AF379292

Test and Evaluation Methods for Human-Machine Interfaces of Automated Vehicles

Test and Evaluation Methods for Human-Machine Interfaces of Automated Vehicles

Editors

Frederik Naujoks
Sebastian Hergeth
Andreas Keinath
Nadja Schömig
Katharina Wiedemann

MDPI • Basel • Beijing • Wuhan • Barcelona • Belgrade • Manchester • Tokyo • Cluj • Tianjin

Editors

Frederik Naujoks
BMW Group
Germany

Sebastian Hergeth
BMW Group
Germany

Andreas Keinath
BMW Group
Germany

Nadja Schömig
Wuerzburg Institute for
Traffic Sciences
Germany

Katharina Wiedemann
Wuerzburg Institute for
Traffic Sciences
Germany

Editorial Office
MDPI
St. Alban-Anlage 66
4052 Basel, Switzerland

This is a reprint of articles from the Special Issue published online in the open access journal *Information* (ISSN 2078-2489) (available at: https://www.mdpi.com/journal/information/special_issues/Automated_Vehicles).

For citation purposes, cite each article independently as indicated on the article page online and as indicated below:

LastName, A.A.; LastName, B.B.; LastName, C.C. Article Title. *Journal Name* **Year**, *Article Number*, Page Range.

ISBN 978-3-03943-198-4 (Hbk)
ISBN 978-3-03943-199-1 (PDF)

Contents

About the Editors

Frederik Naujoks graduated from the University of Wuerzburg in 2010 with a Diploma in Psychology and a PhD in 2015. Between 2011 and 2017, he worked at the Center for Traffic Sciences (IZVW) at the University of Wuerzburg and at the Wuerzburg Institute for Traffic Sciences (WIVW). Since then, he has worked with BMW since 2017. His research focuses on applied psychology topics, such as driver distraction, usability and human-centered design and the evaluation of assisted and automated driving.

Sebastian Hergeth was born in 1986 in Munich, Germany, and obtained his Bachelor of Science in Psychology from Paris Lodron University of Salzburg in 2011, Master of Science in Economic and Organizational Psychology from Ludwig-Maximilians-Universität München in 2013, and PhD in Psychology from Chemnitz University of Technology on the topic of trust in automation in 2016. Since 2016 he is an employee of the BMW Group in Munich. His main research areas include human factors of assisted and automated driving, HMI design and evaluation, method development, trust in automation and driver distraction, as well as exterior human–machine interfaces.

Andreas Keinath is Head of Concept Quality and Usability of the HMI department at BMW Group. He received his PhD in Psychology from Chemnitz University of Technology in 2003. His research focusses on cognitive and applied psychology, as well as automotive systems engineering.

Nadja Schömig finished her studies in psychology at the University of Würzburg in 2003. From 2003 to 2007, she worked at the Centre for Traffic Sciences at the University of Würzburg. In 2008, she started working as a senior researcher at the Würzburg Institute for Traffic Sciences (WIVW). In 2009, she received her PhD in psychology on the topic of driver situation awareness and its measurement. Her main research areas are human factor-related topics in assisted and automated driving, such as HMI design, HMI evaluation methodologies and driver state assessment methodologies (fatigue, distraction).

Katharina Wiedemann has been working at the Würzburg Insitute for Traffic Sciences (WIVW) after finishing her studies in Psychology at the University of Würzburg in 2014. Her research focuses on human factors of assisted and automated driving, HMI design and evaluation, the development of test methods and driver distraction. She is currently finalizing her PhD in psychology about the design of automated vehicle HMIs.

Preface to "Test and Evaluation Methods for Human-Machine Interfaces of Automated Vehicles"

The human-machine interface of automated driving systems (ADS) will play a crucial role in their safe, comfortable and efficient use. For example, the ADS HMI should be capable of efficiently informing the user about the current automated driving mode and the user's responsibilities (e.g., whether the ADS is functioning properly or requesting a transition of control from the ADS to the user). While ADS might allow new and more comfortable seating positions and engagement in nondriving-related tasks that are not allowed in manual driving, these might lower the user's availability for a transfer of control or generate motion sickness. Furthermore, when interacting with other vehicles, ADS might behave differently than manually driven vehicles, which might generate a need for external HMIs or standardized motion patterns for an adequate interaction with non-automated traffic participants. This is only a small proportion of the new challenges for test and evaluation methods of HMIs that arise from the introduction of ADS. Thus, human factor experts need to explore, advance and establish test methods that are able to account for these new challenges in the design of future vehicles.

The articles of this Special Issue analyze developments and new challenges by introducing literature reviews, and analytical as well as experimental approaches to the topics outlined above. The contributions all stem from well-known research institutes and leading practitioners in the field of ADS research. The papers deal with a broad selection of relevant topics, which can be broadly categorized in four clusters:

• Assessing the relationship of automated vehicles and surrounding non-automated traffic: ADS will very likely be introduced into a mixed traffic environment, which means that some road users will be automated, while others will drive manually. Papers [1–4] focus on the impact of automated vehicles on surrounding, non-automated traffic such as pedestrians or cyclists.

• Designing and evaluating external human–machine interfaces (eHMIs): Automated cars may be equipped with eHMIs for communication with other unequipped road users such as pedestrians. Their potential benefits and drawbacks are discussed in the technical and scientific community, but there are currently no available standards for their implementation. Thus, papers [5–9] present empirical studies as well as test protocols for this focus area.

• Evaluating interior HMIs of automated vehicles: As long as vehicles can be driven manually or require manual intervention by their users, the interior HMI will still play a crucial part in their safe and efficient usage. However, guidelines and test methods are only slowly being adapted from those of manual and assisted driving. The next three papers [10–12] investigate methods regarding the assessments of interior HMIs of automated vehicles.

• Evaluating the influence of driver state, driver availability and situational factors on control transitions and the comfort of automated driving: A crucial human factor in the use of automated driving functions is the driver's state, such as the readiness to take over manual driving, mode awareness, fatigue or motion sickness. The driver's state can have an impact both on the safety of control transitions as well as the perceived comfort and acceptance of automated driving. The following papers [13–21] provide empirical studies, as well as theoretical analyses and test protocols on this issue.

This Special Issue brings together research from well-known human factor experts in the field of automated driving. The impressive number of published papers covering a wide range of research topics on test and evaluation methods for automated vehicles HMIs shows the high relevance of this Special Issue. The Special Issue has thus contributed to the promotion and dissemination of these methods within the scientific community, and will hopefully stimulate further research on these topics.

References

1. Fuest, T.; Schmidt, E.; Bengler, K. Comparison of Methods to Evaluate the Influence of an Automated Vehicle's Driving Behavior on Pedestrians: Wizard of Oz, Virtual Reality, and Video. *Information* **2020**, *11*, 291.
2. Fuest, T.; Feierle, A.; Schmidt, E.; Bengler, K. Effects of Marking Automated Vehicles on Human Drivers on Highways. *Information* **2020**, *11*, 286.
3. Feierle, A.; Rettenmaier, M.; Zeitlmeir, F.; Bengler, K. Multi-Vehicle Simulation in Urban Automated Driving: Technical Implementation and Added Benefit. *Information* **2020**, *11*, 272.
4. Krüger, M.; Driessen, T.; Wiebel-Herboth, C.B.; de Winter, J.C.F.; Wersing, H. Feeling Uncertain—Effects of a Vibrotactile Belt that Communicates Vehicle Sensor Uncertainty. *Information* **2020**, *11*, 353.
5. Kaß, C.; Schoch, S.; Naujoks, F.; Hergeth, S.; Keinath, A.; Neukum, A. Standardized Test Procedure for External Human-Machine Interfaces of Automated Vehicles. *Information* **2020**, *11*, 173.
6. Rettenmaier, M.; Schulze, J.; Bengler, K. How Much Space Is Required? Effect of Distance, Content, and Color on External Human–Machine Interface Size. *Information* **2020**, *11*, 346.
7. Kooijman, L.; Riender H.; de Winter, J.C.F. How Do eHMIs Affect Pedestrians' Crossing Behavior? A Study Using a Head-Mounted Display Combined with a Motion Suit. *Information* **2019**, *10*, 386.
8. Eisma, Y.B.; van Bergen, S.; ter Barke, S.M.; Hensen, M.T.T.; Tempelaar, W.J.; de Winter, J.C.F. External Human–Machine Interfaces: The Effect of Display Location on Crossing Intentions and Eye Movements. *Information*, **2020** *11*, 13.
9. Faas, S.M.; Mattes, S.; Kao, A.C.; Baumann, M. Efficient Paradigm to Measure Street-Crossing Onset Time of Pedestrians in Video-Based Interactions with Vehicles. *Information* **2020**, *11*, 360.
10. Albers, S.; Radlmayr, J.; Loew, A.; Hergeth, S.; Naujoks, F.; Keinath, A.; Bengler, K. Usability Evaluation—Advances in Experimental Design in the Context of Automated Driving Human–Machine Interfaces. *Information* **2020**, *11*, 240.
11. Schömig, N.; Wiedemann, K.; Hergeth, S.; Forster, Y.; Muttart, J.; Eriksson, A.; Mitropoulos-Rundus, D.; Grove, K.; Krems, J.; Keinath, A.; Neukum, A.; Naujoks, F. Checklist for Expert Evaluation of HMIs of Automated Vehicles—Discussions on Its Value and Adaptions of the Method within an Expert Workshop. *Information* **2020**, *11*, 233.
12. Wolter, S.; Dominioni, G.C.; Hergeth, S.; Tango, F.; Whitehouse, S.; Naujoks, F. Human–Vehicle Integration in the Code of Practice for Automated Driving. *Information* **2020**, *11*, 284.
13. Wörle, J.; Kenntner-Mabiala, R.; Metz, B.; Fritzsch, S.; Purucker, C.; Befelein, D.; Prill, A. Sleep Inertia Countermeasures in Automated Driving: A Concept of Cognitive Stimulation. *Information* **2020**, *11*, 342.
14. Hollander, C.; Rauh, N.; Naujoks, F.; Hergeth, S.; Krems, J.F.; Keinath, A. Methodological Approach towards Evaluating the Effects of Non-Driving Related Tasks during Partially Automated Driving. *Information* **2020**, *11*, 340.

15. Kurpiers, C.; Biebl, B.; Hernandez, J.M.; Raisch, F. Mode Awareness and Automated Driving—What Is It and How Can It Be Measured? *Information* **2020**, *11*, 277.

16. Forster, Y.; Geisel, V.; Hergeth, S.; Naujoks, F.; Keinath, A. Engagement in Non-Driving Related Tasks as a Non-Intrusive Measure for Mode Awareness: A Simulator Study. *Information* **2020**, *11*, 239.

17. Mühlbacher, D.; Tomzig, M.; Reinmüller, K.; Rittger, L. Methodological Considerations Concerning Motion Sickness Investigations during Automated Driving. *Information* **2020**, *11*, 265.

18. Boelhouwer, A.; van der Beukel, A.P.; van der Voort, M.C.; Verwey, W.B.; Martens, M.H. Supporting Drivers of Partially Automated Cars through an Adaptive Digital In-Car Tutor. *Information* **2020**, *11*, 185.

19. Scharfe, M.S.L.; Zeeb, K.; Russwinkel, N. The Impact of Situational Complexity and Familiarity on Takeover Quality in Uncritical Highly Automated Driving Scenarios. *Information* **2020**, *11*, 115.

20. Metz, B.; Wörle, J.; Hanig, M.; Schmitt, M.; Lutz, A. Repeated Usage of an L3 Motorway Chauffeur: Change of Evaluation and Usage. *Information* **2020**, *11*, 114.

21. Radhakrishnan, V.; Merat, N.; Louw, T.; Lenné M.G.; Romano, R.; Paschalidis, E.; Hajiseyedjavadi, F.; Wei, C.; Boer, E.R. Measuring Drivers' Physiological Response to Different Vehicle Controllers in Highly Automated Driving (HAD): Opportunities for Establishing Real-Time Values of Driver Discomfort. *Information* **2020**, *11*, 390.

Frederik Naujoks, Sebastian Hergeth, Andreas Keinath, Nadja Schömig, Katharina Wiedemann

Editors

Editorial

Editorial for Special Issue: Test and Evaluation Methods for Human-Machine Interfaces of Automated Vehicles

Frederik Naujoks [1,*], Sebastian Hergeth [1], Andreas Keinath [1], Nadja Schömig [2] and Katharina Wiedemann [2]

[1] BMW Group, 80937 Munich, Germany; sebastian.hergeth@bmw.de (S.H.); andreas.keinath@bmw.de (A.K.)
[2] Wuerzburg Institute for Traffic Sciences, D-97209 Veitshöchheim, Germany; nadja.schömig@wivw.de (N.S.); katharina.wiedemann@wivw.de (K.W.)
* Correspondence: frederik.naujoks@bmw.de

Received: 18 August 2020; Accepted: 19 August 2020; Published: 20 August 2020

Abstract: Today, OEMs and suppliers can rely on commonly agreed and standardized test and evaluation methods for in-vehicle human–machine interfaces (HMIs). These have traditionally focused on the context of manually driven vehicles and put the evaluation of minimizing distraction effects and enhancing usability at their core (e.g., AAM guidelines or NHTSA visual-manual distraction guidelines). However, advances in automated driving systems (ADS) have already begun to change the driver's role from actively driving the vehicle to monitoring the driving situation and being ready to intervene in partially automated driving (SAE L2). Higher levels of vehicle automation will likely only require the driver to act as a fallback ready user in case of system limits and malfunctions (SAE L3) or could even act without any fallback within their operational design domain (SAE L4). During the same trip, different levels of automation might be available to the driver (e.g., L2 in urban environments, L3 on highways). These developments require new test and evaluation methods for ADS, as available test methods cannot be easily transferred and adapted. The shift towards higher levels of vehicle automation has also moved the discussion towards the interaction between automated and non-automated road users using exterior HMIs. This Special Issue includes theoretical papers a well as empirical studies that deal with these new challenges by proposing new and innovative test methods in the evaluation of ADS HMIs in different areas.

Keywords: automated driving; human–machine interface; test methods; user studies; evaluation

---

1. Introduction

The human–machine interface (HMI) will play a crucial role in the safe, comfortable and efficient use of automated vehicles. For example, the automated driving system (ADS) HMI should be capable of informing the user about the current mode and minimize confusion about the status of the ADS and the user's current responsibilities (e.g., whether the ADS is functioning properly, ready for use, unavailable for use or requesting a transition of control from the ADS to the user). While ADS might allow new and more comfortable seating positions and engagement in nondriving-related tasks that were not allowed in manual driving, these might lower the user's availability for a transfer of control or generate motion sickness. As the driving task is no longer actively fulfilled by the driver, distraction by nondriving-related tasks might turn into controlled engagement by activating activities that prevent fatigue, generating the need to advance assessment methods for nondriving-related tasks. Furthermore, when interacting with other vehicles, ADS might behave differently than manually driven vehicles, which might generate a need for external HMIs or standardized motion patterns for an adequate interaction with non-automated traffic participants. This is only a small proportion of the new challenges for test and evaluation methods of HMIs that arise from the introduction of ADS. The articles of this Special Issue analyze the developments and new challenges by introducing new test methods about the topics outlined above. Among the submissions received, all of which went through a rigorous peer-review process, 21 papers have been selected for publication. The contributions all stem from well-known research institutes and leading practitioners in the field of ADS research. The papers, which will be described in the following, deal with a broad selection of relevant topics such as the evaluation of the relationship of automated vehicles and surrounding non-automated traffic, external as well as interior human–machine interfaces of automated vehicles and the influence of driver state, driver availability and situational factors on control transitions and comfort of automated driving.

Assessing the relationship of automated vehicles and surrounding non-automated traffic

ADS will very likely be introduced into a mixed traffic environment, which means that some road users will be automated while others will be driven manually. The following papers focus on the impact of automated vehicles on surrounding, non-automated traffic such as pedestrians or cyclists. The first paper "Comparison of Methods to Evaluate the Influence of an Automated Vehicle's Driving Behavior on Pedestrians: Wizard of Oz, Virtual Reality, and Video" by Fuest, Schmidt and Bengler [1] investigates four different methods regarding the communication between automated vehicles and pedestrians. Hence the same study design in four different settings was used. Two video, one virtual reality, and one Wizard of Oz setup was replicated. An automated vehicle approached from the left, using different driving profiles characterized by changing speed to communicate its intention to let the pedestrians cross the road. Participants were asked to recognize the intention of the automated vehicle and to press a button as soon as they realized its intention.

The second paper "Effects of Marking Automated Vehicles on Human Drivers on Highways" by Fuest, Feierle, Schmidt and Bengler [2] presents a simulation study with different highway scenarios each with and without a marked automated vehicle. Common to all scenarios was that the automated vehicles strictly adhered to German highway regulations, and therefore moved in road traffic somewhat differently to human drivers. After each trial, the participants were asked to rate how appropriate and disturbing the automated vehicle's driving behavior was. In addition, objective data, such as the time of a lane change and the time headway were measured.

The third paper "Multi-Vehicle Simulation in Urban Automated Driving: Technical Implementation and Added Benefit" by Feierle, Rettenmaier, Zeitlmeir and Bengler [3] investigates the simultaneous interaction between an automated vehicle (AV) and its passenger, and between the same AV and a human driver of another vehicle. For this purpose a multi-vehicle simulation consisting of two driving simulators, one for the AV and one for the manual vehicle was implemented. This paper analyzes the effect of an automation failure, where the AV first communicates to yield the right of way and then changes its strategy and passes through the bottleneck first, despite oncoming traffic. The research questions the study aims to answer are what methods should be used for the implementation of multi-vehicle simulations with one AV, and is there an added benefit of this multi-vehicle simulation compared to single-driver simulator studies?

The next paper focuses on the communication of surrounding traffic conditions to users of automated vehicles. The paper "Feeling Uncertain—Effects of a Vibrotactile Belt that Communicates Vehicle Sensor Uncertainty" by Krüger, Driessen, Wiebel-Herboth, de Winter and Wersing [4] deals with the design and evaluation of a vibrotactile interface that communicates spatiotemporal information about surrounding vehicles and encodes a representation of spatial uncertainty in a novel way. For the measure of subjective understanding and benefit, a questionnaire, ratings and scores were used, for the objective benefit, the minimum time-to-contact as a measure of safety and gaze distributions as an indicator for attention guidance were computed.

Designing and evaluating external human–machine interfaces (eHMIs)

Automated cars may be equipped with eHMIs for communication with other unequipped road users such as pedestrians. Their potential benefits and drawbacks are discussed in the technical and scientific community, but there are currently no available standards for their implementation. Therefore the first paper "Standardized Test Procedure for External Human-Machine Interfaces of Automated Vehicles", by Kaß, Schoch, Naujoks, Hergeth, Keinath and Neukum [5] presents a standardized test procedure that enables the effective usability evaluation of eHMIs from the perspective of multiple road users. The paper includes a methodological approach to deduce relevant use cases as well as specific usability requirements that should be fulfilled by an eHMI to be effective, efficient, and satisfying. To prove whether an eHMI meets these requirements, a test protocol for the empirical evaluation of an eHMI with a participant study is demonstrated.

To be effective, any message displayed by an automated vehicle to other road users must satisfy legibility requirements based on the dynamics of the road traffic and the time required by the human to process the respective message. Therefore the second paper "How Much Space Is Required? Effect of Distance, Content, and Color on External Human–Machine Interface Size" by Rettenmaier, Schulze and Bengler [6] examines the size requirements of displayed text or symbols regarding eHMIs for ensuring the legibility of a message. Based on a developed eHMI prototype, the influence of content type on content size to ensure legibility from a constant distance, as well as the influence of content type and content color on the human detection range, was investigated.

The third paper "How Do eHMIs Affect Pedestrians' Crossing Behavior? A Study Using a Head-Mounted Display Combined with a Motion Suit" by Kooijmann, Happee and de Winter [7] focuses on the investigation of the effects of eHMIs on participants' crossing behavior. For this purpose, the participants were immersed in a virtual urban environment using a head-mounted display coupled to a motion-tracking suit. The approaching vehicles' behavior (yielding, or nonyielding) and eHMI type (None, Text or Front Brake Lights) were manipulated and the participants could cross the road whenever they felt safe enough to do so. The study shows that the motion suit allows investigating pedestrian behaviors related to bodily attention and hesitation in the context of interacting with automated vehicles.

The fourth paper "External Human–Machine Interfaces: The Effect of Display Location on Crossing Intentions and Eye Movements" by Eisma, van Bergen, Brake, Hensen, Tempelaar and de Winter [8] addresses the effects of the position of the eHMI on the feeling of safety to cross the street. The eHMI showed "Waiting" combined with a walking symbol 1.2 s before the car started to slow down, or "Driving" while the car continued driving. Participants had to press and hold the spacebar when they felt it was safe to cross. After that, the percentages of spacebar presses and the eye-tracking analyses were evaluated.

The last paper regarding the concept of eHMIs "Efficient Paradigm to Measure Street-Crossing Onset Time of Pedestrians in Video-Based Interactions with Vehicles" by Faas, Mattes, Kao and Baumann [9] introduces a methodology to compare eHMI concepts from a pedestrian's viewpoint. Therefore a quantifiable concept that allows participants to naturally step off a sidewalk to cross the street was developed. Hidden force-sensitive resistor sensors recorded their crossing onset time (COT) in response to real-life videos of approaching vehicles in an immersive crosswalk simulation environment.

Evaluating interior HMIs of automated vehicles

As long as vehicles can be driven manually or require manual intervention by their users, the interior HMI will still play a crucial part in their safe and efficient usage. However, guidelines and test methods are only slowly being adapted from those of manual and assisted driving. The next three papers investigate methods regarding the assessments of interior HMIs of automated vehicles. The first one "Usability Evaluation—Advances in Experimental Design in the Context of Automated Driving Human–Machine Interfaces" by Albers, Radlmayr, Löw, Hergeth, Naujoks, Keinath and Bengler [10] aggregates common research methods and findings based on an extensive literature review. These methods and findings are discussed critically, taking into consideration requirements for usability assessments of HMIs in the context of conditional automated driving. The paper concludes with a derivation of recommended study characteristics framing best practice advice for the design of experiments.

The second paper "Checklist for Expert Evaluation of HMIs of Automated Vehicles—Discussions on Its Value and Adaptions of the Method within an Expert Workshop" by Schömig, Wiedemann, Hergeth, Forster, Muttart, Eriksson, Mitropulos-Rundus, Grove, Krems, Keinath, Neukum and Naujoks [11] summarizes the results of a workshop about a checklist method for the evaluation of automated vehicles' HMIs. Within this workshop, members of the human factors community were brought together to discuss the method and to further promote the development of HMI guidelines and assessment methods for the design of HMIs of automated driving systems (ADS). The results will be used to further improve the checklist method and make the process available to the scientific community.

The paper "Human–Vehicle Integration in the Code of Practice for Automated Driving" by Wolter, Dominioni, Hergeth, Tango, Whitehouse and Naujoks [12] deals with a new Code of Practice for automated driving (CoP-AD) as part of the publicly funded European project L3Pilot. It provides developers with a comprehensive guideline on how to design and test automated driving functions, with a focus on highway driving and parking. This paper focuses on the human factors aspects addressed in the CoP-AD, which includes, inter alia, general human factors-related guidelines, mode awareness, trust, and misuse, driver monitoring together with the topic of controllability and the execution of customer clinics, as well as the training and variability of users.

Evaluating the influence of driver state, driver availability and situational factors on control transitions and comfort of automated driving

A crucial human factor in the use of automated driving functions is the driver's state, such as the readiness to take over manual driving, mode awareness, fatigue or motion sickness. The driver's state can have an impact both on the safety of control transitions as well as the perceived comfort and acceptance of automated driving. The following papers provide empirical studies as well as theoretical analyses and test protocols on this issue.

The first one "Sleep Inertia Countermeasures in Automated Driving: A Concept of Cognitive Stimulation" by Wörle, Kenntner-Mabiala, Metz, Fritzsch, Purucker, Befelein and Prill [13] shows the concept and evaluation of a reactive countermeasure against sleep inertia, which could be useful with regard to dual-mode vehicles that allow both manual and automated driving. The so called "sleep inertia counter-procedure for drivers" (SICD), has been developed with the aim to activate and motivate the driver as well as to measure the driver's alertness level. The SICD is evaluated in a study with drivers in a driving simulator.

The second paper "Methodological Approach towards Evaluating the Effects of Non-Driving Related Tasks during Partially Automated Driving" by Hollander, Rauh, Naujoks, Hergeth, Krems and Keinath [14] shows the development of a test protocol for systematically evaluating non driving-related tasks' (NDRT) effects during partially automated driving (PAD). Two generic take-over situations addressing system limits of a given PAD regarding longitudinal and lateral control were implemented to evaluate drivers' supervisory and take-over capabilities while engaging in different NDRTs (e.g., manual radio tuning task). The test protocol was evaluated and refined across the three studies (two simulator and one test track).

The third paper "Mode Awareness and Automated Driving—What Is It and How Can It Be Measured?" by Kurpiers, Biebl, Mejia Hernandez and Raisch [15] introduces a measurement method to assess mode awareness when using automated vehicles. The background of this study is the different responsibility allocation in different automation modes that requires the driver to always be aware of the currently active system and its limits to ensure a safe drive. For that reason, current research focuses on identifying factors that might promote mode awareness. In the method presented by the authors, the behavior aspect is represented by the relational attention ratio in manual, Level 2 and

Level 3 driving as well as the controllability of a system limit in Level 2. The knowledge aspect of mode awareness is operationalized by a questionnaire on the mental model for the automation systems after an initial instruction as well as an extensive enquiry following the driving sequence.

The fourth paper "Engagement in Non-Driving Related Tasks as a Non-Intrusive Measure for Mode Awareness: A Simulator Study" by Forster, Geisel, Hergeth, Naujoks and Keinath [16] describes a driving simulator study, based on the expectation that HMI design and practice with different levels of driving automation influence NDRT engagement. Therefore the participants completed several transitions of control and could engage in an NDRT if they felt safe and comfortable to do so. The NDRT was the Surrogate Reference Task (SuRT) as a representative of a wide range of visual-manual NDRTs. Engagement (i.e., number of inputs on the NDRT interface) was assessed at the onset of a respective episode of automated driving (i.e., after transition) and during ongoing automation (i.e., before subsequent transition).

The fifth paper "Methodological Considerations Concerning Motion Sickness Investigations during Automated Driving" by Mühlbacher, Tomzig, Reinmüller and Rittger [17] discusses methodological aspects for investigating motion sickness in the context of automated driving including measurement tools, test environments, sample, and ethical restrictions. Additionally, methodological considerations guided by different underlying research questions and hypotheses are provided. Selected results from the authors' own studies concerning motion sickness during automated driving which were conducted in a motion-based driving simulation and a real vehicle are used to support the discussion.

The sixth paper "Supporting Drivers of Partially Automated Cars through an Adaptive Digital In-Car Tutor" by Boelhouwer, van den Beukel, van der Voort, Verwey and Martens [18] investigates the effects of a Digital In-Car Tutor (DIT) prototype on appropriate automation use and take-over quality during a driving simulator study. A DIT is proposed to support drivers in learning about, and trying out, their car automation during regular drives. Participants needed to use the automation when they thought that it was safe, and turn it off if they did not. The control group read an information brochure before driving, while the experiment group received the DIT during the first driving session.

The seventh paper "The Impact of Situational Complexity and Familiarity on Takeover Quality in Uncritical Highly Automated Driving Scenarios" by Scharfe, Zeeb and Russwinkel [19] differentiates between the objective complexity and the subjectively perceived complexity of a traffic situation. The aim of the present study was to examine the impact of objective complexity and familiarity on the subjectively perceived complexity and the resulting takeover quality. In a driving simulator study, participants were requested to take over vehicle control in an uncritical situation. Familiarity and objective complexity were varied by the number of surrounding vehicles and scenario repetitions. Subjective complexity was measured using the NASA-TLX; the takeover quality was gathered using the take-over controllability rating (TOC-Rating).

The eighth paper "Repeated Usage of an L3 Motorway Chauffeur: Change of Evaluation and Usage" by Metz, Wörle, Hanig, Schmitt and Lutz [20] investigates changes in drivers' evaluation, in function usage and in drivers' reactions to take-over situations with repeated usage of automated driving functions. Therefore, drivers used a level 3 (L3) automated driving function for motorways during six experimental sessions in a driving simulator study. They were free to activate/deactivate the system as they liked and to spend driving time on self-chosen side tasks. After that the experienced trust and safety, the time spent on side tasks, attention directed to the road and behavioral adaptation was analyzed.

The last paper "Measuring Drivers' Physiological Response to Different Vehicle Controllers in Highly Automated Driving (HAD): Opportunities for Establishing Real-Time Values of Driver Discomfort" by Radhakrishnan, Merat, Louw, Lenné, Romano, Paschalidis, Hajiseyedjavadi, Wei and Boer [21] investigates how driver discomfort was influenced by different types of automated vehicle (AV) controllers, compared to manual driving, and whether this response changed in different road environments, using heart-rate variability and electrodermal activity. The drivers were subjected to manual driving and four AV controllers: two modelled to depict "human-like" driving behavior, one conventional lane-keeping assist controller, and a replay of their own manual drive.

2. Conclusions

This Special Issue brings together research from well-known human factors experts in the field of automated driving. The impressive number of published papers covering a wide range of research topics on test and evaluation methods for automated vehicles HMIs shows the high relevance of this Special Issue. The Special Issue has thus contributed to the promotion and dissemination of these methods within the scientific community and will hopefully stimulate further research on these topics.

Acknowledgments: The guest editors would like to thank the authors for their valuable submissions, the reviewers for their precious and constructive comments. We also thank Helena Opower for proofreading and her input for this editorial.

Conflicts of Interest: The authors declare no conflict of interest.

References

1. Fuest, T.; Schmidt, E.; Bengler, K. Comparison of Methods to Evaluate the Influence of an Automated Vehicle's Driving Behavior on Pedestrians: Wizard of Oz, Virtual Reality, and Video. *Information* **2020**, *11*, 291. [CrossRef]
2. Fuest, T.; Feierle, A.; Schmidt, E.; Bengler, K. Effects of Marking Automated Vehicles on Human Drivers on Highways. *Information* **2020**, *11*, 286. [CrossRef]
3. Feierle, A.; Rettenmaier, M.; Zeitlmeir, F.; Bengler, K. Multi-Vehicle Simulation in Urban Automated Driving: Technical Implementation and Added Benefit. *Information* **2020**, *11*, 272. [CrossRef]
4. Krüger, M.; Driessen, T.; Wiebel-Herboth, C.B.; de Winter, J.C.F.; Wersing, H. Feeling Uncertain—Effects of a Vibrotactile Belt that Communicates Vehicle Sensor Uncertainty. *Information* **2020**, *11*, 353. [CrossRef]
5. Kaß, C.; Schoch, S.; Naujoks, F.; Hergeth, S.; Keinath, A.; Neukum, A. Standardized Test Procedure for External Human-Machine Interfaces of Automated Vehicles. *Information* **2020**, *11*, 173. [CrossRef]
6. Rettenmaier, M.; Schulze, J.; Bengler, K. How Much Space Is Required? Effect of Distance, Content, and Color on External Human–Machine Interface Size. *Information* **2020**, *11*, 346. [CrossRef]
7. Kooijman, L.; Riender, H.; de Winter, J.C.F. How Do eHMIs Affect Pedestrians' Crossing Behavior? A Study Using a Head-Mounted Display Combined with a Motion Suit. *Information* **2019**, *10*, 386. [CrossRef]
8. Eisma, Y.B.; van Bergen, S.; ter Barke, S.M.; Hensen, M.T.T.; Tempelaar, W.J.; de Winter, J.C.F. External Human–Machine Interfaces: The Effect of Display Location on Crossing Intentions and Eye Movements. *Information* **2020**, *11*, 13. [CrossRef]
9. Faas, S.M.; Mattes, S.; Kao, A.C.; Baumann, M. Efficient Paradigm to Measure Street-Crossing Onset Time of Pedestrians in Video-Based Interactions with Vehicles. *Information* **2020**, *11*, 360. [CrossRef]
10. Albers, S.; Radlmayr, J.; Loew, A.; Hergeth, S.; Naujoks, F.; Keinath, A.; Bengler, K. Usability Evaluation—Advances in Experimental Design in the Context of Automated Driving Human–Machine Interfaces. *Information* **2020**, *11*, 240. [CrossRef]
11. Schömig, N.; Wiedemann, K.; Hergeth, S.; Forster, Y.; Muttart, J.; Eriksson, A.; Mitropoulos-Rundus, D.; Grove, K.; Krems, J.; Keinath, A.; et al. Checklist for Expert Evaluation of HMIs of Automated Vehicles—Discussions on Its Value and Adaptions of the Method within an Expert Workshop. *Information* **2020**, *11*, 233. [CrossRef]
12. Wolter, S.; Dominioni, G.C.; Hergeth, S.; Tango, F.; Whitehouse, S.; Naujoks, F. Human–Vehicle Integration in the Code of Practice for Automated Driving. *Information* **2020**, *11*, 284. [CrossRef]
13. Wörle, J.; Kenntner-Mabiala, R.; Metz, B.; Fritzsch, S.; Purucker, C.; Befelein, D.; Prill, A. Sleep Inertia Countermeasures in Automated Driving: A Concept of Cognitive Stimulation. *Information* **2020**, *11*, 342. [CrossRef]
14. Hollander, C.; Rauh, N.; Naujoks, F.; Hergeth, S.; Krems, J.F.; Keinath, A. Methodological Approach towards Evaluating the Effects of Non-Driving Related Tasks during Partially Automated Driving. *Information* **2020**, *11*, 340. [CrossRef]
15. Kurpiers, C.; Biebl, B.; Hernandez, J.M.; Raisch, F. Mode Awareness and Automated Driving—What Is It and How Can It Be Measured? *Information* **2020**, *11*, 277. [CrossRef]

16. Forster, Y.; Geisel, V.; Hergeth, S.; Naujoks, F.; Keinath, A. Engagement in Non-Driving Related Tasks as a Non-Intrusive Measure for Mode Awareness: A Simulator Study. *Information* **2020**, *11*, 239. [CrossRef]

17. Mühlbacher, D.; Tomzig, M.; Reinmüller, K.; Rittger, L. Methodological Considerations Concerning Motion Sickness Investigations during Automated Driving. *Information* **2020**, *11*, 265. [CrossRef]

18. Boelhouwer, A.; van der Beukel, A.P.; van der Voort, M.C.; Verwey, W.B.; Martens, M.H. Supporting Drivers of Partially Automated Cars through an Adaptive Digital In-Car Tutor. *Information* **2020**, *11*, 185. [CrossRef]

19. Scharfe, M.S.L.; Zeeb, K.; Russwinkel, N. The Impact of Situational Complexity and Familiarity on Takeover Quality in Uncritical Highly Automated Driving Scenarios. *Information* **2020**, *11*, 115. [CrossRef]

20. Metz, B.; Wörle, J.; Hanig, M.; Schmitt, M.; Lutz, A. Repeated Usage of an L3 Motorway Chauffeur: Change of Evaluation and Usage. *Information* **2020**, *11*, 114. [CrossRef]

21. Radhakrishnan, V.; Merat, N.; Louw, T.; Lenné, M.G.; Romano, R.; Paschalidis, E.; Hajiseyedjavadi, F.; Wei, C.; Boer, E.R. Measuring Drivers' Physiological Response to Different Vehicle Controllers in Highly Automated Driving (HAD): Opportunities for Establishing Real-Time Values of Driver Discomfort. *Information* **2020**, *11*, 390. [CrossRef]

 *information*

Article

Comparison of Methods to Evaluate the Influence of an Automated Vehicle's Driving Behavior on Pedestrians: Wizard of Oz, Virtual Reality, and Video

Tanja Fuest [1,*], Elisabeth Schmidt [2] and Klaus Bengler [1]

[1] Chair of Ergonomics, Technical University of Munich, 85748 Garching, Germany; bengler@tum.de
[2] BMW Group, New Technologies, 85748 Garching, Germany; elisabeth.schmidt@bmw.de
[*] Correspondence: tanja.fuest@tum.de

Received: 5 May 2020; Accepted: 26 May 2020; Published: 29 May 2020

Abstract: Integrating automated vehicles into mixed traffic entails several challenges. Their driving behavior must be designed such that is understandable for all human road users, and that it ensures an efficient and safe traffic system. Previous studies investigated these issues, especially regarding the communication between automated vehicles and pedestrians. These studies used different methods, e.g., videos, virtual reality, or Wizard of Oz vehicles. However, the extent of transferability between these studies is still unknown. Therefore, we replicated the same study design in four different settings: two video, one virtual reality, and one Wizard of Oz setup. In the first video setup, videos from the virtual reality setup were used, while in the second setup, we filmed the Wizard of Oz vehicle. In all studies, participants stood at the roadside in a shared space. An automated vehicle approached from the left, using different driving profiles characterized by changing speed to communicate its intention to let the pedestrians cross the road. Participants were asked to recognize the intention of the automated vehicle and to press a button as soon as they realized this intention. Results revealed differences in the intention recognition time between the four study setups, as well as in the correct intention rate. The results from vehicle–pedestrian interaction studies published in recent years that used different study settings can therefore only be compared to each other to a limited extent.

Keywords: (automated) vehicle–pedestrian interaction; implicit communication; mixed traffic; virtual reality; Wizard of Oz; video; setup comparison/method comparison

---

1. Introduction

An increasing number of automated functions are being integrated into vehicles, and it is only a question of time before the first conditionally automated vehicles (AVs) [1] are driving on public highways. In the long term, AVs will also travel in urban spaces that are characterized by an increased complexity compared to driving on highways [2]. In both scenarios, in addition to AVs, human road users (HRUs) will continue to participate in the traffic system. For this reason, AVs must not only be able to detect HRUs, but they must also communicate with them to ensure safe and efficient interaction. Explicit and implicit communication already takes place in road traffic today. For example, in terms of explicit communication, human drivers flash their headlights or deploy the horn to communicate their intentions [3]. For AVs, besides the existing communication forms, it is also possible to extend the explicit communication by using external human–machine interfaces (eHMIs) (e.g., [4–11]), such as light strips [6,12] or displays [4,7,13].

However, it is still unknown whether AVs require eHMIs. In addition, it has not yet been fully investigated as to what driving profile AVs should follow, and if these trajectories should differ from situation to situation. The driving profile and eHMI might influence traffic safety, as well as the communication between AVs and other HRUs.

Several studies have already been carried out to investigate the influence of AV markings, eHMIs and driving profiles. Most studies focused on the interaction between AVs and pedestrians, using different methods, e.g., images, videos, virtual reality (VR), or Wizard of Oz (WoZ) vehicles. More recently, driving simulator studies subsequently investigated the interaction between AVs and human drivers. However, the extent of transferability of results between these studies is still unknown.

1.1. Images

One method suggested by researchers for development process for human machine interactions are images. For a comparison of 30 early stage design concepts of eHMIs within a short space of time, images were used [14]. Participants had to rate their understanding of the different concepts. The results presented gave no clear recommendation regarding the concepts, but the conclusion of the paper was that the method is suitable for evaluating design elements at an early stage [14]. The method of presenting photos to participants to evaluate the AV's communication strategies was also used in a preliminary study by [15]. Photos of an approaching vehicle were shown to the participants, who were then asked what they would focus on when crossing the street [15]. The authors found out that pedestrians pay particular attention to the AV's braking behavior before crossing the road [15]. Most participants would even wait for a complete standstill, especially when they did not see a driver in the AV [15]. Reference [16] used images of different vehicles to evaluate which vehicle type is most suitable for a subsequent video-based survey.

To sum up, these references suggest that the image setup can be useful for gleaning initial impressions for subsequent studies and for evaluating early stage design concepts.

1.2. Videos

The subsequent video experiment of [16] was used to evaluate the crossing behavior of participants at an unmarked road, depending on different vehicles driving behavior and the automation state of the vehicle [16]. Again, it was shown that the braking behavior plays an important role in the pedestrians' decision to cross the road independent of the vehicle's automation status or the presence of a driver [16].

Additional eHMIs have a positive impact on the imagined crossing behavior of pedestrians [13]. During the braking process, eHMIs have influenced the subjective feeling of participants that it is safe to cross [17]. The eHMIs should be installed on the roof, windscreen, or grille; however, projections and eHMIs on wheels should be avoided [17].

The video studies presented were used to identify possible differences between different implicit and explicit AV communication forms.

1.3. Virtual Reality

Whereas the participants in the video studies sat in front of a monitor, for VR, participants usually saw the environment, including the AV, through a head-mounted display.

The results from a VR study show that pedestrians react with confusion and mistrust to atypical trajectories compared to conventional trajectories [18]. This gives a first hint that VR is a good tool for evaluating pedestrian–vehicle interaction [18]. Other results illustrate that pedestrians understand the AV's driving behavior and recommend early deceleration when yielding [15]. A hard initial braking with a pitch reduced the time pedestrians need to realize an AV's yielding intention [19]. Moreover, defensive driving strategies led to pedestrians starting to cross at an earlier point in time [19].

In addition, eHMIs enhance the interaction between pedestrians and AVs [4] and improve the perceived safety and comfort of participants introduced to the eHMI, when encountering an AV [20]. However, the vehicle size has a small effect on the perceived safety [4]. Larger vehicles reduce the perceived safety of participants [4]. The authors of [21] integrated display into their AV mimicking eyes looking at the pedestrians. These "eyes" help pedestrians to feel safer crossing the street and make their decision to cross quicker [21]. However, eHMIs do not necessarily have the same advantages in all countries: using an eHMI when yielding helps pedestrians in Germany and the United States to realize the AV's intention; however, this effect is not apparent for those in China [6]. In addition, the results have shown that, across Germany, the United States, and China, eHMIs deteriorate the pedestrians' recognition of the AV's passing intention [6]. Moreover, the implemented test environment had an influence, and especially the sound. The study by [22] showed that a spatial audio enhanced task performance compared to unimodal muting.

In summary, it can be stated that many questions concerning explicit and implicit communication of AVs have been carried out in VR. VR setups were especially advantageous due to the cost-effective implementation of a study design that can be replicated in different countries. In addition, setup is more immersive than video or image setups.

1.4. Wizard of Oz

To investigate the interaction of a user with a computer system that is not yet fully developed, a WoZ approach can be used [23]. In this approach, an investigator—who is hidden from the user—simulates the system [23]. In most WoZ studies that examine the interaction between AVs and pedestrians, seat covers are used to hide the driver from the pedestrians' view, so as to simulate an AV [24–28]. The results of WoZ studies demonstrated that being able to see the driver is not very important for pedestrians [12,25,28]. In the study by [12], only half of the sample recognized the driver; however, when asked directly, they expressed that they felt safer when a driver is present. This result stands in contrast to the results of [28], where the perceived safety was not influenced by being able to see the driver. As a reason for their increased feeling of safety in the study of [12], some participants did not mention the eHMI, but instead mentioned the driving strategy of the AV [12]. This is in line with the results of [8], who stated that pedestrians rely on proven methods, and therefore focus on the driving behavior of vehicles rather than on additional eHMIs. The results also demonstrate that not every eHMI is suitable for communication with pedestrians [12]. The pedestrians did not relate the cyan light bar consisting of 12 LEDs on the roof used in the study to themselves, and could not understand the vehicle's intention as communicated by the eHMI [12].

In recent years, the number of WoZ studies has increased. With the WoZ setup, similar questions were investigated as with the VR setup, but the WoZ method is closer to reality. However, the use of a vehicle, a trained driver, a test track, the objective data measurement, and the safety protocol in WoZ studies are complex and cost-intensive.

1.5. Driving Simulator

While the design of AV communication initially focused on pedestrian–vehicle interaction, current studies also deal with human driver–AV interaction. In order to evaluate the influence of different driving strategies and eHMIs on other drivers, simulator studies have been conducted. Reference [7] examined the potential of eHMIs in bottlenecks and recommends the use of eHMIs due to a reduced passing time compared to a condition without an interface. However, labeling an AV did not have an influence on drivers in a simulation setup [9,11].

Investigated driver–AV interaction via a driving-simulator has the benefit of a risk-free setup, compared to WoZ setups.

1.6. Objectives

With regard to the different results, the question arises as to the method by which the communication of AVs should be investigated to obtain valid results. Furthermore, it is unclear whether the obtained results can be compared with each other and whether recommendations should be derived from the different studies.

To answer the question of comparability, we replicated the same study design in four different setups: two videos, one VR, and one WoZ approach. The video setup was divided in two parts: In the first part, we used videos from the VR setup, and in the second part, we filmed the WoZ vehicle. To the knowledge of the authors, such a thorough method validation has not been contributed to the state of the art yet.

Based on the previous results, we focused on the comparison of AVs' driving behavior without the use of eHMIs. In particular, the results for the video, VR and driving simulator studies revealed a positive impact of eHMIs on pedestrians' intention-recognition and the imagined crossing behavior. This contrasts with the results of the WoZ studies, in which hardly any effects were found for eHMIs, and in which driving behavior likely plays a greater role in the pedestrians' decision to cross the road. Across all methods, it can be seen that the driving behavior has an influence on the crossing behavior of pedestrians.

Images were excluded as a method variant in this study because they do not illustrate vehicle dynamics. The focus was on pedestrian–AV interaction, as this is the focus of most published studies. For this reason, driving simulator studies are not included in the comparison, as they investigate human driver–AV interaction.

2. Materials and Methods

2.1. Procedure

A study plan was implemented in three different setups, namely WoZ, VR, and videos, of both setups. The studies were conducted in Germany, which implicates that the motorized traffic was driving on the right lane. In all setups, participants stood at the roadside, in a shared space. An AV approached from the left, using different driving profiles, characterized by changing speed, to communicate whether the HRU—in this case, a pedestrian—was allowed to go first or should wait. Participants were asked to recognize the AV's intention and to press a button when they thought they realized the intention (intention recognition time, IRT).

In the WoZ study, we used the button of a light barrier system. The vehicle activated the sensors after driving over a determined point and a light flashed, when participants pressed the button. This light was visible to the driver, so that he could accelerate to the original speed. Therefore, the rest of the driving profile did not influence pedestrians [24].

In the VR study, participants were asked to press a button on a remote control, and the simulation stopped simultaneously. Additionally, we tracked the walking movement. In this variant, we replicated all driving profiles and asked participants not to press the button, but to cross the virtual street. However, for safety reasons, we did not ask participants in the WoZ setup to cross the street.

In the video setup, participants saw all trials on a monitor. They were asked to press a key on the keyboard, at the moment they realized the intention, upon which the video disappeared.

After each trial, participants had to answer a small number of questionnaire items in each study setup.

2.2. Apparatus

2.2.1. Wizard of Oz Setup

The WoZ vehicle was a BMW 2 series (F46, 220d xDrive) with automatic transmission and equipped with a speed limiter (Figures 1 and 2). The vehicle was marked as an "automated test vehicle" with two magnetic signs. A non-professional driver drove the vehicle and was hidden from the pedestrians' view by a seat cover (Figure 1). The driver practiced the trajectories, so that there was little deviation with each repetition [24]. We implemented the light barrier system SmartSpeed Pro of the company Fusion Sport, connected to a remote control with one button via Bluetooth, and recorded at a sampling rate of 1000 Hz.

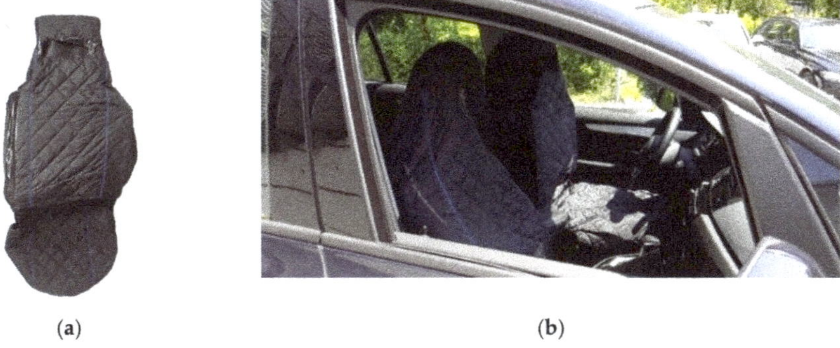

(**a**) (**b**)

Figure 1. Wizard of Oz vehicle: (**a**) seat cover used to hide the driver; (**b**) driver hidden under the seat cover [24].

Figure 2. Wizard of Oz setup.

2.2.2. Virtual Reality Setup

An HTC Vive Pro VR setup with a head-mounted display, two infrared sensors, two trackers, and one remote controller wereused for the VR study setup. All participants held the remote control in their hand, and a tracker was attached to each foot. The simulation software is based on Unity 3D, and a simulated BMW 3 series (F30) was used (Figure 3). The vehicle had no driver, but also no additional markings. The investigator could manipulate the driving behavior by adding a trajectory path and maneuver points. Driving data and the triggering of the button were recorded at 5 Hz. However, no sound was utilized, due to technical reasons.

Figure 3. Virtual reality setup.

2.2.3. Video Setups

We filmed the WoZ vehicle, using a SONY FDR-AX53 with a 26.8 mm wide-angle lens (Figure 2). The camera was mounted on a tripod at a height of 1.61 m, at the same position the participants were standing in the WoZ study. The videos from the VR setup were recorded, using the open-source software OBS (Open Broadcaster Software) studio, and the viewing height was also 1.61 m (Figure 3). The videos were incorporated via HTML, and the survey was accessible from a website. However, since the videos were too large for low internet capacity, most participants watched the videos in the premises of the Chair of Ergonomics (Technical University of Munich) or BMW. The invited participants saw the videos on a 24″ monitor. For all videos, no sound was recorded.

2.3. Study Design and Variables

For all four study setups, almost the same study design was implemented. However, there were some small differences between the study setups:

- For the WoZ setup, participants saw each driving profile twice.
- For the VR setup, we added the condition "walking" instead of a second trial, since we did not let participants cross the road in the WoZ setup for safety reasons. One group of participants started walking when they thought it was safe to cross, and afterward, they were asked to press the button at the moment they realized the AV's intention. The other group started with the IRT condition and walked in the second part of the study. The allocation of participants was randomized.
- In the video setups, each participant saw the two video types, WoZ and VR, in a randomized order.

We randomized two AV intentions: either the *AV goes first*, or to *Let the HRU go first*. For both intentions, an unambiguous and ambiguous driving profile was presented to the participants. To communicate the intentions, altered driving strategies were used that differed in the longitudinal dynamics.

2.3.1. Independent Variables

A within-subject design with two AV intentions (*Let the HRU go first* and *AV goes first*) and—for each of these intentions—an unambiguous and an ambiguous driving profile was implemented. Previous studies showed that the IRT is not sensitive enough to evince differences in driving profiles that are rated very well by humans; thus, we chose highly opposite profiles to apply the IRT [24,25]. All profiles were extracted from human trajectories: In a previous study, participants drove three times, in an unambiguous and ambiguous way, to communicate both intentions to a pedestrian. After each trial, participants rated how satisfied they were with the respective driving profile. We extracted the best rated profiles and defined the specified target trajectories. For the factor "Unambiguity of Driving Profiles" the driver drove either in an understandable or misleading way, to communicate both intentions.

For both intentions, the vehicle accelerated to 28.5 km/h on a 100 m test track. All indicated distances refer to the vehicle's front bumper. If the *AV goes first*, it had a speed of at least 20 km/h when passing the pedestrian. For the second intention, to *Let the HRU go first* the AV decelerated and came to a full stop.

The driving profile *AV goes first, unambiguous* is defined by a constant speed of 28.5 km/h. In contrast, for the profile *AV goes first, ambiguous* the vehicle accelerated to 28.5 km/h and decelerated to 13 km/h after 60 m. After another 32.6 m (7.4 m distance from the pedestrian's position), the vehicle accelerated again (Figure 4).

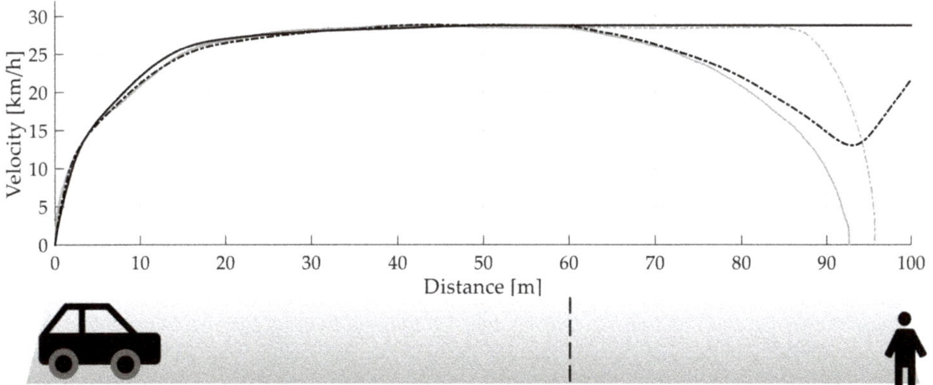

Figure 4. Unambiguous (solid lines) and ambiguous (dashed lines) target trajectory for the intentions *Let the HRU go first* (gray lines) and *AV goes first* (black lines). The vertical dashed line represents the position of the beginning of the time measurement.

For the intention *Let the HRU go first* the vehicle decelerated in two different ways. For the driving profile *Let the HRU go first, unambiguous* the vehicle decelerated by at most 1.5 m/s^2 at a distance of 60 m from the start position. Thus, it started decelerating at the same point as in the driving profile *AV goes first, ambiguous*. The vehicle stopped completely 7.4 m away from the pedestrian—the same point at which the vehicle accelerated in the *AV goes first, ambiguous* profile. In contrast to the smooth deceleration (at max. 1.5 m/s^2) for the unambiguous profile, the vehicle decelerated by at most 4.1 m/s^2 for the driving profile *Let the HRU go first, ambiguous*. The braking process started at 85.2 m from the starting position; hence, the vehicle slowed down in 25.2 m distance to the braking point for the unambiguous driving profile. The vehicle stopped completely after a driving distance of 95.7 m, 4.3 m away from the pedestrian's position.

2.3.2. Dependent Variables

As mentioned, participants pressed a button when they thought they had recognized the vehicle's intention [15,24,25]. We measured the time lapse between the vehicle being at a 40 m distance from the pedestrian's position and the moment at which the participants pressed the button. This time lapse, the IRT, was measured for each trial.

After each trial, participants filled out a five-item questionnaire. This questionnaire was already published in [24] and based on previous studies [15,25]. Based on the IRT, the participants were asked about the vehicle's assumed intention (*Let the HRU go first* or *AV goes first*) and whether they would cross the street at the moment they recognized the intention. Then, pedestrians evaluated their certainty about the vehicle's intention (very uncertain to very certain), the vehicle's driving behavior (very poor to very good), and the perceived criticality of the situation (very critical to very uncritical) on a five-point Likert scale [24]. In the video study, participants were also asked if the video activity had run smoothly from a technical point of view. This item was used to exclude data from the evaluation if videos had frozen during playback.

In order to track the walking movement in the VR setup, we asked participants not to press the button, but to cross the virtual street. The trackers on each foot detected when the participant walked over a virtual line. This line was located about one meter from the participants' starting position. In order to be able to compare the time at the beginning of road crossing with the IRT, the times were synchronized: In both cases, the time measurement started at a 40 m distance from the pedestrian's position. However, the IRT was always independent from the walking movement (Figure 5).

Setup	WoZ	VR	Video WoZ	Video VR
IRT	✓	✓	✓	✓
Start of Road Crossing	-	✓	-	-
Questionnaire Items	✓	✓	✓	✓

Figure 5. Comparison of the different setups.

2.4. Sample

For the VR setup, 37 participants (23 male and 14 female) with a mean age of $M = 27.32$ years ($SD = 9.93$ years) and for the WoZ experiment 34 participants (24 male and 10 female) with a mean age of $M = 40.94$ years ($SD = 21.39$ years) were recruited via BMW and postings at the Technical University of Munich (Table 1). In the video setup, from altogether 46 participants (20 male and 26 female) with a mean age of $M = 30.50$ years ($SD = 11.55$ years), 28 participants were recruited via emailing lists among BMW employees and postings at the Technical University of Munich, and the remaining participants participated online. All participants received compensation; however, in the video setup, participants either received monetary compensation or—the participants who participated online—were entered into a lottery for vouchers for an electronic commerce company.

Table 1. Samples for all study setups.

	WoZ	VR	Video
Sample	N = 34	N = 37	N = 46
ø-age (years)	$M = 40.94, SD = 21.39$ Min. = 17, Max. = 81	$M = 27.32, SD = 9.93$ Min. = 20, Max. = 79	$M = 30.50, SD = 11.55$ Min. = 17, Max. = 67
Sex	$\male = 24$ $\female = 10$	$\male = 23$ $\female = 14$	$\male = 20$ $\female = 26$
Travel as pedestrians in traffic (h per week)	$M = 7.06, SD = 6.33$ Min. = 1, Max. = 25	$M = 8.03, SD = 6.01$ Min. = 1, Max. = 30	$M = 6.57, SD = 5.76$ Min. = 1, Max. = 30

On average, participants travel as pedestrians in traffic $M = 7.06$ h ($SD = 6.33$ h) per week in the WoZ setup, $M = 8.03$ h ($SD = 6.01$ h) per week in the VR setup, and $M = 6.57$ h ($SD = 5.76$ h) per week in the video setup.

2.5. Analysis

The different study setups (WoZ, VR, and video) were compared with a between subject design. However, for the video setups, we had two kinds of videos (video WoZ and video VR) and dependent samples. The samples of the WoZ, VR and the two video setups are independent. We were only interested in the comparison between WoZ and VR; WoZ and video WoZ setup; and VR and the video VR study (Figure 5). Therefore, all outcomes are related to these comparisons. Moreover, as a result of the different nature of the samples (the samples of the two video setups are dependent and the other samples independent), a statistical analysis was not useful for all results and most data were compared descriptively.

For the WoZ setup, we had to exclude three participants, because they did not understand the task. In the VR setup, seven participants did not press the button. Therefore, the IRT was evaluated for only 30 participants; however, subjective data are still described for all 37 participants. For the video setups, we asked participants to answer if the video ran smoothly from a technical point of view. All trials in which participants indicated technical problems were excluded from the evaluation.

Due to the different setups, we had dissimilar maximum values for the IRT: in the WoZ setup, the driving behavior varied from trial to trial because of the human driver [24]. Accordingly, the videos of the WoZ setup are also dependent on the driver. Both videos were cut at the moment the AV came to a complete stop or had passed the pedestrian. The time may vary due to human error, so the lengths of the routes were calculated to specify the maximum IRTs. Therefore, it is not possible to compare the absolute values of the IRT between the different setups. However, we had the participants' answers about the vehicle's assumed intention and if they would cross the street. Both dependent variables are related to the IRT, but can still be evaluated.

3. Results

This section is divided into five subsections. In the first three subsections, the setups are compared with each other with regard to the frequency of misinterpretations of intentions (Section 3.1), the mentioned crossing behavior (Section 3.2), and the time of decision (Section 3.3). In Section 3.4, we analyzed for each setup, separately, whether the unambiguity of driving profiles led to different IRTs and evaluations of driving behavior. In Section 3.5., IRT is compared to the start of crossing behavior for the VR setup.

3.1. Misinterpretations of Intentions

Table 2 presents the misinterpretation rate for the intention *Let the HRU go first*, while Table 3 illustrates the misinterpretation rate for the intention *AV goes first* for all setups. The results of the misinterpretations of intentions for the WoZ study were already published in [24].

For the intention *Let the HRU go first*, we found correct interpretation rates of 100% (WoZ), 97% (VR), 96% (video VR), and 89% (video WoZ) for the unambiguous driving profile. In contrast, the interpretation for the ambiguous driving profile was only correct in 23% (WoZ), 36% (video VR), 39% (video WoZ), and 70% (VR) of all trials.

For the intention *AV goes first*, the results showed a similar outcome. For the unambiguous driving profile, we found correct interpretation rates of 93% (video WoZ), 97% (WoZ and VR), and 98% (video VR). In contrast, the interpretation for the ambiguous driving profile was only correct in 29% (WoZ), 60% (VR), 68% (video WoZ), and 72% (video VR) of all trials.

To sum up, for all methods, the ambiguous driving profiles lead to higher misinterpretation rates, compared to the unambiguous profiles. This effect can especially be seen for the WoZ setup, whereas the effect is more moderate for the VR setup. However, for the video setups we found different results. The misinterpretation rate for the intention *Let the HRU go first* is between the rate for the WoZ and VR setup for both video setups. In contrast, for the intention *AV goes first* the misinterpretation rate is lower than for the WoZ and VR setup for both video setups.

Table 2. Misinterpretations of the intention *Let the HRU go first*.

	WoZ	VR	Video WoZ	Video VR
Unambiguous	0.0% (0) n = 62	2.7% (1) n = 37	11.1% (4) n = 36	4.3% (2) n = 46
Ambiguous	77.4% (48) n = 62	29.7% (11) n = 37	61.5% (24) n = 39	63.6% (28) n = 44

Table 3. Misinterpretations of the intention *AV goes first*.

	WoZ	VR	Video WoZ	Video VR
Unambiguous	3.2% (2) n = 62	2.7% (1) n = 37	7.3% (3) n = 41	2.4% (1) n = 42
Ambiguous	71.0% (44) n = 62	40.5% (15) n = 37	31.7% (13) n = 41	27.9% (12) n = 43

3.2. Mentioned Crossing Behavior

Besides the vehicle's assumed intention, we asked participants if they would cross the street. Tables 4–7 present the mentioned crossing behavior for all four intentions and setups. The tables are subdivided into the correctly or incorrectly recognized intention and the respective mentioned crossing behavior.

In total, 52% of all participants correctly realized the intention for *Let the HRU go first, unambiguous* and would have crossed the road in the WoZ study. This value is higher for the other study setups: 62% for the VR setup, 67% for the video WoZ setup, and 85% for the video VR setup (Table 4). Compared to the ambiguous driving profile, more participants would have crossed the road (Table 5). The tendency for the WoZ and the VR setup is the same: More participants would have crossed the road in the VR setup, as compared to the WoZ setup (Figure 6). Nevertheless, for the unambiguous driving profile, the highest number of participants crossed the road for both video setups, whereas for the ambiguous driving profile, the fewest participants crossed the road for the video setups (Figure 6).

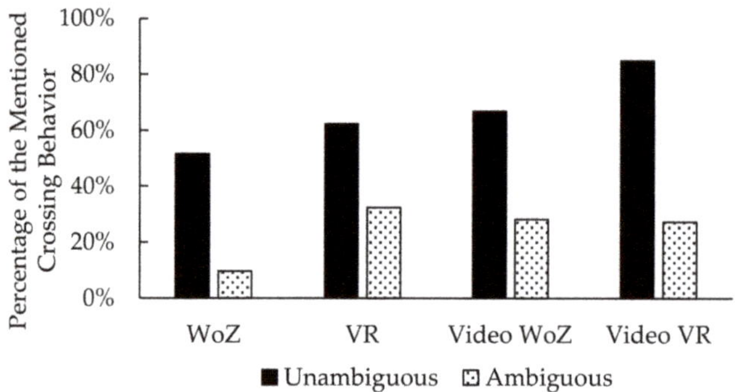

Figure 6. Mentioned crossing behavior for the intention *Let the HRU go first*, for the participants who understood the intention correctly.

For the intention *AV goes first*, it poses a safety risk if participants misunderstand the intention and would still cross the road. That risk is higher for the ambiguous driving profile for all study setups than for the unambiguous driving profile (Figure 7). Especially for the ambiguous driving profile, fewer participants would have crossed the road by mistake in the VR setup (16%), as compared to the WoZ setup (23%). The result for the video WoZ setup had the same tendency as the WoZ setup (WoZ: 23%, video WoZ: 22%; Table 7); in addition, the video VR setup had the same tendency as the VR setup (VR: 16%; video VR: 16%; Table 7). However, for the unambiguous driving profile, the collision risk was comparatively low for all four study setups (Table 6).

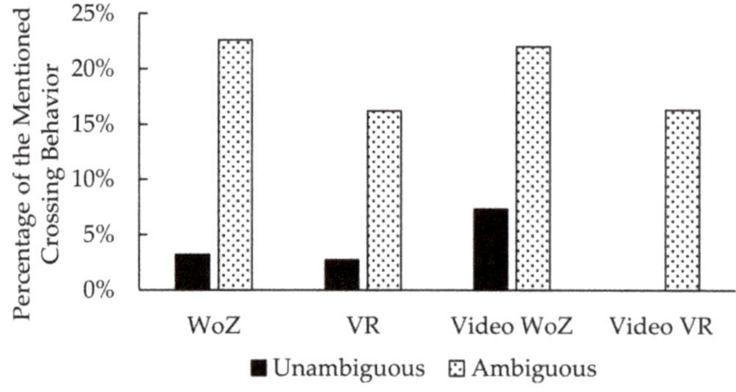

Figure 7. Mentioned crossing behavior for the intention *AV goes first*, for the participants who misunderstood the intention.

Table 4. Mentioned crossing behavior for the intention *Let the HRU go first, unambiguous.*

		WoZ				VR				Video WoZ				Video VR	
		Intention Recognition				Intention Recognition				Intention Recognition				Intention Recognition	
		Correct	False			Correct	False			Correct	False			Correct	False
Crossing	Yes	51.6% (32)	0.0% (0)	Crossing	Yes	62.2% (23)	0.0% (0)	Crossing	Yes	66.7% (24)	0.0% (0)	Crossing	Yes	84.8% (39)	0.0% (0)
	No	48.4% (30)	0.0% (0)		No	35.1% (13)	2.7% (1)		No	22.2% (8)	11.1% (4)		No	10.9% (5)	4.3% (2)

Table 5. Mentioned crossing behavior for the intention *Let the HRU go first, ambiguous.*

		WoZ				VR				Video WoZ				Video VR	
		Intention Recognition				Intention Recognition				Intention Recognition				Intention Recognition	
		Correct	False			Correct	False			Correct	False			Correct	False
Crossing	Yes	9.7% (6)	0.0% (0)	Crossing	Yes	32.4% (12)	0.0% (0)	Crossing	Yes	28.2% (11)	0.0% (0)	Crossing	Yes	27.3% (12)	2.3% (1)
	No	12.9% (8)	77.4% (48)		No	37.8% (14)	29.7% (11)		No	10.3% (4)	61.5% (24)		No	9.1% (4)	61.4% (27)

Table 6. Mentioned crossing behavior for the intention *AV goes first, unambiguous.*

		WoZ				VR				Video WoZ				Video VR	
		Intention Recognition				Intention Recognition				Intention Recognition				Intention Recognition	
		Correct	False			Correct	False			Correct	False			Correct	False
Crossing	Yes	0.0% (0)	3.2% (2)	Crossing	Yes	0.0% (0)	2.7% (1)	Crossing	Yes	0.0% (0)	7.3% (3)	Crossing	Yes	0.0% (0)	0.0% (0)
	No	96.8% (60)	0.0% (0)		No	97.3% (36)	0.0% (0)		No	92.7% (38)	0.0% (0)		No	97.6% (41)	2.4% (1)

Table 7. Mentioned crossing behavior for the intention *AV goes first, ambiguous.*

		WoZ				VR				Video WoZ				Video VR	
		Intention Recognition				Intention Recognition				Intention Recognition				Intention Recognition	
		Correct	False			Correct	False			Correct	False			Correct	False
Crossing	Yes	0.0% (0)	22.6% (14)	Crossing	Yes	0.0% (0)	16.2% (6)	Crossing	Yes	2.4% (1)	22.0% (9)	Crossing	Yes	2.3% (1)	16.3% (7)
	No	29.0% (18)	48.4% (30)		No	59.5% (22)	24.3% (9)		No	65.8% (27)	9.8% (4)		No	69.8% (30)	11.6% (5)

3.3. Time of Decision

For the time of decision, we evaluated how often the participants waited to press the button until the AV came to a complete standstill or passed by for each setup. To analyze this, only correct answers were included. Therefore, *n* varies for the different driving strategies and settings.

The results showed that, for the WoZ setup, only one participant waited until the AV passed by. However, in the other three setups, more participants waited for a complete standstill when faced with the ambiguous driving profile, compared to the unambiguous driving profile (Table 8). For the intention *AV goes first*, more participants waited in the VR and video VR setup for the AV to pass by with the ambiguous driving profile, as compared to the unambiguous profile. Only for the video WoZ setup did more participants wait for a complete standstill when faced with the unambiguous driving profile (Table 8).

Table 8. Percentage and number of participants waited to press the button until the AV came to a complete standstill or passed by, for each setup.

	WoZ	VR	Video WoZ	Video VR
Let the HRU go first, Unambiguous	0.0% (0) n = 62	11.1% (4) n = 36	43.8% (14) n = 32	13.6% (6) n = 44
Let the HRU go first, Ambiguous	0.0% (0) n = 14	42.9% (9) n = 21	73.3% (11) n = 15	87.5% (14) n = 16
AV goes first, Unambiguous	0.0% (0) n = 60	6.1% (2) n = 33	42.1% (16) n = 38	22.0% (9) n = 41
AV goes first, Ambiguous	1.6% (1) n = 18	35.0% (7) n = 20	39.3% (11) n = 28	41.9% (13) n = 31

3.4. Unambiguity of Driving Profiles: Subjective Data and Intention Recognition Time

To evaluate the subjective data and the IRT, we only used correct answers. As we focused only on the comparison between WoZ and VR, WoZ and video WoZ, VR and video VR, and video WoZ and video VR (Figure 5), we calculated planned contrasts between those setups and compared the *p*-values with a Bonferroni-corrected alpha of 0.0125. For the comparison between the independent samples, Mann–Whitney U-tests were calculated, and for the comparison between the two video setups (in which the samples are dependent), Wilcoxon tests were calculated (Figure 8).

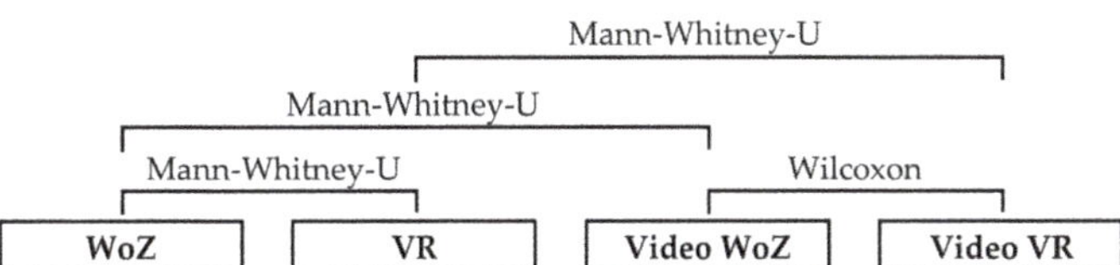

Figure 8. Comparison for the subjective data.

As already published in [24] for the WoZ setup, we also tested whether the driving profiles led to different IRTs and evaluations of driving behavior. Therefore, we used the mean of the repeated measurements for every dependent variable for each driving profile for the results of the WoZ setup. Hence, two non-parametric Wilcoxon tests were calculated for all dependent variables (one for each intention), and we compared the *p*-values with an alpha of 0.05.

3.4.1. Intention Recognition Time

The Wilcoxon tests only revealed significant differences for the intention *Let the HRU go first* for the two video setups. Moreover, the IRT was higher for the ambiguous driving profile for the WoZ, VR and video WoZ setups, whereas for the video VR setup, the IRT was higher for the unambiguous driving profile (Table 9).

However, for the intention *AV goes first*, significant differences for all four setups comparing the unambiguous and ambiguous driving profile were found (Table 9). For all four setups, participants needed more time to correctly interpret the ambiguous driving profile.

Table 9. Median (*Mdn*) of the IRT (measured in seconds), segregated by setup.

	WoZ	VR	Video WoZ	Video VR
Let the HRU go first, Unambiguous	4.1 s	4.5 s	5.3 s	5.6 s
Let the HRU go first, Ambiguous	4.2 s	4.8 s	6.5 s	5.5 s
	$z = -0.62$	$z = -0.45$	$z = -3.26$	$z = -2.80$
	$p = 0.534$	$p = 0.657$	$p = 0.001$	$p = 0.005$
			$r = 0.87$	$r = 0.70$
	$(n = 11)$	$(n = 26)$	$(n = 14)$	$(n = 16)$
AV goes first, Unambiguous	3.3 s	3.8 s	4.7 s	4.4 s
AV goes first, Ambiguous	4.6 s	5.3 s	6.7 s	6.8 s
	$z = -2.85$	$z = -2.82$	$z = -3.75$	$z = -4.72$
	$p = 0.004$	$p = 0.005$	$p \leq 0.001$	$p \leq 0.001$
	$r = 0.86$	$r = 0.81$	$r = 0.74$	$r = 0.88$
	$(n = 11)$	$(n = 12)$	$(n = 26)$	$(n = 29)$

3.4.2. Subjective Decision-Making Reliability

For the intention *Let the HRU go first, unambiguous* ($z = -1.38$, $p = 0.167$), the intention *Let the HRU go first, ambiguous* ($z = -0.14$, $p = 0.892$), and the intention *AV goes first, unambiguous* ($z = -2.35$, $p = 0.019$), we did not find significant differences between the WoZ and VR setups after the Bonferroni correction. However, for the intention *AV goes first, ambiguous*, there was a significantly higher subjective decision-making reliability ($z = -2.84$, $p = 0.004$, $r = 0.45$) for the VR setup (*Mdn* = 5.0), as compared to the WoZ setup (*Mdn* = 3.0; Figure 9).

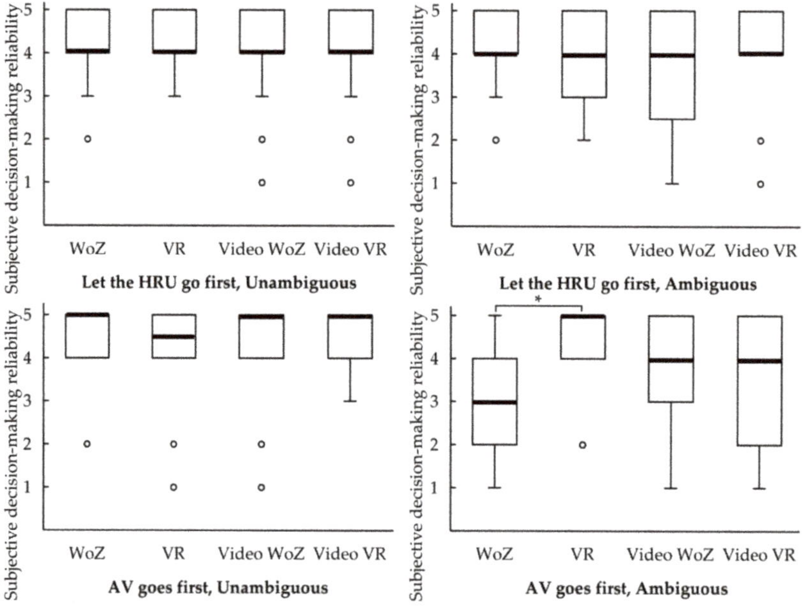

Figure 9. Boxplots for the subjective decision-making reliability (1 = very uncertain; 5 = very certain), segregated by setup (* = $p < 0.0125$).

The comparison between VR and video VR revealed no significant differences for any of the four intentions (*Let the HRU go first, unambiguous*: $z = -0.58$, $p = 0.561$; *Let the HRU go first, ambiguous*: $z = -0.08$, $p = 0.934$; *AV goes first, unambiguous*: $z = -0.05$, $p = 0.963$; *AV goes first, ambiguous*: $z = -1.43$, $p = 0.154$).

In addition, the results for the subjective decision-making reliability of the WoZ and the video WoZ setups revealed no significant differences (*Let the HRU go first, unambiguous*: $z = -1.61$, $p = 0.107$; *Let the HRU go first, ambiguous*: $z = -0.324$, $p = 0.746$; *AV goes first, unambiguous*: $z = -1.39$, $p = 0.163$; *AV goes first, ambiguous*: $z = -2.01$, $p = 0.045$).

We also found no significant differences for the video VR and the video WoZ setups (*Let the HRU go first, unambiguous*: $z = -2.39$, $p = 0.017$; *Let the HRU go first, ambiguous*: $z = -0.33$, $p = 0.740$; *AV goes first, unambiguous*: $z = -0.43$, $p = 0.668$; *AV goes first, ambiguous*: $z = -0.53$, $p = 0.595$).

The boxplots (Figure 9) illustrated that the inter-quartile ranges (IQRs) for the WoZ setup for the intention *Let the HRU go first* are both comparatively small. In contrast, for the intention *AV goes first* the boxplots differ in their IQRs with regard to the unambiguous and the ambiguous driving profile: The range for the ambiguous driving profile is greater than the range for the unambiguous driving profile. The boxplots for the VR setup revealed a different result: the IQRs for the intention *AV goes first* are both small. For the intention *Let the HRU go first* the range is greater for the ambiguous driving profile than for the unambiguous profile. As presented in Section 3.3, more participants in the VR setup waited for a complete standstill or for the vehicle to pass before answering the questions. For both driving strategies, the participants who waited for the complete driving strategy were very confident in their decision (first quartile, median, and third quartile: 5.0). For the other participants, the boxplots are very tall (first quartile: 2.8, median: 4.0, and third quartile: 4.3).

The IQRs for the video setups are relatively small for the unambiguous driving profiles, but comparatively large for the *AV goes first, ambiguous* driving profile. This is comparable with the boxplots from the WoZ setup. However, for the intention *Let the HRU go first, ambiguous*, the IQR for the video WoZ setup is much greater than for the video VR setup and the WoZ setup. For both video setups, the number of participants who waited for the complete driving profile is relatively high (Table 8).

For the intention *Let the HRU go first*, none of the setups showed a significant difference in terms of decision-making reliability between the ambiguous and the unambiguous driving profile. For the WoZ setup, the subjective decision-making reliability revealed a significant difference for the driving profile *AV goes first* between the unambiguous and the ambiguous driving profile (Table 10; the median in Table 10 for the WoZ setup differs from the median in Figure 9, since we used the mean of the repeated measurements for comparison within the setup). The participants were more confident with their decision when the driving profile was unambiguous. This is comparable with the results from both video setups, even if these were not significant.

Table 10. Median (*Mdn*) of the subjective decision-making reliability (1 = very uncertain, 5 = very certain), segregated by setup.

	WoZ	VR	Video WoZ	Video VR
Let the HRU go first, Unambiguous	4.5	4.0	4.0	4.0
Let the HRU go first, Ambiguous	4.0	4.0	4.0	4.0
	$z = -1.21$	$z = -0.83$	$z = -0.98$,	$z = -0.50$
	$p = 0.226$	$p = 0.406$	$p = 0.329$	$p = 0.615$
	$(n = 11)$	$(n = 26)$	$(n = 14)$	$(n = 16)$
AV goes first, Unambiguous	5.0	4.5	5.0	5.0
AV goes first, Ambiguous	3.0	5.0	4.0	4.0
	$z = -2.94$	$z = -0.88$	$z = -1.83$	$z = -1.84$
	$p = 0.003$	$p = 0.377$	$p = 0.068$	$p = 0.066$
	$r = 0.89$	$(n = 22)$	$(n = 26)$	$(n = 29)$
	$(n = 11)$			

3.4.3. Evaluation of Driving Behavior

Just as for the subjective decision-making reliability, the differences for the evaluation of driving behavior showed no significant differences for the intention *Let the HRU go first, unambiguous* ($z = -0.38$, $p = 0.702$) and the intention *Let the HRU go first, ambiguous* ($z = -1.42$, $p = 0.156$). We also found no significant difference for the intention *AV goes first, ambiguous* ($z = -0.34$, $p = 0.734$). However, the participants rated the driving behavior significantly better in the WoZ setup (*Mdn* = 4.0) than in the VR setup (*Mdn* = 4.0) ($z = -4.59$, $p \leq 0.001$, $r = 0.47$) for the intention *AV goes first, unambiguous* (Figure 10).

The comparison between the WoZ and the video WoZ setup showed a significant difference for the intention *Let the HRU go first, unambiguous* ($z = -3.12$, $p = 0.002$, $r = 0.33$). The rating is better for the WoZ setup (*Mdn* = 4.0) than for the video WoZ setup (*Mdn* = 4.0). Moreover, the intention *AV goes first, unambiguous* revealed a significantly better rating for the WoZ setup (*Mdn* = 4.0) than for the video WoZ setup (*Mdn* = 4.0; $z = -4.20$, $p \leq 0.001$, $r = 0.42$). For the intention *Let the HRU go first, ambiguous* ($z = -2.04$, $p = 0.041$), and *AV goes first, ambiguous* ($z = -0.42$, $p = 0.678$) no significant differences were found.

No significant differences for all intentions were found when comparing the VR and video VR setup (*Let the HRU go first, unambiguous*: $z = -1.63$, $p = 0.103$; *Let the HRU go first, ambiguous*: $z = -1.00$, $p = 0.319$; *AV goes first, unambiguous*: $z = -0.08$, $p = 0.936$; *AV goes first, ambiguous*: $z = -0.11$, $p = 0.909$), as well as video WoZ and video VR setups (*Let the HRU go first, unambiguous*: $z = -0.28$, $p = 0.776$; *Let the HRU go first, ambiguous*: $z = -0.14$, $p = 0.890$; *AV goes first, unambiguous*: $z = -1.08$, $p = 0.279$; *AV goes first, ambiguous*: $z = -1.04$, $p = 0.299$).

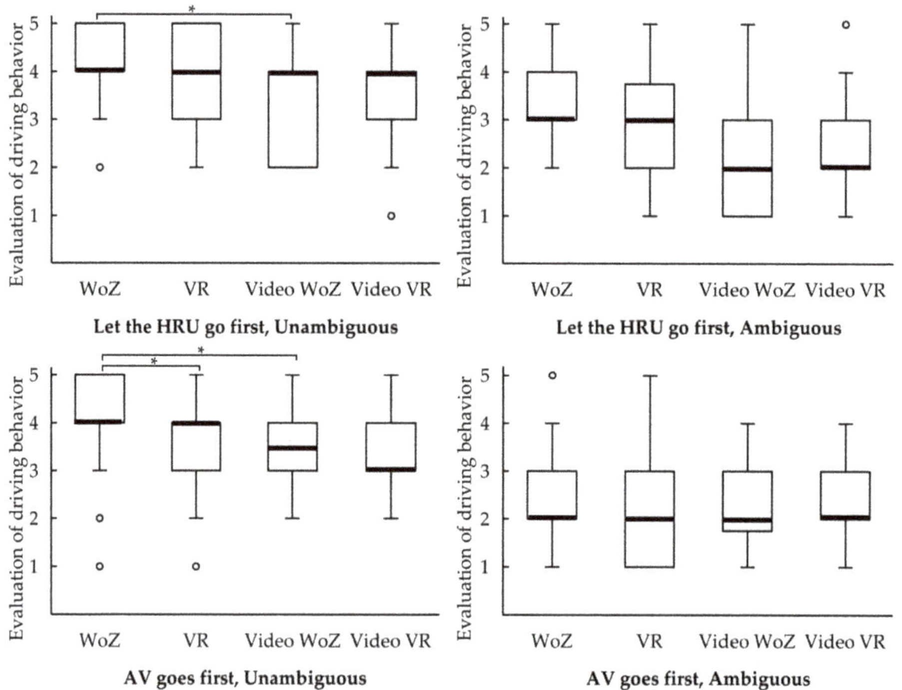

Figure 10. Boxplots for evaluation of driving behavior (1 = very poor, 5 = very good), segregated by setup (* = $p < 0.0125$).[M1] [W2]

The IQRs for all boxplots for the WoZ setups are comparatively small. However, with the exception of the intention *AV goes first, unambiguous*, the IQRs for the VR setup are rather large. For the mentioned intention, very few participants (6%) waited until the vehicle had passed by (Table 8). The large IQRs for both ambiguous driving profiles might have occurred due to those participants who waited to see the entirety of the driving profiles (Figure 11). However, this does not explain the larger IQR for the intention *Let the HRU go first, unambiguous*, because only four participants waited for the complete standstill of the AV (Table 8). In addition, the boxplots for both video setups revealed different IQRs that cannot be explained by the fact that some participants waited. However, all boxplots illustrate that the unambiguous driving profiles tend to be rated better than the ambiguous driving profiles (Figure 10).

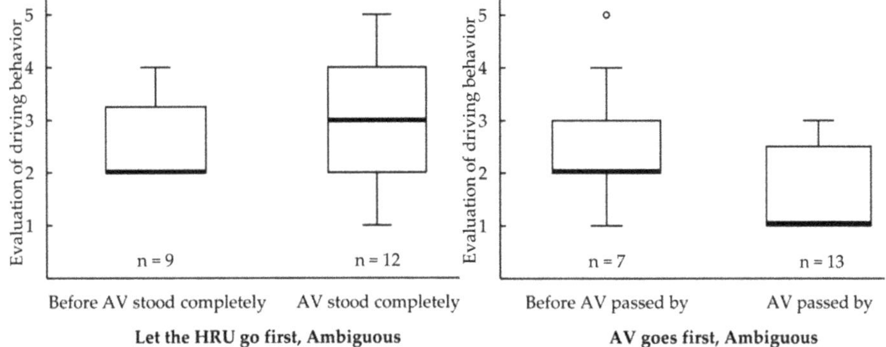

Figure 11. Boxplots for evaluation of driving behavior (1 = very poor, 5 = very good) for the VR setup, segregated by time of decision (before the AV reached standstill or after the AV reached standstill, and before the AV passed by or waited until the AV passed by).[M3] [W4]

The evaluation of the driving behavior showed significant differences for all four setups and both driving strategies (*Let the HRU go first* and *AV goes first*), between the unambiguous and ambiguous driving profiles. The participants rated the unambiguous driving profiles better than the ambiguous driving profiles in all four setups (Table 11). Here, the deviating median listed in the table and boxplots results from using the mean of the repeated measurements for the WoZ setup.

Table 11. Median (*Mdn*) of the evaluation of driving behavior (1 = very poor, 5 = very good), segregated by setup.

	WoZ	VR	Video WoZ	Video VR
Let the HRU go first, Unambiguous	4.5	4.0	4.0	4.0
Let the HRU go first, Ambiguous	3.5	3.0	2.0	2.0
	$z = -2.70$,	$z = -3.79$,	$z = -2.99$	$z = -2.56$
	$p = 0.007$,	$p \leq 0.001$,	$p = 0.003$	$p = 0.011$
	$r = 0.81$	$r = 0.74$	$r = 0.80$	$r = 0.64$
	$(n = 11)$	$(n = 26)$	$(n = 14)$	$(n = 16)$
AV goes first, Unambiguous	4.5	4.0	3.5	3.0
AV goes first, Ambiguous	2.0	2.0	2.0	2.0
	$z = -2.96$	$z = -3.01$	$z = -3.03$	$z = -3.20$
	$p = 0.003$	$p = 0.003$	$p = 0.002$	$p = 0.001$
	$r = 0.89$	$r = 0.64$	$r = 0.59$	$r = 0.59$
	$(n = 11)$	$(n = 22)$	$(n = 26)$	$(n = 29)$

3.4.4. Perceived Criticality

In terms of perceived criticality, no significant differences were revealed between the WoZ and VR setups (*Let the HRU go first, unambiguous*: $z = -0.32$, $p = 0.749$; *Let the HRU go first, ambiguous*: $z = -0.46$, $p = 0.645$; *AV goes first, unambiguous*: $z = -1.44$, $p = 0.151$; *AV goes first, ambiguous*: $z = -0.25$, $p = 0.801$).

However, we found significant differences for the intention *Let the HRU go first, ambiguous* ($z = -2.56$, $p = 0.011$, $r = 0.26$) and the intention *AV goes first, unambiguous* ($z = -2.79$, $p = 0.005$, $r = 0.28$) between the WoZ and the video WoZ setups (Figure 12). For both intentions, the perceived criticality is higher for the WoZ setup (for both intentions: $Mdn = 4.0$), as compared to the video WoZ setup (*Let the HRU go first, ambiguous*: $Mdn = 3.0$; *AV goes first, unambiguous*: $Mdn = 4.0$). For the intention *Let the HRU go first, unambiguous* ($z = -1.44$, $p = 0.151$) and for the intention *AV goes first, unambiguous* ($z = -0.10$, $p = 0.917$), no significant differences were found.

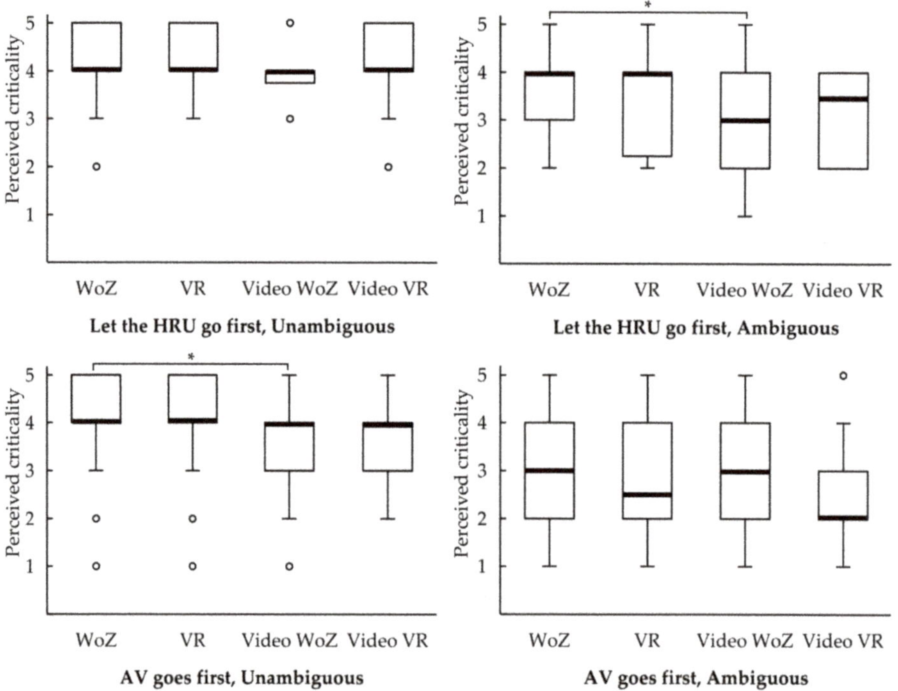

Figure 12. Boxplots for perceived criticality (1 = very critical, 5 = very uncritical), segregated by setups (* = $p < 0.0125$).

Moreover, the VR and video VR setup (*Let the HRU go first, unambiguous*: $z = -1.50$, $p = 0.134$; *Let the HRU go first, ambiguous*: $z = -1.03$, $p = 0.305$; *AV goes first, unambiguous*: $z = -1.14$, $p = 0.255$; *Go first, ambiguous*: $z = -0.43$, $p = 0.595$) revealed no significant differences.

Furthermore, no differences were found for the perceived criticality between the video VR and video WoZ setup (*Let the HRU go first, unambiguous*: $z = 0.00$, $p \geq 0.999$; *Let the HRU go first, ambiguous*: $z = -0.82$, $p = 0.412$; *AV goes first, unambiguous*: $z = -1.89$, $p = 0.059$; *AV goes first, ambiguous*: $z = -0.86$, $p = 0.388$).

All boxplots illustrate that the ambiguous driving profiles tend to be rated more critically than the unambiguous driving profiles (Figure 12). The boxplots for both ambiguous driving profiles showed larger IQRs for all setups compared to the unambiguous driving profiles. The only exception is the boxplot for the intention *AV goes first, ambiguous* for the video VR setup: The IQRs are not larger for the ambiguous driving profile than for the unambiguous driving profile. This is independent of whether the participants waited to see the entirety of the driving profile (IQRs for both groups: first quartile: 2.0, median: 2.0, third quartile: 3.0).

We also evaluated the extent to which the unambiguity influences the perceived criticality for all setups. In all four setups, participants rated the situation to be significantly less critical if the driving profile was unambiguous for both intentions (Table 12). As before, the median in the boxplot differs from the median listed in the table for the WoZ setup, because the mean of the repeated measurements for the comparison was used for the table (Table 12).

Table 12. Median (*Mdn*) of the perceived criticality (1 = very critical, 5 = very uncritical), segregated by setup.

	WoZ	VR	Video WoZ	Video VR
Let the HRU go first, Unambiguous	4.5	4.0	4.0	4.0
Let the HRU go first, Ambiguous	4.0	4.0	3.0	3.5
	$z = -2.41$	$z = -2.98$	$z = -2.57$	$z = -2.23$
	$p = 0.016$	$p = 0.003$	$p = 0.010$	$p = 0.026$
	$r = 0.73$	$r = 0.58$	$r = 0.69$	$r = 0.56$
	(n = 11)	(n = 26)	(n = 14)	(n = 16)
AV goes first, Unambiguous	4.5	4.0	4.0	4.0
AV goes first, Ambiguous	3.0	2.5	3.0	2.0
	$z = -2.82$	$z = -2.42$	$z = -2.02$	$z = -2.98$
	$p = 0.005$	$p = 0.016$	$p = 0.043$	$p = 0.003$
	$r = 0.85$	$r = 0.52$	$r = 0.40$	$r = 0.55$
	(n = 11)	(n = 22)	(n = 26)	(n = 29)

3.5. VR Study: IRT vs. Start of Road Crossing

As mentioned in Section 2.3.2, we asked participants in the VR setup to cross the street instead of pressing a button. Reaction times such as IRTs and the crossing time were not normally distributed. Therefore, two Wilcoxon tests were calculated to evaluate possible differences between the IRT and the crossing time for the intention *Let the HRU go first*.

The results revealed that participants made their decision for the intention *Let the HRU go first, unambiguous* earlier (IRT, *Mdn* = 4.5 s) and waited significantly longer to cross the street (*Mdn* = 7.2 s; $z = -5.09, p \leq 0.001, r = 0.87$). A comparable result was found for the intention *Let the HRU go first, ambiguous* ($z = -3.90, p \leq 0.001, r = 0.76$). Participants made their decision first (IRT, *Mdn* = 4.8 s) and crossed the street later (*Mdn* = 6.9 s). This leads to lower misinterpretation rates for all intentions (Table 13).

Table 13. Misinterpretations of the intentions for the metrics IRT and start of road crossing.

	Let the HRU Go First, Unambiguous	Let the HRU Go First, Ambiguous	AV Goes First, Unambiguous	AV Goes First, Ambiguous
IRT	2.7% (1) n = 37	29.7% (11) n = 37	2.7% (1) n = 37	40.5% (15) n = 37
Start of Road Crossing	2.7% (1) n = 37	2.7% (1) n = 37	2.7% (1) n = 37	5.4% (2) n = 37

Just as with the IRT, there are no significant differences between the unambiguous and the ambiguous driving profiles for the start of road crossing ($z = -0.77, p = 0.442$).

4. Discussion

The aim of the study was to compare different study setups that can be used to evaluate the driving behavior of AVs, in order to be able to give indications as to whether already-conducted studies can be compared with each other. Therefore, we replicated the same study design in four different settings: WoZ, VR, video WoZ, and video VR. In all studies, participants stood at the roadside in a shared space. An AV approached from the left, using different driving profiles, characterized by changing speed as a way of communicating its intention to let the pedestrian cross the road. Participants were asked to recognize the intention of the AV and to press a button as soon as they had realized this intention.

Since the WoZ setup is the closest to reality, the authors assume that the values measured in this setup are the most realistic ones. The other setups were related to the results of the WoZ setup.

The misinterpretation rates for the ambiguous driving profiles were underestimated in VR, video WoZ, and video VR, as compared to the WoZ setup: The misinterpretation rate is lower in those setups. However, differences between unambiguous and ambiguous driving strategies were revealed in all setups, since the misinterpretation rate was higher for ambiguous driving profiles compared to the unambiguous profiles. This coincides with the results of previous studies, employing video, VR, and WoZ setups, where pedestrians refer to differences in driving strategies when crossing the road (e.g., [8,12,15,16,19]).

For the intention *Let the HRU go first*, it was preferable that participants recognize the intention correctly and cross the road before the AV had to come to a standstill. The results for the crossing behavior showed that the proportion of those pedestrians is overestimated in VR, video WoZ, and video VR, as compared to the WoZ setup for the unambiguous and the ambiguous driving profile. While the results for both video setups for the intention *Let the HRU go first, ambiguous* are approximately the same (Δ 1%), there is a rather high discrepancy for the intention *Let the HRU go first, unambiguous* (Δ 18%). This result suggests that the crossing behavior is dependent on the type of video.

As mentioned in the results, it poses a safety risk if participants misunderstand the intention and cross the road for the intention *AV goes first*. As for the misinterpretation rate, all setups detect this risk especially for the ambiguous driving profile. While the risk for the unambiguous driving strategy is assessed almost equally by all setups, the risk was underestimated in the VR setup for the ambiguous driving profile compared to the WoZ setup (WoZ vs. VR: *AV goes first, unambiguous* Δ 1%, *AV goes first, ambiguous* Δ 6%). Just like the results for the intention *Let the HRU go first, unambiguous*, the results for the intention *AV goes first, ambiguous* are also dependent on the choice of video: The video WoZ setup can reproduce the critical crossing rate from the WoZ setup (Δ <1%), and the video VR can reproduce the results from the VR setup (Δ <1%).

The comparison also showed that, in the WoZ setup, only one participant waited to see the whole driving profile; all others had made their decision before this point. In the VR setup, a total of 20% of all participants who correctly realized the intention, waited to make their decision until the end of the driving profile. That rate is higher for the ambiguous driving profile (39%) compared to the unambiguous profile (9%). Therefore, it seems that the perception of the driving profiles is more difficult for participants in a VR setup. However, understanding intentions by using the driving profiles appears to be even more difficult when only seeing videos. Most participants waited until the end of the driving profile (46%) in the video WoZ setup; however, also in the video VR setup, many participants waited to see the whole driving profile (32%). It is possible to differentiate between unambiguous and ambiguous driving profiles with just the results of a VR or a video study, but the results are not transferable to reality, because the pedestrians made their decisions in the WoZ setup at an earlier stage.

The results for the subjective decision-making reliability let no clear statement be made regarding the significance tests. The different IQRs result from participants who waited until the vehicle stood completely or had passed by, depending on the study setup. However, the results for the WoZ setup revealed the greatest IQR for the intention *AV goes first, ambiguous*. In addition, the comparison between the *AV goes first, unambiguous* and *AV goes first, ambiguous* driving profile in the WoZ setup showed the only significant difference across all setups. The results indicate that the *AV goes first, ambiguous* profile leads to the most uncertainties. In contrast, the *AV goes first, unambiguous* profile revealed the shortest IQRs across all setups. A reason could be that, in this driving strategy, the AV does not change its speed.

This can also be seen for the evaluation of the driving profile: in all four setups, the IQRs for the intention *AV goes first, unambiguous* were short. The driving strategy led to clear trends in the evaluations. With one exception, the intention *AV goes first, ambiguous*, the driving strategies were rated better in the WoZ setup. The intention *AV goes first, ambiguous* is rated equally bad in all setups. When looking at the boxplots and the significance tests, it becomes clear that the item can be used to distinguish between unambiguous and ambiguous driving strategies in all settings. This effect can especially be seen for the WoZ setup, because the effect size is greatest for this setup, compared to the other three setups. However, the IQRs for the VR setup to some extent—but especially for the video setups—cannot be explained by the results. This could be due to perception and/or decision artefacts.

The perceived criticality is higher in the WoZ setup for some intentions, as compared to the video WoZ setup. However, there is no clear tendency for the perceived criticality to be systematically underestimated in the video setups or the VR setup. It is possible in all setups to differentiate between the unambiguous and ambiguous driving profiles. However, the effect size is greatest for the WoZ setup.

In addition to the setup comparison, the VR setup was used to check how the IRT metrics differ in terms of the start of road crossing. Results revealed that participants made their decision regarding the AV's intention significantly earlier than they would cross the road. A motor process must be performed for both metrics; however, more time is needed to walk one meter than to press a button. Nevertheless, this does not explain the time difference of 2.7 s between IRT and the start of road crossing for the unambiguous and 2.1 s for the ambiguous driving profile. However, it can be assumed that pedestrians assess the AV's driving behavior at an early stage, but wait until they are certain in their decision before crossing the road. Due to the longer time period, participants saw more of the whole driving profile and made more correct decisions, compared to the IRT metric. However, for the intention *AV goes first, ambiguous*, two persons still crossed the road by mistake. In real-life traffic situations, but also in the WoZ setup, this behavior would probably have led to an accident.

5. Limitations

Even though we tried to replicate the setups as much as possible, there were small differences: In the VR and video setups, for example, no engine sound was presented to the participants. Compared to the results from [22], this might deteriorate the task performance. Furthermore, the environment varied in the WoZ (rather rural) and VR setup (rather urban).

In addition, in the WoZ setup the driver accelerated to the original speed at the moment the participants pressed the button, so that they were not influenced by the remaining driving profile. In the VR setup and both video setups, the video was frozen the moment participants pressed the button. These limitations might have led to differences between the setups.

Although all vehicles were BMWs, a BMW 2 series was used in the WoZ setup, and a BMW 3 series was used in the VR setup. As mentioned, Ref. [4] found a significant effect for different vehicles sizes. However, the authors compared a Smart Fortwo, a BMW Z4, and a Ford F150; therefore, the different sizes of the vehicles were comparatively large compared with our vehicles. In addition, the differences found had only a small effect [4].

Furthermore, there are also weaknesses in the analysis: Equivalence tests should have been carried out instead of significance tests for differences. Unfortunately, the prerequisites were not met, due to the ordinal-scaled data and small sample sizes. For this reason, the authors have limited themselves to report descriptive data for most results.

Methodologically, it was not possible to compare IRT between the studies, because the different times measurements calculating the IRT were not synchronized. We implemented the driving profiles for the VR setup as a replicate from the specification. However, due to the low sampling rate of 5 Hz, differences of a maximum of 200 ms may occur. For the video VR setup, the videos were screened on the monitor, and for the video WoZ setup, a driving throughput was recorded. Due to the cutting of the video sequences, the driving data can no longer be clearly calculated for the respective video. This makes it impossible to use the absolute IRT values for the setup comparison. However, the comparison within the setup is possible, even if the driving profiles themselves are of different lengths

Furthermore, it would have been useful to add a setup in which a programmable vehicle runs the given profiles, since the driving strategies in the WoZ differ for each trial, because a human driver is not able to precisely replicate a given driving profile [24].

6. Conclusions

To sum up, it can be stated that the WoZ setup is a useful approach to evaluate large differences between trajectories. However, small changes in driving behavior cannot be assessed, as a human driver is not able to replicate these [24]. Using the misinterpretation and crossing rate, it is possible to differentiate between unambiguous and ambiguous driving profiles in VR setups. Nevertheless, the collision risk would be underestimated in the VR setup compared to the WoZ setup, because less participants would have crossed the road by mistake in the VR setup. Conclusions as to absolute values are not possible in the VR setup. It is possible to detect a potential ambiguous driving profile when using a video setup. However, the type of video influences, among other things, the collision risk. Additionally, it is possible that perception and decision artefacts will emerge in a video study.

Author Contributions: Conceptualization, T.F., E.S., and K.B.; methodology, T.F..; formal analysis, T.F.; investigation, T.F.; resources, T.F., E.S., and K.B.; data curation, T.F.; writing—original draft preparation, T.F.; writing—review and editing, T.F., E.S., and K.B.; visualization, T.F. and E.S.; supervision, T.F. and K.B. All authors have read and agreed to the published version of the manuscript.

Funding: This research received no external funding.

Acknowledgments: The authors would like to thank Lars Michalowski for support in conducting the Wizard of Oz and the VR study.

Conflicts of Interest: The authors declare no conflict of interest.

References

1. SAE International. *Taxonomy and Definitions for Terms Related to Driving Automation Systems for On-Road Motor Vehicles (J3016)*; SAE International: Warrendale, PA, USA, 2018.
2. Schneemann, F.; Gohl, I. Analyzing driver-pedestrian interaction at crosswalks: A contribution to autonomous driving in urban environments. In Proceedings of the IEEE Intelligent Vehicles Symposium (IV), Gothenburg, Sweden, 19–22 June 2016; pp. 38–43, ISBN 978-1-5090-1821-5.
3. Fuest, T.; Sorokin, L.; Bellem, H.; Bengler, K. Taxonomy of Traffic Situations for the Interaction between Automated Vehicles and Human Road Users. In *Advances in Human Aspects of Transportation. AHFE 2017. Advances in Intelligent Systems and Computing*; Stanton, N.A., Ed.; Springer International Publishing: Cham, Switzerland, 2018; Volume 597, pp. 708–719. [CrossRef]
4. De Clercq, K.; Dietrich, A.; Núñez Velasco, J.P.; de Winter, J.; Happee, R. External Human-Machine Interfaces on Automated Vehicles: Effects on Pedestrian Crossing Decisions. *Hum. Factors* **2019**, *61*, 8. [CrossRef]
5. Burns, C.G.; Oliveira, L.; Thomas, P.; Iyer, S.; Birrell, S. Pedestrian Decision-Making Responses to External Human-Machine Interface Designs for Autonomous Vehicles. In Proceedings of the IEEE Intelligent Vehicles Symposium (IV), Paris, France, 9–12 June 2019; pp. 70–75.
6. Weber, F.; Chadowitz, R.; Schmidt, K.; Messerschmidt, J.; Fuest, T. Crossing the Street Across the Globe: A Study on the Effects of eHMI on Pedestrians in the US, Germany and China. In *HCI in Mobility, Transport, and Automotive Systems*; Krömker, H., Ed.; Springer International Publishing: Cham, Switzerland, 2019; pp. 515–530, ISBN 978-3-030-22665-7.
7. Rettenmaier, M.; Pietsch, M.; Schmidtler, J.; Bengler, K. Passing through the Bottleneck—The Potential of External Human-Machine Interfaces. In Proceedings of the IEEE Intelligent Vehicles Symposium (IV), Paris, France, 9–12 June 2019; pp. 1687–1692. [CrossRef]

8. Clamann, M.; Aubert, M.; Cummings, M.L. Evaluation of Vehicle-to-Pedestrian Communication Displays for Autonomous Vehicles. In Proceedings of the Transportation Research Board 96th Annual Meeting, Washington, DC, USA, 8–12 January 2017.

9. Kühn, M.; Stange, V.; Vollrath, M. Menschliche Reaktion auf hochautomatisierte Fahrzeuge im Mischverkehr auf der Autobahn. In *VDI Tagung Mensch-Maschine-Mobilität 2019—Der (Mit-)Fahrer im 21.Jahrhundert!?* VDI Verlag: Düsseldorf, Germany, 2019; pp. 169–184.

10. Bengler, K.; Rettenmaier, M.; Fritz, N.; Feierle, A. From HMI to HMIs: Towards an HMI Framework for Automated Driving. *Information* **2020**, *11*, 61. [CrossRef]

11. Fuest, T.; Feierle, A.; Schmidt, E.; Bengler, K. Effects of Marking Automated Vehicles on Human Drivers on Highways. *Information* **2020**, *11*, 286. [CrossRef]

12. Hensch, A.-C.; Neumann, I.; Beggiato, M.; Halama, J.; Krems, J.F. Effects of a light-based communication approach as an external HMI for Automated Vehicles—A Wizard-of-Oz Study. *ToTS* **2020**, *10*, 18–32. [CrossRef]

13. Song, Y.E.; Lehsing, C.; Fuest, T.; Bengler, K. External HMIs and Their Effect on the Interaction between Pedestrians and Automated Vehicles. In *Intelligent Human Systems Integration*; Karwowski, W., Ahram, T., Eds.; Springer International Publishing: Cham, Switzerland, 2018; pp. 13–18, ISBN 978-3-319-73887-1.

14. Fridman, L.; Mehler, B.; Xia, L.; Yang, Y.; Facusse, L.Y.; Reimer, B. To Walk or Not to Walk: Crowdsourced Assessment of External Vehicle-to-Pedestrian Displays. In Proceedings of the 98th Annual Transportation Research Board Meeting, Washington, DC, USA, 12–17 January 2019.

15. Fuest, T.; Maier, A.S.; Bellem, H.; Bengler, K. How Should an Automated Vehicle Communicate Its Intention to a Pedestrian?—A Virtual Reality Study. In *Human Systems Engineering and Design II*; Ahram, T., Karwowski, W., Pickl, S., Taiar, R., Eds.; Springer International Publishing: Cham, Switzerland, 2020; pp. 195–201. [CrossRef]

16. Dey, D.; Martens, M.; Eggen, B.; Terken, J. Pedestrian road-crossing willingness as a function of vehicle automation, external appearance, and driving behaviour. *Transp. Res. Part F Traffic Psychol. Behav.* **2019**, *65*, 191–205. [CrossRef]

17. Eisma, Y.B.; van Bergen, S.; ter Brake, S.M.; Hensen, M.T.T.; Tempelaar, W.J.; de Winter, J.C.F. External Human-Machine Interfaces: The Effect of Display Location on Crossing Intentions and Eye Movements. *Information* **2020**, *11*, 13. [CrossRef]

18. Schmidt, H.; Terwilliger, J.; AlAdawy, D.; Fridman, L. Hacking Nonverbal Communication between Pedestrians and Vehicles in Virtual Reality. *arXiv* **2019**, arXiv:1904.01931.

19. Dietrich, A.; Maruhn, P.; Schwarze, L.; Bengler, K. Implicit Communication of Automated Vehicles in Urban Scenarios: Effects of Pitch and Deceleration on Pedestrian Crossing Behavior. In *Human Systems Engineering and Design II*; Ahram, T., Karwowski, W., Pickl, S., Taiar, R., Eds.; Springer International Publishing: Cham, Switzerland, 2020; pp. 176–181, ISBN 978-3-030-27927-1.

20. Böckle, M.-P.; Brenden, A.P.; Klingegård, M.; Habibovic, A.; Bout, M. SAV2P—Exploring the Impact of an Interface for Shared Automated Vehicles on Pedestrians' Experience. In Proceedings of the 9th International Conference on Automotive User Interfaces and Interactive Vehicular Applications Adjunct(Automotive UI '17), Oldenburg, Germany, 24–27 September 2017; Löcken, A., Boll, S., Politis, I., Osswald, S., Schroeter, R., Large, D., Baumann, M., Alvarez, I., Chuang, L., Feuerstack, S., et al., Eds.; ACM Press: New York, NY, USA, 2017; pp. 136–140.

21. Chang, C.-M.; Toda, K.; Sakamoto, D.; Igarashi, T. Eyes on a Car: an Interface Design for Communication between an Autonomous Car and a Pedestrian. In Proceedings of the 9th International Conference on Automotive User Interfaces and Interactive Vehicular Applications (Automotive UI '17), Oldenburg, Germany, 24–27 September2017; Boll, S., Pfleging, B., Politis, I., Large, D., Domnez, B., Eds.; ACM Press: New York, NY, USA, 2017; pp. 65–73.

22. Bernhard, M.; Grosse, K.; Wimmer, M. Bimodal Task-Facilitation in a Virtual Traffic Scenario through Spatialized Sound Rendering. *ACM Trans. Appl. Percept.* **2011**, *8*, 1–22. [CrossRef]

23. Fraser, N.M.; Gilbert, G.N. Simulating speech systems. *Comput. Speech Lang.* **1991**, *5*, 81–99. [CrossRef]

24. Fuest, T.; Michalowski, L.; Schmidt, E.; Bengler, K. Reproducibility of Driving Profiles—Application of the Wizard of Oz Method for Vehicle Pedestrian Interaction. In Proceedings of the IEEE Intelligent Transportation Systems Conference (ITSC), Auckland, New Zealand, 27–30 October 2019; pp. 3954–3959. [CrossRef]

25. Fuest, T.; Michalowski, L.; Träris, L.; Bellem, H.; Bengler, K. Using the Driving Behavior of an Automated Vehicle to Communicate Intentions—A Wizard of Oz Study. In Proceedings of the 21st International Conference on Intelligent Transportation Systems (ITSC), Maui, HI, USA, 4–7 November 2018; pp. 3596–3601. [CrossRef]

26. Currano, R.; Park, S.Y.; Domingo, L.; Garcia-Mancilla, J.; Santana-Mancilla, P.C.; Gonzalez, V.M.; Ju, W. ¡Vamos! Observations of Pedestrian Interactions with Driverless Cars in Mexico. In Proceedings of the 10th International Conference on Automotive User Interfaces and Interactive Vehicular Applications—AutomotiveUI'18, Toronto, ON, Canada, 23–25 September 2018; ACM Press: New York, NY, USA, 2018; pp. 210–220.

27. Rothenbücher, D.; Li, J.; Sirkin, D.; Mok, B.; Ju, W. Ghost Driver: A Field Study Investigating the Interaction Between Pedestrians and Driverless Vehicles. In Proceedings of the 7th International Conference on Automotive User Interfaces and Interactive Vehicular Applications—AutomotiveUI'15, Nottingham, UK, 1–3 September 2015; Burnett, G., Gabbard, J., Green, P., Osswald, S., Eds.; ACM Press: New York, NY, USA, 2015; pp. 44–49.

28. Joisten, P.; Alexandi, E.; Drews, R.; Klassen, L.; Petersohn, P.; Pick, A.; Schwindt, S.; Abendroth, B. Displaying Vehicle Driving Mode—Effects on Pedestrian Behavior and Perceived Safety. In *Human Systems Engineering and Design II*; Ahram, T., Karwowski, W., Pickl, S., Taiar, R., Eds.; Springer International Publishing: Cham, Switzerland, 2020; pp. 250–256, ISBN 978-3-030-27927-1.

 *information*

Article

Effects of Marking Automated Vehicles on Human Drivers on Highways

Tanja Fuest [1,*], Alexander Feierle [1], Elisabeth Schmidt [2] and Klaus Bengler [1]

1   Chair of Ergonomics, Technical University of Munich, 85748 Garching, Germany;
    alexander.feierle@tum.de (A.F.); bengler@tum.de (K.B.)
2   BMW Group, New Technologies, 85748 Garching, Germany; elisabeth.schmidt@bmw.de
*   Correspondence: tanja.fuest@tum.de

Received: 4 May 2020; Accepted: 25 May 2020; Published: 28 May 2020

Abstract: Due to the short range of the sensor technology used in automated vehicles, we assume that the implemented driving strategies may initially differ from those of human drivers. Nevertheless, automated vehicles must be able to move safely through manual road traffic. Initially, they will behave as carefully as human learners do. In the same way that driving-school vehicles tend to be marked in Germany, markings for automated vehicles could also prove advantageous. To this end, a simulation study with 40 participants was conducted. All participants experienced three different highway scenarios, each with and without a marked automated vehicle. One scenario was based around some roadworks, the next scenario was a traffic jam, and the last scenario involved a lane change. Common to all scenarios was that the automated vehicles strictly adhered to German highway regulations, and therefore moved in road traffic somewhat differently to human drivers. After each trial, we asked participants to rate how appropriate and disturbing the automated vehicle's driving behavior was. We also measured objective data, such as the time of a lane change and the time headway. The results show no differences for the subjective and objective data regarding the marking of an automated vehicle. Reasons for this might be that the driving behavior itself is sufficiently informative for humans to recognize an automated vehicle. In addition, participants experienced the automated vehicle's driving behavior for the first time, and it is reasonable to assume that an adjustment of the humans' driving behavior would take place in the event of repeated encounters.

Keywords: marking automated vehicles; automated vehicles—human drivers interaction; mixed traffic; explicit communication; external human-machine interface

---

1. Introduction

BMW has announced that the first highly automated vehicles (AVs) will be integrated into road traffic by 2021 [1]. It can be assumed that, initially, level 3 functions [2] will be available on highways. At the beginning, there will be several situations where the implemented driving strategy of an AV differs from that of a human driver. These include, in particular, situations where anticipatory driving is required, such as waiting for large gaps or reacting to missing traffic signs (e.g., changes in the speed limit). These atypical driving strategies could lead to confusion and distrust by other human road users (HRUs) [3]. One way of counteracting the confusion of HRUs is the clear identification of AVs, e.g., through special marking or additional light signals.

One argument for marking AVs, besides the positive marketing effect, is an increased understanding of larger gap sizes or ambiguous driving strategies [4]. One argument against marking is that the compliant behavior of AVs could lead to unwanted external interference [4]. For example, pedestrians could step onto the road, as they could be sure that the AV will brake [4].

Similar markings already exist for several types of vehicle, e.g., driving-school vehicles [5]. Those vehicles can be marked when they are being used for lessons to draw the attention of the surrounding traffic to the presence of the learner driver [5]. For example, in Austria and New Zealand, when driving at the age of 17, a clearly visible sign must be attached to the vehicle [6,7]. This allows other drivers to adjust their driving behavior to the learner driver and, if necessary, maintain a greater than usual distance from the vehicle, or overtake quickly.

In a study to evaluate the influence of marked AVs on human drivers, drivers encountered an AV that was either marked, not marked, or wrongly marked, in different highway scenarios [8]. The authors asked participants to rate the perceived safety, risk, and how pleasant it was to encounter the AV. Objective driving data were recorded during the simulator study. The results show that human drivers evaluate encounters with AVs independent of the marking [8].

Moreover, the critical gap acceptance and the perceived safety of participants crossing a road in front of an AV is not affected by the vehicle's driving mode (manual vs. automated) [9]. A comparable result was found in the study by Rodríguez Palmeiro [10]: even if participants noticed that the vehicle had an automated-driving sign, and they were subjectively influenced by feeling less safe and more doubtful, the objective behavior of participants did not change [10]. In addition, Faas, Mathis and Baumann [11] recommended providing—as a minimum—information for the pedestrians on the vehicle's status, so as to increase trust, perceived safety and to improve the road user experience.

Although few studies exist that have investigated the marking of AVs as such, there is currently increased research into visual external human-machine interfaces (eHMIs) for AVs, used to communicate explicitly with other HRUs [12]. Even though the focus of eHMIs is on other communication content, they result in additional marking of the AV. Light strips (e.g., [11,13,14]), displays (e.g., [9,15–17]) and projections (e.g., [18]) have primarily been used to communicate intentions to pedestrians (e.g., [13–15]) or human drivers (e.g., [16,19]). Cyan is recommended for eHMIs as it is a highly visible color and has no specific association in road traffic contexts [11,18,20,21]. Therefore, it seems to be well-suited to represent AVs [20].

The current results indicate that eHMIs improve the interaction between pedestrians and AVs [15,17] and increase the perceived safety and comfort of participants [22,23]. However, with regard to pedestrian–AV interaction, projections and eHMIs on wheels should be avoided, whereas eHMIs on roofs, windscreens or grilles work quite well [23]. In addition, it was found that eHMIs are useful for human driver–AV interaction, whereby displays are recommended rather than projections [16].

However, there are also results which indicate that the interpretation of eHMIs by pedestrians is sometimes ambiguous [14] and suggest that pedestrians make their decision to cross the road depending on the AV's driving behavior [24–27].

2. Objectives

When integrating AVs into traffic, communication might differ from situation to situation depending on the communication partner, such as pedestrians or human drivers [28]. As mentioned, there will be situations where the AV's driving strategy differs from that of human drivers. It can be assumed that these driving strategies can only be adapted with improved technology and algorithms. For as long as better technology cannot be implemented, consideration should be given to mark AVs. Such markings can be used by drivers to identify AVs and adapt their driving behavior if necessary. The aim of the study is to investigate whether marking the vehicles with a cyan LED strip in the upper part of the rear window as AVs (Figure 1) results in differences in the drivers' behavior and subjective evaluation in situations where it can be expected that an AV's driving strategy will deviate from that of a human.

Figure 1. Marked automated vehicle with a cyan LED strip in the upper part of the rear window in the driving simulation.

3. Method

3.1. Preliminary Study: Interview with Driving Instructors

In order to obtain an initial impression of the effects of vehicle marking, we posted a question in two Facebook online groups for German driving instructors. We asked for their experience of marking their driving schools' vehicles. We received 53 responses sharing different impressions. Of the 53 responses, it was possible to analyze 40 answers, as the others did not discuss the topic of marking driving school's vehicles. Altogether, 20% of the driving instructors mentioned that they do not experience differences in the behavior of surrounding traffic while driving a marked driving-school vehicle, compared to driving a vehicle without markings. In total, 27.5% are in favor of marking and 52.5% prefer not being identified as a driving-school vehicle. Reasons mentioned for preferring markings are the greater consideration demonstrated by other road users (10%), less honking (12%), and more acceptance from others (5%) (Figure 2). However, other driving instructors perceive less consideration from other road users when they see such markings, along with riskier behavior by the same (for example, not adhering to appropriate distances when cutting in and out during overtaking; 27%). In their opinion, others honk more (3%) if they recognize a driving-school vehicle. Therefore, from their perspective, it is more relaxing (23%) to drive without markings.

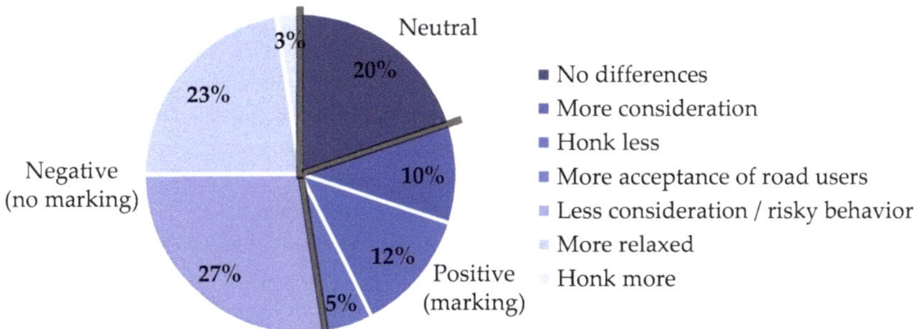

Figure 2. Attitude towards the marking of driving school vehicles.

To evaluate whether marking AVs also leads to differing opinions, we conducted a driving simulation study. The ethics committee of the Technical University of Munich approved this study. The corresponding code is 448/19 S.

3.2. Procedure

After welcoming the participants, they had to sign a declaration of consent. They were then asked to fill out demographic questions on a tablet and take a seat in the driving simulator in order to adjust the driver's seat and mirrors. Participants were introduced to the simulator, and experienced the driving simulation during a familiarization drive. All participants experienced six trials in random order. Each trial consisted of one of three different highway scenarios in which the driver encountered an AV (see Section 3.4.1). After each trial, participants were asked about the surrounding traffic (see Section 3.5.1). At the end of the study, we asked about the attitude towards marking AVs. With the exception of the demographic information, the experimenter gathered the information via oral questions and responses.

3.3. Apparatus

The basis of the static driving simulator (Figure 3) was a BMW 6 series mockup. A 6-channel projection system provided a realistic driving environment, with a refresh rate of 60 Hz. Three projectors were used for the 180° front view, and three projectors for the rear view (side and rear mirrors). We used the driving simulation software SILAB 6.5 of the Würzburg Institute for Traffic Sciences GmbH [29] and logged the driving data with 240 Hz. A 6-channel noise simulation completed the driving simulation. A freely programmable instrument cluster was used as human-machine interface. A tachometer and a speedometer were implemented for displaying driving-relevant information in this study. No additional advanced driver-assistance systems were used.

Figure 3. Driving simulator of the Chair of Ergonomics at the Technical University of Munich [30].

3.4. Independent Variables

We implemented a 3×2 within-subject design with three different scenarios on a three-lane highway (Figure 4), each with and without a marked AV. In all trials, participants started from a highway rest area and drove manually on a highway at a maximum speed of 130 km/h. The participants were instructed to adhere to the German highway regulations, in particular driving in the right lane, except when overtaking. To keep participants in the right lane, we implemented a high traffic density with a speed of 144 km/h in the middle lane at the beginning of all scenarios.

After a short time, an AV appeared in front of the participants in the right lane. The AV was either marked as such or looked like a manual vehicle. In all scenarios, the AV adhered strictly to the highway regulations and stayed in the right lane in front of participants. The appearance of the AV indicated the beginning of one of three different scenarios (Figure 4).

3.4.1. Scenarios

Roadworks

Participants drove through roadworks where a speed limit of 60 km/h was applicable. The scenario started at the end of the roadworks. There was no sign to inform drivers that the 60 km/h limit no longer applied. Therefore, the AV remained at 60 km/h, whilst all vehicles in the other lanes accelerated to 100 km/h (Figure 4a).

Traffic Jam

During the second scenario, a traffic jam occurred on the highway. The vehicles drove at a speed of 30 km/h in the left lane and in the middle lane. The vehicles in the middle lane used the large gaps to cut in front of the AV. The AV had a speed range of 15 to 40 km/h while maintaining a minimum gap of 5 seconds to the vehicles cutting in. Since the ego vehicle drove behind the AV, the participant had to brake in accordance with the AV (Figure 4b).

Lane Change

The AV used the indicator to signal to change lanes from the right to the middle lane in the third scenario. The vehicles on the middle lane were traveling at a speed of 130 km/h, and at 140 km/h in the left lane. However, the gaps between the vehicles on the target track were too small for the AV's algorithm to conduct a lane change and the AV stayed in the right lane. As a result, the AV drove at a varying speed of between 110 and 120 km/h (Figure 4c).

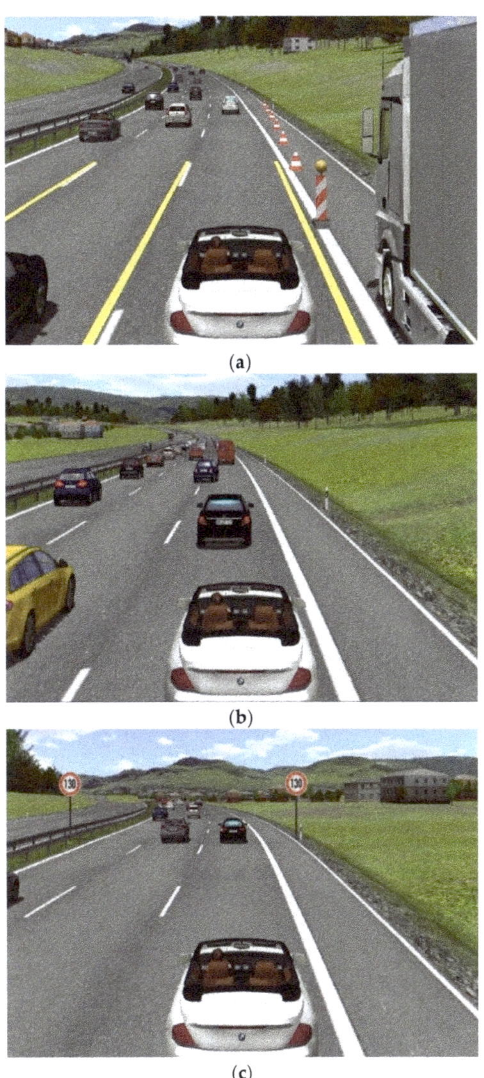

(a)

(b)

(c)

Figure 4. Scenarios implemented in the driving simulation: (**a**) Roadworks, (**b**) Traffic Jam, (**c**) Lane Change.

3.4.2. Marking the AV

In every scenario, a different vehicle type was used, so that participants could not recognize the AV immediately (Figure 5). The vehicle size was kept as constant as possible and the vehicle colors were kept unobtrusive (Figure 5). The participants experienced all scenarios with and without a marked AV. We marked the AV with a cyan LED strip in the upper part of the rear window that was visible to the participant when they followed the AV (Figures 1 and 5) [11,18,20,21]. We used an LED strip since it is less costly and easier to integrate into common commercial vehicles, compared to display or projection systems.

Figure 5. Vehicle types and colors for the AV, with and without a marking.

3.5. Dependent Variables

3.5.1. Subjective Data

The questionnaire was divided into three parts. In the first part, we surveyed demographic information such as age, sex, kilometers driven per year and the attitude towards the development of AVs on a five-point Likert scale (1 = *very positive* to 5 = *very negative*).

The second part of the questionnaire comprised five questions and was repeated after each trial. The first question asked about the surrounding traffic (*Did you notice anything particularly positive or negative about the surrounding traffic?*). With these questions, we aimed to find out whether participants recognized any different driving behavior in scenarios where the AV is not marked. The next questions enquired about conformity to the participants' expectations (*Did the vehicle in front behave as you would have expected?* and *How should the vehicle have behaved to meet your expectations?*). In addition, two further items were rated on a five-point Likert scale to investigate the driving behavior of the vehicle in front. With the first item, we measured, with regard to rationality, the perceived appropriateness of the driving behavior (*How appropriate was the driving behavior of the vehicle in front?*; 1 = *inappropriate* to 5 = *appropriate*) [31]. With the second item, we measured, with regard to emotionality, the perceived disturbance caused by the vehicle in front (*How disturbing was the driving behavior of the vehicle in front?*; 1 = *disturbing* to 5 = *not disturbing*) [31].

After all trials—in order to compare the objective driving data with the subjective perception—we asked the participants how they reacted when the vehicle in front was marked as an AV. In addition, we wanted to find out how people would react in real traffic situations. Therefore, we asked participants how they would behave in real traffic if they were to encounter an AV (which behaved as experienced in the simulation). Finally, we evaluated whether and for what reasons human drivers would like AVs to be marked as such.

3.5.2. Objective Data

We counted the number of lane changes conducted by the participants in overtaking the AV. The time between the start of the scenario and the completion of the lane change was calculated, to assess whether the AV's marking led to earlier overtaking. The lane change was considered to be completed when the vehicle's center of gravity crossed the lane marking.

To evaluate whether participants kept a greater safety gap to the AV, the minimum time headway (THW) of each participant was calculated for the period that the participant followed the AV in the same lane. THW is calculated using the distance of the AV to the human driver ($x_{AV\text{-}EGO}$) and the speed of the driver (v_{EGO}) according to [32], see Equation (1).

$$\text{THW} = \frac{x_{AV-EGO}}{v_{EGO}} \tag{1}$$

3.6. Participants

Altogether, 40 participants were recruited via postings at the Technical University of Munich and received compensation. This sample did not consist of the driving instructors from the preliminary study. Due to simulation sickness, we had to exclude two participants. In total, we analyzed 38 participants (24 male, 14 female) with a mean age of 29.63 years ($SD = 9.58$ years). The participants had had their driver's license for an average of 12.13 years ($SD = 9.37$ years) and drove on average 7997.37 km per year ($SD = 7535.95$ km per year). Their attitude towards AVs was rather positive ($Mdn = 2$). This attitude was based, among other things, on the expectation of increasing road safety, improved traffic flow, and more comfort, but also on personal enthusiasm for the topic (Figure 6).

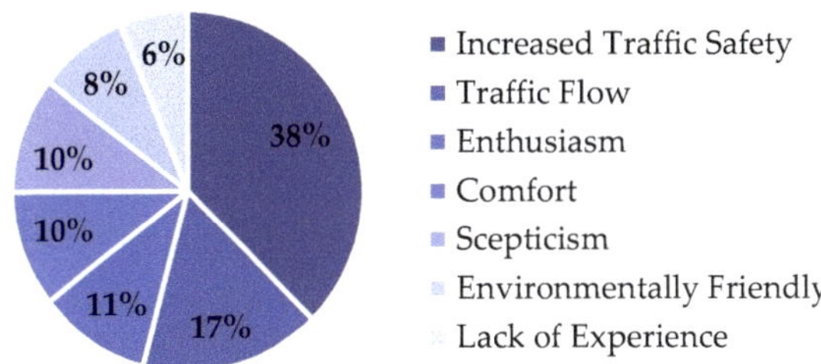

Figure 6. Attitude towards automated vehicles.

3.7. Analysis

We had some lags in the simulation, especially in the *Traffic Jam* scenario. Due to the technical problems, we had to exclude 20 trials from the subjective data and 25 trials from the objective data. In total, 208 trials were analyzable. Data were analyzed using Matlab, SPSS, and Excel. The Bonferroni correction was used for all statistical tests and the p-values were compared with a corrected alpha of 0.017.

The subjective data were ordinal scaled variables. Hence, two non-parametric Wilcoxon tests were calculated for both dependent questionnaire items.

The time elapsed until participants changed lanes in the *Roadworks* scenario is not normally distributed (marking: $W(27) = 0.91$, $p = 0.02$, no marking: $W(24) = 0.88$, $p \leq 0.01$). However, for the *Traffic Jam* scenario, the time elapsed until participants changed lanes is normally distributed (marking: $W(14) = 0.88$, $p = 0.06$, no marking: $W(15) = 0.90$, $p = 0.08$). In addition, the Shaphiro–Wilk test showed no significant departure from normality for the time elapsed until participants changed lanes in the *Lane Change* scenario (marking: $W(16) = 0.95$, $p = 0.47$, no marking: $W(15) = 0.90$, $p = 0.09$). As a result, we calculated one Wilcoxon test for the *Roadworks* scenario, and two t-tests for the other two scenarios.

The Shaphiro–Wilk test showed a significant departure from normality for the THW in the *Roadworks* scenario (marking: $W(37) = 0.78, p \leq 0.001$, no marking: $W(38) = 0.61, p \leq 0.001$). The THW for the *Traffic Jam* scenario (marking: $W(29) = 0.94, p = 0.12$, no marking: $W(27) = 0.92, p = 0.05$), and the THW for the *Lane Change* scenario are normally distributed (marking: $W(36) = 0.97, p = 0.31$, no marking: $W(36) = 0.96, p = 0.13$). Therefore, we calculated one Wilcoxon test for the *Roadworks* scenario, and two Wilcoxon tests for the other two scenarios.

The open questionnaire items were evaluated descriptively.

4. Results

4.1. Subjective Data

We wanted to find out, whether marking an AV influences drivers. However, we found no significant differences for the item *How appropriate was the driving behavior of the vehicle in front?* (*Roadworks*: $z = -0.26, p = 0.79, n = 37$; *Traffic Jam*: $z = -1.00, p = 0.32, n = 25$; *Lane Change*: $z = -0.76, p = 0.45, n = 38$; Figure 7), and for the item *How disturbing was the driving behavior of the vehicle in front?* (*Roadworks*: $z = -0.94, p = 0.35, n = 37$; *Traffic Jam*: $z = -1.36, p = 0.17, n = 25$; *Lane Change*: $z = -0.34, p = 0.73, n = 38$; Figure 8).

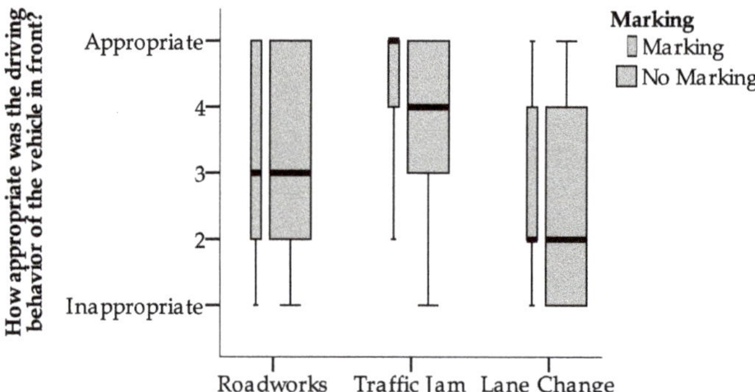

Figure 7. Boxplot for the item *How appropriate was the driving behavior of the vehicle in front?*, segregated by situation and marking.

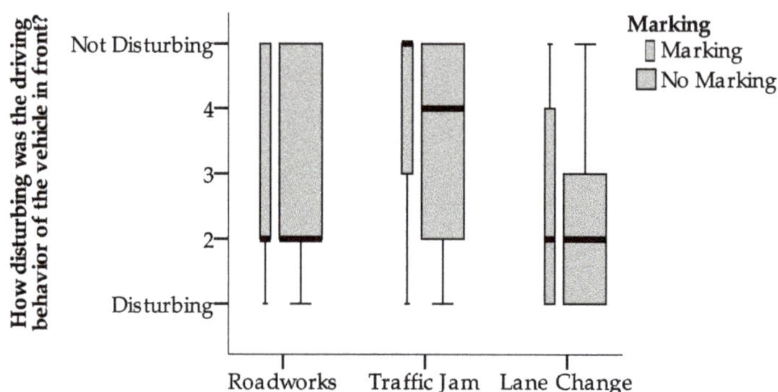

Figure 8. Boxplot for the item *How disturbing was the driving behavior of the vehicle in front?*, segregated by situation and marking.

In addition, the open questions illustrated that marking the vehicle does not affect the perception of the surrounding vehicles. For the *Roadworks* scenario, participants expressed incomprehension that the vehicle in front drove with only 60 km/h even after the roadworks. The vehicle in the *Traffic Jam* scenario was criticized for letting other vehicles merge in front of it. For the last scenario, *Lane Change*, participants mentioned that the vehicle flashed but did not change lanes and that the vehicle lost speed trying to change lanes. For all scenarios, the aspects were named with and without a marking of the vehicle in front.

Moreover, no descriptive differences were found for the expected driving behavior between the scenarios where the AV is marked or not (Table 1). For the *Roadworks* scenario, participants expected the perceived AV's driving behavior in 48% (marking: 46%, no marking: 50%) of all cases. Over 60% of all participants wished that the AV would have accelerated again after the *Roadworks*, regardless of the marking. One participant in the *Roadworks* scenario with the marked vehicle and two participants in the analogous scenario with the unmarked vehicle mentioned that the AV drove as expected because of the lack of the appropriate road sign. For the *Traffic Jam* scenario, nearly 76% of all participants (marking: 81%, no marking: 70%) expected the driving behavior. Those who had expected other driving behavior wished for a smoother driving style without letting as many vehicles merge. For the *Lane Change* scenario, only 26% (marking: 32%, no marking: 21%) of all participants expected the observed AV's driving behavior. Regardless of the marking of the vehicle, 80% wanted the AV to change lane or to switch off the indicators (27%) and accelerate once again (18%).

Table 1. Assessment of the expected driving behavior.

	Roadworks		Traffic Jam		Lane Change	
	Marking	No Marking	Marking	No Marking	Marking	No Marking
Behavior as Expected	45.9%	50.0%	80.6%	69.2%	31.6%	21.1%
Behavior not as Expected	54.1%	50.0%	19.4%	30.8%	68.4%	78.9%

However, even if the marking had no influence on subject's ratings, 66% would prefer AVs to be marked (Figure 9). The other 34% do not want the vehicle to be directly identified as automated (Figure 9). Participants preferring AVs to be marked argued that it is easier to assess the AV's driving behavior (41%) and to adapt their own behavior to the new road user (e.g., greater gaps, increased attention; 22%). Another 15% would like to have marking in order to increase acceptance, and 15% mentioned that the AV is a role model, because it complies with the German highway regulations. In addition, 7% would like markings only as additional information. Reasons mentioned against marking the AV include that it is a normal road user (20%) and should not attract attention (33%). Another 27% mentioned that the potential for abuse is too high due to the marking and 20% were afraid that the uncertainty in road traffic will become too great (Figure 9).

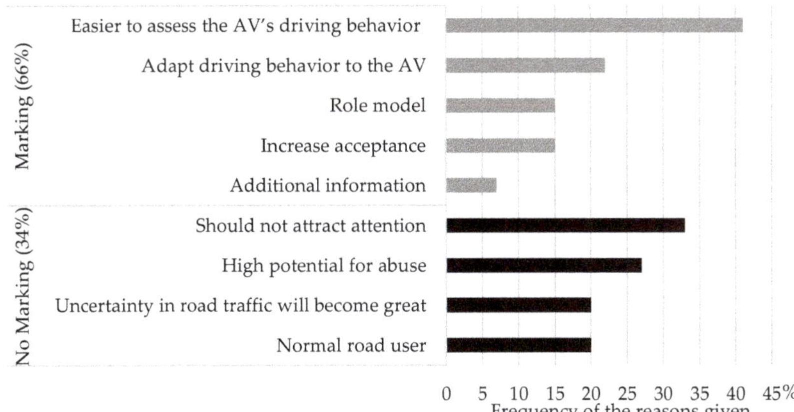

Figure 9. Reasons for or against AV marking.

At the end of the study, we asked participants how they reacted when seeing an AV in the simulation. Altogether, 50% of all participants mentioned that they behaved as usual, whereas others raised their attention levels (10%), drove more carefully (8%), and/or kept a greater distance (5%). Another 8% had higher confidence in the vehicle, because it adheres to German highway regulations (Figure 10).

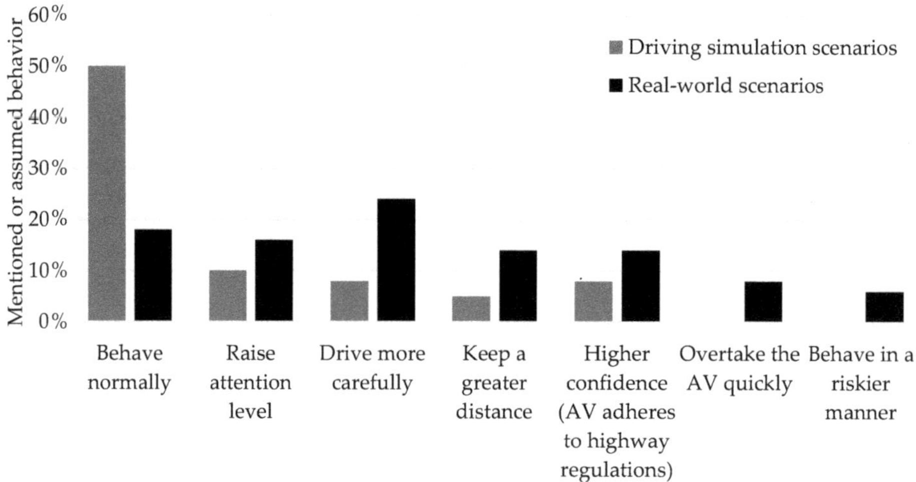

Figure 10. Participants' mentioned behavior in the driving simulation scenarios and assumed behavior in real-world scenarios.

Nevertheless, in real traffic, only 18% of all participants said they would behave "normally". Another 24% would drive more carefully, 16% would raise their attention level, and 14% would keep a greater distance and drive more defensively. Another 14% said they would follow the vehicle and orientate themselves to the driving behavior of the AV, because it adheres to German highway regulations. However, another 8% would overtake the AV quickly and 6% would behave in a more risky manner than usual, because the AV drives in an error-free way (Figure 10).

4.2. Objective Data

Altogether, participants changed lanes in 55% of all trials. The most lane changes happened in the scenario *Roadworks* (marking: 73%, no marking: 63%), followed by *Traffic Jam* (marking: 48%, no marking: 56%), and *Lane Change* (marking: 44%, no marking: 42%; Table 2). However, we found no tendency that marking the AV influences the frequency of lane changes on a descriptive level (Table 2). We also found no significant differences in the time elapsed until the lane change was conducted (*Roadworks*: $z = -1.48$, $p = 0.14$, $n = 21$; *Traffic Jam*: $t(10) = -1.26$, $p = 0.24$; *Lane Change*: $t(8) = -0.21$, $p = 0.84$; Figure 11). In addition, the presence of markings had no significant influence on the THW in any of the three scenarios (*Roadworks*: $z = -0.52$, $p = 0.60$, $n = 37$; *Traffic Jam*: $t(24) = -0.16$, $p = 0.88$; *Lane Change*: $t(34) = -0.54$, $p = 0.59$; Figure 12).

Table 2. Number of lane changes over all trials.

	Roadworks	Traffic Jam	Lane Change
Marking	27 (73.0%)	14 (48.3%)	16 (44.4%)
No Marking	24 (63.2%)	15 (55.6%)	15 (41.7%)

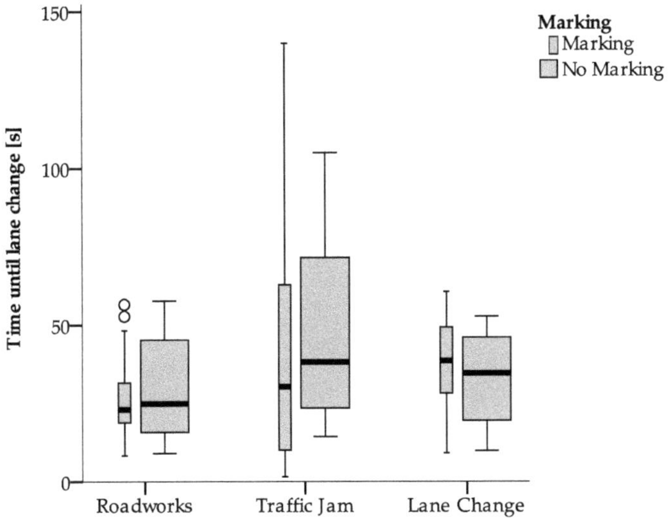

Figure 11. Boxplot for the time until lane change, segregated by situation and marking.

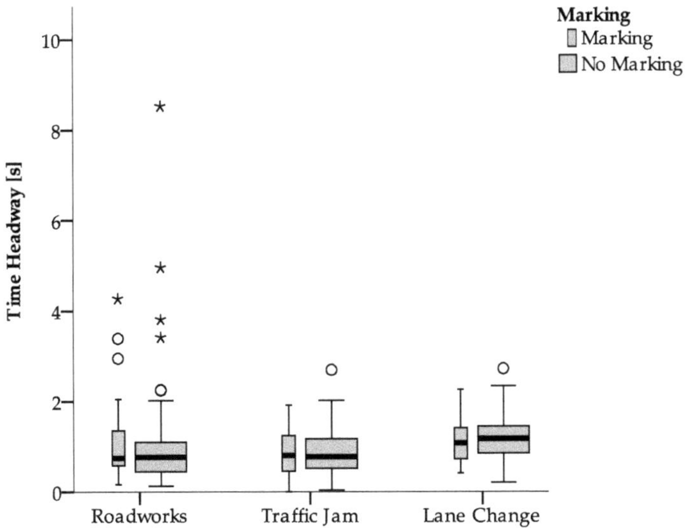

Figure 12. Boxplot for the minimum time headway, segregated by situation and marking.

5. Discussion

The aim of the study was to determine the influence of marking AVs on human drivers in three scenarios, in order to deduce whether markings should be implemented for AVs.

The results illustrate that marking AVs does not influence the driving behavior of human drivers and their subjective rating. This confirms the results of Kühn, Stange and Vollrath [8]. It is possible that the driving behavior itself is sufficiently informative in order to be able to recognize an AV. This is also consistent with the statements of Kühn et al. [8], who mentioned that drivers have a fairly accurate idea of how AVs will behave in situations on highways where interaction with other HRUs is required. Another aspect could be that drivers can deal with ambiguous driving strategies of other drivers and have already learned to compensate for such behavior by, for example, increasing the gap to the vehicle in front or by overtaking. Rodríguez Palmeiro [10] already stated that the driving behavior of the AV is more important than external signs in pedestrians' deciding whether or not to cross the road.

Although no significant differences were found in participants' ratings, the majority of participants preferred the AV to be marked. Due to the marking, they could assess the AV's driving behavior and adapt their own driving behavior accordingly. In addition, most participants mentioned that they behaved normally in the simulation, but when encountering an AV in real traffic, they would behave more carefully with increased attention levels. Therefore, it can be assumed that a study involving encounters with AVs in real traffic might lead to different results.

However, this study only examined the encounter with a single AV in three selected scenarios. The drivers experienced the AV's driving behavior in every scenario for the first time. Therefore, it might be difficult to adapt their driving behavior to the AV without knowing what the AV is going to do next. It is reasonable to assume that an adjustment of human driving behavior would take place in the event of their repeated encounters with AVs. This also explains the participants' preference for marked AVs, as it enables drivers to recognize the AV at an early stage, and adapt their driving behavior accordingly. Therefore, it might be useful to investigate long-term effects in a further study. In addition, the effects of age and gender should be evaluated.

Besides the result that the marking had no influence, we found descriptive differences—dependent on the given scenario—for the number of lane changes and the time participants needed until they changed lanes. The *Roadworks* scenario showed the highest number of lane changes in the shortest time passed. This may be due to the large speed difference of the AV compared to the vehicles in the middle lane (60 to 100 km/h) in relation to the other scenarios (*Traffic Jam*: 15–40 km/h to 30 km/h; *Lane Change*: 110–120 km/h to 130 km/h). Based on these results, more scenarios should be investigated in future studies.

Due to technical issues in the *Traffic Jam* scenario, the simulation did not run smoothly, therefore participants' driving behavior might be influenced. As a result, absolute values can only be interpreted with caution. Nevertheless, the comparison between the scenarios with and without a marking is still possible.

6. Conclusions

As a general conclusion, it can be stated that the marking of an AV made no differences to human drivers in terms of their driving behavior and their subjective ratings. It seems that drivers can compensate for AVs' driving behavior, whereby they do not require the AV to be identified as such. Nevertheless, the participants indicate that they prefer to be able to distinguish AVs from other vehicles. However, this study did not address the long-term effects, which may affect the results, and should be investigated in future studies.

Author Contributions: Conceptualization, T.F., A.F. and E.S.; methodology, T.F. and A.F.; software, A.F.; validation, T.F. and A.F.; formal analysis, T.F. and A.F.; investigation, T.F. and A.F.; resources, T.F., A.F. and K.B.; data curation, T.F.; writing—original draft preparation, T.F. and A.F.; writing—review and editing, T.F., A.F., E.S. and K.B.; visualization, T.F. and A.F.; supervision, K.B. All authors have read and agreed to the published version of the manuscript.

Funding: This research received no external funding.

Acknowledgments: The authors thank Franz Daisenberger for conducting the study.

Conflicts of Interest: The authors declare no conflict of interest.

References

1. BMW Group. BMW Group's Driver Assistance and Autonomous Driving Development Department under New Leadership. Alejandro Vukotich Takes Over at the Helm, Elmar Frickenstein to Retire after Handover Phase. Available online: https://www.press.bmwgroup.com/africa-dom-easteurope/article/detail/T0288264EN/bmw-group%E2%80%99s-driver-assistance-and-autonomous-driving-development-department-under-new-leadership-alejandro-vukotich-takes-over-at-the-helm-elmar-frickenstein-to-retire-after-handover-phase?language=en (accessed on 3 May 2019).

2. SAE International. *Taxonomy and Definitions for Terms Related to Driving Automation Systems for On-Road Motor Vehicles (J3016)*; SAE International: Warrendale, PA, USA, 2018.

3. Schmidt, H.; Terwilliger, J.; AlAdawy, D.; Fridman, L. Hacking Nonverbal Communication between Pedestrians and Vehicles in Virtual Reality. *arXiv* **2019**, arXiv:1904.01931.

4. Färber, B. Communication and Communication Problems between Autonomous Vehicles and Human Drivers. In *Autonomous Driving*; Springer: Berlin/Heidelberg, Germany, 2016; pp. 125–144.

5. Federal Ministry of Justice and Consumer Protection. Durchführungsverordnung zum Fahrlehrergesetz. Available online: https://www.gesetze-im-internet.de/fahrlg2018dv/BJNR000210018.html (accessed on 27 May 2020). (In German)

6. Federal Ministry for Digital and Economic Affairs. L17—Ausstattung des Ausbildungsfahrzeugs. Available online: https://www.oesterreich.gv.at/themen/dokumente_und_recht/fuehrerschein/1/2/Seite.040112.html (accessed on 16 January 2020). (In German)

7. NZ Transport Agency. Conditions of a Learner Licence. Available online: https://www.nzta.govt.nz/driver-licences/getting-a-licence/licences-by-vehicle-type/cars/learners-licence/conditions-of-a-learner-licence/ (accessed on 28 January 2020).

8. Kühn, M.; Stange, V.; Vollrath, M. Menschliche Reaktion auf hochautomatisierte Fahrzeuge im Mischverkehr auf der Autobahn. In *VDI Tagung Mensch-Maschine-Mobilität 2019—Der (Mit-) Fahrer im 21. Jahrhundert!?* VDI Verlag: Düsseldorf, Germany, 2019; pp. 169–184. (In German)

9. Joisten, P.; Alexandi, E.; Drews, R.; Klassen, L.; Petersohn, P.; Pick, A.; Schwindt, S.; Abendroth, B. Displaying Vehicle Driving Mode—Effects on Pedestrian Behavior and Perceived Safety. In *Human Systems Engineering and Design II*; Ahram, T., Karwowski, W., Pickl, S., Taiar, R., Eds.; Springer International Publishing: Cham, Switzerland, 2020; pp. 250–256, ISBN 978-3-030-27927-1.

10. Rodríguez Palmeiro, A. Interaction between pedestrians and Wizard of Oz automated vehicles. Master's Thesis, Delft University of Technology, Delft, The Netherlands, 2017.

11. Faas, S.M.; Mathis, L.-A.; Baumann, M. External HMI for self-driving vehicles: Which information shall be displayed? *Transp. Res. Part F Traffic Psychol. Behav.* **2020**, *68*, 171–186. [CrossRef]

12. Bengler, K.; Rettenmaier, M.; Fritz, N.; Feierle, A. From HMI to HMIs: Towards an HMI Framework for Automated Driving. *Information* **2020**, *11*, 61. [CrossRef]

13. Weber, F.; Chadowitz, R.; Schmidt, K.; Messerschmidt, J.; Fuest, T. Crossing the Street across the Globe: A Study on the Effects of eHMI on Pedestrians in the US, Germany and China. In *Human-Computer Interaction in Mobility, Transport, and Automotive Systems*; Krömker, H., Ed.; Springer International Publishing: Cham, Switzerland, 2019; pp. 515–530, ISBN 978-3-030-22665-7.

14. Hensch, A.-C.; Neumann, I.; Beggiato, M.; Halama, J.; Krems, J.F. How Should Automated Vehicles Communicate?—Effects of a Light-Based Communication Approach in a Wizard-of-Oz Study. In *Advances in Human Factors of Transportation*; Stanton, N., Ed.; Springer International Publishing: Cham, Switzerland, 2020; pp. 79–91, ISBN 978-3-030-20502-7.

15. Song, Y.E.; Lehsing, C.; Fuest, T.; Bengler, K. External HMIs and Their Effect on the Interaction between Pedestrians and Automated Vehicles. In *Intelligent Human Systems Integration*; Karwowski, W., Ahram, T., Eds.; Springer International Publishing: Cham, Switzerland, 2018; pp. 13–18, ISBN 978-3-319-73887-1.

16. Rettenmaier, M.; Pietsch, M.; Schmidtler, J.; Bengler, K. Passing through the Bottleneck—The Potential of External Human-Machine Interfaces. In Proceedings of the 2019 IEEE Intelligent Vehicles Symposium (IV), Paris, France, 9–12 June 2019; pp. 1687–1692, ISBN 978-1-7281-0560-4.

17. De Clercq, K.; Dietrich, A.; Velasco, J.P.N.; De Winter, J.C.F.; Happee, R. External Human-Machine Interfaces on Automated Vehicles: Effects on Pedestrian Crossing Decisions. *Hum. Factors* **2019**, *61*, 1353–1370. [CrossRef] [PubMed]

18. Dietrich, A.; Willrodt, J.-H.; Wagner, K.; Bengler, K. Projection-Based External Human Machine Interfaces–Enabling Interaction between Automated Vehicles and Pedestrians. In Proceedings of the DSC Europe 2018 VR. Driving Simulation & Virtual Reality Conference & Exhibition, Antibe, France, 5–7 September 2018; pp. 43–50, ISBN 978-2-85782-734-4.

19. Rettenmaier, M.; Albers, D.; Bengler, K. After you?!—Use of external human-machine interfaces in road bottleneck scenarios. *Transp. Res. Part F Traffic Psychol. Behav.* **2020**, *70*, 175–190. [CrossRef]

20. Werner, A. New Colours for Autonomous Driving: An Evaluation of Chromaticities for the External Lighting Equipment of Autonomous Vehicles. *Colour Turn* **2019**. [CrossRef]

21. SAE International. *Automated Driving System (ADS) Marker Lamp*; SAE International: Warrendale, PA, USA, 2019.
22. Böckle, M.-P.; Brenden, A.P.; Klingegård, M.; Habibovic, A.; Bout, M. SAV2P—Exploring the Impact of an Interface for Shared Automated Vehicles on Pedestrians' Experience. In Proceedings of the 9th International Conference on Automotive User Interfaces and Interactive Vehicular Applications Adjunct (Automotive UI '17), Oldenburg, Germany, 24–27 September 2017; Löcken, A., Boll, S., Politis, I., Osswald, S., Schroeter, R., Large, D., Baumann, M., Alvarez, I., Chuang, L., Feuerstack, S., et al., Eds.; ACM Press: New York, NY, USA, 2017; pp. 136–140, ISBN 9781450351515.
23. Eisma, Y.; Van Bergen, S.; Ter Brake, S.; Hensen, M.; Tempelaar, W.; De Winter, J.C.F. External human—Machine interfaces: The effect of display location on crossing intentions and eye movements. *Information* **2019**, *11*, 13. [CrossRef]
24. Clamann, M.; Aubert, M.; Cummings, M.L. Evaluation of Vehicle-to-Pedestrian Communication Displays for Autonomous Vehicles. In Proceedings of the 96th Annual Transportation Research Board Meeting, Washington, DC, USA, 8–12 January 2017.
25. Fuest, T.; Michalowski, L.; Traris, L.; Bellem, H.; Bengler, K. Using the Driving Behavior of an Automated Vehicle to Communicate Intentions—A Wizard of Oz Study. In Proceedings of the 2018 21st International Conference on Intelligent Transportation Systems (ITSC), Maui, HI, USA, 4–7 November 2018; pp. 3596–3601. [CrossRef]
26. Fuest, T.; Maier, A.S.; Bellem, H.; Bengler, K. How Should an Automated Vehicle Communicate Its Intention to a Pedestrian?—A Virtual Reality Study. In *Human Systems Engineering and Design II*; Ahram, T., Karwowski, W., Pickl, S., Taiar, R., Eds.; Springer International Publishing: Cham, Switzerland, 2020; pp. 195–201. [CrossRef]
27. Fuest, T.; Michalowski, L.; Schmidt, E.; Bengler, K. Reproducibility of Driving Profiles—Application of the Wizard of Oz Method for Vehicle Pedestrian Interaction. In Proceedings of the 2019 IEEE Intelligent Transportation Systems Conference (ITSC), Auckland, New Zealand, 27–30 October 2019; pp. 3954–3959. [CrossRef]
28. Fuest, T.; Sorokin, L.; Bellem, H.; Bengler, K. Taxonomy of Traffic Situations for the Interaction between Automated Vehicles and Human Road Users. In *Advances in Human Aspects of Transportation. AHFE 2017. Advances in Intelligent Systems and Computing*; Stanton, N.A., Ed.; Springer International Publishing: Cham, Switzerland, 2018; Volume 597, pp. 708–719. [CrossRef]
29. Würzburg Institute for Traffic Sciences GmbH. Driving Simulation and SILAB. Available online: https://wivw.de/en/silab (accessed on 14 April 2020).
30. Technical University of Munich. Static Driving Simulator. Available online: https://www.mw.tum.de/en/lfe/research/labs/static-driving-simulator/ (accessed on 14 April 2020).
31. Surges, F. Einfluss Hochautomatisiert Fahrender Fahrzeuge auf das Fahrverhalten und Die Einstellungen Manueller Fahrer im Mischverkehr; BASt (Bundesanstalt für Strassenwesen (Federal highway research institute))-Project: F1100.4318007. Available online: https://trid.trb.org/view/1576671 (accessed on 27 May 2020). (In German)
32. Wachenfeld, W.; Winner, H. Do Autonomous Vehicles Learn. In *Autonomous Driving*; Maurer, M., Gerdes, J.C., Lenz, B., Winner, H., Eds.; Springer: Berlin/Heidelberg, Germany, 2016; pp. 451–471.

 *information*

Article

Multi-Vehicle Simulation in Urban Automated Driving: Technical Implementation and Added Benefit

Alexander Feierle *,†, **Michael Rettenmaier** *,†, **Florian Zeitlmeir and Klaus Bengler**

Chair of Ergonomics, Technical University of Munich, 85748 Garching, Germany; florian.zeitlmeir@tum.de (F.Z.); bengler@tum.de (K.B.)

* Correspondence: alexander.feierle@tum.de (A.F.); michael.rettenmaier@tum.de (M.R.); Tel.: +49-89-289-15335 (A.F.); +49-89-289-15419 (M.R.)

† These authors contributed equally to this work.

Received: 11 April 2020; Accepted: 16 May 2020; Published: 19 May 2020

Abstract: This article investigates the simultaneous interaction between an automated vehicle (AV) and its passenger, and between the same AV and a human driver of another vehicle. For this purpose, we have implemented a multi-vehicle simulation consisting of two driving simulators, one for the AV and one for the manual vehicle. The considered scenario is a road bottleneck with a double-parked vehicle either on one side of the road or on both sides of the road where an AV and a simultaneously oncoming human driver negotiate the right of way. The AV communicates to its passenger via the internal automation human–machine interface (HMI) and it concurrently displays the right of way to the human driver via an external HMI. In addition to the regular encounters, this paper analyzes the effect of an automation failure, where the AV first communicates to yield the right of way and then changes its strategy and passes through the bottleneck first despite oncoming traffic. The research questions the study aims to answer are what methods should be used for the implementation of multi-vehicle simulations with one AV, and if there is an added benefit of this multi-vehicle simulation compared to single-driver simulator studies. The results show an acceptable synchronicity for using traffic lights as basic synchronization and a distance control as the detail synchronization method. The participants had similar passing times in the multi-vehicle simulation compared to a previously conducted single-driver simulation. Moreover, there was a lower crash rate in the multi-vehicle simulation during the automation failure. Concluding the results, the proposed method seems to be an appropriate solution to implement multi-vehicle simulation with one AV. Additionally, multi-vehicle simulation offers a benefit if more than one human affects the interaction within a scenario.

Keywords: multi-vehicle simulation; mixed traffic; human–machine interface; automated driving

---

1. Introduction

A current research focus in the context of automated driving is human–machine interface (HMI) design. In urban areas, which are characterized by a high number of objects [1], a high number of vulnerable road users [2], and high information density [3], the automated vehicle (AV) must be able to clearly communicate with the passenger and the surrounding human road user [4]. The only way to investigate the simultaneous communication via the automation HMI (aHMI) and the external HMI (eHMI) [4] is by conducting a multi-vehicle simulation. This requires a human road user, such as a human driver, who perceives the eHMI and a passenger in the AV who perceives information from the aHMI.

A scenario of particular interest is the bottleneck scenario in urban areas [5] where communicating via eHMIs has the potential to enhance traffic efficiency and safety [6]. Partially automated driving systems (ADS) are already state of the art. Nevertheless, the current operation design domain (ODD) in partially automated driving is limited to highways, since these are characterized by a lower complexity compared to urban areas. As the driver must still monitor the ADS and must be able to take over vehicle guidance at any time without a request to intervene [7], it could be assumed that such systems will be realized sooner than systems with a higher level of driving automation in urban areas. Therefore, this study addresses the interaction between a human driver and a partially AV and its passenger in bottleneck scenarios in urban areas.

Compared to investigations with fixed programmed road users, multi-vehicle simulations should generate a more realistic driving behavior [8]. With regard to partially automated driving, there may be an added benefit, especially when the passenger of the AV has to take over vehicle guidance again. For this purpose, a controlled interaction scenario must be achieved, which is a special challenge of multi-vehicle simulation [9]. Therefore, this publication aims at the realization and evaluation of the technical implementation of such a multi-vehicle study. Additionally, a multi-vehicle experiment has been conducted to compare the results with a single-driver simulation to identify added benefits using multi-vehicle simulation.

2. State of Research

2.1. Previous Studies on Multi-Agent Simulation

Multi-agent simulation is a useful tool for analyzing the interaction of various road users in the same environment. It permits the measurement of the parameters of each individual participant as well as the objectification of the behavior within a group of several drivers, e.g., in platoons [10]. Additionally, the multi-agent simulation retains the single-agent simulation's benefits of being controllable and accurate and enriches the experiments with a more realistic traffic flow environment [11,12]. Thus, the multi-vehicle simulation increases the ability of both driving and traffic simulation [11]. It enables the investigation of social interaction [13] and the analysis of advanced driver assistance systems affecting several drivers [14]. A classification of previous research can be made according to the characteristics of the road users involved.

Lehsing, Benz, and Bengler [15] investigated the interaction between a human driver and a pedestrian in a pedestrian crossing scenario. In half of the encounters a confederate controlled the pedestrian, resulting in a more human-like behavior since he was able to react to the participants' driving behavior. In the other half of the encounters the pedestrian's behavior was programmed. The authors state that the approach of physically linking both simulators is a meaningful method in traffic research since it raises the validity of investigations in human–human interaction [15].

In contrast to the driver-pedestrian interaction there were studies researching the interaction between several human drivers, which could be clustered in experiments investigating safety-critical situations and experiments researching the interaction and cooperation between several road users. Hancock and de Ridder [16] used the multi-vehicle simulation to investigate the participants' avoidance responses at the brink of a collision. The authors emphasize the value of multi-vehicle simulation because it analyzes critical situations in a safe and efficient manner. Moreover, the method provided similar avoidance responses compared to real-world investigations [16]. Yasar, Berbers, and Preuveneers [17] also used the multi-vehicle simulation to investigate safety critical situations at intersections by analyzing the incident rate and the participants' driving behavior affected by a voice-based command system and the presence of traffic lights. Will [18] found a decrease in the criticality of encounters between a human driver and a motorcyclist due to a system supporting the interaction at intersections.

Aside from conducting multi-vehicle simulations to investigate safety critical situations, the method was also used to analyze the interaction or cooperation of different human drivers. The method was used to realize the presence of multiple participants in a platoon of four vehicles to identify parameters describing the behavior of different drivers within the platoon as well as the behavior of the platoon as a whole [19,20]. Moreover, Heesen, Baumann, Kelsch, Nause, and Friedrich [21] conducted a multi-vehicle study to examine the effect of a cooperative lane change assistant on possible conflicts on motorways. Results of the experiment show that drivers consider the other driver's possible actions when requesting to cooperate. In addition, the capability to anticipate affects the willingness to cooperate [21]. Sun, Ma, Li, and Niu [11] confirm the positive effect of multi-vehicle simulation on the behavior in lane change maneuvers to be consistent with the data of field observations. Further research including multi-vehicle simulation was applied, e.g., the evaluation of dynamic speed guidance strategies [22] or the analysis of the "rubbernecking" phenomenon, consisting of a driver slowing down due to an accident on the opposite side of the road [23].

Furthermore, multi-vehicle simulations are used to analyze the subjective feeling of human drivers. Rittger, Mühlbacher, Maag, and Kiesel [24] found that the usage of a traffic light assistant could raise the feeling of bothering other road users and it induces anger in participants without an assistant. Additionally, the participants' knowledge of the presence of another real human in the same simulation influences the participants' sensation [25] and the willingness to cooperate [21].

The implementation of AVs and the associated investigation of the interaction between AVs and other road users enlarge the application of multi-agent simulation. Bazilinskyy, Kooijman, Dodou, and de Winter [26] analyzed the interaction between an AV communicating via an eHMI, a human driver, and a pedestrian at a T-intersection with a zebra-crossing. The authors concluded that the multi-agent simulation is a promising tool to research interaction in traffic in the future.

2.2. Implementation of Multi-Agent Simulation

One challenge in conducting a multi-agent simulation is to induce the interaction in a controlled manner [9]. In the case that the interaction does not occur in the simulation, there is no added benefit of multi-agent simulation [8]. The following possibilities to realize the participants' coordination were used to avoid the insufficiently synchronized encounters of several participants.

Schindler and Köster [27] used the implementation of detours, dynamically modified speed adjustments, and the manipulation of the participants' speedometer to synchronize the participants' encounter. Another possibility is the dynamic change of the route length [16,27] or to have one interaction partner as a confederate [15]. The confederate knows about the experimental condition and is able to react to the driving behavior of the other participant. Moreover, the instruction of participants could be used to enable a synchronized interaction [24,28]. In the simulation, the implementation of road sections where the participants have to follow programmed traffic and the control of implemented traffic lights are methods to enable coordinated interaction in a multi-agent simulation [27,29].

3. Objectives

One challenge of multi-vehicle simulation is that the participants have to approach the investigated scenario at the same time in order to ensure controlled interaction. Various publications have already taken up this challenge. However, all these studies investigated scenarios without automated road users. Since the present work investigates the interaction between an AV and a human driver at bottlenecks, new opportunities arise to achieve the synchronous arrival of both road users via the ADS and its implemented longitudinal control. This creates new challenges in terms of reproducibility and comprehensibility. Therefore, this publication aims at the technical implementation and evaluation of such a method with an automated road user in a multi-vehicle simulation. Hence, a multi-vehicle simulation was conducted. The results are compared with the results of a single-driver study on eHMI design to identify the relevant use cases where multi-vehicle simulation offers an added benefit. The objectives of this study lead to the following research questions (RQ):

- RQ1: What methods should be used for multi-vehicle simulations with one automated vehicle to ensure synchronicity?
- RQ2: What is the added benefit of a multi-vehicle simulation with one automated vehicle compared to single-driver simulations?

4. Technical Implementation

4.1. Basic Synchronization

After analyzing synchronization methods used in research for multi-vehicle simulation studies (see Section 2.2), we decided to synchronize the AV and the human driver via a traffic light control. The basic synchronization with traffic lights enables the compensation of large time differences and has a low space requirement in the simulation environment. Figure 1 shows the basic synchronization we implemented in the simulation. For the manual vehicle, a speed limit of 30 km/h was applied directly after the traffic lights. For the AV, the speed limit of 30 km/h was set at the beginning of the interaction phase. When approaching the bottleneck the traffic light in front of the human driver shows red and the human driver has to wait at the stop line. The AV arrives at the other traffic light with a delay (Δt) due to course design. During the approach the AV passes a trigger point at the course which causes the traffic lights in front of the AV to switch from green to red so that the AV decelerates to a standstill in front of its stop line. Subsequently, both traffic lights switch from red to green. Since both traffic lights have the same distance to the road bottleneck and due to the simultaneous change of the traffic light's state, the AV and the human driver are basically synchronized when entering the scenario.

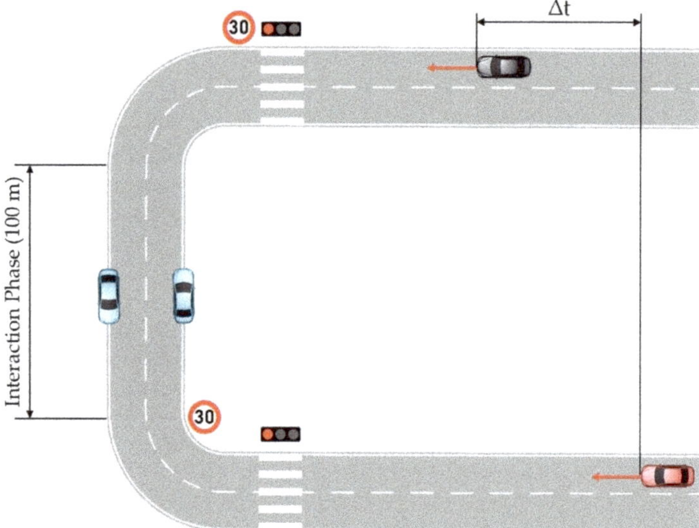

Figure 1. Basic synchronization of the AV (red vehicle in the lower part) and human driver (black vehicle in the upper part) via the traffic light control. The route does not correspond to the real course in the simulation and is shown schematically.

4.2. Detail Synchronization

After the basic synchronization has compensated large time differences, both vehicles start from a standstill after the traffic lights have turned green. A distance difference may already occur while waiting in front of the traffic lights if the human driver comes to a standstill with a different distance to the traffic lights than the AV. According to Rettenmaier, Albers, and Bengler [30] the interaction phase was defined in a radius of 50 m around the road bottleneck (see Figure 1). After passing the green traffic light, distance differences (Δd) would occur without detail synchronization, where the AV adapts to the behavior of the human driver. These differences would result due to the different speed profiles. In the case of large distance differences, there would be no interaction because the passing of the bottleneck would be regulated by the earlier arrival of one of the vehicles [31]. In order to achieve a high degree of synchronicity when the interaction phase is reached, the automated longitudinal control of the AV is used to adapt to the behavior of the human driver.

The automated driving system is realized by using simulation state data. The longitudinal control of the automation during free driving without a front vehicle or traffic light consists of a PID control, which receives speed settings as input. An acceleration is generated as output of the PID control, which is transferred to the internal vehicle dynamics using a single-track model of the driving simulation software SILAB. Here, several implementation opportunities to adapt to the behavior of the manual vehicle exist (Figure 2):

- Implementation of a PID controller which controls the speed difference of both vehicles and has the acceleration as an output (Method 1)
- Implementation of a PID controller which controls the distance difference of both vehicles to the road bottleneck and has the acceleration as an output (Method 2)
- Transmitting the acceleration of the manual vehicle directly to the AV's internal driving dynamics in SILAB (Method 3)
- Transmitting the pedals' positions of the manual vehicle directly to the AV's internal driving dynamics in SILAB (Method 4)

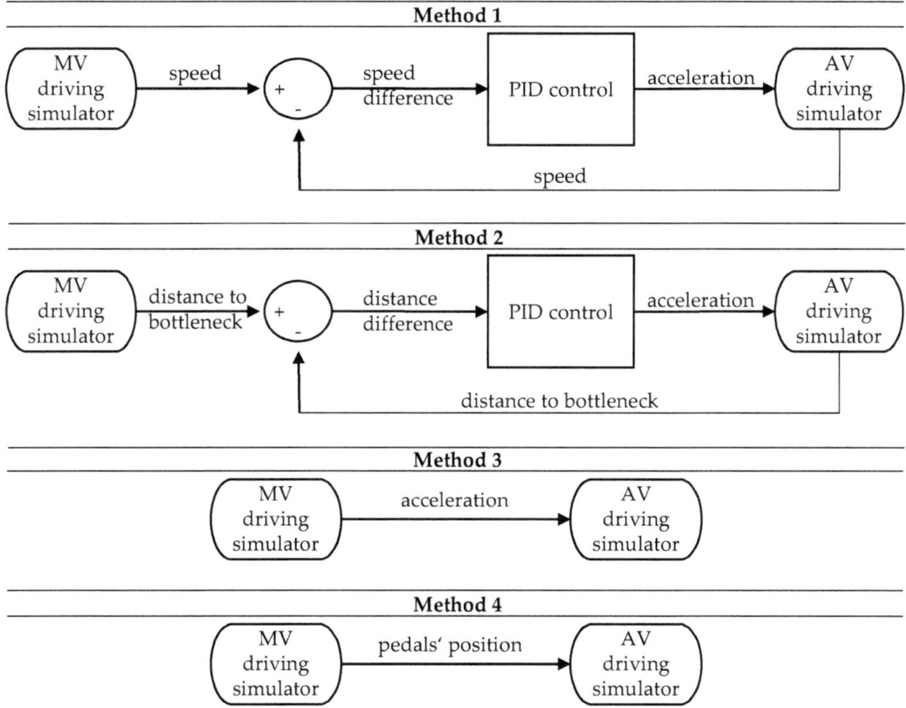

Figure 2. Block diagrams of the proposed methods to adapt the AV to the behavior of the manual vehicle (MV).

In order to analyze which of these methods is most appropriate, speed profiles for a simulation of the manual vehicle are required. To exclude influences of lateral steering on the longitudinal dynamics, the scenario (Figure 1) was implemented on a straight track instead of a u-shaped one. Subsequently, three different speed profiles were implemented using a cruise control (Figure 3). The different speed profiles are intended to represent different human driver types (offensive, neutral, defensive). Nevertheless, these synthetic profiles cannot represent a human driver exactly, so they are only suitable for a first pre-test.

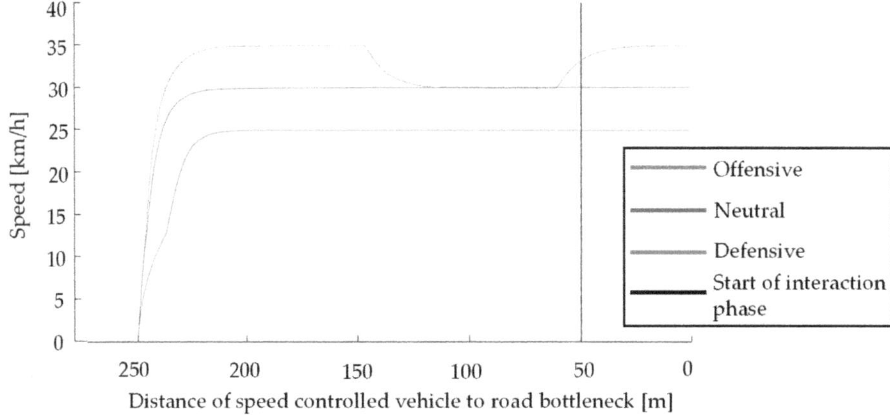

Figure 3. Three different implemented speed profiles (offensive, neutral, defensive) using a cruise control to simulate the manual vehicle during the pre-test.

Negative values for the distance differences of both vehicles to the road bottleneck mean that the manual vehicle reached the interaction phase first. Method 1 using the speed as input resulted in a mean (M) difference of −8.26 m with a standard deviation (SD) of 7.61 m. Using the distance difference as an input in Method 2 led to M = −0.79 m (SD = 1.12 m) difference. Method 3 using the acceleration of the manual vehicle did not lead to any interaction scenarios due to implementation issues. Method 4 using the pedals' positions as input led to the smallest average difference of M = −0.51 m (SD = 0.63 m). However, since the same pedals in terms of hardware and software were not installed in both simulators, a factor was required to convert the pedal values. This factor was also dependent on the lateral dynamics, so that it was not possible to configure this factor for the u-shaped track and we had to reject this method. Due to the smaller resulting differences for Method 2 compared to Method 1, we used the distance difference as an input for a separate PID controller to do the detail synchronization. In order to enable the PID control of the AV to compensate for the distance difference, the AV's speed limit of 50 km/h should be maintained until the start of the interaction phase.

For standardized conditions of the interactions, the speed profiles of the AV should be as identical as possible during all encounters within the interaction phase. For this purpose, a further pre-test was carried out in which the detail synchronization was switched off before the interaction phase so that the automated longitudinal guidance could be adjusted to 30 km/h. Again, the three synthetic speed profiles (offensive, neutral, defensive) were used on a straight course, while the distance of the switch-off to the road bottleneck was varied. The longitudinal control needs about 40 m to compensate for a speed difference of 5 km/h to the target speed of 30 km/h. Therefore, the distance of the switch-off was varied in 10 m steps between 80 m and 120 m to the road bottleneck. The start of the interaction phase (distance of 50 m) was used as reference. The mean distance differences with the standard deviation between the AV and implemented manual vehicle to the road bottleneck, respectively, are shown in Table 1.

Table 1. Mean and standard deviation of the distance differences of the AV and the implemented manual vehicle using each speed profile (offensive, neutral, defensive) once (n = 3). The distance to the road bottleneck when the detail synchronization was switched off was varied.

Switch-Off Distance [m]	M [m]	SD [m]
50	−0.79	1.11
80	0.21	4.49
90	0.76	5.31
100	1.57	6.11
110	2.19	7.14
120	2.65	8.19

The earlier the switch-off is performed, the more the mean distance difference, and especially its standard deviation, increases. Thus, an earlier switch-off point leads to a reduction in synchronicity. In contrast, an early switch-off of the detail synchronization leads to a constant speed profile in the interaction phase and thus to a corresponding reproducibility of the AV's speed profile. At speeds of less than 30 km/h of the AV during detail synchronization, the AV would subsequently accelerate to 30 km/h after switching off the detail synchronization in front of the bottleneck. This could lead to a lack of comprehensibility by the passenger, which in turn could result in passenger intervention. Thus, synchronicity, reproducibility, and comprehensibility must be taken into account when designing the detail synchronization (Figure 4). It is not possible to guarantee the desired interaction scenarios with the human driver by simultaneously fulfilling these three attributes. Therefore, one of the criteria had to be neglected in the design and either a limited synchronicity, a limited reproducibility, or a limited comprehensibility had to be accepted (Figure 4).

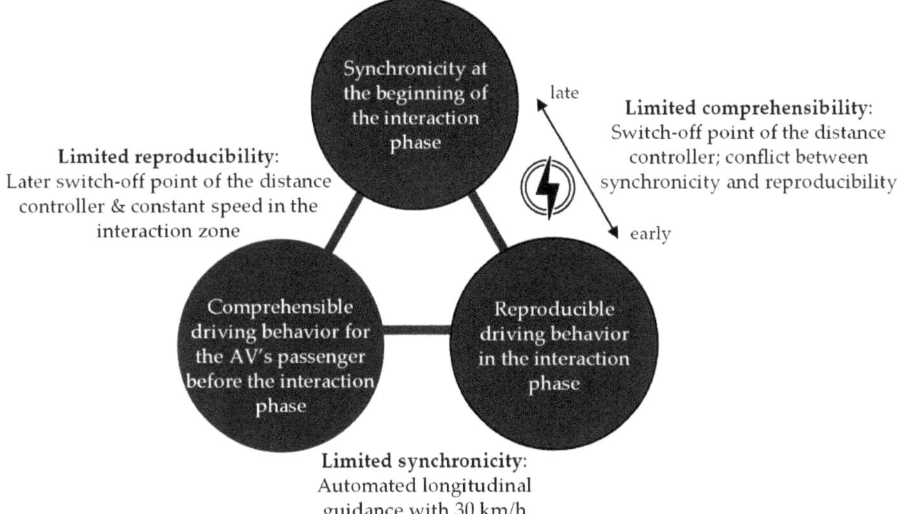

Figure 4. Effect of the switch-off point of the distance controller on the attributes synchronicity, reproducibility, and comprehensibility for detail synchronization. The switch-off point results in a trade-off between these attributes in a way that a simultaneous fulfilling of all attributes cannot be guaranteed.

For the investigation of the interaction at bottlenecks and possible automation failures, we considered the highest possible synchronicity and reproducibility as most important, so that the vehicles arrive at the bottleneck simultaneously and the AV has the target speed of 30km/h at the beginning of the interaction phase. Low synchronicity or reproducibility could limit the validity of the experimental setting and may lead to many excluded datasets. Therefore, we decided to use a limited comprehensibility. Switching off the detail synchronization 80 m in front of the bottleneck represents the best compromise between synchronicity and reproducibility (Table 1). A final pre-test with two participants and three runs each showed a distance difference of $M = -2.6$ m ($SD = 8.2$ m). We considered reaching the 30 km/h before the start of the interaction phase and having a distance difference of less than one vehicle length as a good reproducibility and synchronicity for our approach to use it for our experimental setting.

4.3. Course Design

Figure 5 presents the two course modules (Module I and Module II) we used in our study from the bird's eye view including the navigation details which supported both participants as they passed through the respective module on the intended route. Each participant drives through an individually designed urban route consisting of different streets and intersections. Since the participants are separated by a row of houses during entry into and exit from the module, they encounter each other only once per module at the road bottleneck. The size of the modules results in an average transit time of five minutes per module. The basic and detail synchronization occurs in the area around the bottleneck. The straight section on which the interaction takes place is 300 m long. Each traffic light of basic synchronization is 250 m apart of the bottleneck. The access to the interaction section consists of a slight bend so that the participants are not able to see each other while waiting at the respective traffic lights. For the manual vehicle, the speed limit at the interaction section was set to 30 km/h directly after the corresponding traffic light. The 30 km/h speed limit of the AV was set 50 m in front of the bottleneck. On the remaining course the speed limit was 50 km/h.

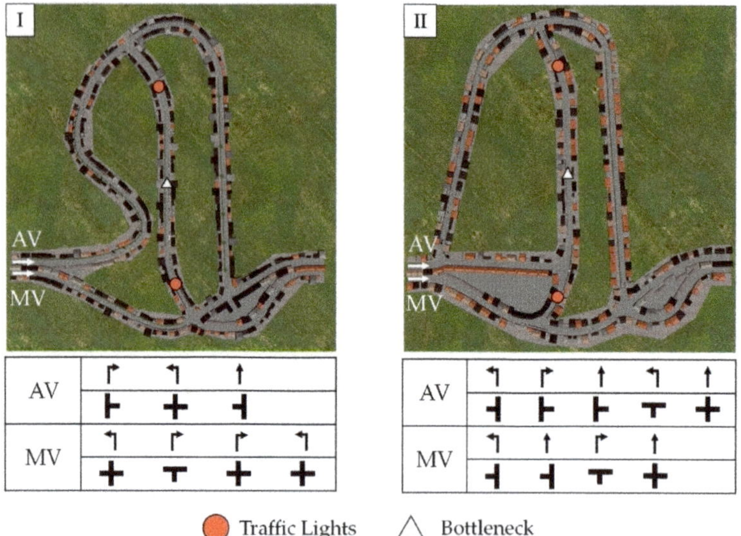

Figure 5. Course design consisting of two modules (Module I and Module II) the participants passed through during the experiment. Additionally, the navigation through the modules of the AV and the manual vehicle (MV) is presented.

5. Multi-Vehicle Study

5.1. Sample

Twenty-six participants took part in this study resulting in 13 participant pairs. The participants were comprised of 31% women and 69% men. The mean age of the participants was $M = 27.50$ years with a standard deviation of $SD = 8.99$ years. They possessed their driver's license for $M = 10.08$ years ($SD = 8.93$ years) and evaluated their previous knowledge of automated driving on a 5-point Likert scale from "very low" to "very high" with a median of 4 (= high). A statistical evaluation showed no differences between automated and manual vehicle groups. The requirement for participation in this experiment was a valid driver's license.

5.2. Experimental Design

The multi-vehicle study consisted of a 2 (message) × 2 (bottleneck type) repeated measures design. The first factor message (within-subject) represented the AV's intention. It contained the factor levels AV yields the right of way and AV insists on the right of way. The second factor bottleneck type (within-subject) consisted of the levels bottleneck narrowed on both sides and bottleneck narrowed on one side. Additionally, we implemented an automation failure where the AV first communicated to yield the right of way at a bottleneck narrowed only on the AV's side. Thirty meters in front of the bottleneck the AV failed to detect the oncoming human driver. Therefore, it stopped communicating by switching off the eHMI and started to pass through the bottleneck despite the oncoming human driver. Each participant pair experienced the Use Cases 1-4 once in a permuted order followed by Use Case 5 with the automation failure at the end of the experimental drive (Table 2).

Table 2. Five different Use Cases the participants passed through.

	AV Insists on Right of Way	AV Yields Right of Way	Automation Failure
Bottleneck narrowed on both sides	Use Case 1 (Module I)	Use Case 3 (Module I)	-
Bottleneck narrowed on one side	Use Case 2 (Module II)	Use Case 4 (Module II)	Use Case 5 (Module II)

5.3. Driving Simulators

The study took place in the two modular driving simulators at the Chair of Ergonomics of the Technical University of Munich (Figure 6). Both simulators offer a 120° horizontal field of view on three 55-inch screens with Ultra-HD resolution. While the rearview mirror is integrated in the view of the middle screen, two additional displays visualize the side mirrors. An additional display behind the steering wheel serves as a freely programmable instrument cluster (IC). In the AV setup, an LED-strip was positioned where the bottom of the windshield would be. In addition, the AV setup was equipped with a motion platform. Four D-BOX actuators generated pitch and roll movements, which provided participants with improved feedback about the behavior of the AV. Sound systems in both simulators generated engine and environmental sounds. We used the driving simulation software SILAB 6.0 from the Würzburg Institute of Traffic Sciences [32]. A data collection rate of 240 Hz and a refresh rate of 60 Hz was used. The partially automated driving system of the AV had to be activated by a button on the steering wheel. The automated driving system could be deactivated at any time using the same button or by braking, accelerating, or steering. The simulators are located in different rooms and were networked via LAN cable.

(a)   (b)

Figure 6. Modular driving simulators. (**a**) Manual vehicle setup; (**b**) Automated vehicle setup with blue LED-strip.

5.4. HMI Design

5.4.1. Human–Machine Interface of the Manual Vehicle

We used an instrument cluster (IC) and head-up display (HUD) for the HMI of the manual vehicle. Both HMI elements presented navigation and speed information. No other information, such as from driver assistance systems, was implemented in the manual vehicle's HMI.

5.4.2. Automation Human–Machine Interface

The aHMI [4] consisted of an instrument cluster (IC), a head-up display (HUD), and an LED strip. The aHMI should provide information about current and planned maneuvers in addition to the system status to the passenger when monitoring a partial automated driving system [33–35]. The LED-strip was mounted at the bottom of the windshield since this is an often used position in the context of automated driving [35–39]. When the ADS was available, the LED-strip illuminated white and after activation, the LED-strip illuminated blue [40]. For displaying the current and planned maneuver, the IC and HUD were used. The IC display (Figure 7) has been further modified from the adaptive concept of Feierle, Bücherl, Hecht, and Bengler [41]. The current speed is displayed on the left part of the IC, while the system status is displayed on the right and at the bottom as part of an automation scale. Central to the display is the indication of the planned and current maneuvers of the vehicle as well as the traffic sign recognition. Above this, as an extension of the road, is the navigation display. The visualization of the maneuvers regarding the investigated bottleneck scenarios depending on the oncoming traffic, are shown in Figure 8.

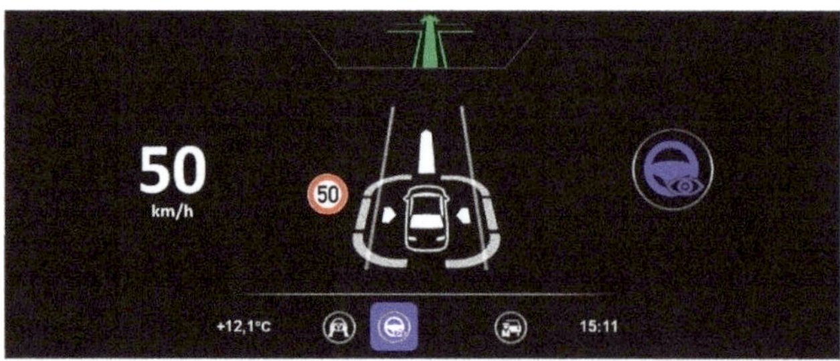

Figure 7. Visualization of the instrument cluster, modified from Feierle et al. (2020) [41].

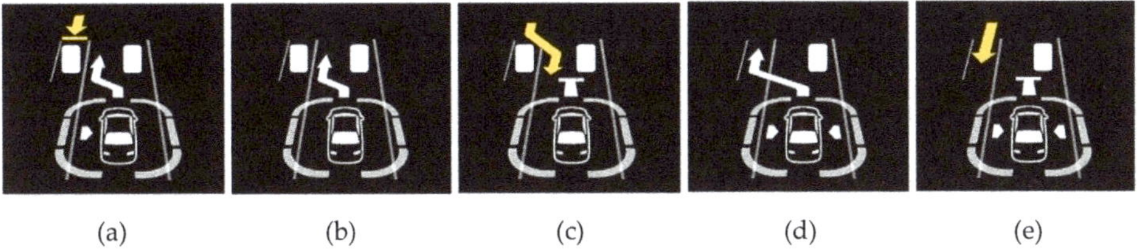

(a)   (b)   (c)   (d)   (e)

Figure 8. Visualization of the maneuver in the IC during the bottleneck scenarios: (**a**) bottleneck narrowed on both sides, AV insists on the right of way; (**b**) bottleneck narrowed on both sides, no oncoming traffic, AV passes; (**c**) bottleneck narrowed on both sides, AV yields the right of way; (**d**) bottleneck narrowed on one side, no upcoming traffic, AV passes; (**e**) bottleneck narrowed on one side, AV yields the right of way.

The HUD is based on the concept of Feierle, Beller, and Bengler [42]. The display (Figure 9) is divided into three sections. Speed information is located at the left section, system status, and driving maneuvers are shown in the middle section, and the right section shows the navigation information.

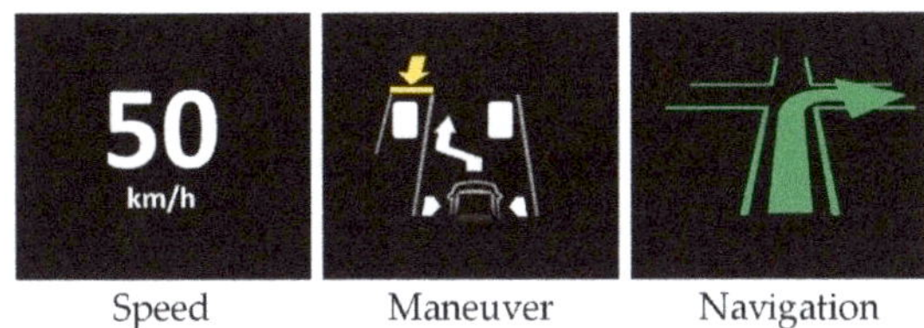

Speed   Maneuver   Navigation

Figure 9. Head-up display showing speed, maneuver, and navigation information when the AV insists on the right of way in the road bottleneck scenario narrowed on both sides. The black background is transparent in the driving simulation.

5.4.3. External Human–Machine Interface

The eHMI [4] consisted of a display mounted at the front of the vehicle, since its message is visible for the human driver, especially for long distances like in the road bottleneck scenario [6]. The design of the eHMI (Figure 10) was developed by Rettenmaier et al. [30]. The eHMI uses an arrow to indicate which negotiation partner can pass through the bottleneck first. With the green arrow the AV communicates to yield the right of way to the human driver. The orange arrow indicates that the AV insists on the right of way. Both arrows are animated with a frequency of 1 Hz building up in the direction the bottleneck may be passed through first. Additionally, with the arrows the eHMI design includes the contour of the road represented by two gray lines [30].

AV yields the right of way

AV insists on the right of way

Figure 10. External HMI used in the study. In the upper part of the picture the AV indicates to yield the right of way to the human driver. In the lower part the AV communicates to insist on the right of way. The illustrated scenario is the road bottleneck narrowed on both sides of the road [30].

5.5. Experimental Track and Bottleneck Scenarios

The experimental track consisted of a route network in an urban area with several intersections and connecting roads. The scenario examined in the study is the road bottleneck scenario composed of the simultaneous encounter of a human driver and an AV approaching from the opposite direction. The scenario varies the bottleneck type and the right of way. Figure 11 presents the five resulting Use Cases the participants passed through during the experimental drive. The scenario is subdivided into the approaching phase in which both participants approach the bottleneck until the start of the interaction phases starting 50 m in front of the bottleneck. In the interaction phase the AV switches on its eHMI and it starts communicating to yield the right of way or to insist on it. If the AV yielded the right of way it stopped (S) 13 m in front of the obstacle. In Use Case 1 and Use Case 3 the bottleneck was constricted on both sides of the road due to two double-parked vehicles. In Use Case 2 and Use Case 4 there was only one obstacle, either on the human driver's side of the road or in the AV's lane. Use Case 5 represents the implemented automation failure when the AV first communicates to yield the right of way at the bottleneck narrowed on one side. Then the AV changes the strategy 30 m in front of the bottleneck and demonstrates insisting on the right of way. The 30 m results from adding the travel distance within a one second reaction time ($x = 8.33$ m), the braking distance with a deceleration of -2 m/s^2 ($y = 17.35$ m), and the stopping distance to the middle of the bottleneck ($z = 4$ m). After oncoming traffic is initially detected, the AV changes the communication strategy by switching off the eHMI due to losing the detection of the oncoming human driver during the passage. The speed limit in the interaction phase was set to 30 km/h for both participants.

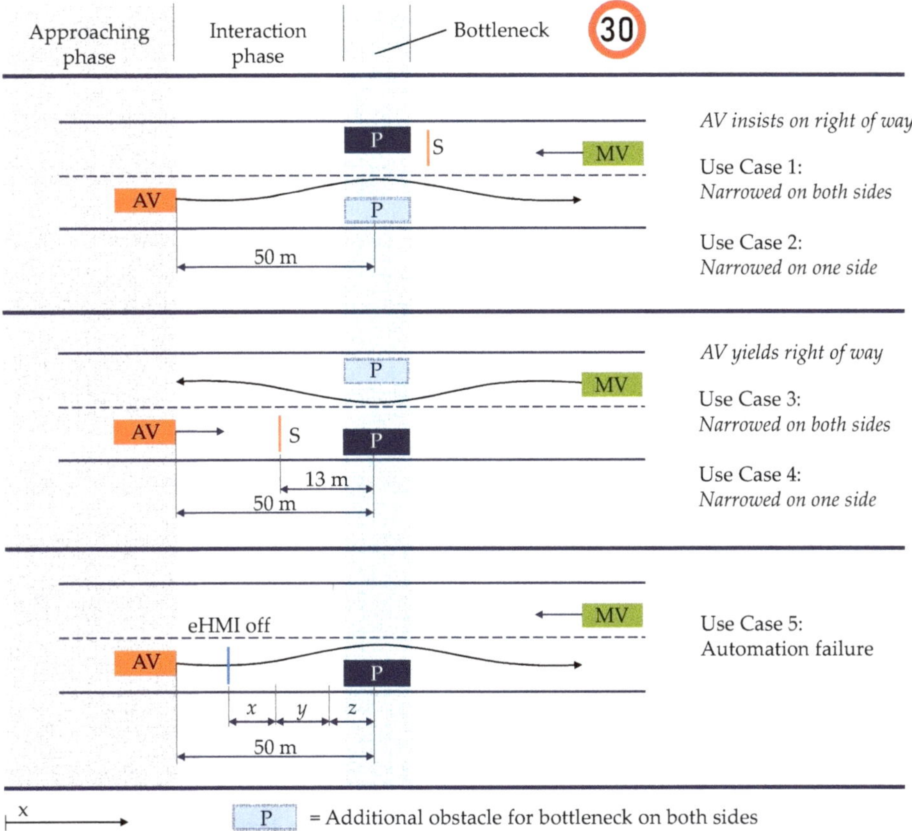

Figure 11. Different bottleneck scenarios the participants passed through during the experimental drive. The scenarios are located in the interaction phase with a speed limit of 30 km/h. In the interaction phase the AV communicates to the driver of the manual vehicle (MV) via the eHMI either to yield the right of way or to insist on it [30].

5.6. Procedure

During the experiment there were two experimenters, one for each participant. Welcoming and introducing the participants was conducted separately by the experimenters to avoid the influence of gender effects, sympathy/antipathy, or social similarity between the participants. After reading the safety instructions and the participant information the participants consented to the experiment. Subsequently, the participants filled in a demographic questionnaire including the age, gender, experience with automated driving, and the possession of their driver's license. Afterwards the participants received the instruction. The participants acting as the human drivers in the simulation were instructed about manual driving with navigation instructions and were informed that there would be interactions with an AV. Moreover, the human drivers were also made aware of the presence of another human in the AV in the simulation. The AV's passengers were instructed about partially automated driving, its capacities, and about the obligation of monitoring the driving scene. Additionally, the AV's passengers also received information about the presence of a human driver in the same simulation, since this awareness could positively affect the willingness to cooperate [21].

Subsequently, both participants completed an introductory drive (duration: 10 min) in the multi-vehicle simulation. The human drivers had the opportunity to familiarize themselves with the simulator's driving behavior and the navigation information. The AV's passengers got acquainted with the driving automation including the oversteering of the same. Afterwards, the experimental drive (duration: 25 min) was conducted consisting of passing through the Use Cases 1–4 in a permuted order followed by the experience of the automation failure in Use Case 5. The experiment concluded by both participants filling out a questionnaire and having an oral interview referring to the automation failure they experienced.

5.7. Measures and Analysis

We used the differences in distance and in time to arrival (TTA) of the two simulated vehicles to the bottleneck to assess the synchronicity and the driving profiles resulting from the methodology. Both metrics were calculated once the first of the two vehicles reached the interaction phase. For this purpose, six of the 65 possible encounters had to be discarded due to the intervention of participants in the AV before reaching the interaction phase.

To determine traffic efficiency and safety, we excluded the data of three participant pairs due to technical issues within the interaction phase. The traffic efficiency was operationalized by means of participants' passing times. This metric was defined as the time that elapsed from the manual driver's entrance to the interaction phase (50 m in front of the bottleneck) until passing the AV 15 m behind the bottleneck. The crash rate was used to assess the controllability of the automation failure. Additionally, as a further metric the time to collision (TTC) was calculated when the passenger of the AV took over control of the vehicle guidance. Based on the small sample size in multi-vehicle simulation and the large difference in sample size compared to single driver simulations, we refrained from a statistical evaluation and we descriptively analyzed the data.

6. Results

6.1. Technical Implementation

Figure 12 shows the distances of the manual vehicles and AVs to the bottleneck (blue line) as a result of the distance control. The angle bisector (orange line) represents the distances for an ideal synchronization, if the implemented control does not result in any delay. It can be seen that both vehicles start at different distances from the bottleneck after the traffic light turns green. At the beginning the manual vehicles approach the bottleneck faster than the AVs, resulting in a vertical rise in the curves. Therefore, the distance control results in an offset as the initial accelerations of the human drivers cannot be compensated for quickly enough. The maximum deviations occur between 250 m and 200 m. From 200 m to the bottleneck, the control is more successful in compensating for the difference in distance, which brings the curves closer to ideal synchronicity again, whereby in some cases an offset remains until 80 m before the bottleneck. The deviation increases again directly before the interaction phase. This may be due to the switch of the synchronization mode to the longitudinal control independent of the human drivers' behavior. In most cases, the manual vehicle reaches the interaction phase first since the human drivers show higher speed than 30 km/h in most cases.

The differences in distance (Δd) (Figure 13) result in $M = -5.70$ m ($SD = 4.06$ m) which corresponds to a difference in TTA of $M = -0.34$ s ($SD = 1.10$ s). A negative difference in distance and TTA mean an earlier arrival of the manual vehicle at the interaction phase.

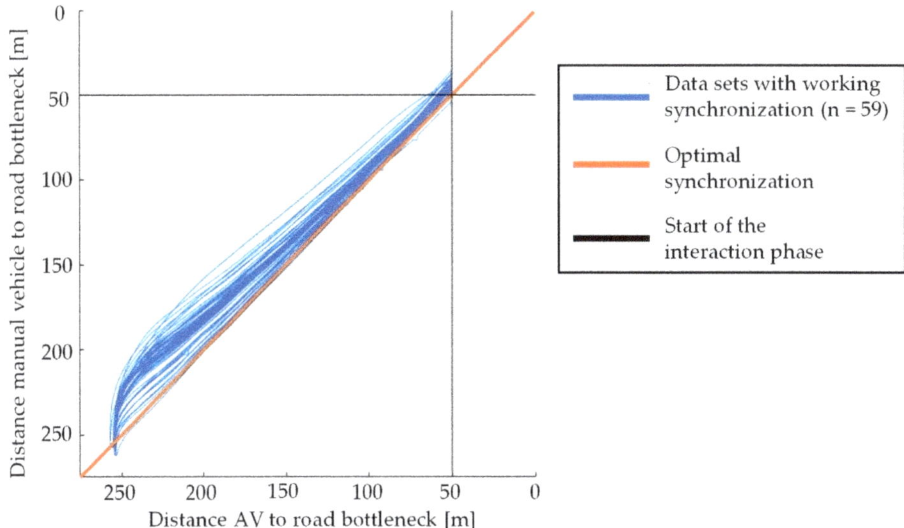

Figure 12. Distances of the manual vehicles plotted over the distances of the AVs to the bottleneck during the detail synchronization phase. The angle bisector visualizes the distances for an ideal synchronization.

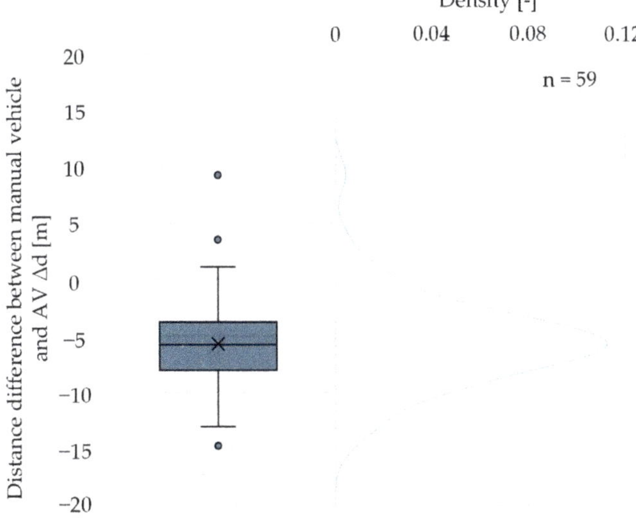

Figure 13. Distance differences and its relative frequency of the manual vehicles and AVs when one of them has already reached the interaction phase. Negative values mean that the manual vehicle arrived at the road bottleneck first.

6.2. Multi-Vehicle Study

6.2.1. Human Driving Behavior

Figure 14 shows the participants' passing time in the case that the AV yielded the right of way divided by the data of the single-driver simulation [30] (21 data sets) and the data of the multi-vehicle simulation (10 data sets). One data set ($n = 9$) was removed in multi-vehicle simulation due to an intervening participant in the AV. Table 3 contains the descriptive data. At the bottleneck narrowed on one side the passing time is similar in both studies with an average difference of 158 ms. At the bottleneck narrowed on both sides the participants in the single-driver study needed on average 465 ms more than in the multi-vehicle simulation.

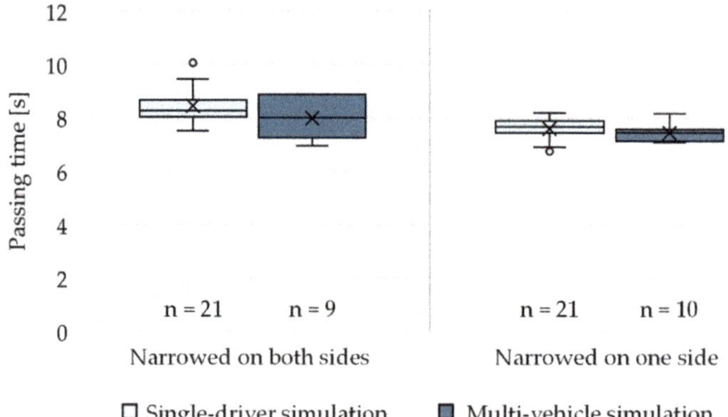

Figure 14. Participants average passing time in the case that the AV yields the right of way to the oncoming human driver divided by the bottleneck type. The data of the single-driver simulation are derived from Rettenmaier et al. [30].

Table 3. Descriptive data of the participants' passing time. The data of the single-driver simulation are derived from Rettenmaier et al. [30].

	Bottleneck Narrowed on Both Sides	Bottleneck Narrowed on One Side
Single-driver simulation M (SD) [ms]	8445 (1405) (n = 21)	7598 (495) (n = 21)
Multi-vehicle simulation M (SD) [ms]	7980 (740) (n = 9)	7440 (310) (n = 10)

6.2.2. Effect of Automation Failure

In the multi-vehicle simulation from ten trials four crashes occurred during the automation failure, where the human driver crashed with the AV and its passenger. These encounters are characterized by a late intervening AV's passenger (TTC: 0.37 s, 0.65 s, 0.90 s, 0.94 s). The change in the aHMI was not detected by the AV's passenger in all four cases. Moreover, switching off the AV's eHMI was only detected by one manual driving participant. The other three participants did not detect that the eHMI was deactivated. In contrast to the 40% crash rate of the multi-vehicle simulation, the single-driver simulation showed a crash rate of 95% [30].

In six trials no crash occurred. These encounters include faster interventions of the AV's passenger braking to standstill (TTC: 1.31 s, 1.83 s, 2.06 s, 2.32 s, 2.44 s, 2.73 s). The AV's passengers stated that no oncoming traffic was detected permanently (once), that they had noticed the change in the aHMI (three times), or that they could not give any information about the aHMI during the automation failure (two times). None of the six human drivers that did not crash noticed that the eHMI was switched off. They stated that the eHMI continuously communicated to yield the right of way.

7. Discussion

7.1. Technical Implementation

The results of the detail synchronization show that at the beginning the distance of the manual vehicle to the road bottleneck is decreasing faster than the distance of the AV to the bottleneck.

According to the driving data, this is due to a slower acceleration of the AV. Thus, the manual vehicle accelerates strongly in the beginning and quickly reaches the maximum permitted speed. This fact can be attributed to the accelerator pedal in the manual driving simulator setup, which has a lower resistance than one of the Sensowheel pedals in the automated driving simulator setup. The resulting distance difference is compensated by the distance control in the detail synchronization with the distance passed. This can only be achieved by increasing the speed of the AV compared to the manual vehicle. Nevertheless, the allowed 50 km/h on the AV side during detail synchronization were rarely reached before the interaction phase. None of the participants reported that the speed was below the maximum speed, so different speed regulations seem to be a good way to compensate for differences in distance. Since the synchronicity increases with distance traveled, extending the distance of the detail synchronization could provide an improvement. A further adjustment of the PID controller could additionally provide improved synchronicity with a lower deviation. In addition, a modification of the control loop, e.g., by a two-cascade control, would be thinkable. However, the inconsistent setpoint changes caused by the driving behavior of the human driver make it difficult to minimize the control deviation with the proposed possible improvements. In particular, switching off the detail synchronization 80 m before the bottleneck leads to an increase in asynchrony directly before the interaction phase. The absolute value of the resulting mean distance difference ($M = 5.7$ m) only moderately exceeds the AV's length (4.68 m), which we consider a tolerable deviation. Previous multi-agent studies including two manual road users lacked in inducing the intended interactions in a controlled manner in half of the recorded interactions in Will [18] and between 30% and 43% in Hancock and de Ridder [16]. Compared to these studies, the synchronization of the AV and the manual driver in this paper succeeded in all cases without an intervening AV's passenger. Therefore, the proposed method appears to be valid to implement a multi-vehicle simulation with one AV.

7.2. Multi-Vehicle Study

7.2.1. Human Driving Behavior

The AV supports the human driver to efficiently pass through the bottleneck scenarios by communicating to yield the right of way. The enhancement in traffic efficiency is reflected in the human drivers' short passing times. In comparison to the passing times of the single-driver simulation [30], the ones of the multi-vehicle study are similar or even slightly faster. This could be attributable to the fact that the AV arrived at the bottleneck a little later, which is an indication to yield the right of way in real world traffic [31]. However, as there are no clear tendencies, we state that the synchronization of both participants was implemented with sufficient accuracy and that there is no major influence by the variance of synchronization. Thus, the multi-vehicle simulation has, apart from the complex implementation, no disadvantage compared with the single-driver simulation when investigating the interaction of an AV with a manual driving participant.

7.2.2. Effect of Automation Failure

The multi-vehicle simulation resulted in a lower crash rate compared to the single-driver study [30]. However, the automation failure in this paper resulted in four crashes of the AV and the human driver, which means that the implemented scenario was too critical to be resolved by the participants. Only one participant noticed that the AV switched off its eHMI. Switching off the eHMI to communicate that the AV changes its strategy and passes through the bottleneck is insufficient. As already shown in the single-driver simulation [30] the AV has to communicate the changing driving strategy more saliently by displaying at least the message of the AV's actual status. The increased stimulus would result in faster reaction times by the participants [43] and could lower the crash rate. Additionally, only 30% of the AVs' passengers noticed the change in the IC or HUD. Here, a salient presentation of the planned maneuver by an augmented reality HUD and the resulting shift of the visual attention to the relevant driving environment could offer added value for future investigations [42].

In summary, participants were used to a perfect working automated system due to the previous encounters. During the automation failure, participants were not attentive enough since it was hardly possible for humans to monitor for unlikely abnormalities [44]. Therefore, we state that the AV's internal and external communication must be reliable and the AV must not change its strategy.

7.2.3. Is Multi-Vehicle Simulation Beneficial?

If a study deals with the interaction of a perfectly working AV with its passenger or surrounding road users, there is no benefit of multi-vehicle simulation compared to the single-driver simulation because the results show no clear descriptive tendencies. It makes no difference to the human drivers' driving behavior whether there is a real passenger in the AV or whether the AV is implemented within a single-driver simulation because in both cases the AV is programmed and the passenger has no influence on the AV's behavior. We state that in scenarios where only one human negotiation partner affects the interaction it is sufficient to use single-driver simulation, thus avoiding the additional effort of the multi-vehicle simulation.

If research deals with the interaction of two human negotiation partners like after the take-over of the AV's passenger during the automation failure, there is a benefit for multi-vehicle simulation. The results show that the AV's passenger lowered the crash rate by intervening in the multi-vehicle simulation. The take-over including the timing and the braking behavior of the AV's passenger is barely possible to implement in the single-driver simulation.

7.3. Limitations

A statistical analysis between the data of the multi-vehicle simulation and the single-driver simulation is not reasonable since the sample size in the present study was too small. Nevertheless, descriptively analyzing the data shows similar results in driving behavior in multi-vehicle and in single-driver simulation. Moreover, the sample was young and an above-average number of male participants attended. It will be useful to conduct future experiments with an age- and gender-balanced sample.

Since the human drivers' driving behavior differed, the synchronization and thus the arrival at the bottleneck was not completely simultaneous in each trial in the way that the human driver reached the interaction phase first. This fact could have affected the participants' passing times. The variance in manual driving behavior had the additional effect of the AV sometimes demonstrating incomprehensible driving behavior to compensate for the difference in distance. However, this problem did not disturb any participant.

8. Conclusions and Future Work

Based on the successful synchronization of the AV and manual vehicle in this study, we recommend a traffic light control for basic synchronization and a distance control for detail synchronization for future investigations using multi-vehicle simulation. The multi-vehicle simulation compared to a single-driver simulation revealed an added benefit for the automation failure scenario by realizing a more human-like interaction of two potential reacting and acting participants.

Single-driver studies seem to be appropriate to enable a worst-case consideration without an intervening AV's passenger, for example, in automation failure scenarios. To investigate more realistic regular interactions between several road users further multi-vehicle simulation studies should be conducted. We suggest conducting a large-scaled study addressing several scenarios (e.g., bottlenecks, intersections, roundabouts) to allow a deeper comparison with single-driver studies and a simultaneous investigation of AV's internal and external communication. Furthermore, future multi-agent simulation studies should not be limited to motorized road users, but should also address vulnerable road users such as cyclists and pedestrians.

Author Contributions: Conceptualization, A.F., M.R., and F.Z.; methodology, A.F., M.R., and F.Z.; software, A.F., M.R., and F.Z.; validation, A.F., M.R., and F.Z.; formal analysis, A.F., M.R., and F.Z.; investigation, A.F., M.R., and F.Z.; resources, A.F. and M.R.; data curation, A.F., M.R., and F.Z.; writing—original draft preparation, A.F. and M.R.; writing—review and editing, A.F. and M.R.; visualization, A.F. and M.R.; supervision, K.B. All authors have read and agreed to the published version of the manuscript.

Funding: This research was funded by the German Federal Ministry of Economics and Energy within the project @CITY: Automated Cars and Intelligent Traffic in the City, grant number 19A17015B. The authors are solely responsible for the content.

Conflicts of Interest: The authors declare no conflict of interest. The funders had no role in the design of the study; in the collection, analyses, or interpretation of data; in the writing of the manuscript, or in the decision to publish the results.

References

1. Götze, M. Entwicklung und Evaluation eines integrativen MMI Gesamtkonzeptes zur Handlungsunterstützung für den urbanen Verkehr. Ph.D. Thesis, Technical University of Munich, Munich, Germany, 2018.

2. Lüke, S.; Fochler, O.; Schaller, T.; Regensburger, U. Stauassistenz und-automation. In *Handbuch Fahrerassistenzsysteme: Grundlagen, Komponenten und Systeme für aktive Sicherheit und Komfort*; Winner, H., Hakuli, S., Lotz, F., Singer, C., Eds.; Springer Fachmedien Wiesbaden: Wiesbaden, Germany, 2015; pp. 995–1007. [CrossRef]

3. Rittger, L.; Götze, M. HMI Strategy—Recommended Action. In *UR:BAN Human Factors in Traffic*; Bengler, K., Drüke, J., Hoffmann, S., Manstetten, D., Neukum, A., Eds.; Springer Fachmedien Wiesbaden: Wiesbaden, Germany, 2018; pp. 119–150. [CrossRef]

4. Bengler, K.; Rettenmaier, M.; Fritz, N.; Feierle, A. From HMI to HMIs: Towards an HMI Framework for Automated Driving. *Information* **2020**, *11*, 61. [CrossRef]

5. Habibovic, A.; Andersson, J.; Malmsten Lundgren, V.; Klingegård, M.; Englund, C.; Larsson, S. External Vehicle Interfaces for Communication with Other Road Users. In *Road Vehicle Automation*; Springer: Cham, Germany, 2018; pp. 91–102. [CrossRef]

6. Rettenmaier, M.; Pietsch, M.; Schmidtler, J.; Bengler, K. Passing through the Bottleneck—The Potential of External Human-Machine Interfaces. In Proceedings of the IEEE Intelligent Vehicles Symposium (IV), Paris, France, 9–12 June 2019; pp. 1687–1692. [CrossRef]

7. SAE International. *Taxonomy and Definitions for Terms Related to Driving Automation Systems for On-Road Motor Vehicles*; SAE International: Warrendale, PA, USA, 2018; p. 3016.

8. Mühlbacher, D.; Preuk, K.; Lehsing, C.; Will, S.; Dotzauer, M. Multi-Road User Simulation: Methodological Considerations from Study Planning to Data Analysis. In *UR:BAN Human Factors in Traffic: Approaches for Safe, Efficient and Stress-Free Urban Traffic*; Bengler, K., Drüke, J., Hoffmann, S., Manstetten, D., Neukum, A., Eds.; Springer Fachmedien Wiesbaden: Wiesbaden, Germany, 2018; pp. 403–418. ISBN 978-3-658-15418-9.

9. Oeltze, K.; Dotzauer, M. Towards A Best Practice for Multi-Driver Simulator Studies. In Proceedings of the Workshop on Practical Experiences in Measuring and Modeling Drivers and Driver-Vehicle Interactions at AutomotiveUI'15, Nottingham, UK, 1–3 September 2015.

10. Oeltze, K.; Schießl, C. Benefits and challenges of multi-driver simulator studies. *IET Intell. Transp. Syst.* **2015**, *9*, 618–625. [CrossRef]

11. Sun, J.; Ma, Z.; Li, T.; Niu, D. Development and application of an integrated traffic simulation and multi-driving simulators. *Simul. Model. Pract. Theory* **2015**, *59*, 1–17. [CrossRef]

12. Mühlbacher, D.; Rittger, L.; Maag, C. Real vs. Simulated Surrounding Traffic—Does It Matter. In Proceedings of the Driving Simulation Conference 2014, Paris, France, 4–5 September 2014.

13. Mühlbacher, D. The Multi-Driver Simulation: A Tool to Investigate Social Interactions Between Several Drivers. In *UR:BAN Human Factors in Traffic: Approaches for Safe, Efficient and Stress-free Urban Traffic*; Bengler, K., Drüke, J., Hoffmann, S., Manstetten, D., Neukum, A., Eds.; Springer Fachmedien Wiesbaden: Wiesbaden, Germany, 2018; pp. 379–391. ISBN 978-3-658-15418-9.

14. Maag, C. Emerging Phenomena During Driving Interactions. In *Co-Evolution of Intelligent Socio-Technical Systems*; Mitleton-Kelly, E., Ed.; Springer: Berlin/Heidelberg, Germany, 2013; pp. 185–218. [CrossRef]

15. Lehsing, C.; Benz, T.; Bengler, K. Insights into Interaction—Effects of Human-Human Interaction in Pedestrian Crossing Situations using a linked Simulator Environment. *IFAC-PapersOnLine* **2016**, *49*, 138–143. [CrossRef]

16. Hancock, P.A.; de Ridder, S.N. Behavioural accident avoidance science: Understanding response in collision incipient conditions. *Ergonomics* **2003**, *46*, 1111–1135. [CrossRef] [PubMed]
17. Yasar, A.-U.-H.; Berbers, Y.; Preuveneers, D. Computational Analysis of Driving Variations on Distributed Multiuser Driving Simulators. In Proceedings of the Second IASTED Africa Conference on Modelling and Simulation, Gaborone, Botswana, 8–10 September 2008; Science and technology innovation for sustainable development. Ogwu, F.J., Ed.; International Association of Science and Technology for Development: Calgary, AB, Canada, 2008. ISBN 9780889867642.
18. Will, S. A New Approach to Investigate Powered Two Wheelers' Interactions with Passenger Car Drivers: The Motorcycle—Car Multi-Driver Simulation. In *UR:BAN Human Factors in Traffic: Approaches for Safe, Efficient and Stress-free Urban Traffic*; Bengler, K., Drüke, J., Hoffmann, S., Manstetten, D., Neukum, A., Eds.; Springer Fachmedien Wiesbaden: Wiesbaden, Germany, 2018; pp. 393–402. ISBN 978-3-658-15418-9.
19. Mühlbacher, D.; Zimmer, J.; Fischer, F.; Krüger, H.-P. The multi-driver simulator—A new concept of driving simulation for the analysis of interactions between several drivers. In *Human Centred Automation*; de Waard, D., Gérard, L., Onnasch, L., Wiczorek, R., Manzey, D., Eds.; Shaker Publishing: Maastricht, The Netherlands, 2011; pp. 147–158.
20. Mühlbacher, D.; Krüger, H.-P. The effect of car-following on lateral guidance during cognitive load—A study conducted in the multi-driver simulation. In Proceedings of the 2nd International Conference on Driver Distraction and Inattention 2011, Gothenburg, Sweden, 5–7 September 2011.
21. Heesen, M.; Baumann, M.; Kelsch, J.; Nause, D.; Friedrich, M. Investigation of Cooperative Driving Behaviour during Lane Change in a Multi-Driver Simulation Environment. In Proceedings of the HFES Europe Chapter Conference Toulouse, Toulouse, France, 10–12 October 2012.
22. Niu, D.; Sun, J. Eco-driving Versus Green Wave Speed Guidance for Signalized Highway Traffic: A Multi-vehicle Driving Simulator Study. *Procedia-Soc. Behav. Sci.* **2013**, *96*, 1079–1090. [CrossRef]
23. Gajananan, K.; Nantes, A.; Miska, M.; Nakasone, A.; Prendinger, H. An Experimental Space for Conducting Controlled Driving Behavior Studies based on a Multiuser Networked 3D Virtual Environment and the Scenario Markup Language. *IEEE Trans. Hum. -Mach. Syst.* **2013**, *43*, 345–358. [CrossRef]
24. Rittger, L.; Mühlbacher, D.; Maag, C.; Kiesel, A. Anger and bother experience when driving with a traffic light assistant: A multi-driver simulator study. In Proceedings of the Human Factors and Ergonomics Society Europe Chapter 2014 Annual Conference, Lisbon, Portugal, 8–10 October 2014.
25. Friedrich, M.; Nause, D.; Heesen, M.; Keich, A.; Kelsch, J.; Baumann, M.; Vollrath, M. Validation of the MoSAIC-Driving Simulator—Investigating the impact of a human driver on cooperative driving behavior in an experimental simulation setup. *Proc. Hum. Factors Ergon. Soc. Annu. Meet.* **2013**, *57*, 2052–2056. [CrossRef]
26. Bazilinskyy, P.; Kooijman, L.; Dodou, D.; de Winter, J.C.F. Coupled Simulator for Research on the Interaction between Pedestrians and (Automated) Vehicles. 2019. Available online: https://www.researchgate.net/publication/338118077_Coupled_simulator_for_research_on_the_interaction_between_pedestrians_and_automated_vehicles (accessed on 19 May 2020).
27. Schindler, J.; Köster, F. A Dynamic and Model-Based Approach for Performing Successful Multi-Driver Studies. Proceedings of DSC Europe, Paris, France, 7–9 September 2016.
28. Mühlbacher, D. Die Pulksimulation als Methode zur Untersuchung verkehrspsychologischer Fragestellungen. Ph.D. Thesis, University of Würzburg, Würzburg, Germany, 2013.
29. Will, S. Die—vernetzte—Fahrsimulation zur Untersuchung des Fahr-und Interaktionsverhaltens von Motorradfahrern. 2016. Available online: https://docplayer.org/22386886-Sim-die-vernetzte-fahrsimulation-zur-untersuchung-des-fahr-und-interaktionsverhaltens-von-motorradfahrern.html (accessed on 19 May 2020).
30. Rettenmaier, M.; Albers, D.; Bengler, K. After you?!—Use of external human-machine interfaces in road bottleneck scenarios. *Transp. Res. Part F Traffic Psychol. Behav.* **2020**, *70*, 175–190. [CrossRef]
31. Rettenmaier, M.; Requena Witzig, C.; Bengler, K. Interaction at the Bottleneck—A Traffic Observation. In *Human Systems Engineering and Design II*; Ahram, T., Karwowski, W., Pickl, S., Taiar, R., Eds.; Springer International Publishing: Cham, Germany, 2020; pp. 243–249. [CrossRef]
32. Würzburg Institute for Traffic Sciences GmbH. Driving Simulation and SILAB. Available online: https://wivw.de/en/silab (accessed on 14 December 2019).

33. Drüke, J.; Semmler, C.; Bendewald, L. The "HMI tool kit" as a Strategy for the Systematic Derivation of User-Oriented HMI Concepts of Driver Assistance Systems in Urban Areas. In *UR:BAN Human Factors in Traffic*; Bengler, K., Drüke, J., Hoffmann, S., Manstetten, D., Neukum, A., Eds.; Springer Fachmedien Wiesbaden: Wiesbaden, Germany, 2018; pp. 53–74. [CrossRef]

34. van den Beukel, A.P.; van der Voort, M.C. Design Considerations on User-Interaction for Semi-Automated Driving. In Proceedings of the FISITA 2014 World Automotive Congress, Maastricht, The Netherlands, 2–6 June 2014.

35. Othersen, I. Vom Fahrer zum Denker und Teilzeitlenker: Einflussfaktoren und Gestaltungsmerkmale nutzerorientierter Interaktionskonzepte für die Überwachungsaufgabe des Fahrers im teilautomatisierten Modus. *AutoUni-Schr.* **2016**. [CrossRef]

36. Yang, Y.; Götze, M.; Laqua, A.; Dominioni, G.C.; Kawabe, K.; Bengler, K. A method to improve driver's situation awareness in automated driving. In Proceedings of the Human Factors and Ergonomics Society Europe Chapter 2017 Annual Conference, Rome, Italy, 28–30 September 2017.

37. Yang, Y.; Karakaya, B.; Dominioni, G.C.; Kawabe, K.; Bengler, K. An HMI Concept to Improve Driver's Visual Behavior and Situation Awareness in Automated Vehicle. In Proceedings of the 2018 IEEE Intelligent Transportation Systems Conference (ITSC), Maui, HI, USA, 4–7 November 2018; pp. 650–655. [CrossRef]

38. Feldhütter, A.; Härtwig, N.; Kurpiers, C.; Hernandez, J.M.; Bengler, K. Effect on Mode Awareness When Changing from Conditionally to Partially Automated Driving. In Proceedings of the 20th Congress of the International Ergonomics Association (IEA 2018), Florence, Italy, 26–30 August 2018; Volume VI: Transport Ergonomics and Human Factors (TEHF), Aerospace Human Factors and Ergonomics. Bagnara, S., Tartaglia, R., Albolino, S., Alexander, T., Fujita, Y., Eds.; Springer International Publishing: Cham, Germany, 2019; pp. 314–324. [CrossRef]

39. Feierle, A.; Danner, S.; Steininger, S.; Bengler, K. Information Needs and Visual Attention during Urban, Highly Automated Driving—An Investigation of Potential Influencing Factors. *Information* **2020**, *11*, 62. [CrossRef]

40. Feierle, A.; Holderied, M.; Bengler, K. Evaluation of Ambient Light Displays for Requests to Intervene and Minimal Risk Maneuvers in Highly Automated Urban Driving. In Proceedings of the 2020 IEEE Intelligent Transportation Systems Conference (ITSC), Rhodes, Greece, 20–23 September 2020.

41. Feierle, A.; Bücherl, F.; Hecht, T.; Bengler, K. Evaluation of Display Concepts for the Instrument Cluster in Urban Automated Driving. In *Human Systems Engineering and Design II*; Ahram, T., Karwowski, W., Pickl, S., Taiar, R., Eds.; Springer International Publishing: Cham, Germany, 2020; pp. 209–215. [CrossRef]

42. Feierle, A.; Beller, D.; Bengler, K. Head-Up Displays in Urban Partially Automated Driving: Effects of Using Augmented Reality. In Proceedings of the 2019 IEEE Intelligent Transportation Systems Conference (ITSC) 2019, Auckland, New Zealand, 27–30 October 2019; pp. 1877–1882. [CrossRef]

43. Werneke, J.; Vollrath, M. Signal evaluation environment: A new method for the design of peripheral in-vehicle warning signals. *Behav. Res. Methods* **2011**, *43*, 537–547. [CrossRef] [PubMed]

44. Bainbridge, L. Ironies of automation. *Automatica* **1983**, *19*, 775–779. [CrossRef]

 *information*

Article

Feeling Uncertain—Effects of a Vibrotactile Belt that Communicates Vehicle Sensor Uncertainty

Matti Krüger [1,†], Tom Driessen [2,†], Christiane B. Wiebel-Herboth [1], Joost C. F. de Winter [2,*] and Heiko Wersing [1]

[1] Honda Research Institute Europe, 63073 Offenbach am Main, Germany; matti.krueger@honda-ri.de (M.K.); christiane.wiebel@honda-ri.de (C.B.W.-H.); heiko.wersing@honda-ri.de (H.W.)
[2] Department of Cognitive Robotics, Faculty of Mechanical, Maritime and Materials Engineering, Delft University of Technology, 2628 CD Delft, The Netherlands; t.driessen@tudelft.nl
*   Correspondence: j.c.f.dewinter@tudelft.nl
†   These authors contributed equally to this work.

Received: 31 May 2020; Accepted: 30 June 2020; Published: 6 July 2020

Abstract: With the rise of partially automated cars, drivers are more and more required to judge the degree of responsibility that can be delegated to vehicle assistant systems. This can be supported by utilizing interfaces that intuitively convey real-time reliabilities of system functions such as environment sensing. We designed a vibrotactile interface that communicates spatiotemporal information about surrounding vehicles and encodes a representation of spatial uncertainty in a novel way. We evaluated this interface in a driving simulator experiment with high and low levels of human and machine confidence respectively caused by simulated degraded vehicle sensor precision and limited human visibility range. Thereby we were interested in whether drivers (i) could perceive and understand the vibrotactile encoding of spatial uncertainty, (ii) would subjectively benefit from the encoded information, (iii) would be disturbed in cases of information redundancy, and (iv) would gain objective safety benefits from the encoded information. To measure subjective understanding and benefit, a custom questionnaire, Van der Laan acceptance ratings and NASA TLX scores were used. To measure the objective benefit, we computed the minimum time-to-contact as a measure of safety and gaze distributions as an indicator for attention guidance. Results indicate that participants were able to understand the encoded uncertainty and spatiotemporal information and purposefully utilized it when needed. The tactile interface provided meaningful support despite sensory restrictions. By encoding spatial uncertainties, it successfully extended the operating range of the assistance system.

Keywords: spatiotemporal displays; sensory augmentation; reliability display; uncertainty encoding; automotive hmi; human-machine cooperation; cooperative driver assistance; state transparency display

---

1. Introduction

Modern cars are equipped with sensor systems that surpass human perception in various ways. For example, camera systems may offer continuous 360-degree vision and Lidar can provide vision in the dark. Advanced driver assistance systems use these sensor capabilities by providing the driver with supportive information (e.g., lane departure warning, blind-spot detection, navigation) or by taking over control (e.g., adaptive cruise control, automated lane-keeping). However, the reliability of sensory systems may degrade due to changes in the environment. For example, the accuracy of Lidar measurements tends to decrease in the rain [1], and car manufacturers warn about reduced reliability of sensors in tunnels (e.g., Reference [2] (p. 96)). Since drivers cannot be expected to have an understanding of the functioning (or the mere existence) of these sensor systems, they may benefit from the availability of information on sensor reliability. An assistance system could assess such measures of uncertainty by itself, where the level of uncertainty may be based on signal variance or the disagreement between different sensor signals. A system that would share information on sensor uncertainty could help drivers adjust their level of trust in the automation to appropriate levels [3]. This approach is in line with a cooperative automation framework, which challenges designers to regard assistance functions as cooperative partners or team agents, rather than as tools, for example, References [4–8]. Among ten challenges to make automation a team player, Klein et al. [6] (p. 93) listed the team agent's ability to "make pertinent aspects of their status and intentions obvious to their teammates". Communicating system uncertainty might be one step in this direction.

1.1. Related Work

Drivers have been found to show safer behavior when being given appropriate supplementary information about the traffic environment (see e.g., References [9–11], but also Reference [12] for potential adverse effects). Several studies in the automotive context have further investigated the potential of reliability displays, especially for automated driving. Most attempts to communicate system uncertainty have focused on visual displays [13–18]. Variants of such displays include function-specific versus function-unspecific uncertainty encodings or different types of implicit and explicit visualization. Qualitative displays, for example, have illustrated uncertainty through icons, while quantitative displays have incorporated multiple levels or continuous measures of uncertainty using graphs and scales. Beller et al. [13] used an emoji-like icon showing a confused face reaching out with open palms to indicate system uncertainty in a driving simulator experiment. Helldin et al. [15] investigated the impact of visualizing assistance uncertainty on drivers' trust by displaying a visualization of assistance competence (*SAE* level 2 [19]) in a driving simulation with varying weather conditions. The amount of machine confidence was displayed by means of seven empty bars that filled up as confidence increased, in a similar way to mobile phone status bars displaying signal quality. Kunze et al. [16] designed an anthropomorphic reliability display for a simulated *SAE* level 3 automated vehicle. They made a visual display showing a peak from a heartbeat graph that lit up according to a simulated heartbeat frequency between 50 bpm (high reliability) and 140 bpm (low reliability). In addition to the graph, a numeric value of the current machine heart rate was visible.

Uncertainty communication has been shown to be beneficial. Previous work has found improved safety measures [13] and faster take-over times [15,16,20], as well as accompanying changes in gaze behavior [15,16,20]. Furthermore, it was found that drivers showed a more appropriate trust calibration [13,15,18] and gave higher acceptance ratings for such systems [13] compared to baseline conditions. Also, system comprehension [13] and situation awareness [13] were shown to be improved due to uncertainty communication. However, the deployment of the visual modality as a feedback channel has also been subject to criticism. One disadvantage of visual uncertainty communication is that the driver's visual modality might not be continuously available for input as other activities compete for visual attention. When observing the road or engaging in non-driving tasks, drivers may neglect continuous visual displays [21]. This might become especially problematic in automated driving, where the driver is likely to be engaged in a non-driving task. Thus, the use of visual displays for communicating uncertainty carries the risk of disuse or an increase in perceptual workload [16,20].

Recent studies have investigated the use of touch [22], olfaction [23], as well as peripheral vision to share measures of system uncertainty with the driver. In particular, a driving simulator study by Kunze et al. [22] investigated different variants of vibrotactile feedback in a car seat to communicate increases or decreases in the global uncertainty of an automated vehicle for initiating a takeover by the driver. They showed that encodings of uncertainty increase were more intuitive to users than encodings of uncertainty decrease. Moreover, changes in amplitude and rhythm of the vibrotactile feedback were rated highest. The authors did not investigate the effect of the tactile uncertainty feedback on objective measures and recommended that it should still be examined whether people can make use of the feedback and respond to it appropriately. In another study, Kunze et al. [20] coupled a peripheral awareness display with vibrotactile feedback in order to communicate different levels of global system uncertainty in an automated driving simulator experiment. However, they only used the vibrotactile feedback to communicate the highest level of system uncertainty. Results showed that driver workload was significantly lower compared to a visual display condition that needed focal visual attention for the uncertainty communication to be perceived. In addition, they found that users had a more appropriate attention distribution and showed better take-over performance.

Apart from its potential for reliability communication, vibrotactile interfaces have been identified as promising elements of user interfaces [24] and particularly applicable in the context of driver assistance [25] such as for driving- [26–30] or navigation support [31–39]. In addition, also advanced tactile encodings of relevant information such as spatial distances [40–46], directions [32,47–52] and spatio-temporal measures [53,54] have been investigated.

Auspicious reports from these studies let us conclude that vibrotactile feedback is a promising candidate for uncertainty communication in the automotive context and should be investigated in greater detail. To our knowledge, no study so far has investigated tactile communication of system uncertainty relating to individual sensing and signaling about other traffic participants. Here we extend previous research by investigating a previously presented vibrotactile driving assistance system [53,54], augmented with an uncertainty communication functionality.

1.2. Current Study

The main goal of this study is to evaluate driving experience and performance with a driving assistance system that communicates safety-relevant information and additionally conveys its uncertainty about this information. Using a driving simulation environment, we test how the tactile encoding of one dimension of system uncertainty affects the driver's perception of the system in terms of its usefulness and satisfaction and how it affects perceived workload. In addition, we explore whether such a signal influences measures of driving safety and gaze-based attention.

We extend a vibrotactile driving assistance interface that has been shown before to successfully support a driver in gaining a better understanding of the environment through sensory augmentation [53,54]. The tactile assistance provides two types of information—temporal distances and the directions of objects that are on a collision trajectory with the ego-vehicle. The extension introduced here consists of further encoding uncertainty in the tactile stimuli about the directions of objects that are directly approaching. We refer to this uncertainty as directional or spatial uncertainty. Because the underlying assistance system provides information about both direction and temporal distance, also temporal uncertainty, that is, uncertainty about temporal distances can exist. This dimension of uncertainty is not investigated here and the system is marginalized to have full temporal certainty in this study.

We expect that the effect of directional uncertainty communication will be moderated by the driver's own certainty about the directions of potential collision objects. More specifically, we propose the following hypotheses:

Hypotheses 1 (H1). *Understanding. Drivers perceive and understand directional uncertainty encoded in tactile stimuli which communicate spatiotemporal distances of approaching vehicles.*

Hypotheses 2 (H2). *Subjective Benefit. Drivers utilize complementary uncertainty information in tactile stimuli for their subjective benefit.*

Hypotheses 3 (H3). *Disturbance. Drivers are not disturbed by receiving redundant uncertainty information.*

Hypotheses 4 (H4). *Safety. Signaling complementary uncertainty information leads to higher objective safety.*

We here understand *subjective benefit* as a term that subsumes impressions of usefulness, satisfaction and reduced workload and *objective safety* as an expression of safety derived from driving data such as the the smallest predicted time-to-contact to any vehicle that is on a collision trajectory with the ego-vehicle (i.e., the minimal time-to-contact, see Sections 2.3 and 2.5.5.4). *Complementary uncertainty* information is here defined as information that augments uncertain human perception. *Redundant uncertainty* information is defined as information that is already fully covered by more certain human perception. *Disturbance* should be understood as the opposite of benefit and would be expressed in lower scores on the subjective measures and lower performance on the objective measures. For this study, we created conditions that enable us to induce both machine and human sensory uncertainty and thereby determine how complementary or how redundant the encoded uncertainty information becomes.

2. Materials and Methods

2.1. Participants

Fourteen drivers (1 female) between 21 and 41 years old ($M = 29.1$, $SD = 5.4$) participated in the study. All participants reported that they had (corrected-to) normal vision and held a valid driving license for an average of 11 years. All participants gave their written informed consent before taking part in the study.

2.2. Experimental Setup

The experiment was conducted in a static driving simulator (Figure 1) with controls for steering, braking, and accelerating. Gear-shifting/transmission was set to automatic mode. Three display panels (50 inch diagonal, 1080p each, 60 Hz) presented the driving scenario and the remaining parts of the interior (dashboard, instrument cluster, mirrors), using the SILAB 5.1 driving simulation software developed by the WIVW GmbH (Würzburg Institute for Traffic Sciences, Germany). Participants wore a 120 Hz monocular eye-tracker (Pupil Labs GmbH [55]). In addition, participants wore a waist belt (feelSpace GmbH [56]) containing 16 equally spaced vibromotors (between 4.9 and 7.5 cm depending on the size of the belt). In particular, the belt contains eccentric rotation mass motors that can have a maximum amplitude of 2.2 g and a frequency spectrum of 50–240 Hz (0.45–3.3 V) triggered with a 50 ms latency. Frequency and amplitude were set to scale approximately linearly with voltage. Four belt sizes were used in the experiment to ensure a good fit for all participants. The firmware of the belt interface was customized for the experiment.

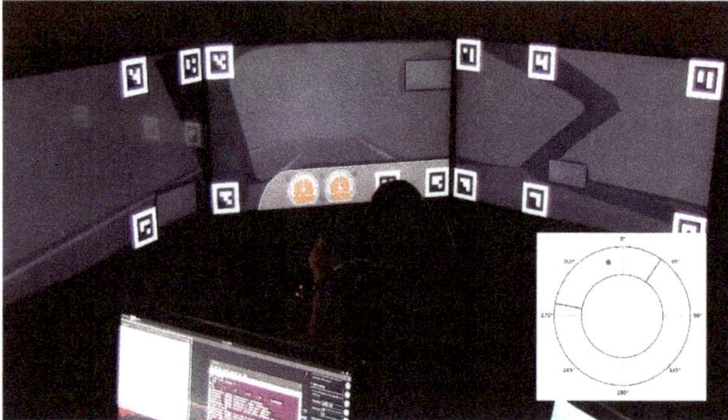

Figure 1. Driving simulator setup in the foggy tunnel scenario. The experimenter screen (bottom left) shows a visualization of the tactile stimuli. In this visualization (magnified in the white box on the right side) the location of a dark dot corresponds to the current direction communicated via a tactile stimulus and the size of the dot indicates the intensity of the respective stimulus. Black bars mark the boundaries between which stimuli oscillate dependent on the current range of spatial uncertainty. This visualization was not available to participants.

2.3. Stimuli

The tactile communication was implemented with a signaling mode similar to the interface used in the experiments by Krüger et al. [53,54]. Two information dimensions about approaching objects were encoded in the tactile stimuli. First, the direction of approaching objects relative to the ego-vehicle was encoded in a mapping of stimulus location on the belt. That is, stimulus location signaled from which lane(s) and lane segments (i.e., center front/back, left front/back, right front/back) vehicles were approaching by activating pre-defined vibromotors that were corresponding to the direction of the lane and segment. In previous studies [53,54], we have found a circular arrangement of actuators, as provided by the feelSpace belt, to be suitable for intuitive signaling of direction information. Nevertheless, other arrangements may also be suitable and could be preferred when working with specific design constraints. Six out of the 16 vibromotors were chosen to realize such mapping (Figure 2). The vibromotors for directional lane encoding were distributed according to the schematic shown in Figure 2. Thereby we chose to set distances between dorsal actuators to be larger than those for the front direction due to differences in spatial discriminability between dorsal and ventral regions [47,57]. A similar direction encoding with eight actuators but no varied treatment of ventral and dorsal regions has, for instance, been successfully employed before by Van Erp et al. [32].

Second, the temporal proximity to the approaching object was encoded in the stimulus intensity. We defined the temporal proximity as the complement of the time to collision (TTC) towards a surrounding object that is on a collision track with the ego-vehicle within a fixed temporal range. Assuming that an object b is moving behind an object a along the same path and trajectory with velocities V_a and V_b and a and b are distance D_{ab} apart, the TTC between a and b is given by:

$$TTC = \begin{cases} \frac{D_{ab}}{V_b - V_a}, & \text{if } V_b > V_a \\ \infty, & \text{otherwise.} \end{cases} \tag{1}$$

For the left and right lanes, we simplified TTC computation by calculating the L^2 norm of a vector consisting of the respective hypothetical (i.e., assuming already being on the respective lane) longitudinal TTC (TTC_{Long}) and the time to lane crossing (TLC) for the respective lane according to Equation (2). The TLC is derived as a TTC that is based on the lateral velocity relative to the lane and the distance to the lane boundary.

$$TTC_{L/R} = \left(TTC_{Long}^2 + TLC_{L/R}^2 \right)^{\frac{1}{2}}. \tag{2}$$

The TTC defines the time it would take until a collision occurred if two objects maintained their current velocities and direction of travel. In the present experiment, we decided to make the stimulus intensity correspond to the complement of the TTC for a temporal range between zero and nine seconds. Stimulus onset occurred whenever the TTC between the ego-vehicle and a surrounding object dropped below a threshold (θ) of nine seconds. This value was chosen as a compromise between the goal of maximizing the range of intensity coding and the need to keep stimuli in a range that can still be perceived by the participants as relevant. Stimulus intensity at onset was set to the smallest perceivable intensity identified by the experimenter, and increased linearly as the TTC dropped. If the TTC was zero (a collision), stimulus intensity reached its maximum, which was equal to the maximum intensity provided by tactile interface. Accordingly, close temporal proximities were signaled with more intense vibration and vice versa.

$$\text{Intensity} = max \left(\frac{\theta - TTC}{\theta}, 0 \right). \tag{3}$$

The tactile interface can give exact signals about the location and temporal proximity of an approaching object as long as the vehicle has precise knowledge about the location and velocity of this object. We refer to this signal as the precise signal, which served as a baseline.

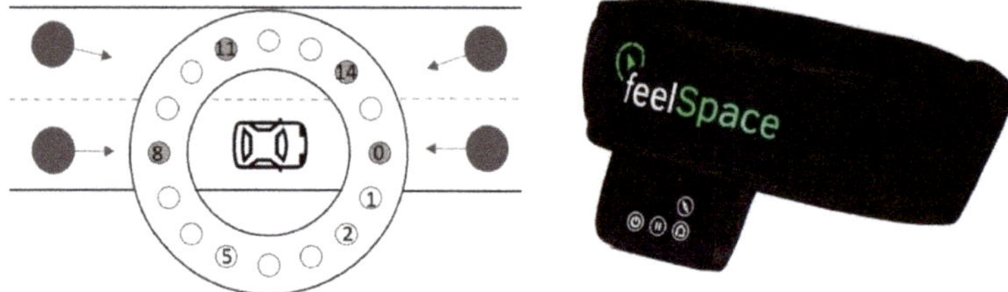

Figure 2. (**Left**) Schematic of the belt in an example situation where from every left and center lane direction an object (large gray dot) is approaching with a time to collision (TTC) value under nine seconds. Vibromotors nr. 0, 14, 8 and 11 (small grey dots) would activate in this case. If the ego-vehicle drove on the left lane, the activations would occur at vibromotor 0, 2, 5 and 8. Note that the selected vibromotors on the rear were spaced two instead of one gap apart to account for differences in spatial discriminability between dorsal and ventral regions [47,57]. (**Right**) Photograph of the tactile waist belt (© feelSpace GmbH).

2.3.1. Uncertainty Communication

In addition to the precise signal, a second signaling mode was realized to communicate the machine's uncertainty about an exact object direction to the user. We refer to this signal as the uncertainty communication. For the uncertainty communication, the encoding of temporal proximity was identical to the precise signal; only the location encoding was varied. The rationale behind the uncertainty communication was that, due to the environmental changes, the vehicle's sensory system may be unable to measure precise object locations (the exact lane), but could still signal the *presence* of an approaching object from either front or back, without specifying the ego- or a neighboring lane. In order to convey this information to the user, the direction of approach for a vehicle was no longer signaled by one unique stimulus location, but through a dynamic vibration pattern traveling over a specific range that represented the overlap between the two lanes on which a vehicle might appear. Upon stimulus onset, neighboring vibromotors were successively activated in the clock- or counter-clockwise direction, creating a tactile illusion of *apparent motion* [24]. The initial vibromotor position and direction was chosen randomly from the available vibromotors within the respective uncertainty range.

Figure 3A shows a schematic of the uncertainty signal. The stimulus development is illustrated by the pointer oscillating between the two borders with a constant frequency (1.0 Hz, from start-to-start point). The next vibromotor activated at the same instance that its predecessor switched off (Figure 3B). The pointer continued to bounce between these borders until either one of two events occured: (1) the TTC became greater than nine seconds, in which case the signal disappeared, or (2) a reliable estimate of the current lane of the approaching vehicle became available. In the latter case, the width of the range converged to one, conveying the same unique direction as in the precise signal condition. We also experimented with other representations of uncertainty, such as synchronously activating multiple actuators in the uncertainty range. However, such variants which employ co-activation of nearby actuators can produce side effects like the funneling illusion [58] and a perceived stimulus intensity increase [59]. Because such effects would interfere with the encoding of information in stimulus direction and intensity, we favored the described method of sequential activation.

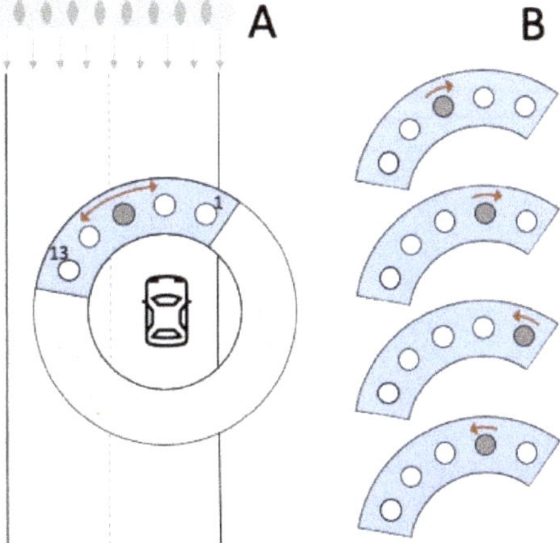

Figure 3. Uncertainty signal for an object approaching from the front on a two-lane road (**A**). Grey dots indicate possible locations of the approaching vehicle as signaled by the system. The stimulus *traveled* between the borders and bounced back in the other direction as it hit one of the borders (**B**). The width of the range was chosen to be between the vibromotors that were allocated for the static signal (Figure 2) plus one extra vibromotor on each side. Thus, in the example in this image, the signal bounced between vibromotors 13 and 1.

2.4. Experimental Design

Independent Variables

Two factors were systematically varied in the experiment in order to evaluate the proposed uncertainty communication system. First, we varied the availability of uncertainty communication (on vs. off). Second, we varied the perceptual uncertainty in the different scenarios between human and machine (machine certain-human uncertain (MC-HU), machine uncertain-human certain (MU-HC), both uncertain (MU-HU)). The uncertainty manipulation was realized through contextual conditions in the driving scenarios that aimed at independently modulating the uncertainty of the vehicle's observations and the uncertainty of the human's observations. Machine uncertainty was introduced by means of driving through (a) a foggy tunnel and (b) rain. Both situations would decrease sensor reliability and increase machine uncertainty. Human uncertainty was provoked by driving through (a) a foggy tunnel and (b) a foggy road. The foggy tunnel thus served as the joint uncertainty condition, in which both the human and the machine suffered from limited sensory input. Since the goal of this study was to examine the effects of uncertainty communication in human-machine cooperation, we decided to omit a condition in which both the human and the machine would be certain. In the foggy road scenario, the machine had an accurate estimate of the position of vehicles at any distance away from it, and it could always communicate the precise signal. Therefore, uncertainty communication (uc) was only available in the foggy tunnel and rain scenarios. Participants drove through these scenarios twice: once without (MU-HU, MU-HC) and once with the uncertainty communication functionality enabled (MU-HU-uc, MU-HC-uc). In case the uncertainty communication was disabled, the vibrotactile interface provided a precise signal only as soon as the approaching car entered a visible range (see Section 2.5 for details). In case the uncertainty communication was enabled, the vibrotactile interface communicated the uncertain signal whenever the defined threshold of a TTC lower than nine seconds to an approaching object was reached. This resulted in a total amount of five experimental conditions, the characteristics of which are summarized in Figure 4.

	Human Uncertain: **HU**			Human Certain: **HC**		
	Scene	Sensor Range	Human Vision	*Scene*	Sensor Range	Human Vision
Machine Certain: **MC**	*foggy road*	inf.	33m			
Machine Uncertain + Unc. Comm.: **MU-uc**	*foggy tunnel*	Unc. > 33 m	33m	*rainy road*	Unc. > 33 m	inf.
Machine Uncertain: **MU**	*foggy tunnel*	33m	33m	*rainy road*	33 m	inf.

Figure 4. Overview of five experimental conditions with corresponding ranges for human vision and machine sensors. Colors are assigned to individual conditions to facilitate condition mapping of the results. For machine uncertain conditions (blue and green), the light colors mark conditions without uncertainty communication while their dark counterparts indicate uncertainty communication.

2.5. Procedure

The study was structured into five experimental and two familiarization blocks. The two familiarization blocks had the purpose of introducing the participants to the driving simulator and the tactile interface. The first familiarization procedure was carried out according to guidelines specified by Hoffmann and Buld [60]. This procedure aimed at reducing the probability of causing simulator sickness by gradually increasing exposure to virtual accelerations. The second familiarization scenario allowed the driver to explore the direction and temporal proximity encoding provided by the tactile interface in a scenario where the machine was certain (precise signal). In the five experimental blocks, the participant's task was to maintain a speed of 120 km/h where possible and avoid collisions with other vehicles. All scenarios consisted of a straight two-lane highway. To rule out potential learning effects, the order in which experimental conditions were conducted varied between participants. Half of the participants started with the two uncertainty communication conditions and half without. Foggy scenarios and rain scenarios were alternated. Before the uncertainty communication conditions, participants were verbally instructed by the experimenter about the machine limitations as follows—"In this section, you will drive through rain/a tunnel. Therefore, the vehicle is less certain about the locations of vehicles that are further away". The following sections further detail the design of the scenarios. Conceptually each scenario followed the same structure: To maintain an objective speed of 120 km/h the driver had to detect and overtake slower cars on the left or right lane from the front, and avoid faster cars that approached at 160 km/h from the rear, possibly changing lanes for overtaking.

2.5.1. Familiarization—System Exploration Scenario

The scenario consisted of a two-lane highway on a sunny day. Participants were not informed about the functionality of the tactile interface and were asked to maintain a speed of 120 km/h where possible. Since vehicles on the passing lane were designed to drive faster than the target speed, the task was most easily fulfilled by driving on the rightmost lane. However, vehicles on the right lane that were trailed by the ego-vehicle would occasionally slow down, forcing the participant to either overtake via the left lane or brake to avoid a collision. These instances ensured that the time to collision between the ego-vehicle and its surrounding vehicles dropped below the threshold value of nine seconds, causing exposure to the tactile stimuli (the precise signal). After five minutes of driving, participants were asked to park their car on the emergency lane, and the system exploration scenario was stopped. Participants were then asked what they thought the tactile stimuli communicated, and they were informed about the true nature of the assistance function. This scenario was similar to the experimental scenario by Krüger et al. [53,54], who found that participants were able to develop an intuitive understanding of the stimuli within four minutes of system exposure. Similarly rapid user understanding times for directional tactile displays were described by Cassinelli et al. [40] and Hogema et al. [61].

2.5.2. Experimental Block-Foggy Road: Machine Certain, Human Uncertain (MC-HU)

The foggy road scenario was simulated as a night-time scenario, designed to make the human uncertain by inserting a dense fog field and disabled lights of surrounding traffic. The fog was parameterized to limit the look-ahead distance to about 33 m (Figure 5), corresponding to a look-ahead time of about one second assuming the driver drove at the target speed. A temporal distance of one second has been suggested as the threshold below which a driving situation can be considered critical [62,63]. We assumed that this look-ahead distance would induce uncertainty in drivers, as they would need to be continuously prepared for the occurrence of a critical situation.

Figure 5. Visibility in the foggy scenarios. Vehicles disappear at a distance of approximately 33 m.

Machine observations were not affected by the mist or darkness, so a precise signal was communicated for vehicles driving at any distance away from the ego-vehicle. The experimenter triggered the onset of a target vehicle approaching the ego-vehicle according to a fixed script. This approach allowed for an easy verification that participants were driving at the approximate target speed, which was a prerequisite for the correct situation development. When a command was given, the target vehicle started approaching behind the fog barrier from one of the four possible lane directions (front, front-left, rear, rear-left). Vehicles coming from the rear were driving at a speed of 160 km/h. Vehicles in the front were driving at 80 km/h. As a consequence, the target vehicle would overtake (or be overtaken by) the ego-vehicle, assuming that the participant kept driving around the target speed of 120 km/h. Vehicles that approached from the rear on the right lane were programmed to change lanes and overtake the ego-vehicle at a distance of 30 m. After the target vehicle had passed and disappeared into the fog again, and the experimenter confirmed that the participant was driving at the target speed, the next target vehicle was launched. This procedure was carried out 14 times. Directions from which cars approached were pseudo-randomized.

2.5.3. Experimental Block-Foggy Tunnel: Machine and Human Uncertain (MU-HU)

The foggy tunnel scenario was identical to the foggy road (MC-HU) scenario, except for the addition of a tunnel that ran for the entire course and a change in *sensor reliability* such that vehicles outside a 33 m radius from the ego-vehicle could at most be signaled via uncertainty communication as described in Section 2.3.1. Limitations of the look-ahead distance were the same as in the foggy road condition (33 m, 1 s) for the human. For comparability reasons, traffic definitions were identical to the foggy road scenario (MC-HU).

2.5.4. Experimental Block-Rain: Machine Uncertain, Human Certain (MU-HC)

The rain scenario consisted of a straight road on a rainy day. The rain was visually present, though at an intensity at which it did not have much influence on the driver's visual perception. The reliability of the machine was said to be negatively affected by the rain, in the same manner as it was in the foggy tunnel scenario. That is, the look-ahead distance of the machine for precise direction identification and signaling was limited to 33 m. Because the driver's field of view was not obstructed, the traffic setup had to be organized in a different way compared to the fog conditions. The altered traffic profile for the rain scenario is explained in Figure 6.

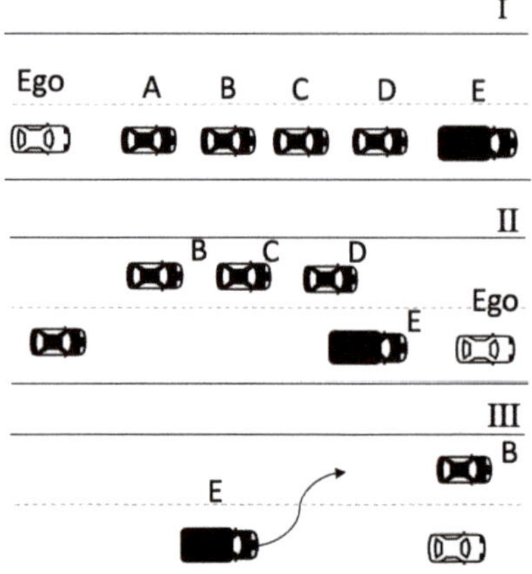

Figure 6. Traffic definition in the rain (machine uncertain-human certain condition (MU-HC)) scenarios. Five vehicles were driving on the right lane at 80 km/h, spaced 250 m apart (I). The ego-vehicle could maintain the target speed (120 km/h) by overtaking the vehicles. When the front truck (E) was overtaken, a trigger point was activated that made the trailing cars B, C, and D switch to the left lane, and adjust their speed to 160 km/h (II). This resulted in B, C and D eventually overtaking the ego-vehicle from the rear. When D passed the ego-vehicle (III), the leading vehicle (A) accelerated to 160 km/h, and it changed to the left lane if it came within a distance of 30 m of the ego-vehicle.

2.5.5. Dependent Measures

As dependent variables, we recorded subjective measures concerning the usefulness, satisfaction and perceived workload in the different experimental conditions, as well as the overall understanding and experience. In addition, we were interested in objective measures which express effects on peoples' gazing behavior and their performance in a driving task.

We used three questionnaires for the subjective evaluation. These were used to gain insights into the subjective experiences which the different experimental conditions induced and see whether the conditions were correctly perceived and understood.

2.5.5.1. Task Load, Usefulness, Satisfaction

After each experimental condition, the NASA Task Load Index (Raw-TLX, [64]) assessment was conducted. Usefulness and satisfaction ratings were obtained using the Van der Laan acceptance scale [65].

2.5.5.2. Understanding and Experience

Furthermore, after every experimental block, participants were asked to rate a number of statements on a 5-point Likert scale (strongly disagree to strongly agree). These statements were included to check if (a) the modulation of human perceptual confidence through environmental factors was successful, (b) the participants had understood the machine's level of uncertainty, and (c) participants experienced that the machine expressed its level of uncertainty.

2.5.5.3. Gaze Distributions

The front gaze ratio was computed as the ratio of the number of gaze points in the front window versus the total amount of gaze points in the mirrors and windshield (Equation (4)). A higher front gaze ratio indicates that the driver allocated more attention towards the front; a lower front gaze ratio indicates that the user allocated more attention towards the rear. By means of this measure, we aimed at evaluating whether the uncertainty communication caused shifts in visual attention towards the direction of the presented signal.

$$\text{front gaze ratio} = \frac{\text{gaze count on windshield}}{\text{gaze count on windshield} + \text{mirrors}}. \tag{4}$$

2.5.5.4. Trial Safety

Trial safety was operationalized as the *Minimum Time-to-Contact* (MTTC) recorded in each trial in any direction. The MTTC can be understood as a conservative measure of safety that only takes into account the smallest recorded TTC and thus indicates how dangerous a trial became at the most (see e.g., References [20,54]).

2.5.5.5. Trial Definition

We restricted the analysis of gaze distributions and safety to specific periods of interest which we refer to as trials. A trial occurred for every vehicle that overtook or was overtaken by the ego vehicle. The starting point of a trial was set to the moment where time to passing (TTP) of a surrounding vehicle dropped below nine seconds. Here, we defined the TTP as the time it would take until two vehicles would pass each other if they would maintain their current velocities. The TTP can be understood as a TTC (see Equation (1)) without the requirement for being on a collision trajectory. We set the end point of a trial to the moment at which the ego-vehicle and the other vehicle passed each other.

2.6. Analysis

We split the analysis of the data into three parts—(1) custom questionnaire data, (2) subjective data on perceived workload as well as on perceived system acceptance in terms of usefulness and satisfaction, and (3) objective behavioral and performance data, including gaze distribution results and measures of trial safety. To rule out potential confounds, we only ran statistical tests between experimental conditions that shared the same traffic profiles. While the differences in traffic profiles prevented comparisons between fog and rain conditions, this design choice did not impair the investigation of our research hypotheses. It allowed us to prioritize internal validity through the implementation of scenarios that contained credible sources of uncertainty for each environmental condition.

Statistical analysis was carried out using the *scipy* python library. Plots were generated using the python packages *matplotlib* and *seaborn*.

2.6.1. Custom Questionnaire Data—H1 (Understanding)

Custom questionnaire data for all conditions were analyzed descriptively based on median responses and interquartile ranges. According to H1, we expected participants to indicate understanding of the uncertainty encoding stimuli.

2.6.2. Acceptance and Workload—H2 (Subjective Benefit) and H3 (Disturbance)

Usefulness and satisfaction scores were obtained by mapping subsets of Van Der Laan Questionnaire responses to two respective scales in the [−2, 2] range (see [65]). Figure 7 illustrates the outcome that we would expect for usefulness, satisfaction and workload under our research hypotheses H2 and H3. We expected usefulness and satisfaction to be higher in human uncertain (HU) conditions with uncertainty communication than when omitting the information. We further assumed that an advantage of the machine certain (MC—red) over the uncertainty communication (dark blue) condition should exist due to the higher information gain achievable by precise signals. On the other hand, for cases with higher human certainty (HC—green) we would expect information from an uncertainty communication to be redundant and therefore to cause no advantage over an omission of signals in the uncertainty range. However, under H3 also no disadvantage from redundant uncertainty communication was assumed.

For workload, measured as the NASA Task Load Index (Raw-TLX [64]), the expected relationship would be reversed because we define the relationship between workload and benefit as inverse, that is, a high workload reflects low benefit whereas a low workload can indicate higher benefit.

We compared scores of human uncertain conditions (MC-HU, MU-HU-uc, MU-HU—red, blue) using Friedmann tests and post-hoc one-sided Wilcoxon signed rank tests with Bonferroni adjusted alpha levels for repeated testing. As there were only two human certain conditions (MU-HC-uc, MU-HC—green) we directly compared scores for these conditions using Wilcoxon signed rank tests with Bonferroni adjusted alpha levels.

| | Human Uncertain: **HU** | | | | Human Certain: **HC** | | | |
	Scene	Usefulness	Satisfaction	Workload	*Scene*	Usefulness	Satisfaction	Workload
Machine Certain: **MC**	*foggy road*	++	++	lowest				
Machine Uncertain + Unc. Comm.: **MU-uc**	*foggy tunnel*	+	+	low	*rainy road*	0	0	low
Machine Uncertain: **MU**	*foggy tunnel*	0	0	highest	*rainy road*	0	0	low

Figure 7. Predicted outcome of subjective evaluations according to our research hypotheses when assuming successful experimental manipulations. Usefulness and satisfaction: Symbols +,0 are used to illustrate the predicted valuation. Relative workload predictions were given verbally. For machine uncertain conditions (blue and green), the light colors mark conditions without uncertainty communication. Their dark counterparts indicate uncertainty communication.

2.6.3. Gaze Distribution and Safety—H4 (Safety)

Figure 8 illustrates the outcome that we would expect for safety and gaze guidance under H4. While gaze guidance is not directly subsumed in the *benefit* term, here we understand it as a behavioral indicator for an influence on peoples' information sampling which relates to our second and fourth hypotheses. The assistance system primes relevant regions of interest through tactile stimuli which may prompt users to shift their gaze accordingly in order to acquire additional information or visual confirmation. Under H2 and H4 we would therefore expect gaze guidance to be observable for conditions in which the system can provide novel information, that is, machine certain (MC—red) and human uncertain with uncertainty communication (MU-HU-uc—dark blue) conditions. In contrast, according to H3 this should not be the case for cases in which human uncertainty is equal or lower than machine uncertainty (light blue and green).

Prior to gaze distribution analysis, we filtered the data to only include trials in which vehicles approached from behind. As driving requires frontal visual attention at most times, especially with low visibility conditions, a comparison of front gaze ratios is more meaningful for situations in which safety-relevant events take place behind the ego vehicle. Due to the presence of outliers and a violation of the normality assumption, we compared front gaze ratios of human uncertain conditions (MC-HU, MU-HU-uc, MU-HU—red, blue) using Friedmann tests and post-hoc one-sided Wilcoxon signed rank tests with Bonferroni adjusted alpha levels for repeated testing. As there were only two human certain conditions (MU-HC-uc, MU-HC—green) we directly compared front gaze ratios for these conditions using one-sided Wilcoxon signed rank tests with Bonferroni adjusted alpha levels.

For the analysis of safety we focused on human uncertain conditions and trials in which vehicles approached from the front right lane because these trials required corrective actions by the driver to ensure safety. In line with H4 we expected safety to be highest in the machine certain (MC—red) condition, lowest in the absence of >33 m signaling (MU-HU—light blue) and intermediate with uncertainty communication enabled (MU-HU-uc—dark blue). MTTC scores (see Section 2.5.5.4) were calculated for each trial and mean MTTC scores per participant and condition were compared using a Friedmann test and post-hoc one-sided Wilcoxon signed rank tests with Bonferroni adjusted alpha levels for repeated testing.

	Human Uncertain: **HU**			Human Certain: **HC**		
	Scene	Gaze Guidance	Safety	*Scene*	Gaze Guidance	Safety
Machine Certain: **MC**	*foggy road*	High	Highest			
Machine Uncertain + Unc. Comm.: **MU-uc**	*foggy tunnel*	High	Medium	*rainy road*	Low	
Machine Uncertain: **MU**	*foggy tunnel*	Low	Lowest	*rainy road*	Lowest	

Figure 8. Predicted outcome of behavioral measures according to our research hypotheses when assuming successful experimental manipulations through the introduced conditions. For machine uncertain conditions (blue and green), the light colors mark conditions without uncertainty communication. Their dark counterparts indicate uncertainty communication.

3. Results

3.1. Subjective Reports

3.1.1. Custom Questionnaire—H1 (Understanding)

Response distributions to the eight Likert items that were used in our customized questionnaire are shown in Figure 9 for each experimental condition. For human uncertain conditions, participants strongly indicated weather conditions as a cause for feeling unconfident whereas other road users had a smaller influence and belt signals were not negatively affecting confidence. For human certain conditions, none of these three factors reduced confidence. These ratings suggest that our experimental manipulation of human uncertainty through different weather conditions was successful. Statements 4 and 5 targeted the understanding of the tactile stimuli and the machine uncertainty state. In support of H1, participants generally identified system uncertainty when present (MU), especially with uncertainty communication (uc) and correctly indicated its absence (MC). This suggests that the state transparency achieved by the uncertainty communication supported system state understanding. The last three statements were included for an estimate on which modalities the participants relied during the different conditions. Reliance on own capabilities and visual sensing was highest in the human certain conditions (HC). For human uncertain conditions (HU), reliance on the tactile stimuli was high, especially for the machine certain (MC) and machine uncertain + communication (MU-HU-uc) conditions. This was no longer the case when uncertainty communication was disabled (MU-HU). In support of the H2 and H3, this suggests that participants utilized tactile stimuli depending on system reliability and their own confidence state. In summary, participant responses suggest that the experimental manipulations worked as intended and induced different levels of congruency between human and machine perceptual uncertainty.

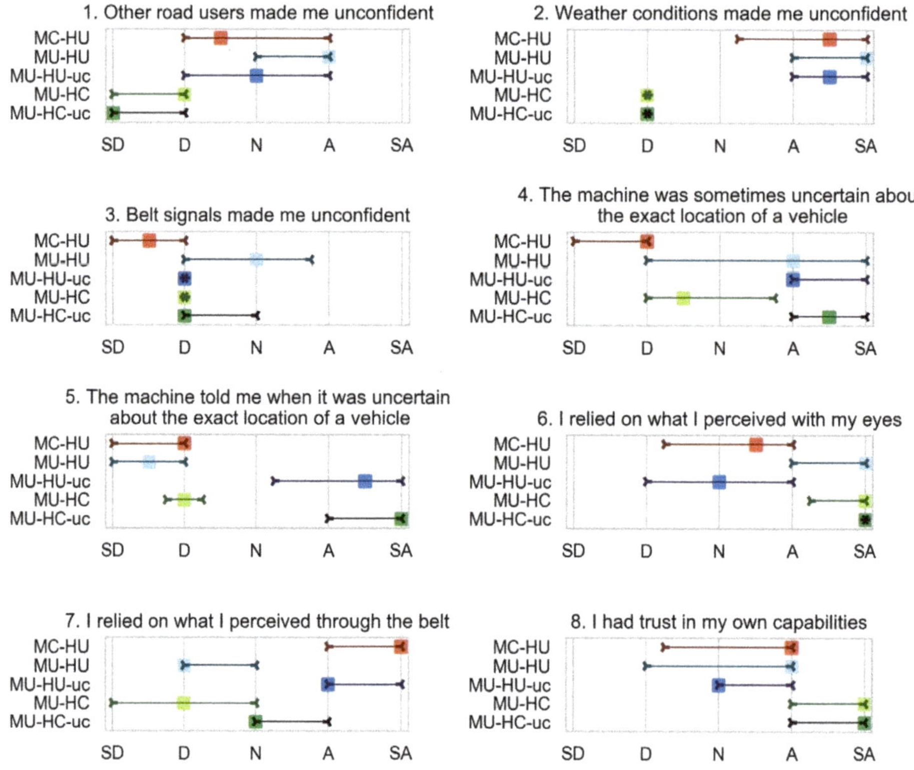

Figure 9. Median agreement ratings (square) and 25th and 75th percentiles on a custom 5-point Likert scale questionnaire. SD = Strongly Disagree, D = Disagree, N = Neutral, A = Agree, SA = Strongly Agree.

3.1.2. Usefulness and Satisfaction—H2 (Subjective Benefit) and H3 (Disturbance)

An overview of the usefulness and satisfaction scores that were obtained in each experimental condition can be found in Figure 10b. As expected, the overall highest score was found for the machine certain and human uncertain condition (MC-HU). The overall lowest score was obtained for the machine uncertain-human certain condition (MU-HC). We were interested in comparing conditions within a given level of human certainty, that is a comparison between the three human uncertain conditions (HU—red and blue) and between the two human certain conditions (HC—green).

The human uncertain conditions (MC-HU, MU-HU-uc, MU-HU) differed significantly for usefulness, $\chi^2(2) = 20.87$, $p < 0.001$ ($<\alpha = 0.025$), as well as for the satisfaction scores, $\chi^2(2) = 16.62$, $p < 0.001$ ($<\alpha = 0.025$). Post-hoc comparisons revealed that usefulness was rated significantly higher with uncertainty communication enabled (MU-HU-uc—dark blue) than disabled (MU-HU—light blue), MU-HU-uc vs. MU-HU: $w = 0.0$, $p < 0.001$ ($<\alpha = 0.008$) where w denotes the sum of the ranks of the differences above zero (In contrast to test statistics of many parametric tests, a small value for w is therefore a strong indicator for consistent and significant differences). Similarly, usefulness in the machine certain condition (MC—red) was rated significantly higher than in the machine uncertain condition without uncertainty communication (MU-HU), MC-HU vs. MU-HU: $w = 0.0$, $p < 0.001$ ($<\alpha = 0.008$). However, there was no significant difference in usefulness ratings between the machine certain (MC-HU) and the uncertainty communication condition (MU-HU-uc), MC-HU vs. MU-HU-uc: $w = 32.0$, $p = 0.289$ ($>\alpha = 0.008$). The same pattern of results was observed for the satisfaction ratings, MU-HU-uc vs. MU-HU: $w = 10.5$, $p = 0.004$ ($<\alpha = 0.008$), MC-HU vs. MU-HU: $w = 0.0$, $p < 0.001$ ($<\alpha = 0.008$), MC-HU vs. MU-HU-uc: $w = 34.5$, $p = 0.219$ ($>\alpha = 0.008$).

		Human Uncertain: **HU**				Human Certain: **HC**		
	Scene	Usefulness	Satisfaction	Workload	*Scene*	Usefulness	Satisfaction	Workload
Machine Certain: **MC**	*foggy road*	1.41 (0.54)	0.82 (0.64)	39.6 (14.5)				
Machine Uncertain + Unc. Comm.: **MU-uc**	*foggy tunnel*	1.37 (0.50)	0.62 (0.76)	50.5 (11.1)	*rainy road*	0.16 (0.50)	-0.04 (0.76)	24.8 (9.1)
Machine Uncertain: **MU**	*foggy tunnel*	-0.28 (0.92)	-0.23 (0.56)	57.7 (14.3)	*rainy road*	-0.68 (0.67)	0.03 (0.38)	22.1 (10.6)

(**a**)

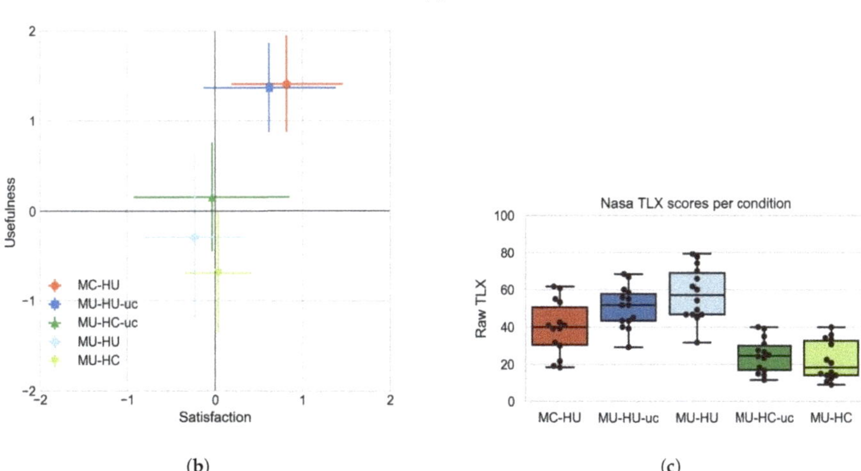

(**b**) (**c**)

Figure 10. Results of subjective measures for different conditions. Conditions are visually represented by distinct colors. For machine uncertain conditions (blue and green), the light colors mark conditions without uncertainty communication. Their dark counterparts indicate uncertainty communication. (a) Mean usefulness, satisfaction, and NASA TLX scores for each condition. Standard deviations are shown in brackets. Asterisks indicate statistically significant differences between conditions linked by brackets; (b) Mean usefulness and satisfaction scores of the assistance functionality in MC-HU (Foggy Road), MU-HU-uc (Foggy Tunnel), MU-HC-uc (Rain), MU-HU (Foggy Tunnel, no UC), MU-HC (Rain, no UC). Error bars display the standard deviation; (c) NASA Raw TLX scores per condition. Scores of individual questions were averaged to obtain the overall RTLX score in the range [0,100].

These results support the prediction driven by H2 that usefulness and satisfaction ratings should be higher with enabled than disabled uncertainty communication. However, contrary to our assumption, no advantage of the machine certain (MC-HU) over the uncertainty communication (MU-HU-uc) condition, reflecting a difference in potential information gain, could be confirmed.

Also for the human certain conditions (HC–green), we found that usefulness was rated as significantly higher with uncertainty communication enabled (MU-HC-uc) than disabled (MU-HC), MU-HC-uc vs. MU-HC: $w = 16.5$, $p = 0.012$ ($<\alpha = 0.05$). For satisfaction ratings, the differences between human certain conditions were not significant, MU-HC-uc vs. MU-HC: $w = 21.0$, $p = 0.429$ ($>\alpha = 0.05$). While average satisfaction ratings were somewhat neutral for both conditions, the usefulness of a late-supporting system was negatively judged. Average neutral usefulness ratings for the uncertainty communication condition support our predictions made under H3, presumably because it was neither needed nor disturbing.

3.1.3. Workload—H2 (Subjective Benefit) and H3 (Disturbance)

NASA TLX workload ratings (Figure 10c) differed significantly between human uncertain conditions (MC-HU, MU-HU-uc, MU-HU), $\chi^2(2)$ = 11.66, p = 0.003 ($<\alpha$ = 0.05). Post-hoc comparisons revealed that workload was rated significantly lower with uncertainty communication enabled (MU-HU-uc—dark blue) than disabled (MU-HU—light blue), MU-HU-uc vs. MU-HU: w = 14.0, p = 0.008 ($<\alpha$ = 0.016). Also in the machine certain condition (MC—red), workload was rated significantly lower than in the machine uncertain condition without uncertainty communication (MU-HU), MC-HU vs. MU-HU: w = 1.0, p = 0.001 ($<\alpha$ = 0.016). These results confirm the prediction that workload should be reduced when enabling uncertainty communication and thus support H2. However, differences in subjective workload between the machine certain (MC-HU) and the uncertainty communication condition (MU-HU-uc) were not significant, MC-HU vs. MU-HU-uc: w = 19.0, p = 0.032 ($>\alpha$ = 0.016). In contrast to H2, an assumed advantage of the machine certain (MC-HU) over the uncertainty communication (MU-HU-uc) could therefore not be confirmed.

For the human certain conditions (HC—green), workload ratings were comparably low and did not differ significantly between conditions with uncertainty communication enabled (MU-HC-uc—dark green) and disabled (MU-HC—light green), MU-HC-uc vs. MU-HC: w = 31.0, p = 0.310 ($>\alpha$ = 0.05). When contrasted with results from the human uncertain (HU) conditions, the low averages and the lack of difference in satisfaction and workload between the two human certain (HC) conditions may be seen as support for H3. However, due to the use of different driving profiles, a formal comparison of differences would not be valid.

3.2. Gaze Distribution—H2 (Subjective Benefit) and H4 (Safety)

Figure 11b shows the ratio of gaze points on the front (front window) divided by front+back (front window + mirrors). Front gaze ratios differed significantly between human uncertain conditions (MC-HU, MU-HU-uc, MU-HU) for trials in which vehicles approached from the back, $\chi^2(2)$ = 16.0, $p < 0.001$ ($<\alpha$ = 0.05). Post-hoc comparisons revealed that the front gaze ratios were significantly lower with uncertainty communication enabled (MU-HU-uc—dark blue) than disabled (MU-HU—light blue), MU-HU-uc vs. MU-HU: w = 0.0, $p < 0.001$ ($<\alpha$ = 0.016). Also in the machine certain condition (HC—red), front gaze ratios were significantly lower than in the machine uncertain condition without uncertainty communication (MU-HU), MC-HU vs. MU-HU: w = 2.0, $p < 0.001$ ($<\alpha$ = 0.016). Differences in front gaze ratios between the machine certain (MC-HU) and the uncertainty communication condition (MU-HU-uc) were not significant, MC-HU vs. MU-HU-uc: w = 14.0, p = 0.007 ($<\alpha$ = 0.016 but $w > w_{critical}$ = 12).

Between human certain conditions (MU-HC, MU-HC-uc—green), differences between front gaze ratios could not be regarded as significant for trials in which vehicles approached from the back, MU-HC vs. MU-HC-uc: w = 24.0, p = 0.037 ($>\alpha$ = 0.016 and $w > w_{critical}$ = 12). These findings indicate an increased overt attention guidance for conditions in which the assistance can provide novel relevant information. They are therefore in line with our predictions (see Figure 8) made under H2 and H4.

For comparison, for situations in which vehicles approached from the front (Figure 11c), the gaze distributions substantially shifted to the front (MU-HC: M = 0.92, SD = 0.05; MU-HC-uc: M = 0.91, SD = 0.06; MU-HU: M = 0.97, SD = 0.02; MU-HU-uc: M = 0.96, SD = 0.04; MC-HU: M = 0.94, SD = 0.07) across all conditions. Differences between uncertainty communication and no uncertainty communication diminished, as stimuli with uncertain direction encoding only drew attention to front regions.

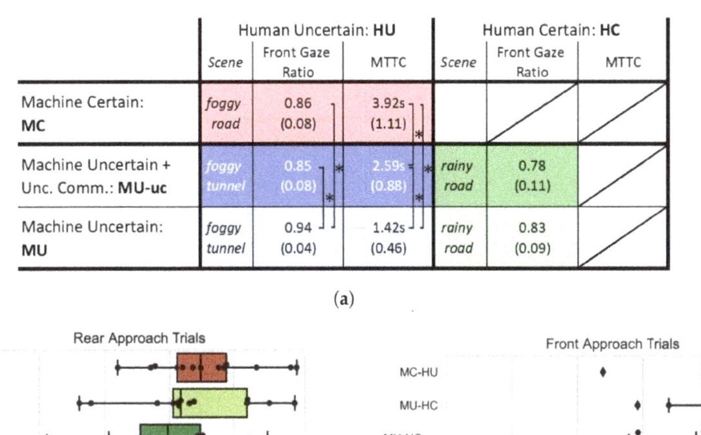

	Human Uncertain: **HU**			Human Certain: **HC**		
	Scene	Front Gaze Ratio	MTTC	*Scene*	Front Gaze Ratio	MTTC
Machine Certain: **MC**	*foggy road*	0.86 (0.08)	3.92s (1.11)			
Machine Uncertain + Unc. Comm.: **MU-uc**	*foggy tunnel*	0.85 (0.08)	2.59s (0.88)	*rainy road*	0.78 (0.11)	
Machine Uncertain: **MU**	*foggy tunnel*	0.94 (0.04)	1.42s (0.46)	*rainy road*	0.83 (0.09)	

(**a**)

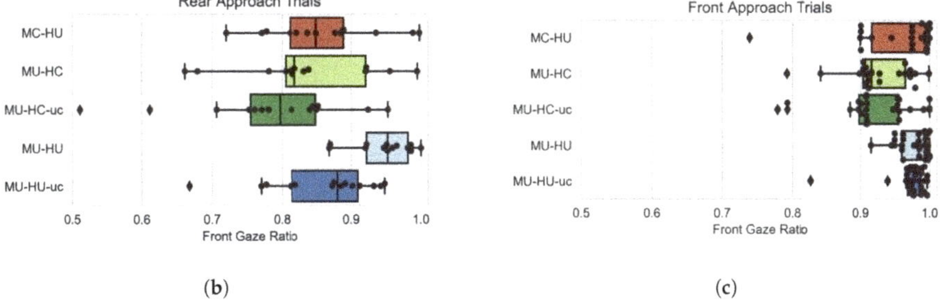

(**b**)   (**c**)

Figure 11. Results of objective measures for different conditions. Conditions are visually represented by distinct colors. For machine uncertain conditions (blue and green), the light colors mark conditions without uncertainty communication. Their dark counterparts indicate uncertainty communication. (**a**) Mean front gaze ratios and MTTC scores for each applicable condition. Standard deviations are shown in brackets. Asterisks indicate statistically significant differences between conditions linked by brackets; (**b**) Gaze ratio for conditions in which the machine was uncertain and for trials in which vehicles were approaching from the rear. Lower values indicate more gazing towards the mirrors. Due to failed eye tracking recordings, $n = 13$ (instead of 14) for all conditions; (**c**) Gaze ratio for conditions in which the machine was uncertain and for trials in which vehicles were approaching from the front.

3.3. Trial Safety—H4 (Safety)

Figure 12 displays the MTTC scores for human uncertain conditions. We only considered the data of the human uncertain (HU—blue and red) conditions for statistical tests. MTTCs differed significantly between human uncertain conditions (MC-HU, MU-HU-uc, MU-HU), $\chi^2(2) = 24.14$, $p < 0.001$ ($<\alpha = 0.05$). We found that the MTTCs were significantly higher for the MU-HU-uc condition ($M = 2.59$ s, $SD = 0.88$) than for the MU-HU condition ($M = 1.24$ s, $SD = 0.46$); $w = 4.0$, $p = 0.001$ ($<\alpha = 0.016$). Furthermore, driving safety in terms of MTTC was also significantly higher in the MC-HU condition ($M = 3.92$ s, $SD = 1.11$) than in the MU-HU-uc condition, $w = 7.0$, $p = 0.002$ ($<\alpha = 0.016$) and the MU-HU condition, $w = 0.0$, $p < 0.001$ ($<\alpha = 0.016$). In poor visibility conditions (MU), imprecise tactile direction signaling (MU-HU-uc) appears superior to a variant only capable of signaling specific, reliable observations within a substantially constrained spatial range (MU-HU). In accordance with H4, participants thus seem to have taken advantage of the information available in the tactile stimuli to adjust their driving behavior for achieving higher safety.

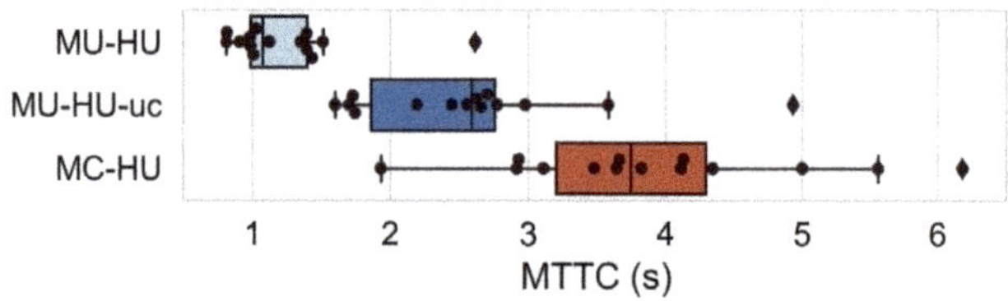

Figure 12. Minimum Time-to-Contact (MTTC) scores for human-uncertain conditions ($n = 14$).

4. Discussion

In the present driving simulator study, we investigated the effects of a novel approach to encode spatial uncertainty in the stimuli of a vibrotactile assistance system. We aimed at evaluating the influence of the uncertainty communication on subjective measures indicative of perceived usefulness, satisfaction, and workload, as well as on behavioral measures, that is, driving safety and gaze allocation. We assumed that any effect of the uncertainty communication would be influenced by the relation of spatial uncertainty in human perception and the assistance system. Therefore, we experimentally varied the driving scenarios to simulate machine uncertainty (tunnel + fog, rain) and to induce human uncertainty (fog, tunnel + fog). We found that our suggested uncertainty communication mode was understood by participants and had significant effects on both subjective and objective behavioral measures. Thereby the utility of the system seemed to depend on the driver's perceptual confidence state. In our experiment, the uncertainty communication was regarded as beneficial and had a measurable influence on driver behavior in cases where the human driver was uncertain as well.

4.1. Signal Understanding and Experiment Validation

A prerequisite to this study was that our environmental scenario manipulations had the effect that we intended. Data from our custom questionnaire indicate that this was indeed the case. Participants reported that they felt uncertain due to the weather conditions and agreed that they relied more on the belt signal than on their own perception in the human uncertain conditions. Furthermore, participants experienced higher workload in the human uncertain conditions compared to the human certain conditions.

Besides, we were interested in the participants' subjective agreement on understanding the manipulation of machine uncertainty and the respective uncertainty communication signal. This was important to further validate our experimental procedure and the design of our uncertainty signal. Participants indicated that they had understood when the machine was uncertain and that they understood the meaning of the signal. Interestingly, they seemed to have noticed the machine uncertainty more strongly in the conditions in which the uncertainty communication was enabled, which suggests that this feature helped to make the machine state more transparent. Taken together, in support of hypothesis H1 (Understanding), these results indicate that our experimental manipulations were valid and that participants seemed to have an appropriate understanding of the uncertainty communication.

An important difference between earlier studies that have demonstrated successful communication of uncertainty (e.g., References [13,16,23]) and the work presented here, is that we relied on an *implicit* representation of uncertainty in the tactile modality: The uncertainty component was encoded within the spatiotemporal signaling functionality of our vibrotactile interface. Instead of explicitly stating that "I am uncertain", the machine agent implicitly communicates uncertainty by being less specific in its display of the location of objects. We argue that the distinction between *implicit* and *explicit* uncertainty communication may be useful for the future design of reliability displays. Implicit uncertainty communication is characterized by an increase in ambiguity or vagueness, or a decrease in specificness of presented information. One example of implicit uncertainty communication that we encountered in the literature is by Finger and Bisantz [14], who added distortions to an image to make it increasingly difficult to specify the underlying image.

4.2. Uncertainty Signaling in Human Uncertain Conditions

In terms of behavioral adaptations and user acceptance, we found substantial differences in the results between the human certain and the human uncertain conditions. In particular, in case of both human and system uncertainty, uncertainty communication was perceived as significantly more useful and satisfying compared to the no uncertainty communication conditions. Uncertainty communication also yielded significantly lower workload, increased driving safety and more strongly guided gaze behavior, indicating that more attention was allocated towards the direction of the uncertainty signal. These results support hypotheses H2 (Subjective Benefit) and H4 (Safety) by showing that the vibrotactile uncertainty communication had beneficial effects on driving comfort and safety.

In the human uncertain conditions, the uncertainty communication signal was not perceived as significantly different from the precise signal in terms of perceived usefulness and satisfaction, as well as in perceived workload. This is somewhat surprising as one might think that participants would naturally value the accessibility of the full information that is provided by the precise signal more than the more ambiguous uncertainty information signal. Overall, this outcome indicates that making the vehicle's perceptual state transparent is appreciated by participants. Our results suggest that users are still satisfied with the directional cues and recognize the usefulness of the uncertainty signals, despite the lower quality in terms of information specificity. However, in case of driving safety, we observed a significant advantage of the precise signal over the uncertainty communication signal. That is, we observed the safest driving behavior in terms of MTTC scores in conditions where the machine's sensory capabilities were unaffected by the environment.

We conclude that the precise signal was appropriately used by participants to acquire a more accurate understanding of the direction of surrounding objects.This finding is in line with the reports by Krüger et al. [53,54], who found that participants rapidly gained an understanding of vibrotactile stimuli and presented safer driving behavior using the same vibrotactile assistance with a precise signal mode compared to driving without.

4.3. Uncertainty Signaling in Human Certain Conditions

Analysis of the eye-tracking data revealed that visual attention distributions were affected significantly by the uncertainty signaling in scenarios in which human visibility was limited (human uncertain conditions), but not in the human certain conditions. Furthermore, usefulness and satisfaction ratings showed neutral ratings in the human certain conditions. In agreement with hypothesis H3 (Disturbance) this suggests that there is no direct disadvantage but also no benefit in sharing observations continuously when the human is confident.

For successful human-machine cooperation [7,8] or teaming, a human mental representation of system uncertainty may not be enough. When the machine also has a representation of human confidence in different environments, it allows the machine to decide under what conditions to provide support to the user. However, such a selective and presumably personalized communication could induce confusion when violating a user's assumptions on what the machine is communicating. In this example, it might not even be possible for a user to unambiguously distinguish between cases in which the machine is not providing stimuli because it has not detected a potential collision event and cases in which it has selectively disabled communication because it could confirm that the user has a sufficient scene understanding. Selectively deactivating systems that implicitly encode the absence of issues through an absence of stimuli could therefore be problematic but may be an important challenge to tackle in the design of future driving assistance systems.

4.4. Limitations

Despite the relatively small sample size, the results show clear statistical significance and accordingly provide support for the benefits of uncertainty communication. A limitation of the current study is that the sample (technically schooled, 13/14 male) was not balanced to be representative of a diverse population. Consequently, inferences are restricted to mostly male drivers younger than 42 years. It is well known that age is associated with sensory and cognitive decline [66]. However, prior work on sensory integration [67] and proximity alerting [68] suggests a positive relationship between age and multimodal facilitation effects such as reaction time shortening. Future work should investigate whether such a relationship also exists with our system. Another limitation comes from the restriction to highly challenging situations for cases with human uncertainty. An advantage of the fast succession of safety-critical situations is that it ensured exposure to the functionality of the device, which currently only provides stimuli when operating outside a safety margin. This means that in safe conditions the system does not produce any stimuli. The fact that the system proved its usefulness in challenging situations can be seen as a strength. However, we do not know if the observed effects would remain with less frequent system activation under more common traffic conditions. Future work could address this issue by implementing easier scenarios where a participant encounters fewer safety-critical event(s) for an overall longer exposure time.

4.5. Conclusions

Taken together, the study yields new insights about the communication of directional uncertainty for a driving assistance system in the tactile modality. We found that an implicit encoding of spatial uncertainty in a vibrotactile interface was easily understood and used by participants, and that its impact on drivers depended on the drivers' sense of certainty. Importantly, in case the human driver was uncertain, the uncertainty communication signal was perceived as equally useful and satisfying as a precise signal of the assistance system. Along with previous literature, our findings stress the value and importance of communicating appropriate information and making machine states transparent to the user. Our results suggest that the tactile modality is a suitable candidate for communicating such information to the user unobtrusively and intuitively while potentially circumventing the risks and challenges which an additional utilization of the visual modality would introduce.

Author Contributions: Conceptualization, M.K., T.D., C.B.W.-H. and J.C.F.d.W.; Methodology, M.K., T.D., C.B.W.-H. and J.C.F.d.W.; Software, T.D. and M.K.; Validation, M.K. and J.C.F.d.W.; Formal analysis, T.D. and M.K.; Investigation, T.D.; Resources, H.W.; Data curation, T.D. and M.K.; Writing–original draft preparation, T.D. and M.K.; Writing–review and editing, M.K., C.B.W.-H., J.C.F.d.W., H.W. and T.D.; Visualization, T.D. and M.K.; Supervision, M.K., C.B.W.-H., J.C.F.d.W. and H.W.; Project administration, C.B.W.-H. and H.W.; Funding acquisition, H.W.; All authors have read and agreed to the published version of the manuscript.

Funding: This research was funded by Honda Research Institute Europe GmbH.

Conflicts of Interest: The authors declare no conflict of interest.

References

1. Hasirlioglu, S.; Kamann, A.; Doric, I.; Brandmeier, T. Test methodology for rain influence on automotive surround sensors. In Proceedings of the 2016 IEEE 19th International Conference on Intelligent Transportation Systems (ITSC), Rio de Janeiro, Brazil, 1–4 November 2016; pp. 2242–2247. [CrossRef]
2. Audi. *A8 Owner's Manual*; Audi AG: Ingolstadt, Germany, 2018.
3. Lee, J.D.; See, K.A. Trust in Automation: Designing for Appropriate Reliance. *Hum. Factors J. Hum. Factors Ergon. Soc.* **2004**, *46*, 50–80. [CrossRef] [PubMed]
4. Flemisch, F.O.; Bengler, K.; Bubb, H.; Winner, H.; Bruder, R. Towards cooperative guidance and control of highly automated vehicles: H-Mode and Conduct-by-Wire. *Ergonomics* **2014**, *57*, 343–360. [CrossRef] [PubMed]
5. Hoc, J.M. Towards a cognitive approach to human-machine cooperation in dynamic situations. *Int. J. Hum. Comput. Stud.* **2001**, *54*, 509–540. [CrossRef]

6. Klein, G.; Woods, D.; Bradshaw, J.; Hoffman, R.; Feltovich, P. Ten Challenges for Making Automation a "Team Player" in Joint Human-Agent Activity. *IEEE Intell. Syst.* **2004**, *19*, 91–95. [CrossRef]

7. Krüger, M.; Wiebel, C.B.; Wersing, H. From Tools Towards Cooperative Assistants. In Proceedings of the 5th International Conference on Human Agent Interaction (HAI '17), Bielefeld, Germany, 17–20 October 2017; ACM Press: New York, NY, USA, 2017; pp. 287–294. [CrossRef]

8. Sendhoff, B.; Wersing, H. Cooperative Intelligence—A Humane Perspective. In Proceedings of the 1st IEEE International Conference on Human-Machine Systems (ICHMS 2020), Rome, Italy, 7–9 September 2020; IEEE: New York, NY, USA, 2020.

9. Caird, J.K.; Chisholm, S.; Lockhart, J. Do in-vehicle advanced signs enhance older and younger drivers' intersection performance? Driving simulation and eye movement results. *Int. J. Hum.-Comput. Stud.* **2008**, *66*, 132–144. [CrossRef]

10. Naujoks, F.; Neukum, A. Timing of in-vehicle advisory warnings based on cooperative perception. In Proceedings of the Human Factors and Ergonomics Society Europe Chapter Annual Meeting, Torino, Italy, 16–18 October 2013.

11. Ali, Y.; Zheng, Z.; Haque, M.M. Connectivity's impact on mandatory lane-changing behaviour: Evidences from a driving simulator study. *Transp. Res. Part Emerg. Technol.* **2018**, *93*, 292–309. [CrossRef]

12. Naujoks, F.; Totzke, I. Behavioral adaptation caused by predictive warning systems—The case of congestion tail warnings. *Transp. Res. Part Traffic Psychol. Behav.* **2014**, *26*, 49–61. [CrossRef]

13. Beller, J.; Heesen, M.; Vollrath, M. Improving the Driver-Automation Interaction. *Hum. Factors J. Hum. Factors Ergon. Soc.* **2013**, *55*, 1130–1141. [CrossRef]

14. Finger, R.; Bisantz, A.M. Utilizing graphical formats to convey uncertainty in a decision-making task. *Theor. Issues Ergon. Sci.* **2002**, *3*, 1–25. [CrossRef]

15. Helldin, T.; Falkman, G.; Riveiro, M.; Davidsson, S. Presenting system uncertainty in automotive UIs for supporting trust calibration in autonomous driving. In Proceedings of the 5th International Conference on Automotive User Interfaces and Interactive Vehicular Applications (AutomotiveUI '13), Eindhoven, The Netherlands, 27–30 October 2013; ACM Press: New York, NY, USA, 2013; pp. 210–217. [CrossRef]

16. Kunze, A.; Summerskill, S.J.; Marshall, R.; Filtness, A.J. Automation transparency: Implications of uncertainty communication for human-automation interaction and interfaces. *Ergonomics* **2019**, 1–16. [CrossRef]

17. Noah, B.E.; Gable, T.M.; Chen, S.Y.; Singh, S.; Walker, B.N. Development and Preliminary Evaluation of Reliability Displays for Automated Lane Keeping. In Proceedings of the 9th International Conference on Automotive User Interfaces and Interactive Vehicular Applications (AutomotiveUI '17), Oldenburg, Germany, 24–27 September 2017; ACM Press: New York, NY, USA, 2017; pp. 202–208. [CrossRef]

18. Faltaous, S.; Baumann, M.; Schneegass, S.; Chuang, L.L. Design Guidelines for Reliability Communication in Autonomous Vehicles. In Proceedings of the 10th International Conference on Automotive User Interfaces and Interactive Vehicular Applications, Toronto, ON, Canada, 23–25 September 2018; pp. 258–267.

19. SAE International On-Road Automated Driving Committee. *Taxonomy and Definitions for Terms Related to Driving Automation Systems for On-Road Motor Vehicles*; Technical report SAE J3016; SAE International: Warrendale, PA, USA, 2016.

20. Kunze, A.; Summerskill, S.J.; Marshall, R.; Filtness, A.J. Conveying Uncertainties Using Peripheral Awareness Displays in the Context of Automated Driving. In Proceedings of the 11th International Conference on Automotive User Interfaces and Interactive Vehicular Applications, Utrecht, The Netherlands, 22–25 September 2019; pp. 329–341.

21. Cohen-Lazry, G.; Borowsky, A.; Oron-Gilad, T. The effects of continuous driving-related feedback on drivers' response to automation failures. In Proceedings of the Human Factors and Ergonomics Society Annual Meeting, Nantes, France, 20 October 2017. SAGE Publications Sage CA: Los Angeles, CA, 2017; Volume 61, pp. 1980–1984.

22. Kunze, A.; Summerskill, S.J.; Marshall, R.; Filtness, A.J. Preliminary Evaluation of Variables for Communicating Uncertainties Using a Haptic Seat. In Proceedings of the 10th International Conference on Automotive User Interfaces and Interactive Vehicular Applications (AutomotiveUI '18), Toronto, ON, Canada, 23–25 September 2018; ACM Press: New York, NY, USA, 2018; pp. 154–158.

23. Wintersberger, P.; Dmitrenko, D.; Schartmüller, C.; Frison, A.K.; Maggioni, E.; Obrist, M.; Riener, A. S(C)ENTINEL. In Proceedings of the 24th International Conference on Intelligent User Interfaces (IUI '19), Marina del Ray, CA, USA, 19–22 March 2019; ACM Press: New York, NY, USA, 2019; pp. 538–546. [CrossRef]

24. Van Erp, J.B.F. Guidelines for the use of vibro-tactile displays in human computer interaction. In Proceedings of the Eurohaptics, Edinburgh, UK, 8–10 July 2002; Volume 2002, pp. 18–22.

25. Petermeijer, S.M.; Cieler, S.; De Winter, J.C.F. Comparing spatially static and dynamic vibrotactile take-over requests in the driver seat. *Accid. Anal. Prev.* **2017**, *99*, 218–227. [CrossRef] [PubMed]

26. Ho, C.; Tan, H.Z.; Spence, C. Using spatial vibrotactile cues to direct visual attention in driving scenes. *Transp. Res. Part Traffic Psychol. Behav.* **2005**, *8*, 397–412. [CrossRef]

27. Spence, C.; Ho, C. Tactile and Multisensory Spatial Warning Signals for Drivers. *IEEE Trans. Haptics* **2008**, *1*, 121–129. [CrossRef] [PubMed]

28. Fitch, G.M.; Hankey, J.M.; Kleiner, B.M.; Dingus, T.A. Driver comprehension of multiple haptic seat alerts intended for use in an integrated collision avoidance system. *Transp. Res. Part Traffic Psychol. Behav.* **2011**, *14*, 278–290. [CrossRef]

29. Meng, F.; Spence, C. Tactile warning signals for in-vehicle systems. *Accid. Anal. Prev.* **2015**, *75*, 333–346. [CrossRef] [PubMed]

30. Petermeijer, S.M.; de Winter, J.C.; Bengler, K.J. Vibrotactile displays: A survey with a view on highly automated driving. *IEEE Trans. Intell. Transp. Syst.* **2016**, *17*, 897–907. [CrossRef]

31. Nagel, S.K.; Carl, C.; Kringe, T.; Märtin, R.; König, P. Beyond sensory substitution—Learning the sixth sense. *J. Neural Eng.* **2005**, *2*, R13. [CrossRef]

32. Van Erp, J.B.F.; Veen, H.A.H.C.V.; Jansen, C.; Dobbins, T. Waypoint Navigation with a Vibrotactile Waist Belt. *ACM Trans. Appl. Percept.* **2005**, *2*, 106–117. [CrossRef]

33. Smets, N.J.; te Brake, G.M.; Neerincx, M.A.; Lindenberg, J. Effects of mobile map orientation and tactile feedback on navigation speed and situation awareness. In Proceedings of the 10th International Conference on Human Computer Interaction with Mobile Devices and Services, Amsterdam, The Netherlands, 2–5 September 2008; ACM: New York, NY, USA, 2008; pp. 73–80.

34. Srikulwong, M.; O'Neill, E. A comparative study of tactile representation techniques for landmarks on a wearable device. In Proceedings of the SIGCHI Conference on Human Factors in Computing Systems, Vancouver, BC, Canada, 7–12 May 2011; ACM: New York, NY, USA, 2011.

35. Zelek, J.; Hobein, M. Wearable Tactile Navigation System. US20130218456, 22 August 2013.

36. Steltenpohl, H.; Bouwer, A. Vibrobelt: Tactile Navigation Support for Cyclists. In Proceedings of the 2013 International Conference on Intelligent User Interfaces, Santa Monica, CA, USA, 19–22 March 2013; ACM: New York, NY, USA, 2013; pp. 417–426.

37. Prasad, M.; Taele, P.; Goldberg, D.; Hammond, T.A. Haptimoto: Turn-by-turn haptic route guidance interface for motorcyclists. In Proceedings of the SIGCHI Conference on Human Factors in Computing Systems, Toronto, ON, Canada, 26 April–1 May 2014; ACM: New York, NY, USA, 2014.

38. Schirmer, M.; Hartmann, J.; Bertel, S.; Echtler, F. Shoe me the way: A shoe-based tactile interface for eyes-free urban navigation. In Proceedings of the 17th International Conference on Human-Computer Interaction with Mobile Devices and Services, Copenhagen, Denmark, 24–27 August 2015; ACM: New York, NY, USA, 2015; pp. 327–336.

39. Dobbelstein, D.; Henzler, P.; Rukzio, E. Unconstrained pedestrian navigation based on vibro-tactile feedback around the wristband of a smartwatch. In Proceedings of the 2016 CHI Conference Extended Abstracts on Human Factors in Computing Systems, San Jose, CA, USA, 7–12 May 2016; ACM: New York, NY, USA 2016, pp. 2439–2445.

40. Cassinelli, A.; Reynolds, C.; Ishikawa, M. Augmenting spatial awareness with Haptic Radar. In Proceedings of the 2006 10th IEEE International Symposium on Wearable Computers, Montreux, Switzerland, 11–14 October 2006; IEEE: Piscataway, NJ, USA, 2006, pp. 61–64. [CrossRef]

41. Cardin, S.; Thalmann, D.; Vexo, F. A wearable system for mobility improvement of visually impaired people. *Vis. Comput.* **2007**, *23*, 109–118. [CrossRef]

42. Franz, M.; Zeidler, A.; dos Santos Rocha, M.; Klein, C. Vibro-Tactile Space-Awareness. In Proceedings of the Tenth International Conference on Ubiquitous Computing, Seoul, Korea, 21–24 September 2008; pp. 117–120.

43. Riener, A.; Ferscha, A. Raising awareness about space via vibro-tactile notifications. In Proceedings of the European Conference on Smart Sensing and Context, Zurich, Switzerland, 29–31 October 2008; Springer: Berlin, Germany, 2008; pp. 235–245.

44. Morrell, J.; Wasilewski, K. Design and evaluation of a vibrotactile seat to improve spatial awareness while driving. In Proceedings of the 2010 IEEE Haptics Symposium, Waltham, MA, USA, 25–26 March 2010; pp. 281–288. [CrossRef]

45. de Barros, P.G.; Lindeman, R.W. Performance effects of multi-sensory displays in virtual teleoperation environments. In Proceedings of the 1st Symposium on Spatial User Interaction (SUI '13), Los Angeles, CA, USA, 20–21 July 2013; ACM Press: New York, NY, USA, 2013; p. 41. [CrossRef]

46. Berning, M.; Braun, F.; Riedel, T.; Beigl, M. ProximityHat: A head-worn system for subtle sensory augmentation with tactile stimulation. In Proceedings of the International Symposium on Wearable Computers, Osaka, Japan, 7–11 September 2015; ACM: New York, NY, USA, 2015.

47. Tsukada, K.; Yasumura, M. Activebelt: Belt-type wearable tactile display for directional navigation. In Proceedings of the International Conference on Ubiquitous Computing, Nottingham, UK, 7–10 September 2004; Springer: Berlin, Germany, 2004, pp. 384–399.

48. Murata, A.; Kemori, S.; Moriwaka, M.; Hayami, T. Proposal of Automotive 8-directional Warning System That Makes Use of Tactile Apparent Movement. In Proceedings of the International Conference on Digital Human Modeling and Applications in Health, Safety, Ergonomics and Risk Management, Las Vegas, NV, USA, 21–26 July 2013; pp. 98–107.

49. Telpaz, A.; Rhindress, B.; Zelman, I.; Tsimhoni, O. Haptic Seat for Automated Driving: Preparing the Driver to Take Control Effectively. In Proceedings of the 7th International Conference on Automotive User Interfaces and Interactive Vehicular Applications (AutomotiveUI '15), Nottingham, UK, 1–3 September 2015; ACM: New York, USA, 2015; pp. 23–30. [CrossRef]

50. Boll, S.; Asif, A.; Heuten, W. Feel your route: A tactile display for car navigation. *IEEE Pervasive Comput.* **2011**, *10*, 35–42. [CrossRef]

51. Pielot, M.; Henze, N.; Heuten, W.; Boll, S. Evaluation of continuous direction encoding with tactile belts. In Proceedings of the International Workshop on Haptic and Audio Interaction Design, 31 August–1 September 2008; Springer: Berlin, Germany, 2008; pp. 1–10.

52. Pielot, M.; Boll, S. Tactile Wayfinder: Comparison of tactile waypoint navigation with commercial pedestrian navigation systems. In Proceedings of the International Conference on Pervasive Computing, Helsinki, Finland, 17–20 May 2010; Springer: Berlin, Germany, 2010, pp. 76–93.

53. Krüger, M.; Wersing, H.; Wiebel-Herboth, C.B. Approach for Enhancing the Perception and Prediction of Traffic Dynamics with a Tactile Interface. In Proceedings of the 10th International Conference on Automotive User Interfaces and Interactive Vehicular Applications (AutomotiveUI '18), Toronto, ON, Canada, 23–25 September 2018; ACM Press: New York, NY, USA, 2018; pp. 164–169. [CrossRef]

54. Krüger, M.; Wiebel-Herboth, C.B.; Wersing, H. The Lateral Line: Augmenting Spatiotemporal Perception with a Tactile Interface. In Proceedings of the AHs '20: Augmented Humans International Conference, Kaiserslautern, Germany, 16–17 March 2020; ACM Press: New York, NY, USA, 2020. [CrossRef]

55. Kassner, M.; Patera, W.; Bulling, A. Pupil: An Open Source Platform for Pervasive Eye Tracking and Mobile Gaze-based Interaction. In Proceedings of the 2014 ACM International Joint Conference on Pervasive and Ubiquitous Computing Adjunct Publication (UbiComp '14 Adjunct), Seattle, WA, USA, 13–17 September 2014; ACM Press: New York, NY, USA, 2014; pp. 1151–1160. [CrossRef]

56. Acik, A.; Bernhard, B.; Dombrowe, I.; Kringe, T.; Märtin, R.; Carl, C.; Honey, C.; Kabisch, C.; Jansen, L.; Lörken, C.; et al. *FeelSpace—Report of a Study Project*; Technical Report; University of Osnabrück: Osnabrück, Germany, 2005.

57. Van Erp, J., Tactile navigation display. In *Haptic Human-Computer Interaction, Proceedings of the First International Workshop Glasgow, UK, 31 August–1 September 2000*; Springer: Berlin/Heidelberg, Germany, 2001; pp. 165–173.

58. Kerdegari, H.; Kim, Y.; Stafford, T.; Prescott, T.J. Centralizing bias and the vibrotactile funneling illusion on the forehead. In Proceedings of the International Conference on Human Haptic Sensing and Touch Enabled Computer Applications, Versailles, France, 24–26 June 2014; Springer: Berlin, Germany, 2014; pp. 55–62.

59. Cholewiak, R.W. Spatial factors in the perceived intensity of vibrotactile patterns. *Sens. Process.* **1979**, *3*, 141–156.

60. Hoffmann, S.; Krüger, H.P.; Buld, S. Vermeidung von Simulator Sickness anhand eines Trainings zur Gewöhnung an die Fahrsimulation. *VDI Berichte* **2003**, *1745*, 385–404.

61. Hogema, J.; De Vries, S.; Van Erp, J.; Kiefer, R. A Tactile Seat for Direction Coding in Car Driving: Field Evaluation. *IEEE Trans. Haptics* **2009**, *2*, 181–188. [CrossRef] [PubMed]

62. Hayward, J.C. Near miss determination through use of a scale of danger. In Proceedings of the 51st Annual Meeting of the Highway Research Board, Washington, DC, USA, 17–21 January 1972; pp. 24–34.

63. Van der Horst, A.R.A. *A Time-Based Analysis of Road User Behaviour in Normal and Critical Encounters*; Delft University of Technology: Delft, The Netherlands, 1991.

64. Hart, S.G. Nasa-Task Load Index (NASA-TLX); 20 Years Later. *Proc. Hum. Factors Ergon. Soc. Annu. Meet.* **2006**, *50*, 904–908. [CrossRef]

65. Van Der Laan, J.D.; Heino, A.; De Waard, D. A simple procedure for the assessment of acceptance of advanced transport telematics. *Transp. Res. Part Emerg. Technol.* **1997**, *5*, 1–10. [CrossRef]

66. Verhaeghen, P.; Salthouse, T.A. Meta-analyses of age–cognition relations in adulthood: Estimates of linear and nonlinear age effects and structural models. *Psychol. Bull.* **1997**, *122*, 231. [CrossRef]

67. Laurienti, P.J.; Burdette, J.H.; Maldjian, J.A.; Wallace, M.T. Enhanced multisensory integration in older adults. *Neurobiol. Aging* **2006**, *27*, 1155–1163. [CrossRef]

68. Kramer, A.F.; Cassavaugh, N.; Horrey, W.J.; Becic, E.; Mayhugh, J.L. Influence of age and proximity warning devices on collision avoidance in simulated driving. *Hum. Factors* **2007**, *49*, 935–949. [CrossRef]

 *information*

Article

Standardized Test Procedure for External Human–Machine Interfaces of Automated Vehicles

Christina Kaß [1,*], Stefanie Schoch [1], Frederik Naujoks [2], Sebastian Hergeth [2], Andreas Keinath [2] and Alexandra Neukum [1]

[1] Würzburg Institute for Traffic Sciences GmbH, 97209 Veitshöchheim, Germany; schoch@wivw.de (S.S.); neukum@wivw.de (A.N.)
[2] BMW Group, 80937 Munich, Germany; frederik.naujoks@bmw.de (F.N.); sebastian.hergeth@bmw.de (S.H.); andreas.keinath@bmw.de (A.K.)
* Correspondence: kass@wivw.de

Received: 28 February 2020; Accepted: 23 March 2020; Published: 24 March 2020

Abstract: Research on external human–machine interfaces (eHMIs) has recently become a major area of interest in the field of human factors research on automated driving. The broad variety of methodological approaches renders the current state of research inconclusive and comparisons between interface designs impossible. To date, there are no standardized test procedures to evaluate and compare different design variants of eHMIs with each other and with interactions without eHMIs. This article presents a standardized test procedure that enables the effective usability evaluation of eHMI design solutions. First, the test procedure provides a methodological approach to deduce relevant use cases for the evaluation of an eHMI. In addition, we define specific usability requirements that must be fulfilled by an eHMI to be effective, efficient, and satisfying. To prove whether an eHMI meets the defined requirements, we have developed a test protocol for the empirical evaluation of an eHMI with a participant study. The article elucidates underlying considerations and details of the test protocol that serves as framework to measure the behavior and subjective evaluations of non-automated road users when interacting with automated vehicles in an experimental setting. The standardized test procedure provides a useful framework for researchers and practitioners.

Keywords: eHMI; standardized test procedure; use cases; test protocol; automated driving

---

1. Introduction

With the introduction of automated vehicles into mixed traffic environments, drivers may be (temporarily) allowed to engage in non-driving-related tasks while driving. As a consequence, the drivers of automated vehicles will often be unavailable for communication while their vehicle is interacting with non-automated road users. To face this change and to ensure safe interactions, there is a broad acceptance among practitioners and researchers that in some situations, automated vehicles may need to replace the informal communication of human drivers (such as gestures and eye contact) with external human–machine interfaces (eHMIs) [1,2]. Currently, eHMI systems represent a completely new and immature technology. Before introducing such a new technological system to the market and to the traffic environment, it is important to carefully determine its usability.

Since 2017, a large body of research has been investigating the impact of different eHMI approaches on the subjective evaluations and behavior of non-automated road users. Previously studied eHMI approaches basically differed with regard to the content of communication (e.g., maneuver intention, automation status, and request for action) [3] and concrete interface design solutions (e.g., the position and modality of the signal) [4–8]. Results are inconclusive regarding the benefit of using an eHMI to signal maneuver intentions of automated vehicles. In some studies, communicating the maneuver intention of the automated vehicle increased the subjective ratings of interaction partners in comparison to interactions without an eHMI [9–12]. In other studies, such eHMI concepts did not have any impact on pedestrians' perceived trust and safety [1] or even had a negative effect on pedestrians' workload during interactions with automated vehicles [13]. Moreover, it is still unclear whether communicating the vehicle's automation status with eHMI systems improves the subjective experiences of interaction partners. On the one hand, eHMI systems that signaled the automation status with light-emitting diode (LED) strips had a positive effect on pedestrians' emotional experience [11] and perceived safety [12] compared to interactions without an eHMI. On the other hand, other studies did not reveal an impact of communicating the automation status on pedestrians' perceived stress [14] and perceived safety [1], such as on cyclists' reported behavior [15]. Furthermore, previous studies have offered contradictory findings on the effect of eHMI signals on the behavioral decisions of non-automated road users. eHMI concepts that communicated the vehicle's intention to stop [10] or gave a concrete request for action ("Walk!" or "Ok") [16] increased pedestrians' willingness to cross the road in a shared space compared to interactions without an eHMI. In addition, two studies found that pedestrians needed less time to make their decision to cross or not cross the road with than without an eHMI [6,17]. However, the results of [14,18] revealed that pedestrians focused to a higher degree on vehicle speed and distance to the vehicle when making crossing decisions than on eHMI signals. Deb et al. [19] found that a verbal warning saying "safe to cross" shortened the time pedestrians needed to cross the street compared to no eHMI, while different visual eHMI concepts had no effect on crossing time.

Overall, although extensive human factors research has been carried out on eHMIs, a systematic understanding of the usability of different eHMI concepts is still lacking. Previous research has used very different methodological approaches and has had methodological limitations. Methodological limitations include a lack of behavioral measurements [8,9], small sample sizes [12,13], and vague result reports [4,20]. In some studies, participants evaluated the eHMI after they had received a thorough briefing and explanation of the signal meanings [9,11,12]. In other studies, participants reported their subjective ratings of the situation even though some of them had not even perceived the eHMI [1,14]. Another limitation pertains to the research environments used in previous studies. Commonly used methods such as the Wizard of Oz technique, virtual reality (VR) pedestrian simulators, and video or photo studies use only simplified behavioral measurements, resulting in limited external validity. For example, participants were instructed to simply report their behavior [15], to press a button [10], or to take only one step forward to indicate their intention to cross [14]. The outlined methodological differences and limitations render comparisons of different eHMI variants impossible. Therefore, results are inconclusive regarding the required content of communication (e.g., maneuver intention, automation status, and detection feedback), interface design requirements (e.g., modality, position, and text or symbols), the operational design domain for eHMIs (e.g., urban environment, crosswalks, and intersections), and the role of the interaction partner (e.g., pedestrian, cyclist, or manual driver).

Furthermore, most studies have not provided an explanation for the selection of the use case under investigation. The majority of studies have examined interactions in urban areas in a low speed range where communication was required to negotiate the right of way. The most frequently investigated use cases so far have been interactions with pedestrians at crosswalks [4,6,9,12,17–19,21–23] or crossing situations with an ambiguous right of way, e.g., shared spaced or parking areas [1,8,10–14,16,18,20,24,25]. While prior work has already developed frameworks to derive use cases to test the in-vehicle HMIs of automated driving systems [26–29], there has been very limited research on taxonomies for use cases of eHMIs [30].

To date, there are no standardized test procedures to assess the usability of eHMIs of automated vehicles. There is no consensus on relevant use cases, evaluation requirements, and proper experimental designs yet. To advance the development of eHMIs, there is a necessity to standardize the evaluation process of eHMIs. Standardized test procedures allow for reliable and meaningful conclusions and enable comparisons between different studies and interface designs. Standardized methods already exist for other research areas of traffic psychology, e.g., for the evaluation of the in-vehicle

HMIs of vehicles with automated driving systems [31] or to measure the eyes-off-road time as an indicator of distraction potential when interacting with in-vehicle information systems [32,33]. In their review article on the current state of research on eHMIs, Rouchitsas and Alm [34] declared that the "standardization of relevant procedures is a fundamental requirement for effective interface evaluations and meaningful comparisons. Therefore, future conceptual and empirical work in the field should primarily be concerned with producing standardized procedures for evaluating and comparing different implementations"(p. 10). The present article provides a response to this request. We propose a newly developed methodological framework that standardizes the usability evaluation process of eHMIs. This standardized test procedure consists of three parts:

1. Definition of relevant use cases: The selection of relevant use cases represents the basis for a test procedure to evaluate the usability of eHMIs. We developed a methodology to deduce relevant use cases for a given eHMI from an exhaustive set of all possible use cases.
2. Definition of usability requirements: We define the usability requirements of an eHMI according to the International Organization for Standardization (ISO9241-11) [35]. Thus, to ensure the usability of an eHMI, it needs to be effective, efficient, and satisfying. To be able to evaluate whether an eHMI meets these requirements, we derived appropriate parameters and criteria for each requirement.
3. Test protocol for empirical studies: The test protocol provides an experimental framework to empirically evaluate a given eHMI with a user study. We outline the methodological details of the test protocol, e.g., sample, test environment, and instruction.

2. Methods and Results

2.1. Definition of Use Cases

Prior research has mainly focused on vehicle–pedestrian interactions at crosswalks or at ambiguous crossing points in urban environments at a low speed. However, this only represents a limited selection of the possible use cases of an eHMI. To evaluate the usability of an eHMI in a standardized way, it is important that study participants encounter the eHMI with a set of relevant use cases. Thus, the definition of relevant use cases is the core of each evaluation process [31], as it ensures that the test procedure generates meaningful and comparable results. Fuest, Sorokin, Bellem, and Bengler [30] published a taxonomy of traffic situations that intends to serve as a basis to assess the communication between automated vehicles and human road users. Their taxonomy provides an overview of attributes and associated value facets that are considered to influence implicit and explicit communication in traffic, e.g., the attribute "right of way" with the value facets automated vehicle, human road user, or undefined. To define a traffic situation, one can choose and combine attributes and value facets that are relevant for the research question at hand. The combination of all listed value facets results in 373,248 situations. The authors do not provide an instruction how to deduce relevant use cases. Furthermore, the taxonomy lacks attributes that specify the approach direction of the interaction partners and the currently executed driving maneuver of the automated vehicle. Therefore, we developed a new methodological approach to deduce relevant use cases for a given eHMI.

We used a multi-stage gradual methodological approach that claims to consider an exhaustive set of use cases of an eHMI. These use cases are subsequently reduced step-by-step by applying different filters. More specifically, the collection and combination of use cases and their specifications alternate with stepwise reductions of use cases based on redundancies and theoretical and practical considerations. Figure 1 illustrates an overview of the procedure of this approach.

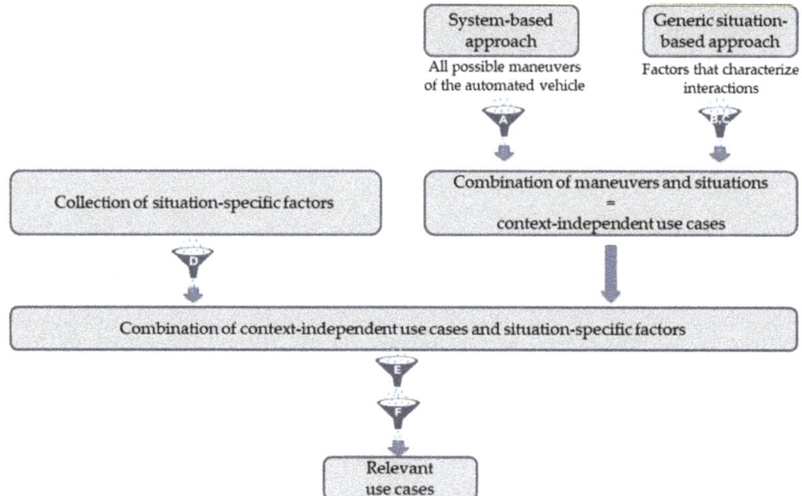

Figure 1. Overview of the methodological approach to select relevant use cases of an external human–machine interface (eHMI).

2.1.1. Defining a Use Case of an eHMI

The basis of the approach was the definition of a use case of an eHMI. A use case of an eHMI is defined as a situation where an automated vehicle and at least one non-automated road user intend to "occupy the same region of space at the same time in the near future" [36]. This situation requires the interactive behavior of at least one involved road user to avoid a potential traffic conflict. Interactive behavior signifies that the road user adapts its initially planned behavior to the anticipated behavior of the other road user, e.g., by changing speed or trajectory. Traffic conflicts arise when "two or more road users approach each other in space and time to such an extent that a collision is imminent if their movements remain unchanged" [37]. The use of eHMIs as communication aids of automated vehicles can potentially support non-automated road users in understanding and anticipating the interactive behavior of the automated vehicle. From this, the users can draw conclusions for their own interactive behavior.

2.1.2. System-Based Approach

The system-based approach was used to collect all possible driving maneuvers that an automated vehicle can execute. Driving maneuvers were divided into lateral and longitudinal maneuvers. Lateral maneuvers consist of driving straight ahead, turning (left, right), and changing the lane (left, right). When the vehicle is in motion, longitudinal maneuvers are keeping a constant speed, decelerating, and accelerating. When at a standstill, longitudinal maneuvers are keeping a constant speed (0 km/h), starting to drive forward, and reversing. Filter A (Figure 1) reduced the number of collected maneuvers based on the assumption that an eHMI should be only used in situations in which it adds benefit to conventional lighting. Consequently, all lateral maneuvers and the reversing maneuver were filtered, as they can be signaled by the turn signal and the reversing light. In principle, acceleration, deceleration, and keeping a constant speed can be perceived by other road users by observing the automated vehicle. However, these cues are often very subtle, and eHMIs could support the perception by signaling these maneuver intentions prior to action execution. In conclusion, the resulting maneuvers are keeping a constant speed (while driving or at standstill), accelerating (while driving or at standstill), and decelerating (while driving).

2.1.3. Generic Situation-based Approach

The generic situation-based approach considered all factors that characterize interactions between traffic participants. This approach is generic because it does not consider the context in which a situation takes place, e.g., urban context, highway, intersection, or parking area. The first factor represents the intended moving direction of the interaction partner, which may be in the opposite direction to the automated vehicle, at a crossing angle to the automated vehicle, laterally approaching the automated vehicle in the same direction, and driving in the same direction as the automated vehicle (see first row of Figure 2). The second factor represents the position of the interaction partner relative to the automated vehicle (see second row of Figure 2). A combination of these two factors leads to certain combinations that would never result in traffic conflicts between the two traffic participants (see definition of eHMI use cases), e.g., when the interaction partner is located next to the automated vehicle while driving in the opposite direction. Filter B (Figures 1 and 2) was used to reduce those combinations that cannot lead to traffic conflicts. The remaining combinations represent situations that would result in traffic conflicts without the interactive behavior of at least one involved road user (see third row of Figure 2). We hypothesized that the driving direction of the interaction partner (left or right) and the exact start position of the interaction partner in a merging situation do not lead to relevant differences between the resulting use cases. These redundant situations are indicated by blue boxes in Figure 2. Thus, Filter C (Figures 1 and 2) filtered out these redundant situations. The resulting three generic situations are shown in the bottom row of Figure 2: The interaction partner approaches the automated vehicle frontally, orthogonally from the side, or merges in front of the automated vehicle with a lateral approach direction. Figure 3 illustrates possible ways to implement these three situations in a driving simulation with a cyclist as the interaction partner.

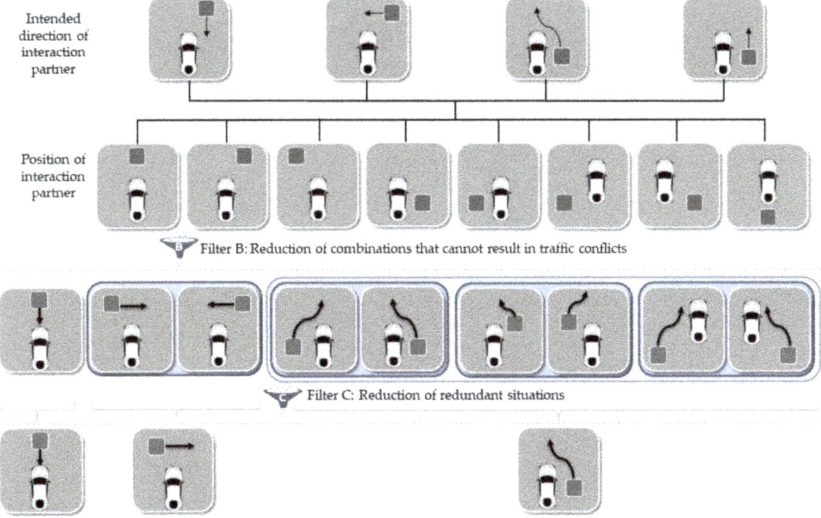

Figure 2. Generic situation-based approach with two filters. Grey squares represent an arbitrary interaction partner, and the white vehicles represent automated vehicles. Blue boxes indicate redundant scenarios.

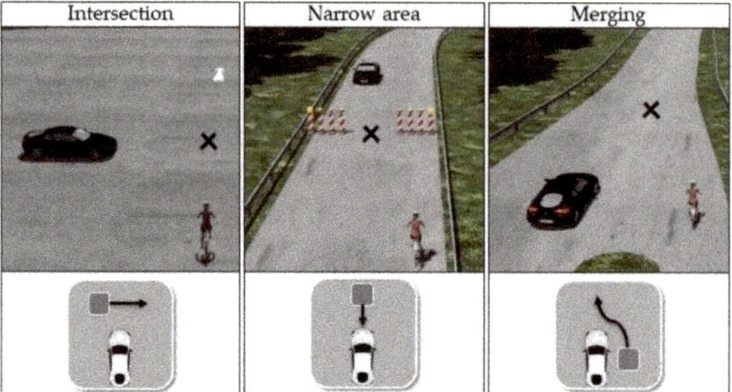

Figure 3. Implementation of the three derived situations in a driving simulation with a cyclist as the interaction partner. The arrows indicate the trajectories of the automated vehicle and the cyclist, and the black cross represents their virtual crossing point.

2.1.4. Combination of Maneuvers and Situations: Context-Independent Use Cases

In the next step, the remaining maneuvers of the automated vehicle and generic situations were combined (see Figures 1 and 4). The resulting context-independent use cases are illustrated in Figure 4. For example, the interaction partner could approach the automated vehicle orthogonally from the side while the automated vehicle keeps a constant speed.

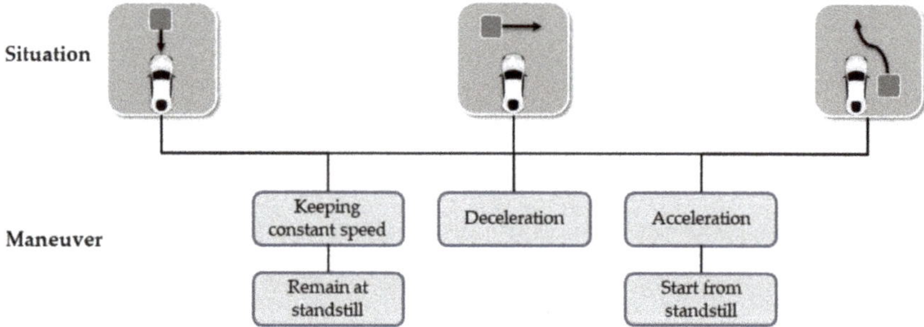

Figure 4. Combination of situations and maneuvers of the automated vehicle, resulting in context-independent use cases. Grey squares represent an arbitrary interaction partner, and the white vehicles represent automated vehicles.

2.1.5. Collection of Situation-Specific Factors

In order to ensure an exhaustive set of use cases, we collected a set of all situation-specific factors that could potentially influence an interaction between the automated vehicle and its interaction partner. Following the procedure by Fuest et al. [30], we assigned value facets to the collected factors. Table 1 presents the collected situation-specific factors and their value facets. Filter D was used to reduce certain value facets or complete factors (filtered factors and value facets are marked by [2] in Table 1). This reduction was based on the guideline that the use cases should be used to evaluate the usability of an eHMI. Accordingly, if we expected that a certain factor and/or its corresponding value facets would not lead to different requirements for an eHMI, they were not further considered. The following paragraph elucidates the collected situation-specific factors and the application of Filter D.

The type of road can be either urban, rural, or a highway [38]. We assumed that the usability of an eHMI would not differ depending on the type of road on which an interaction partner experiences the system. For example, a certain eHMI signal should have the same usability during a merging maneuver regardless of whether the maneuver takes place on an urban or rural road. Independent of the type of road, an eHMI must be able to communicate whether it is letting the interaction partner merge or whether he/she must brake and merge behind the vehicle. Furthermore, use cases can take place in different traffic environments, such as at intersections, in parking areas, or somewhere on the road. It was hypothesized that the system perception and interpretation and, thus, the requirements for an eHMI will not change depending on the traffic environment. Regardless of communicating at an intersection or on the road, the eHMI must signal if the automated vehicle will let the interaction partner cross or not. Thus, by applying Filter D, the factors of the type of road and traffic environment were not further considered as situation-specific factors.

Table 1. Situation-specific factors, their value facets, and the application of Filter D.

Situation-Specific Factor	Value Facets
Type of road	Urban [2] Rural [2] Highway [2]
Traffic environment	Intersection [2] Parking [2] On the road [2]
Right of way [1]	Automated vehicle [2] Interaction partner [2] Undefined
Type of interaction partner [1]	Motorized Non-motorized
Automation level	0 [2] 1 [2] 2 [2] 3 4 [3] 5 [3]
Visibility conditions	Normal Bad [2]
Speed of automated vehicle at beginning of interaction [1]	0 km/h 30 km/h 50 km/h 130 km/h
Speed of interaction partner at beginning of interaction [1]	0 km/h 4.4 km/h 17.5 km/h 30 km/h 50 km/h 130 km/h
Distance between automated vehicle and interaction partner at beginning of interaction	X meters

[1] These factors are based on the taxonomy by Fuest et al. [30]. [2] These factors and value facets are filtered by Filter D. [3] These value facets were combined by Filter D.

The right of way can be either assigned to the automated vehicle (e.g., green traffic light), to the interaction partner (e.g., crosswalk), or can be undefined [30]. To test the usability of an eHMI, the eHMI should be the only mean that influences the interaction between the automated vehicle and the non-automated road user. Thus, we decided to filter those value facets in which clear traffic rules determine the right of way for one of the interaction partners. The use cases for the eHMI test procedure should take place in a traffic environment without right-of-way rules.

The type of interaction partner can be either motorized vehicles (cars, powered two-wheelers, and trucks) or non-motorized vulnerable road users ((VRUs) such as pedestrians and cyclists). To date, there have been no studies that systematically compare the impact of eHMI signals on the interaction of automated vehicles with different types of road users. Interactions with motorized vehicles are usually more dynamic (higher velocities) than with VRUs. The drivers of motorized vehicles often have another visual perspective on the automated vehicle than VRUs. However, prior research and technical developments have suggested that automated vehicles will primarily use vehicle-to-X (V2X) technology to communicate with manual car drivers. With this technology, automated vehicles can send messages directly to the in-vehicle displays of manually-driven vehicles, e.g., about their intent, their willingness to cooperate, or requests of cooperative behavior of the human driver [39]. Thus, with V2X communication, automated vehicles do not necessarily need an eHMI to communicate with the human drivers of manually-driven vehicles. Furthermore, unsuccessful interactions usually have more severe consequences for VRUs than for the drivers of motorized vehicles. Compared to pedestrians, interactions with cyclists are evaluated to be more critical because they move at higher speeds, and, thus, interactions evolve more dynamically [15]. These differences might lead to different requirements for an eHMI when the automated vehicle interacts with different types of interaction partners. Principally, the use cases should be experienced from the perspective of a manual car driver (as the most common representative of a motorized interaction partner), as well as from the perspective of a cyclist (as the worst-case representative of a VRU). However, due to V2X technology as another communication aid between automated and manually-driven vehicles, we recommend to primarily focus on use cases with VRUs as interaction partners.

The automation level represents a further potentially relevant factor. The categorization published by the Society of Automotive Engineers defines six automation levels [40]. Driving on automation levels 0–2 does not represent a use case of an eHMI, as the human driver is responsible to monitor the driving environment and must remain attentive. Thus, the driver him- or herself can still communicate with other road users. On automation levels 3–5, the driver is allowed to engage in non-driving-related tasks as soon as the automated driving system is activated. The system makes decisions about upcoming driving maneuvers and could communicate these to other road users via an eHMI. In general, levels 3–5 can be considered as a single use case because the requirements for an eHMI do not differ. In comparison to levels 4 and 5, however, an automated driving system at level 3 could potentially hand over control to the driver during an interaction situation when a system limit is reached. A so-called take over situation results in the additional requirement that the interaction partner needs to understand that the previous eHMI signal might be no longer be valid once the driver has taken control. As a consequence, a takeover situation during an interaction with an automated vehicle at level 3 should be considered as an additional special use case.

Visibility conditions might influence the perceptibility of an eHMI and, thus, might lead to different requirements of an eHMI. However, these requirements rather relate to the pure visibility of eHMI signals in different visibility conditions than to the usability of the system. Thus, Filter D neglects different visibility conditions. Use cases to test the usability of an eHMI should take place under normal visibility conditions.

As additional factors, the speed of both interaction partners determines how fast an interaction builds up and develops. This might lead to different requirements of the eHMI with regard to the degree of detail and the required velocity of communication. For example, it is conceivable that the communication with eHMI signals must be faster when the driving speed is higher. Additionally, a more detailed eHMI signal could be more useful at a low speed than at a high speed. The speed of the automated vehicle at the beginning of an interaction depends on the used automated driving system and its operational design domain. According to the taxonomy by Fuest et al. [30], 0 km/h represents a vehicle at a standstill, 30 km/h is considered as a low speed range, 50 km/h is considered as an urban speed range, and 130 km/h is the permissible maximum speed in most European countries. These different speeds of the automated vehicle should be considered as use cases within the scope of the operational design domain of the respective automated driving system. The speed of the interaction partner at the beginning of an interaction depends on the type of interaction partner. Motorized vehicles could theoretically approach the automated vehicle at many different speeds. We assumed an average speed of 4.4 km/h for pedestrians [30,41] and 17.5 km/h for cyclists [30,42]. Additionally, speeds of 0 km/h (interaction partner at standstill), 30 km/h (low speed range), 50 km/h (urban speed range), and 130 km/h (maximum speed) should be considered. The reduction of these value facets depends on the type of interaction partner.

The distance between automated vehicle and interaction partner at beginning of interaction depends on their current speed. Based on the initial speed of both interaction partners, the prerequisite that both interaction partners should theoretically arrive at their "virtual crossing" point at the same time (see Figure 3) and a certain predefined time for the interaction partner to perceive and interpret the eHMI to make a behavioral decision, and to execute an action, one can calculate the distance of the interaction partners at the beginning of the interaction. For example, the use case represents the situation in which the interaction partner (cyclist = 17.5 km/h) approaches the automated vehicle (low speed range of 30 km/h) frontally. If we assume a time interval of 4 s, the cyclist drives 12.25 m and the vehicle drives 20.75 m in this time until they reach the virtual crossing point. Thus, the total initial distance must be 33 m. Based on this procedure, the distance does not represent an independent factor but results from other factors.

2.1.6. Combination of Context-Independent Use Cases and Situation-Specific Factors

In a next step, the context-independent use cases and remaining situation-specific factors were combined (Figure 1). However, there were still 864 possible combinations to deduce use cases. Filter E deleted implausible use cases from the full use-case set (Figure 1). This reduction was based on an analysis of realistic and unrealistic combinations of the type of interaction partner, speed, situation, and maneuver of the automated vehicle. For example, when the automated vehicle is at standstill, it can only remain at standstill or start from standstill (lower part of Figure 4). A deceleration maneuver is not possible. Other examples are realistic speeds for the three situations (upper part of Figure 4). An initial speed of 130 km/h for those situations in which the interaction partner approaches the automated vehicle frontally or orthogonally is not realistic for any of the interaction partners.

2.1.7. Deduction of Relevant Use Cases

In the last step, Filter F serves to select those use cases that are relevant for testing the usability requirements defined in Section 2.2 with the eHMI and the automated driving system under investigation. For example, we would like to test the usability of an eHMI of an "urban pilot" with the following specifications: The operational design domain of the system is in urban areas with a speed range between 0 and 30 km/h. If the system detects another road user within a radius of 60 meters, it will not accelerate due to safety reasons. Furthermore, the eHMI signal for keeping a constant speed is the same when the vehicle is at standstill or is moving. When deducing the relevant test cases from the use case set, these specifications further reduce the number of relevant test cases. Filter F can be applied to test different eHMI variants of automated driving systems with varying specifications.

The advantage of this methodological approach is that it provides a reproducible and clear procedure to select relevant use cases to test the usability of any given eHMI. The present set of use cases represents all scenarios that are relevant to test the usability of eHMIs during interactions with automated vehicles. It needs to be noticed that controllability or misuse tests might need different procedures for reducing and selecting relevant use cases. Furthermore, it must be emphasized that this method can and will not cover all conceivable use cases and situations—in particular, sound adaptations will be required for corner cases. Accordingly, researchers and practitioners who want to use this method will have to take care when they apply this method, thus extending and strengthening its validity.

2.2. Usability Requirements, Parameters, and Criteria

Prior research on eHMIs has not yet provided consensus on specific requirements for the usability of eHMIs. For the evaluation of the in-vehicle HMIs of automated driving systems, the National Highway Traffic Safety Administration (NHTSA) has defined minimum requirements that must be fulfilled by an HMI [43]. However, there are no published standardized requirements to assess the usability of eHMIs.

In order to define evaluation requirements, it is important to recall the initial considerations for the development of eHMIs. There were concerns that interactions between automated vehicles and other road users could result in difficulties and dangerous situations because the driver/passenger will not be available for informal communication [18]. Therefore, automated vehicles must ensure safe and efficient interactions with other road users [3]. The implementation of eHMIs is one possible way to support non-automated road users during interactions with automated vehicles. An alternative or complementary approach is to informally communicate driving behavior and intentions to other road users by developing appropriate driving strategies of automated vehicles [44,45]. In order to justify the implementation of an eHMI, it must have advantages for interaction partners compared to automated vehicles without an eHMI. At least, it should not deteriorate the quality of interaction. Thus, the basic requirement for an eHMI is its usability. According to the usability definition by ISO 9241-11 [35], the usability of a system is determined by its effectiveness, efficiency, and satisfaction. To be effective, an eHMI must support the non-automated road user in choosing an accurate behavioral decision during interactions with automated vehicles. An eHMI improves the interaction partner's efficiency if it has a positive effect on the time and mental effort required for a successful interaction. To be satisfying, the interaction partner must perceive the use of the eHMI as pleasant. This is relevant to facilitate its use and acceptance. As such, we defined three usability requirements for an eHMI:

1. The eHMI must be effective.
2. The eHMI must be efficient.
3. The eHMI must be satisfying.

The test procedure needs to differentiate between eHMIs that fulfill or do not fulfill these requirements. To decide whether a certain eHMI meets the defined requirements, it is necessary to define parameters for each requirement. These parameters are used to make the respective usability requirement measurable. The following paragraphs define specific parameters for each usability requirement (effectiveness, efficiency, and satisfaction) and propose methods for how to assess these parameters. To finally decide whether an eHMI is compliant with the respective requirement, it is necessary to define a pass/fail criterion for each parameter. In sum, only when an eHMI passes the specified criteria of all parameters per requirement does it fulfil the specific usability requirement as a whole.

Such parameters can be assessed by behavioral or self-reported measures. Behavioral measures can indicate if and how fast the interaction partner is able to understand the eHMI signal and if they are able to deduce correct behavioral decisions. However, there is a certain guess probability that the interaction partner makes the correct decision by chance (e.g., to either continue driving or to brake/stop). Furthermore, the driving behavior of the automated vehicle serves as an additional indicator for the interaction partner to make an appropriate behavioral decision. Thus, correct behavioral decisions of the interaction partner cannot be exclusively explained by their correct understanding of the eHMI signal. Additionally, self-reported measures are necessary to assess whether the interaction partner understands the eHMI signal correctly or not. On the other hand, self-reported measures alone would be insufficient because it must be ensured that a correct system understanding leads to correct behavior. Therefore, we propose a combination of both behavioral and self-reported measures.

Compared to an interaction without an eHMI, an eHMI should improve the effectiveness and efficiency of an interaction. At least, it should not deteriorate the interaction. To assess this difference between interactions with and without an eHMI, a baseline condition without an eHMI is required.

With this methodological approach, relative criteria can be used to assess the effectiveness and efficiency of an eHMI. However, certain parameters require an absolute instead of a relative criterion. For example, an eHMI should completely prevent the safety-critical behavior of interaction partners. Thus, the investigation should not focus on the question of whether there are less safety-critical situations with than without an eHMI. Instead, it is most important that there are no safety-critical situations with an eHMI at all (absolute criterion). Additionally, to evaluate the satisfaction with an eHMI, absolute criteria appear to be more appropriate than relative criteria.

2.2.1. Parameters and Criteria to Prove the Effectiveness of an eHMI

The effectiveness of an eHMI can be assessed by the parameters system comprehension and the correctness of behavioral decision. To measure system comprehension without giving participants the possibility to additionally consider the observed driving behavior of the automated vehicle as a confounding factor, we propose the occlusion method (see Section 2.3.2 for a detailed explanation). After the view on the automated vehicle has been occluded, participants need to answer the open-ended question "What will the automated vehicle do next?" The experimenter categorizes the answer as either correct or incorrect. The occlusion method does not allow for a comparison with the baseline condition because the screen is blanked before participants can deduce the vehicle's intention from its driving behavior. An absolute criterion can be used to evaluate the system comprehension. We propose a criterion of 85% correct answers for each use case. The appropriate indicators to assess the correctness of the behavioral decision depend on the driving maneuver of the automated vehicle in the respective use case. When the automated vehicle decelerates, the correctness of the behavioral decision can be measured by the minimal speed of the interaction partner during the interaction. The eHMI can be considered as being effective if the interaction partners reduce their speed to a significantly lower extent with an eHMI than without an eHMI (relative criterion). No or only slight reductions of speed would demonstrate that the eHMI supported interaction partners in predicting the unobserved behavior of the automated vehicle prior to real time. When the automated vehicle keeps a constant speed or accelerates, the interaction partner must reduce his or her speed or wait to prevent a safety-critical situation. Continued driving or walking represent incorrect behavioral decisions. However, the correctness of the behavioral decision should be assessed by an absolute criterion with a pass-fail logic. The relevant criterion is the resulting minimum distance between the automated vehicle and the interaction partner. A minimum distance that falls below one meter can be considered as a safety-critical distance. Following the guidelines of the RESPONSE Code of Practice [46], 20 of 20 participants need to pass the defined criterion to support the assumption that 85% of the population would also pass the criterion.

2.2.2. Parameters and Criteria to Prove the Efficiency of an eHMI

To measure the efficiency of an eHMI, we propose the parameters mental workload, time to cross, and visual attention. Mental workload can be assessed by a self-reported measure. After each interaction, the participant answers the question "How high was your mental workload during the interaction with the automated vehicle?" on a 7-point Likert scale ranging from very low to very high. Using a relative criterion, the mental workload should be significantly lower with than without an eHMI. To measure if the eHMI supported the efficiency of the interaction in a timely manner, the time between the first visual contact with the automated vehicle and the crossing of the virtual crossing point (see Figure 3) can be compared with and without the eHMI. The time to cross should be significantly shorter with than without the eHMI (relative criterion). To determine whether the eHMI improved the efficiency of the interaction with regard to the required visual attention, the proportion of visual attention towards the automated vehicle during the interaction should be significantly lower with than without the eHMI (relative criterion). Visual attention can be measured by eye tracking, head tracking, or by video coding.

2.2.3. Parameters and Criteria to Prove the Satisfaction with an eHMI

The satisfaction with the eHMI can be determined by the parameters satisfaction, attitude toward use, behavioral intention, and preference. All parameters are measured by items after participants have encountered all use cases with an eHMI. Table 2 includes a list of proposed items and the respective scales. All parameters use an absolute criterion. With regard to satisfaction, attitude toward use, and behavioral intention, at least 85% of all participants must choose a positive judgement (ratings between 5 and 7 on a 7-point Likert scale). To assess the preference, participants need to decide whether they would prefer interactions with automated vehicles with or without an eHMI in the future. To pass the relative criterion, a significantly higher proportion of participants must prefer future interaction with an eI IMI to interactions without an cHMI.

Table 2. Parameters and items to assess the satisfaction with an eHMI.

Parameter	Item	Scale	Reference
Satisfaction	Overall, how satisfied were you with the signals of the automated vehicle?	7-point Likert: very dissatisfied (1), neither nor (4), very satisfied (7)	Self-formulated
Attitude toward use	The interaction with the system is a wise idea.	7-point Likert: strongly disagree (1), neither nor (4), strongly agree (7)	Technology acceptance model [1]
Behavioral intention	Given that I had access to such signals when interacting with automated vehicles, I predict that I would use them.	7-point Likert: strongly disagree (1), neither nor (4), strongly agree (7)	Technology acceptance model [1]
Preference	In the future, would you prefer to interact with automated vehicles with or without signals?	Binary scale: with; without	Self-formulated

[1] Item adapted from [47].

The proposed requirements, parameters, and criteria contribute to the standardization of test procedures for evaluating the usability of eHMIs. Together with the definition of use cases, these standardized requirements form the basis for reliable eHMI evaluations and allow for meaningful comparisons between different eHMI variants and the results of different studies. Overall, this contribution will support the definition of design requirements for optimal interface specifications. The selection of the parameters can be adapted to the respective research questions and selected use cases.

2.3. Test Protocol

To evaluate the usability of an eHMI, it is important that users interact with the system in a standardized manner. We developed a test protocol for the empirical evaluation of eHMIs with a user study. The test protocol provides a proper experimental design to systematically investigate the usability of eHMIs. The objective is to prove whether a certain eHMI meets the usability requirements defined in Section 2.2. For this purpose, the test protocol defines a methodological procedure to observe and measure users' behavior and subjective evaluations during specified use cases and experimental conditions. The following sections elucidate the methodological details of the test protocol and underlying considerations.

2.3.1. Test Environment

The test environment must allow for controlled, standardized, and economic testing in a safe environment. At the same time, participants should encounter realistic scenarios to guarantee external validity. Furthermore, it is important that the parameters defined in Section 2.2. can be measured. Thus, the test environment must enable behavioral measurements, the observation of participants' behavior and the communication between experimenter and participants for interim questions. Additionally, a realistic implementation of an eHMI is important. Prior research mainly used methods such as VR pedestrian simulators with head-mounted displays [7], desktop computers to demonstrate photos or videos [23], or the Wizard of Oz technique [14]. These test environments often do not enable the dynamic development of interactions. This leads to limitations of external validity, limited use case selection, and limited possibilities to measure behavioral data. We recommend the use of high-fidelity driving simulators to investigate interactions with motorized interaction partners (cars, trucks, powered-two wheelers) or VRUs (cyclists). The chosen simulator should include a realistic mock-up; active intervention options for braking, accelerating, and steering; and the possibility to implement the eHMI. To investigate interactions with pedestrians, VR pedestrian simulators remain the most suitable test environment. However, it is important that the pedestrian simulator provides enough of a physical environment to enable dynamic interactions and possibilities to measure dynamic pedestrian behavior, e.g., by using a motion suit [6].

2.3.2. Procedure and Instruction

The procedure of the test protocol is shown in Figure 5. The instruction informs participants that the study investigates interactions between automated vehicles and manual drivers/cyclists/pedestrians. They are told that automated driving systems perform the entire dynamic driving task, at least in a specific operational design domain. Thus, the car driver can perform tasks other than driving. Furthermore, participants are informed that the experimental drive will take place on a simulated test track without right of way rules. The latter information is very important to ensure ambiguous interaction situations. The instruction at the beginning of the study does not include any information about the eHMI.

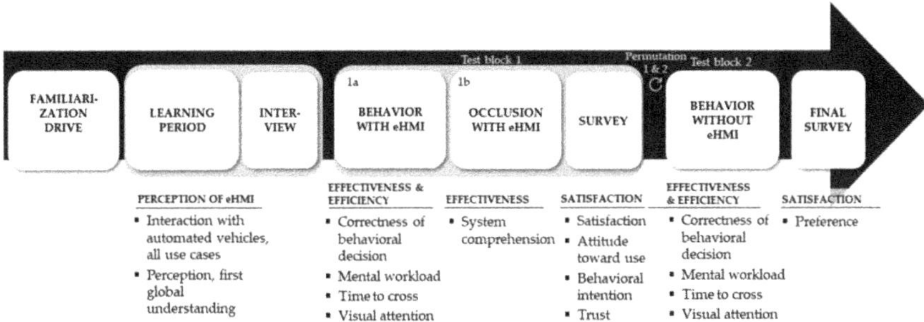

Figure 5. Procedure with measured parameters.

After a short familiarization with the respective simulator (about 5 min) without any interactions with an automated vehicle, participants go through a learning period. They already encounter all use cases in which they interact with automated vehicles that use the tested eHMI. The learning period serves as the opportunity to learn to associate the eHMI signals with the subsequently executed driving maneuver of the automated vehicle. After the leaning period, a short interview is conducted. The experimenter asks the following questions:

1. Did you notice anything while interacting with the automated vehicle?
2. Did you see the signals of the automated vehicle? Please describe the signals.
3. What was the meaning of the signals?

The participants' answers to these questions indicate the perceptibility and visibility of the eHMI signals. Furthermore, the questions serve to assess a first, global understanding of the eHMI. Independent of the answers of the respective participant, the experimenter explains at the end of the interview that the study aims to investigate signals that automated vehicles use to communicate with other road users. This information is important to achieve a common basis for all participants for the subsequent test blocks. The experimenter emphasizes to not being able to give any advice or help during the experimental drive as the objective is to investigate whether the signals are comprehensive and helpful.

Thereupon, participants either first experience the test block with the eHMI (Test Block 1) or without the eHMI (Test Block 2). The sequence of the test blocks should be counterbalanced to control for transition and learning effects. Test Block 1 consists of three parts that should be encountered in the same recommended sequence (see Figure 5). In Test Block 1a, participants encounter all use cases with the eHMI while behavioral data (driving data and visual attention) are constantly recorded. Additionally, participants verbally indicate their mental workload after each interaction (see Section 2.2). The scale to measure mental workload should be located somewhere in the simulator that is visible to the participants. In Test Block 1b, the occlusion test block serves to measure system comprehension. Therefore, participants experience all use cases once again. With the occlusion method, the simulation screen is blanked during each interaction at predefined points in time. This method was adapted from [48] and intends to achieve an open outcome of the situation. The screen should be blanked when the eHMI already signaled the subsequent intention (or communication content in general) but before the automated vehicle has already started to execute the signaled maneuver. After some seconds (e.g., 5 s), the screen shows the last scene again while the automated vehicle has been removed in the meantime. To prevent simulation sickness, it is recommended to automatically brake down the participant to a standstill while the screen is blanked. The suggested open-ended question "What will the automated vehicle do next?" can be adapted to the communication content of the tested eHMI. After the occlusion test block, participants answer a survey that includes different items to measure satisfaction with the eHMI (see Table 2, except for the preference item). In Test Block 2, participants encounter all use cases without the eHMI while behavioral data are recorded and they indicate their mental workload after each interaction. At the end of the study, participants finally evaluate their preference for future interactions with automated vehicles with or without an eHMI. After each test block, participants have the opportunity to take a break. At the end of the experiment, the experimenter thoroughly debriefs the participant.

To control for transition effects between the different use cases, it is recommended to permutate the sequence of the use cases to three different sequences. Thus, the use cases of each test block (1a, 1b, and 2) are encountered in different sequences (A, B, and C). Furthermore, a certain test block should not be experienced in the same sequence by all participants, e.g., each participant experiences Test Block 2 in sequence C. Therefore, the different sequences of use cases should be additionally counterbalanced between the three test blocks. The sequence of the use cases in the learning period can be the same for all participants. In conclusion, the outlined considerations require an equal division of the participants in six different groups. Table 3 shows an exemplary experimental design with six different experimental groups that differ according to the sequence of Test Blocks 1 and 2 and the sequence of use cases in the different test blocks.

Table 3. Example of the experimental design with different sequences of test blocks and use cases.

	Group 1 (Test Block 1 → Test Block 2)			Group 2 (Test Block 2 → Test Block 1)		
	Group 1.1	Group 1.2	Group 1.3	Group 2.1	Group 2.2	Group 2.3
1.	TB 1a in Seq. A	TB 1a in Seq. B	TB 1a in Seq. C	TB 2 in Seq. A	TB 2 in Seq. B	TB 2 in Seq. C
2.	TB 1b in Seq. B	TB 1b in Seq. C	TB 1b in Seq. A	TB 1a in Seq. B	TB 1a in Seq. C	TB 1a in Seq. A
3.	TB 2 in Seq. C	TB 2 in Seq. A	TB 2 in Seq. B	TB 1b in Seq. C	TB 1b in Seq. A	TB 1b in Seq. B

Note. TB = Test block, Seq. = Sequence of use cases in the test block.

2.3.3. Sample

To deduce reliable conclusions from the experimental data, the sample size should be sufficiently large. In reference to RESPONSE [46], at least 20 test persons should take part in the study. The target population of persons who will interact with automated vehicles in the future is very broad. Accordingly, people of all ages, nationalities, educational levels, body heights, and so forth should be eligible for studies that test the effects of eHMIs. To achieve a representative age distribution, NHTSA [43] proposed different age groups of $n = 5$ each, 18–24, 25–39, 40–54, and older than 54 years. Beyond these age groups, it is important to examine the effects of eHMIs on children's behavior and comprehension [21]. Dependent on the interaction partner under investigation, participants may need to fulfill further specific prerequisites. For example, participants in a bicycle simulator study should ride a bike on a regular basis, and participants in a driving simulator study should hold a driver's license.

3. Discussion

Due to a great variety of methodological approaches and methodological limitations, the current state of research on the usability of eHMIs does not allow to draw general conclusions. The standardization of test procedures is, thus, a fundamental prerequisite to effectively evaluate and compare different eHMI design variants. Therefore, the aim of the present article was to outline a standardized test procedure that allows for the systematic investigation of the usability of eHMIs. We have proposed a methodological framework that consists of a method to deduce relevant use cases, a definition of specific usability requirements and appropriate parameters, and a test protocol for the empirical evaluation of an eHMI.

The definition of relevant use cases provides the basis of the test procedure to ensure meaningful and comparable results. To make reliable conclusions, the usability of an eHMI must be proved in use cases previously defined as relevant. Prior studies on eHMIs have often used only one randomly selected use case. The proposed multi-stage gradual methodological approach presented in this article claims to consider all theoretically possible use cases of an eHMI. Using a variety of theoretical and practical considerations, the approach finally results in a set of use cases that are relevant to evaluate the usability of an eHMI. The intersection scenario represents the use case that has been studied most often in previous work on eHMIs [1,3–5,7,8,11–23]. To the best of our knowledge, there has only been one study that used a narrow area as a use case of an eHMI so far [49], and there has been no study that has examined a merging scenario. Thus, the approach to define use cases provides new perspectives for future research on eHMIs. Researchers can easily apply the proposed procedure to select use cases for the eHMI and automated driving system under investigation. All stages and filters before Filter F can be taken as default. Therefore, the selection process can be entered at Filter F. At this point, users can select those use cases that are relevant for the eHMI and automated driving system at hand. A potential limitation of the presented methodological approach to select use cases is that it only considers use cases in which automated vehicles interact with one non-automated road user. In principle, an eHMI that exclusively communicates information about the automated vehicle, such as its status and intentions, should always have the same usability, independent from the number of non-automated road users with which it is currently interacting. With this content of communication, it is not relevant whether only one pedestrian or three pedestrians and two cyclists need to understand the meaning of the eHMI signal and, thus, make decisions about their subsequent behavior. However, if an eHMI directly addresses its message to a specific road user, interactions with more than one non-automated road user quickly become very complex and require an extended approach to deduce use cases. For example, many previous studies have examined eHMI signals that tell pedestrians to "walk," "go ahead," or "don't walk" [8,16,22], that project green arrows [8] or crosswalks [17] on the road surface in front of the vehicle or show a green pedestrian in the windscreen [22]. If another traffic participant feels addressed by such an eHMI signal that was initially directed to another road user, the situation can become very critical. Therefore, we highly recommend not to use eHMI signals that ask a particular road user to take any specific action. As a result, the methodological approach presented in this paper provides an appropriate tool to deduce use cases for eHMIs that communicate information about the automated vehicle itself rather than communicating requests for action to other road users.

The definition of evaluation requirements constitutes an additional prerequisite to standardize the evaluation process of eHMIs. Following the ISO definition of usability [35], we derived three requirements: An eHMI must render the communication of automated vehicles with non-automated road-users effective, efficient, and satisfying. By defining specific parameters and criteria for each usability requirement, the test procedure can differentiate between eHMIs that fulfill or do not fulfill these requirements. Further work is necessary to evaluate the discriminatory power of the proposed parameters. It might be possible that some parameters can better differentiate between eHMIs that meet or do not meet the requirements than others. With increasing experience based on future empirical studies, the specific measurement methods of the parameters can be adapted and extended, e.g., the selection of appropriate items to measure the satisfaction parameters. The parameters could be supplemented by further parameters and the criteria could be adapted if necessary. For example, following the controllability guidelines of the RESPONSE code of practice [46], it is also justified to aim at a system comprehension rate of 100%. The guideline requires that 20 of 20 participants pass the predefined criterion and give the correct answer. However, it must be emphasized that the proposed requirements, parameters, and criteria focus on the usability testing of eHMIs. To prove the controllability of eHMIs, the test procedure needs to be adapted. Nevertheless, part of our test procedure already addresses controllability testing, as the criterion for the minimal distance to the automated vehicle has a pass-fail logic and does not even allow for one fail event in 20 subjects.

The test protocol provides a proper experimental design to systematically evaluate eHMI variants with user studies in a standardized way. The test protocol provides several advantages. First, results of studies that are conducted in accordance with the test protocol allow for reliable conclusions regarding whether the tested eHMI can fulfill the defined usability requirements. Second, the results of different studies that followed the test protocol allow for comparisons between the tested interface designs. Thus, the test protocol constitutes a basis to derive optimal interface specifications based on comparisons of different studies. Another major advantage of the test protocol is that it enables the measurement of an eHMI's usability without confounding factors. As there are no right of way rules and the sequence of the test blocks with and without eHMIs is counterbalanced, different behavioral decisions of the interaction partners in the different test blocks can be explained by the usability of the tested eHMI. Similarly, the occlusion method ensures that comprehension measurements are also essentially based on the comprehensibility of the eHMI. To compare two or more eHMI variants with each other, the test protocol can be adapted and extended. Test Block 1 can be repeated with an additional eHMI variant with the same group of subjects as a repeated-measures design. However, it is very important to always compare participants' behavior with an eHMI with their behavior during interactions without an eHMI in Test Block 2. Moreover, the inclusion of a further test block requires the permutation of the three test blocks and a random distribution of the participants to the resulting sequences. Alternatively, the test protocol allows for the comparison of different eHMI variants that were examined in different studies with different samples. As prerequisites, the samples must be comparable and the studies must select the same use cases.

The next step is the application of the test procedure for the usability evaluation of different eHMI design variants and automated driving systems with different specifications. With increasing experience, the method can be iteratively refined and improved. In turn, the standardized evaluation procedure will become a valuable tool for the scientific and technical community. The standardized test procedure can serve as a basis to establish best practices in the field of communication between automated vehicles and non-automated road users.

Author Contributions: Conceptualization, F.N., S.H., A.K. and A.N.; methodology, C.K. and S.S.; writing—original draft preparation, C.K.; writing—review and editing, S.H., F.N. and S.S.; supervision, A.K. and A.N.; project administration, C.K., S.S., F.N. and S.H. All authors have read and agreed to the published version of the manuscript.

Funding: This research received no external funding.

Acknowledgments: We thank Stefanie Ebert, Thomas Stemmler, and Florian Fischer for their technical support.

Conflicts of Interest: The authors declare no conflict of interest.

References

1. Hensch, A.-C.; Neumann, I.; Beggiato, M.; Halama, J.; Krems, J.F. How Should Automated Vehicles Communicate?—Effects of a Light-Based Communication Approach in a Wizard-of-Oz Study. In *Proceedings of the AHFE 2019 International Conference on Human Factors in Transportation, Washington, DC, USA, 24–28 July 2019*; Springer: Berlin, Germany; pp. 79–91. [CrossRef]
2. Merat, N.; Louw, T.; Madigan, R.; Wilbrink, M.; Schieben, A. What externally presented information do VRUs require when interacting with fully Automated Road Transport Systems in shared space? *Accid. Anal. Prev.* **2018**, *118*, 244–252. [CrossRef] [PubMed]
3. Schieben, A.; Wilbrink, M.; Kettwich, C.; Madigan, R.; Louw, T.; Merat, N. Designing the interaction of automated vehicles with other traffic participants: Design considerations based on human needs and expectations. *Cogn. Technol. Work* **2019**, *21*, 69–85. [CrossRef]
4. Mahadevan, K.; Somanath, S.; Sharlin, E. Communicating awareness and intent in autonomous vehicle-pedestrian interaction. In Proceedings of the 2018 CHI Conference on Human Factors in Computing Systems, Montréal, QC, Canada, 21–26 April 2018; pp. 1–12. [CrossRef]
5. Eisma, Y.; Van Bergen, S.; Ter Brake, S.; Hensen, M.; Tempelaar, W.; De Winter, J. External human-machine interfaces: The effect of display location on crossing intentions and eye movements. *Information* **2020**, *11*, 13. [CrossRef]
6. Kooijman, L.; Happee, R.; de Winter, J.C.F. How do eHMIs affect pedestrians' crossing behavior? A study using a head-mounted display combined with a motion suit. *Information* **2019**, *10*, 386. [CrossRef]
7. Otherson, I.; Conti-Kufner, A.S.; Dietrich, A.; Maruhn, P.; Bengler, K. Designing for Automated Vehicle and Pedestrian Communication: Perspectives on eHMIs from Older and Younger Persons. In Proceedings of the Human Factors and Ergonomics Society Europe Chapter 2018 Annual Conference, Berlin, Germany, 8–10 October 2018; pp. 135–148.
8. Ackermann, C.; Beggiato, M.; Schubert, S.; Krems, J.F. An experimental study to investigate design and assessment criteria: What is important for communication between pedestrians and automated vehicles? *Appl. Ergon.* **2019**, *75*, 272–282. [CrossRef] [PubMed]
9. Böckle, M.-P.; Brenden, A.P.; Klingegård, M.; Habibovic, A.; Bout, M. SAV2P: Exploring the Impact of an Interface for Shared Automated Vehicles on Pedestrians' Experience. In Proceedings of the 9th International Conference on Automotive User Interfaces and Interactive Vehicular Applications Adjunct, Oldenburg, Germany, 24–27 September 2017; pp. 136–140.
10. De Clercq, K.; Dietrich, A.; Núñez Velasco, J.P.; de Winter, J.; Happee, R. External Human-Machine Interfaces on Automated Vehicles: Effects on Pedestrian Crossing Decisions. *Hum. Factors* **2019**, *61*, 1353–1370. [CrossRef] [PubMed]
11. Lagstrom, T.; Malmsten Lundgren, V. AVIP-Autonomous Vehicles Interaction with Pedestrians. Master's Thesis, Chalmers University of Technology, Gothenburg, Sweden, 2015.
12. Habibovic, A.; Lundgren, V.M.; Andersson, J.; Klingegard, M.; Lagstrom, T.; Sirkka, A.; Fagerlonn, J.; Edgren, C.; Fredriksson, R.; Krupenia, S.; et al. Communicating Intent of Automated Vehicles to Pedestrians. *Front. Psychol.* **2018**, *9*, 1336. [CrossRef] [PubMed]
13. Gruenefeld, U.; Weiß, S.; Löcken, A.; Virgilio, I.; Kun, A.L.; Boll, S. VRoad: Gesture-based interaction between pedestrians and automated vehicles in virtual reality. In Proceedings of the 11th International Conference on Automotive User Interfaces and Interactive Vehicular Applications: Adjunct Proceedings, Utrecht, The Netherlands, 22–25 September 2019; pp. 399–404.
14. Rodríguez Palmeiro, A. Interaction between Pedestrians and Wizard of Oz Automated Vehicles. Master's Thesis, Technical University Delft, Delft, The Netherlands, 2017.
15. Hagenzieker, M.P.; Van der Kint, S.; Vissers, L.; Van Schagen, I.N.G.; De Bruin, J.; Van Gent, P.; Commandeur, J.J. Interactions between cyclists and automated vehicles: Results of a photo experiment. *J. Transp. Saf. Secur.* **2020**, *12*, 94–115. [CrossRef]

16. Song, Y.E.; Lehsing, C.; Fuest, T.; Bengler, K. External HMIs and their effect on the interaction between pedestrians and automated vehicles. In *Intelligent Human Systems Integration*; Karwowski, W., Ahram, T., Eds.; Springer: Cham, Switzerland, 2018; Volume 722, pp. 13–18. [CrossRef]

17. Dietrich, A.; Willrodt, J.-H.; Wagner, K.; Bengler, K. Projection-Based External Human Machine Interfaces-Enabling Interaction between Automated Vehicles and Pedestrians. In Proceedings of the Driving Simulation Conference 2018 Europe VR, Antibes, France, 5–7 September 2018.

18. Clamann, M.; Aubert, M.; Cummings, M.L. Evaluation of vehicle-to-pedestrian communication displays for autonomous vehicles. In Proceedings of the 96th Annual Transportation Research Board Meeting, Washintgon, DC, USA, 8–12 January 2017.

19. Deb, S.; Strawderman, L.J.; Carruth, D.W. Investigating pedestrian suggestions for external features on fully autonomous vehicles: A virtual reality experiment. *Transp. Res. Part F Traffic Psychol. Behav.* **2018**, *59*, 135–149. [CrossRef]

20. Li, Y.; Dikmen, M.; Hussein, T.G.; Wang, Y.; Burns, C. To cross or not to cross: Urgency-based external warning displays on autonomous vehicles to improve pedestrian crossing safety. In Proceedings of the 10th International Conference on Automotive User Interfaces and Interactive Vehicular Applications, Toronto, ON, Canada, 23–25 September 2018; pp. 188–197. [CrossRef]

21. Deb, S.; Carruth, D.W.; Fuad, M.; Stanley, L.M.; Frey, D. Comparison of Child and Adult Pedestrian Perspectives of External Features on Autonomous Vehicles Using Virtual Reality Experiment. In *AHFE 2019: Advances in Human Factors of Transportation*; Stanton, N., Ed.; Springer: Cham, Switzerland, 2019; Volume 964, pp. 145–156. [CrossRef]

22. Fridman, L.; Mehler, B.; Xia, L.; Yang, Y.; Facusse, L.Y.; Reimer, B. To Walk or Not to Walk: Crowdsourced Assessment of External Vehicle-to-Pedestrian Displays. 2017. Available online: https://arxiv.org/abs/1707.02698 (accessed on 24 March 2020).

23. Yang, S. Driver Behavior Impact on Pedestrians' Crossing Experience in the Conditionally Autonomous Driving Context. Student's Thesis, School of Computer Science and Communication, Stockholm, Sweden, 2017.

24. Löcken, A.; Golling, C.; Riener, A. How Should Automated Vehicles Interact with Pedestrians? A Comparative Analysis of Interaction Concepts in Virtual Reality. In Proceedings of the 11th International Conference on Automotive User Interfaces and Interactive Vehicular Application, Utrecht, The Netherlands, 22–25 September 2019; pp. 262–274. [CrossRef]

25. Petzoldt, T.; Schleinitz, K.; Banse, R. Potential safety effects of a frontal brake light for motor vehicles. *IET Intell. Transp. Syst.* **2018**, *12*, 449–453. [CrossRef]

26. Naujoks, F.; Hergeth, S.; Wiedemann, K.; Schömig, N.; Keinath, A. Use cases for assessing, testing, and validating the human machine interface of automated driving systems. In Proceedings of the Human Factors and Ergonomics Society Annual Meeting, Philadelphia, PA, USA, 1–5 October 2018; pp. 1873–1877. [CrossRef]

27. Gold, C.; Naujoks, F.; Radlmayr, J.; Bellem, H.; Jarosch, O. Testing scenarios for human factors research in level 3 automated vehicles. In *AHFE 2017: Advances in Human Aspects of Transportation*; Stanton, N., Ed.; Springer: Cham, Switzerland, 2017; Volume 597, pp. 551–559. [CrossRef]

28. McCall, R.; McGee, F.; Meschtscherjakov, A.; Louveton, N.; Engel, T. Towards a taxonomy of autonomous vehicle handover situations. In Proceedings of the 8th International Conference on Automotive User Interfaces and Interactive Vehicular Applications, Ann Arbor, MI, USA, 24–26 October 2016; pp. 193–200. [CrossRef]

29. Lu, Z.; Happee, R.; Cabrall, C.D.; Kyriakidis, M.; de Winter, J.C. Human factors of transitions in automated driving: A general framework and literature survey. *Transp. Res. Part F Traffic Psychol. Behav.* **2016**, *43*, 183–198. [CrossRef]

30. Fuest, T.; Sorokin, L.; Bellem, H.; Bengler, K. Taxonomy of traffic situations for the interaction between automated vehicles and human road users. In *AHFE 2017: Advances in Human Aspects of Transportation*; Stanton, N., Ed.; Springer: Cham, Switzerland, 2017; Volume 597, pp. 708–719. [CrossRef]

31. Naujoks, F.; Hergeth, S.; Wiedemann, K.; Schömig, N.; Forster, Y.; Keinath, A. Test procedure for evaluating the human–machine interface of vehicles with automated driving systems. *Traffic Inj. Prev.* **2019**, *20*, 146–151. [CrossRef] [PubMed]

32. Alliance of Automobile Manufacturers. *Statement of Principles, Criteria and Verification Procedures on Driver Interactions with Advanced In-Vehicle Information and Communication Systems Including*; Alliance of Automobile Manufacturers: Washington, DC, USA, 2006.

33. National Highway Traffic Safety Administration. *Visual-Manual NHTSA Driver Distraction Guidelines for In-Vehicle Electronic Devices*; Department of Transportation: Washington, DC, USA, 2014.

34. Rouchitsas, A.; Alm, H. External Human-Machine Interfaces for Autonomous Vehicle-to-Pedestrian Communication: A Review of Empirical Work. *Front. Psychol.* **2019**, *10*, 2757. [CrossRef] [PubMed]

35. International Organization for Standardization. *Ergonomics of Human-System Interaction—Part 11: Usability: Definitions and Concepts*; International Organization for Standardization: Geneva, Switzerland, 2018; ISO 9241-11.

36. Markkula, G.; Madigan, R.; Nathanael, D.; Portouli, E.; Lee, Y.M.; Dietrich, A.; Billington, J.; Schieben, A.; Merat, N. Defining Interactions: A Conceptual Framework for Understanding Interactive Behaviour in Human and Automated Road Traffic. 2020. Available online: https://doi.org/10.31234/osf.io/8w9z4 (accessed on 24 February 2020).

37. Amundsen, F.H.; Hydén, C. *Proceedings of the First Workshop on Traffic Conflicts*; Institute of Transport Economics: Oslo, Norway, 1977.

38. Forschungsgesellschaft für Straßen- und Verkehrswesen. AG 2 Straßenentwurf. 2018. Available online: https://www.fgsv.de/gremien/strassenentwurf.html (accessed on 24 February 2020).

39. Kraft, A.-K.; Maag, C.; Baumann, M. How to support cooperative driving by HMI design? *Transp. Res. Interdiscip. Perspect.* **2019**, *3*. [CrossRef]

40. SAE International. Taxonomy and Definitions for Terms Related to Driving Automation Systems for on-Road Motor Vehicles (No. J3016). 2018. Available online: https://saemobilus.sae.org/content/j3016_201806 (accessed on 24 February 2020).

41. Federal Highway Administration. *Manual on Uniform Traffic Control Devices*; Federal Highway Administration: Washington, DC, USA, 2003.

42. Panis, L.I.; De Geus, B.; Vandenbulcke, G.; Willems, H.; Degraeuwe, B.; Bleux, N.; Mishra, V.; Thomas, I.; Meeusen, R. Exposure to particulate matter in traffic: A comparison of cyclists and car passengers. *Atmos. Environ.* **2010**, *44*, 2263–2270. [CrossRef]

43. National Highway Traffic Safety Administration. *Federal Automated Vehicles Policy 2.0*; Department of Transportation: Washington, DC, USA, 2017.

44. Fuest, T.; Michalowski, L.; Träris, L.; Bellem, H.; Bengler, K. Using the Driving Behavior of an Automated Vehicle to Communicate Intentions-A Wizard of Oz Study. In Proceedings of the 2018 21st International Conference on Intelligent Transportation Systems (ITSC), Maui, HI, USA, 4–7 November 2018; pp. 3596–3601. [CrossRef]

45. Ackermann, C.; Beggiato, M.; Bluhm, L.-F.; Löw, A.; Krems, J.F. Deceleration parameters and their applicability as informal communication signal between pedestrians and automated vehicles. *Transp. Res. Part F Traffic Psychol. Behav.* **2019**, *62*, 757–768. [CrossRef]

46. Response Consortium. Code of Practice for the Design and Evaluation of ADAS; A Prevent Project; Response: 2006; Volume 3. Available online: https://www.acea.be/uploads/publications/20090831_Code_of_Practice_ADAS.pdf (accessed on 24 March 2020).

47. Venkatesh, V.; Morris, M.G.; Davis, G.B.; Davis, F.D. User acceptance of information technology: Toward a unified view. *MIS Q.* **2003**, *27*, 425–478. [CrossRef]

48. Kaß, C.; Schmidt, G.J.; Kunde, W. Towards an assistance strategy that reduces unnecessary collision alarms: An examination of the driver's perceived need for assistance. *J. Exp. Psychol. Appl.* **2018**, *25*, 291–302. [CrossRef] [PubMed]

49. Rettenmaier, M.; Pietsch, M.; Schmidtler, J.; Bengler, K. Passing through the Bottleneck-The Potential of External Human-Machine Interfaces. In Proceedings of the 2019 IEEE Intelligent Vehicles Symposium (IV), Paris, France, 9–12 June 2019; pp. 1687–1692. [CrossRef]

 *information*

Article

How Much Space Is Required? Effect of Distance, Content, and Color on External Human–Machine Interface Size

Michael Rettenmaier *, Jonas Schulze and Klaus Bengler

Chair of Ergonomics, Technical University of Munich, 85748 Garching, Germany; schulze.jonas@mytum.de (J.S.); bengler@tum.de (K.B.)
* Correspondence: michael.rettenmaier@tum.de

Received: 3 May 2020; Accepted: 1 July 2020; Published: 3 July 2020

Abstract: The communication of an automated vehicle (AV) with human road users can be realized by means of an external human–machine interface (eHMI), such as displays mounted on the AV's surface. For this purpose, the amount of time needed for a human interaction partner to perceive the AV's message and to act accordingly has to be taken into account. Any message displayed by an AV must satisfy minimum size requirements based on the dynamics of the road traffic and the time required by the human. This paper examines the size requirements of displayed text or symbols for ensuring the legibility of a message. Based on the limitations of available package space in current vehicle models and the ergonomic requirements of the interface design, an eHMI prototype was developed. A study involving 30 participants varied the content type (text and symbols) and content color (white, red, green) in a repeated measures design. We investigated the influence of content type on content size to ensure legibility from a constant distance. We also analyzed the influence of content type and content color on the human detection range. The results show that, at a fixed distance, text has to be larger than symbols in order to maintain legibility. Moreover, symbols can be discerned from a greater distance than text. Color had no content overlapping effect on the human detection range. In order to ensure the maximum possible detection range among human road users, an AV should display symbols rather than text. Additionally, the symbols could be color-coded for better message comprehension without affecting the human detection range.

Keywords: automated driving; external human–machine interface; interface size; legibility

1. Introduction

The process of introducing automated vehicles (AVs) into road traffic is progressing. In urban areas in particular, a gradual change is taking place towards mixed traffic, including AVs, human drivers, cyclists, and pedestrians. From automation level 2 and higher, the system sustains lateral and longitudinal vehicle motion control [1], which could directly impact the nature of the interactions between the AV and road users in the surroundings. One approach for enabling communication of AVs with their environments is to use external human–machine interfaces (eHMIs). These are displays mounted on the surface of the vehicle [2,3], light strips [4–6], and projections on the road [7,8]. These devices enable AVs to indicate, for instance, their status, perception, or intention [9] in relevant scenarios, such as at intersections, in parking lots, in narrow spaces, or in merging traffic [10,11]. Current research is almost exclusively devoted to the question of what content these interfaces should display in order for them to be comprehensible to pedestrians [12] or human drivers [2]. Based on a comprehensible eHMI design, the interaction is comfortable, efficient, and safe if the human interaction partner has enough time to perceive and process the eHMI content and act accordingly.

The dynamics of road traffic and the time required by the receiver result in a certain lead time within which the AV has to communicate its message. In turn, a minimum content size is required in which the AV has to display its message.

For the purpose of dimensioning the eHMI, this paper makes reference to the road bottleneck scenario from two previous studies [2,7], with obstacles on both sides of the road due to double-parked vehicles. In this scenario, an AV and a simultaneously oncoming human driver negotiate the right of way within a 30 km/h speed limit zone. The AV displays its message to the human driver at a distance of 100 m. Rettenmaier, Pietsch, and Bengler [7] recommend that in such a bottleneck scenario an AV should communicate via a display mounted on the front of the vehicle, in order for the interaction to be efficient and safe. Front-mounted displays are particularly suitable for communication purposes in straight-approach scenarios [13]. Owing to the high dynamics and relative speeds of the AV and the human driver when approaching the road bottleneck, the resulting required eHMI size exceeds that which would be needed for interactions in a tighter space. Thus, the determined size is also suitable for communicating with pedestrians in road crossing scenarios in which the AV's communication commences at a shorter distance between the AV and the pedestrian, for instance, 45 m [4] or 50 m [12]. Despite all its positive potentials, one disadvantage of communicating via displays is that the content size must be large to be viewed at a distance [14]. However, there was no research found that deals with the question of how large text or symbols need to be with respect to content and color in order for them to be legible from a particular distance. As there are as yet no standards governing the design of eHMIs, this paper investigates the size that displayed text or symbols must have, in order for them to render a message legibly in a bottleneck scenario.

2. Objectives

The present study aims to determine the content size required to render distinct communication at a certain distance for different content types. An additional aim is to examine the influence of content color and content type on the human detection range, which we defined as the distance from which a certain content size is legible. For this purpose, we developed an eHMI prototype (Section 3) including a package space analysis (Section 3.1), ergonomic requirements (Section 3.2), the selection of hardware and software (Section 3.3), and the presentation of the final prototype (Section 3.4). We conducted a study involving 30 participants (Section 4) to analyze the effects of distance, content type, and content color on the required content size, and we set up the following research questions (RQs):

RQ1: Is there any difference in the required content size for it to be legible at a certain distance depending on the content type?
RQ2: Is there any difference in the human detection range depending on the content type?
RQ3: Is there any difference in the human detection range depending on the content color?

3. Development of External HMI Prototype

3.1. Package Space Analysis of Existing Vehicle Models

An AV communicates with an oncoming human driver via its external display. For this reason, the vehicle front is the only surface suitable for displaying information. This surface can be divided into the bumper, radiator grille, headlights, hood, windshield, and rear of the side mirrors. The area of the side mirrors is small and incoherent, while the projection area of the hood is small in the vertical plane. Moreover, the AV's passenger must be able to use the windshield for monitoring the driving scene, while the function of the headlights is to illuminate the road ahead. For these reasons, we considered the bumper and the radiator grille as suitable areas for implementing the eHMI, as the radiator grille is no longer required for engine cooling in an electric vehicle. It is also a suitable area for displaying messages from the AV in straight approach scenarios [13].

We selected three car models to represent each of the six vehicle categories of the Commission for European Communities (mini cars, small cars, large cars, executive cars, luxury cars, and sport utility cars) [15]. The selection was based on the new registration data published by the German Federal Motor Transport Authority for the month of June 2019 [16]. The vehicle's front dimensions were determined by digital measurement of the official dimensions given in a technical drawing and subdividing this area into individual sections for the radiator grille and the bumper (Figure 1). The scale of the technical drawing was recorded, while the pixel size and, in turn, the size of the defined sections were calculated using a digital pixel meter. The potential eHMI size dimensions were determined as the minimum height (*H*) of the radiator grille and the bumper together (Mercedes C-Class: 459 mm) and the minimum width (*W*) of the radiator grille (VW Up: 772 mm) of all car models.

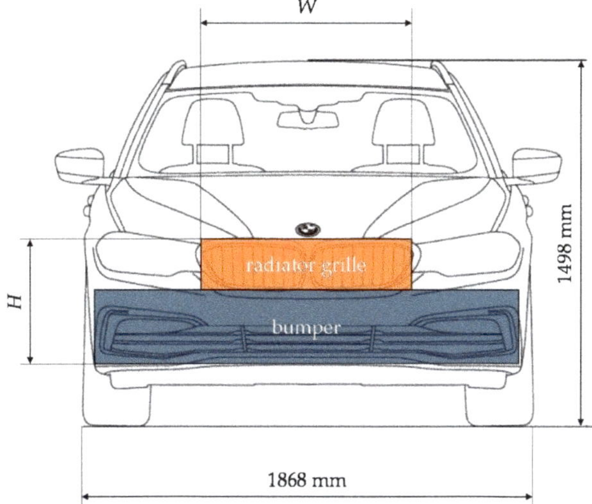

Figure 1. Dimensioning of the vehicle front using the technical drawing [17] of a BMW 5 Touring model as an example. We divided the front into separate radiator grille and bumper sections.

3.2. Ergonomic Requirements

Due to the complexity of the driving task during manual driving, it is necessary that all pertinent information is easily and comfortably legible for drivers. Similarly, the eHMI must be legible at all times of day and night. During the day, the required luminance of the display varies between 1000 cd/m^2 [18] and 5000 cd/m^2 [19] for outdoor use. At night, the eHMI must not be so bright as to dazzle nearby road users. Therefore, the display luminance must be adjustable so as not to impair the eye's adaptability to changes in light levels [20]. Another requirement is that the eHMI should display bright text and symbols on a dark background and not the other way around since this display mode is suitable for day and night use [21]. The contrast ratio of the display between the text/symbol and the background should be 5:1 at high brightness and at least 3:1 at common brightness levels [18].

The symbols on the display should have a minimum visual angle of 20 min of arc (MOA). The minimum visual angle of text written in Latin letters must be 16 MOA and 20–22 MOA for comfortable reading. Moreover, the ratio between letter height and letter width should be 0.7:1–0.9:1. The line width of sans-serif fonts should be 10–17% of the letter height, and there should be a space of one line width between letters [20].

The letter or symbol height requirements specify the minimum display height. The number of letters in a word limit the minimum display width.

3.3. Hardware and Software

The prototype consists of 12 outdoor light-emitting diode (LED) modules made by Coreman Technology Co. [22]. Each red-green-blue (RGB) LED matrix measures 256 × 128 mm with 62 × 32 pixels and a pixel distance of 4 mm. The minimum visual angle of 20 MOA at a distance of 100 m, as considered in the bottleneck scenario [2,7], has a matrix height $H = 582$ mm. The available package space dimensions are $W = 772$ mm and $H = 459$ mm. The 4×3 matrix layout has a size of $W = 768$ mm and $H = 512$ mm, with a resolution of 192×128 pixels. This represents a good compromise between the theoretically required space and the available space. The luminance of each module is higher than 6000 cd/m^2, resulting in a maximum illuminance of 2358 cd when the whole matrix illuminates in full brightness in white. This exceeds the limit value of 1200 cd [23] prescribed for road traffic. Since fewer than 50% of the pixels illuminate for displaying symbols and letters, the illuminance is lower than the legally required threshold.

The working temperature of the module is between −30 °C and +55 °C. The LED matrix is controlled by a Raspberry Pi 4 Computer Model B with 2 GB of memory and a quad-core 64-bit processor with a frequency of 1.5 Hz [24]. The prototype uses the official operating system Raspbian, based on Debian GNU/Linux. The LED matrix is controlled by a laptop via a remote desktop connection. The LED matrix is controlled by an open source C++ library [25]. It is, therefore, able to display pictures, texts, and animations [26].

3.4. Final eHMI Prototype

Figure 2 shows the final eHMI prototype. The LED modules are screw-fitted to a frame made from aluminum sheets. The prototype satisfies the visual angle requirements of 20 MOA at a distance of 88 m between display and participant pursuant to DIN EN ISO 9241-303 [20] with a display size of 768×512 mm. This eHMI display distance is less than the 100 m used in the previous studies [2,7], on which the present investigation is based, but it would provide the human driver in the bottleneck scenario sufficient time to interact comfortably with the AV [27].

| Text in green | E in white | Arrow in red |

Figure 2. The external human–machine interface (eHMI) prototype developed and evaluated in the present investigation. The content colors do not match the real colors due to the display angle and camera distortion.

4. Evaluation of External HMI Prototype

4.1. Sample

Thirty participants took part in the experiment. As no data were discarded, there were 30 valid data sets in the study. The age of the sample was $M = 31.07$ years ($SD = 12.54$ years). The age span ranged from 18 years to 69 years. Nineteen participants were male and 11 were female. Eighteen participants had a visual impairment, which was corrected in 17 cases in the course of the experiment and not corrected in one case. Additionally, there was one participant with red-green deficiency. We refrained from excluding these two data sets from the analysis, as persons with visual impairments also participate in real road traffic. The eye test [28,29] resulted in a visual acuity of $M = 1.47$ ($SD = 0.37$). The visual acuity ranged from 0.8 to 2.0. The participants were recruited at the Technical University of Munich and did not receive an expense allowance.

4.2. Display Content

Figure 3 shows the three different content types displayed by the eHMI prototype during experiment 1 and experiment 2 (Figure 4). The text fulfills the ergonomic requirements (Section 3.2). In experiment 1, we chose to display four letters, since this number was easily readable from a distance of 88 m in a pre-test. In experiment 2, the eHMI displayed five letters (E, P, C, F, D). In both experiments, the eHMI displayed cryptic chunks of letters, so that it was hardly possible to guess the sequence of letters. To avoid the effect of varying legibility for different letters, the display showed the same letters for each participant, but in a randomized order. The letters were derived from one row of the Snellen chart. In addition to text, the prototype also displayed two types of symbols. The arrow and the "E" from the E chart were visualized in four degrees of rotation (0°, 90°, 180°, 270°) such that the limbs of the E and the arrow tip were pointing up, down, to the left, or to the right. The content size is defined throughout this article as the height of the text or the height of the arrow and the E in the orientation given in Figure 3. Even if the arrow is rotated by 90°, its size is the distance from the end of the arrow to its tip.

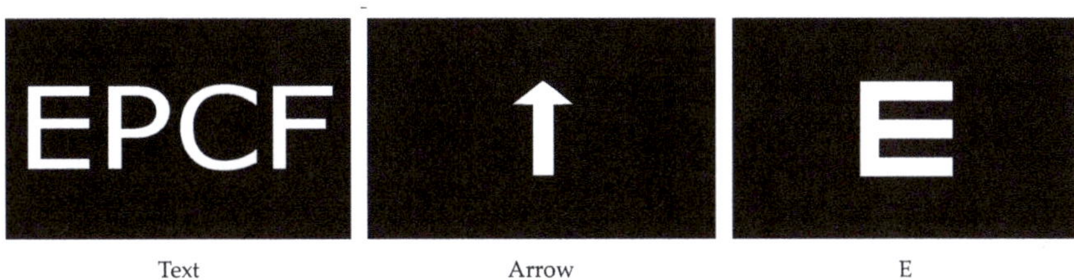

Text　　　　　　　　　　Arrow　　　　　　　　　　E

Figure 3. The three different content types displayed by the eHMI in the present study.

4.3. Experimental Design

It was necessary to conduct two experiments (Figure 4) in order to obtain answers to the research questions. In experiment 1, the participants were at a constant distance of 88 m to the prototype. This distance corresponds to the recommended visual angle of 20 MOA for a prototype height of 512 mm [20]. Following a pre-test, the symbols were scaled to six sizes (ranging from 80 mm to 230 mm), while the text was scaled to five different sizes (from 80 mm to 200 mm) (Table 1) for determining the size required for it to be legible at a distance of 88 m. In experiment 1, the prototype displayed the message in white (R = 255, G = 255, B = 255), since this represents the highest contrast to the LED matrix.

Table 1. Content sizes used in experiment 1. The distance from which the respective size has a visual angle of 20 min of arc (MOA) is presented according to DIN EN ISO 9241-303 [20].

Size (mm)	80	110	140	170	200	230
Distance (m)	13.75	18.91	24.06	29.22	34.38	39.53

Experiment 2 analyzed the effect of content type and content color on the human detection range. The size of symbols and texts was set to 164 mm, which made it possible to display five letters on the prototype. Additionally, texts and symbols were displayed in the colors white (R = 255, G = 255, B = 255), red (R = 255, G = 0, B = 0), and green (R = 0, G = 255, B = 0). We decided to use red and green in addition to white, as they are already familiar in the context of traffic as an indication of either yielding or insisting on the right of way. In experiment 2, the participants approached the eHMI prototype from a distance of 150 m. The participants stopped at a distance X from the eHMI as soon as the content type became legible and thus their detection range was attained.

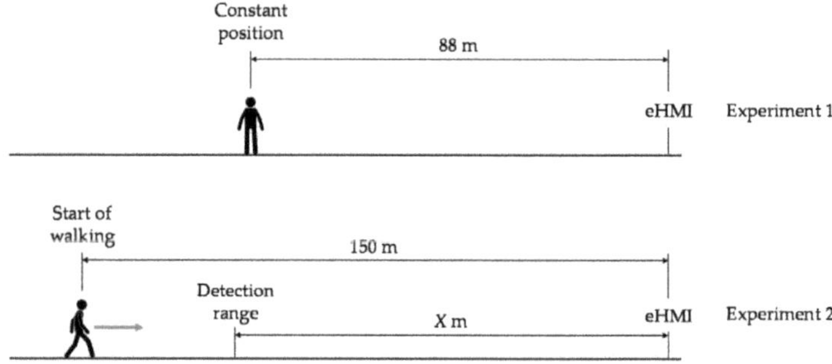

Figure 4. Experiment 1 evaluated the required content size for it to be legible from a distance of 88 m. Experiment 2 analyzed the human detection range (X) depending on content type and content color.

The participants performed experiment 1 and experiment 2 in a permuted order (Figure 5). In experiment 1, the participants read the text (in five different sizes), the arrow (in six different sizes), and the E (in six different sizes) in a permuted order. The text segment of the experiment also displayed two distracting text blocks, in which letters appeared twice, after the first text and after the third text, such that the participants could not assume that the respective letters only appeared once within a text. The data from these two distractor texts were not considered in the evaluation.

In experiment 2, the participants approached the prototype displaying the text, arrow, and E three times each. In each of the three parts, the message was displayed once in white, red, and green.

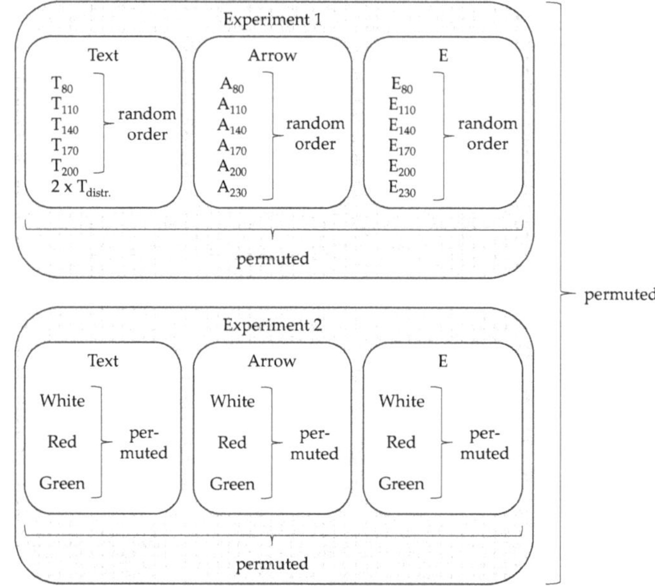

Figure 5. Experimental design dividing the study into two experiments. Both the experiments and the different content types within each experiment were presented in a permuted order.

4.4. Procedure

Once they had been duly informed about the experiment, the participants gave their written consent to take part in the study. They then filled in a demographic questionnaire, which included questions on age and gender. The participants were also asked to indicate whether they had any visual impairment or color vision deficiency. Afterwards, they underwent eye testing using the software FrACT 3.10.2 [29], which displayed the Landolt-C on a computer monitor. The participants had to discern in which of the eight possible positions the Landolt-C opening appeared. The distance between monitor and participant and the number of trials can be configured in the software. The participants then received the instructions for the study, after which experiment 1 and experiment 2 were conducted in a permuted order. The participants were not subject to time limits when identifying the displayed items. Prior to the experiments, the illuminance was measured directly at the eHMI prototype because of the possibility of ambient illumination affecting contrast requirements [30]. The average illuminance was $M = 2812$ lx ($SD = 1092$ lx), with a range of 132 lx to 5483 lx. The total duration of the experiment was about 45 min.

4.5. Dependent Variables

The correctness of the text and symbol identification was evaluated in both experiments. The text was correctly identified and was considered legible if the participant read the sequence of letters in the right order. The arrow and the E were considered legible if the respective symbol and its orientation were correctly identified. In experiment 2, the participants additionally had to state the content color for correct identification. In experiment 1, the content size required for legibility at a distance of 88 m was calculated from the correctly identified content data, while the human detection range from which content of a certain size became legible was investigated in experiment 2.

Experiment 1 collected subjective data regarding the legibility of content, the concentration required for identifying the content, and the participants' confidence in having correctly identified the content, each on a 5-point Likert scale (Table 2). Experiment 2 collected subjective data regarding the participants' confidence in having identified the eHMI content correctly.

Table 2. The three items used to collect subjective data.

	Item	5-Point Likert Scale
Legibility:	Please rate the legibility of the displayed text (symbol).	*Very poor* to *very good*
Concentration:	Please rate the degree of concentration required to read (identify) the text (symbol).	*Very high* to *very low*
Confidence:	How sure are you that you have read (identified) the text (symbol) correctly?	*Very unsure* to *very sure*

4.6. Statistical Analysis

Data preparation was performed with Excel and the statistical analysis was conducted using the software JASP [31]. In experiment 1, since the data were not normally distributed, we applied a Friedman test to analyze the content size required for legibility from a constant distance of 88 m. Post hoc comparisons were conducted using Wilcoxon tests and a Bonferroni correction was applied. The effect size of the Friedman test was classified using Kendall's W (small effect: $W = 0.1$; medium effect: $W = 0.3$; large effect: $W = 0.5$). In the case of the Wilcoxon tests, we classified the effect sizes with the Pearson moment correlation r (small effect: $r = 0.1$; medium effect: $r = 0.3$; large effect: $r = 0.5$) [32].

In experiment 2, we chose to conduct three ANOVAs to evaluate the effect of both content type and content color. The assumption of sphericity (Mauchly's test: $p > 0.05$) was always fulfilled. In both cases, we performed a Bonferroni correction. We refrained from analyzing content type and content color within a single ANOVA, as there were values missing for the text, which would have resulted in the exclusion of nine participants in the analysis as a whole. Our approach allowed the data of these participants to be at least partially incorporated into the statistical analysis. We rated the effect sizes by applying η_p^2 (small effect: $\eta_p^2 = 0.01$; medium effect: $\eta_p^2 = 0.06$; large effect: $\eta_p^2 = 0.14$) for the ANOVA and Cohen's benchmark *d* (small effect: $d = 0.2$; medium effect: $d = 0.5$; large effect: $d = 0.8$) for the post-hoc comparisons [32].

5. Results

5.1. Experiment 1

5.1.1. Effect of Content Size

Table 3 shows the absolute number and the percentage of correct identifications according to content size. All participants usually recognized the three largest sizes, regardless of the content. The only exception was one participant who could not identify the orientation of the arrow at a size of 170 mm. At a size of 110 mm and 80 mm, the number of correct identifications of the text was considerably lower than the number of correct identifications of the arrow and the E.

Table 3. Correct identification in absolute and relative terms ($n = 30$).

	Size (mm)					
	230	200	170	140	110	80
Text	-	30 (100%)	30 (100%)	28 (93%)	20 (67%)	4 (13%)
Arrow	30 (100%)	30 (100%)	29 (97%)	29 (97%)	27 (90%)	16 (53%)
E	30 (100%)	30 (100%)	30 (100%)	29 (97%)	26 (87%)	15 (50%)

Figure 6 shows the content size from which the participants could correctly identify the contents. The text was identified correctly at a size of $Mdn = 110$ mm. The arrow ($Mdn = 95$ mm) and the E ($Mdn = 95$ mm) could be identified at a smaller size. The Friedman test reveals a significant effect of content type on the required content size ($X^2 = 14.59$, $p < 0.001$, Kendall's $W = 0.549$). The post-hoc comparisons using Wilcoxon tests (Table 4) show significant differences between the text and the arrow and between the text and the E, each with a large effect.

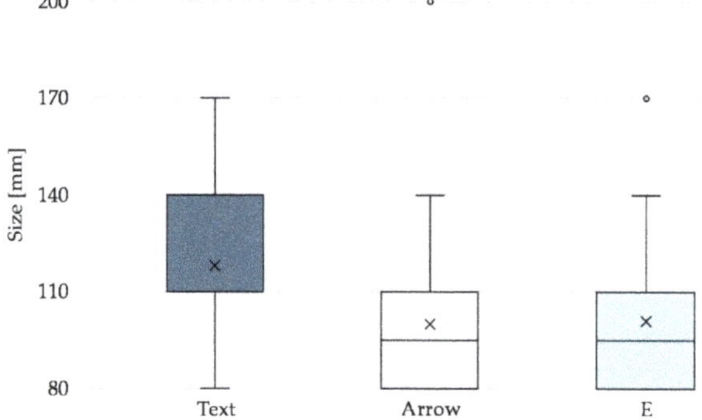

Figure 6. Content size from which the text and symbols were correctly identified ($n = 30$).

Table 4. Post-hoc comparisons using Wilcoxon tests.

		W	*p*	*r*
Text	Arrow	29.00	0.006	0.695
Text	E	27.00	0.002	0.743
Arrow	E	43.00	0.884	0.055

Note: A Bonferroni correction was applied, and the corrected level of significance was set to $\alpha = 0.0167$.

5.1.2. Subjective Results

Table 5 contains the participants' subjective ratings of legibility, concentration, and confidence on a 5-point Likert scale. In the case of legibility and concentration, the two biggest content sizes include high ratings of *Mdn* = 4 and *Mdn* = 5. The two smallest content sizes produce low ratings of *Mdn* = 2 and *Mdn* = 1. As for their confidence in identifying the display content, the participants gave high ratings for the biggest four content sizes and considerably lower ones for the two smallest content sizes.

Table 5. Subjective participant ratings on a 5-point Likert scale with regard to legibility, concentration, and confidence ($n = 30$).

	Size (mm)					
	230	200	170	140	110	80
Legibility:	Please rate the legibility of the displayed text (symbol). (1 = very poor, 5 = very good)					
Text	-	5	4	3.5	2	1
Arrow	4	4	3.5	3	2	1
E	5	5	4	3	2	1
Concentration:	Please rate the degree of concentration required to read (identify) the text (symbol). (1 = very high, 5 = very low)					
Text	-	4	4	3	2	1
Arrow	4	4	3	3	2	1
E	5	4	4	3	2	1
Confidence:	How sure are you that you have read (identified) the text (symbol) correctly? (1 = very unsure, 5 = very sure)					
Text	-	5	5	4	2	1
Arrow	5	5	4	4	3	1
E	5	5	5	4	2.5	1

5.2. Experiment 2

Effect of Content Type and Content Color

Figure 7 shows the detection range from which the participants were able to identify the eHMI content for each content color. The text implies the smallest distance to the prototype for all three colors (Table 6). Table 7 contains the three ANOVAs, one for each color, to evaluate the effect of the content type. For all colors, there were significant effects of the content type on the detection range with large effect sizes. Post-hoc comparisons for the color white (Table 8) reveal significant differences with a medium effect between the text and the arrow and a large effect between the text and the E. The analysis of the red content indicates a significant difference between the text and the E, with a medium effect. The post-hoc comparison of the green content shows significant differences between all three content types with a large effect between the text and the E and medium effect sizes between the text and the arrow as well as between the arrow and the E.

We analyzed the influence of content color by conducting three ANOVAs (Table 9). For text, there was a significant difference with regard to the color, with a large effect. Post-hoc comparisons (Table 10) reveal a significant difference in the distance between the colors white and red and a significant difference between the colors red and green, each with a medium effect.

Table 6. Descriptive data giving the distance from which the display content could be identified divided by content type and content color.

	White	Red	Green
Text, *M* (*SD*)	110.04 m (20.84 m), $n = 27$	116.85 m (19.96 m), $n = 26$	108.71 m (18.93 m), $n = 24$
Arrow, *M* (*SD*)	125.17 m (24.71 m), $n = 30$	125.37 m (23.14 m), $n = 30$	121.03 m (26.04 m), $n = 30$
E, *M* (*SD*)	123.70 m (20.61 m), $n = 30$	128.90 m (19.15 m), $n = 30$	124.30 m (20.81 m), $n = 30$

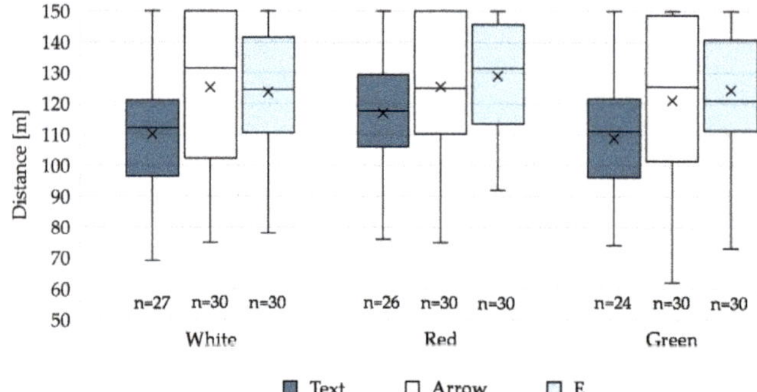

Figure 7. Distance from which the display content could be identified correctly divided by content type and content color.

Table 7. Statistics for the ANOVAs conducted to evaluate the effect of content type with respect to content color.

	F	df	p	η_p^2
White ($n = 27$)	10.704	2, 52	<0.001	0.292
Red ($n = 26$)	5.713	2, 50	0.006	0.186
Green ($n = 24$)	19.267	2, 46	<0.001	0.456

Note: A Bonferroni correction was applied, and the corrected level of significance was set to $\alpha = 0.0167$.

Table 8. Post-hoc comparisons analyzing the content type.

		p_{bonf}	Cohen's d
		White	
Text	Arrow	0.003	0.707
Text	E	<0.001	0.854
Arrow	E	1.000	0.034
		Red	
Text	Arrow	0.086	0.456
Text	E	0.007	0.666
Arrow	E	0.841	0.216
		Green	
Text	Arrow	0.009	0.677
Text	E	<0.001	1.306
Arrow	E	0.035	0.558

Table 9. Statistics for the ANOVAs conducted to evaluate the effect of content color with respect to content type.

	F	df	p	η_p^2
Text ($n = 21$)	5.859	2, 40	0.006	0.227
Arrow ($n = 30$)	1.145	2, 58	0.325	0.038
E ($n = 30$)	1.943	2, 58	0.152	0.063

Note: A Bonferroni correction was applied, and the corrected level of significance was set to $\alpha = 0.0167$.

Table 10. Post hoc comparisons analyzing the text color.

		p_{bonf}	Cohen's d
White	Red	0.046	0.579
White	Green	1.000	0.097
Red	Green	0.006	0.770

6. Discussion

6.1. Effect of Content Type

An increase in content size increases the legibility of the display content regardless of the content type, reflected by the higher numbers of correct identifications from a distance of 88 m, as well as by the participants' higher legibility ratings. Moreover, the concentration required for identifying the content decreases and the confidence in identifying it increases. An increase in text or symbol size leads to the display content taking up more space in the total area of the prototype. Since the brightness of each LED was the same within a color scheme in all trials, the use of larger texts or symbols results in a greater number of illuminated LEDs and thus higher luminance of the message. Additionally to the larger visual angle with large content sizes, with an increase in luminance, there is also a rise in the participants' visual acuity [33], showing that larger content sizes result in increasing legibility.

Text and symbols should be at least 140 mm high to be legible from a distance of 88 m. The participants rated their confidence in identifying the display content as sure (Mdn = 4) for all content types. Moreover, the percentage of correct identifications drops considerably with smaller content sizes. For safety-critical interactions with AVs at a road bottleneck, the oncoming human driver must always be able to identify the message with confidence. Moreover, in real traffic interactions, environmental factors such as vehicle body movements, as well as the driving activity itself, distract the driver from focusing on the eHMI. We can therefore state that the AV should display its message in a slightly larger size than the minimum value. We recommend a value of between 170 mm (6.64 MOA) and 200 mm (7.81 MOA), as these sizes resulted in participants feeling very confident in identifying the display content. For this content size, a display width of 768 mm was sufficient for displaying different symbols and small blocks of text comprising four to five letters, such as "WALK", "GO", "OK", and "STOP", as proposed in several studies [12,34,35].

According to the standard DIN EN ISO 9241-303 [20], 170 mm is the size that should be used at distances of less than 29 m, while the content size for a distance of 88 m should be 512 mm to comply with a recommended visual angle of 20 MOA. However, according to our findings, a content size of 6.64 MOA to 7.81 MOA is sufficient for good legibility. This result underlines the importance of new international standards for future eHMI development. The transferability of findings from guidelines on technology, task, and environment-independent performance specifications and recommendations [20] is not applicable.

Symbols require a smaller size than text for them to be legible, which coincides with the findings of Kline, Ghali, Kline, and Brown [36]. Moreover, symbols of equal size were legible over longer distances than text. The prototype displayed the symbols individually and not surrounded by other elements. The letters within the text did not stand alone and were not delimited from each other, which complicated the correct identification of individual letters. In addition, it can be assumed that the contours of texts and symbols were blurred by the haze effect [20], which depends, among other things, on the relative atmospheric humidity [37]. Even though the haze effect affects symbols and text equally, the contours of text tend to merge in letters that are close together. The impact of haze and the small distances between multiple letters resulted in the text being misread in 13 attempts in experiment 2. The blurred delimitation of individual letters led to confusion, for instance, between the letters C, O, and G, as well as between F and P. In contrast to text identification, participants expected the symbol to be displayed, which means that the symbol type was already identified and only its orientation had to be determined. For safety-critical AV–human driver interaction at road bottlenecks, these findings imply that standalone symbols should be used for communication in order to achieve the most accurate identification and the greatest possible legibility of the AV's message. Moreover, taking into account the comments of the participants, it can be concluded that if using arrows for communication, the arrow tips should be designed more distinctly to improve identification of its orientation. This is reflected in the lower legibility rating of the arrow compared with the distinct orientation of the E for sizes greater than 170 mm.

6.2. Effect of Content Color

The statistical analysis showed that the effect of color was significant for displaying text, in a way that the color red was found to be readable from greater distances, although this color had the lowest contrast ratio. There was no significant effect of symbol color on the human detection range. This finding may be due to the fact that contrast and luminance are confounded variables [30] and thus human visual performance varies with different ratios of contrast and luminance [38]. The red light may have affected the contrast–luminance ratio between several letters in favor of better legibility. All in all, we can state that the influence of color was negligible, which corresponds with the findings of Lin [39], who showed that the color of letters has no significant effect on the visual performance of text identification on TFT-LCD monitors.

We recommend the use of symbols for AV communication (Section 6.1). As the factor of color has no effect on the human detection range, we are free to use red, green or white in an eHMI design in order to attain good legibility. Moreover, the display provides color fidelity at viewing angles of less than 140°, and humans are able to perceive the colors red and green in an area of 65° and 60° respectively [40]. Therefore, in straight approach scenarios like the AV–human driver interaction at a road bottleneck, it is possible to communicate via color and, at the same time, there is no risk of reducing the human detection range. This fact enables coding of AV messages via colors, leading to faster reaction times if the color meets the expectation of the human interaction partner [41]. Red and green are familiar from traffic in the context of yielding or insisting on the right of way. As an example, when texts are green, participants perceive a higher level of safety to cross the street [42], while using symbols in green to communicate to yield the right of way at a road bottleneck enables an efficient and safe passage for the human driver [2].

6.3. Limitations

The sample taking part in the study consisted mainly of young participants between the ages of 25 years and 30 years. This means that a considerable proportion of human drivers were not represented. Elderly people, in particular, are more likely to suffer from vision deficiency such as impaired contrast sensitivity [43], which can influence the results of the experiments. A future study should therefore use an age-balanced sample.

Moreover, in contrast to interactions at road bottlenecks, the participants identified the display content without sitting in a vehicle. Thus, the investigation did not take into account the potential influence of the windshield on the legibility of the display. Additionally, vehicle body movements and dirt can impair the eHMI's legibility in real traffic. A further limitation is that the absence of any driving activity means that participants can devote their full attention to the display. To counteract these effects, we did not recommend a content size of 140 mm for display legibility, but calculated a range of 170 mm–200 mm for use in eHMI designs.

The experiments were conducted on dry winter days. Thus, the analysis did not consider the influence of summer light conditions or rainfall. Before conducting the experiments, we measured the illuminance. Initial analysis indicated an effect of illuminance on the human detection range such that an increase in illuminance led to an increase in range. We refrained from presenting this result in the present paper, because in addition to illuminance, there are several other parameters, such as luminance distribution, light color, and glare [44], which characterize real-life lighting conditions, while haze [37] and thus legibility are affected by air humidity and fog. Therefore, we could not assign the effect only to illuminance. To investigate the influence of individual factors, these need to be isolated and examined in a controlled environment in future work.

7. Conclusions

Content type significantly influences the required display size, with a large effect. Symbols can be displayed in a smaller size than text for them to be legible from a constant distance. Moreover, symbols can be identified at a greater distance than text, which means that in the same scenario the human interaction partner has more time to perceive and process an AV's message in the form of a symbol. In the bottleneck scenario, we state that the height of the display content should be 170 mm (6.64 MOA) to 200 mm (7.81 MOA), as this leads to very good legibility at a distance of 88 m and the majority of the participants were able to identify the smaller content in experiment 2 from even greater distances. In addition, this recommendation considers potential environmental influences that may negatively affect legibility.

Regardless of the display content, we did not find a content overlapping effect of color on the human detection range. The influence of color was only significant when displaying text. In conclusion, we state that in order to ensure the widest possible range of AV communication, the colors investigated in this study are suitable for displaying simple symbols without running the risk of negatively influencing legibility. Therefore, color coding in addition to the symbol shape can be employed in the interests of good legibility and communicating AV messages more clearly.

Author Contributions: Conceptualization, M.R. and J.S.; data curation, M.R. and J.S.; formal analysis, M.R. and J.S.; funding acquisition, K.B.; investigation, M.R. and J.S.; methodology, M.R. and J.S.; project administration, M.R.; software, J.S.; supervision, K.B.; validation, M.R. and J.S.; visualization, M.R.; writing—original draft, M.R.; writing—review and editing, M.R. All authors have read and agreed to the published version of the manuscript.

Funding: The German Federal Ministry of Economics and Energy funded this research within the project @City: Automated Cars and Intelligent Traffic in the City, grant number 19A17015B.

Conflicts of Interest: The authors declare no conflicts of interest.

References

1. SAE International. *Taxonomy and Definitions for Terms Related to Driving Automation Systems for On-Road Motor Vehicles (J3016)*; SAE International: Warrendale, PA, USA, 2018.
2. Rettenmaier, M.; Albers, D.; Bengler, K. After you?!—Use of external human-machine interfaces in road bottleneck scenarios. *Transp. Res. Part F* **2020**, *70*, 175–190. [CrossRef]
3. Clamann, M.; Aubert, M.; Cummings, M.L. Evaluation of Vehicle-to-Pedestrian Communication Displays for Autonomous Vehicles. In Proceedings of the 96th Annual Transportation Research Board Meeting, Washington DC, USA, 8–12 January 2017.
4. Habibovic, A.; Lundgren, V.M.; Andersson, J.; Klingegård, M.; Lagström, T.; Sirkka, A.; Fagerlönn, J.; Edgren, C.; Fredriksson, R.; Krupenia, S.; et al. Communicating Intent of Automated Vehicles to Pedestrians. *Front. Psychol.* **2018**, *9*, 1336. [CrossRef] [PubMed]
5. Faas, S.M.; Baumann, M. Yielding Light Signal Evaluation for Self-driving Vehicle and Pedestrian Interaction. In *Human Systems Engineering and Design II*; Ahram, T., Karwowski, W., Pickl, S., Taiar, R., Eds.; Springer International Publishing: Cham, Switzerland, 2019; pp. 189–194.
6. Faas, S.M.; Baumann, M. Light-Based External Human Machine Interface: Color Evaluation for Self-Driving Vehicle and Pedestrian Interaction. *Proc. Hum. Factors Ergon. Soc. Annu. Meet.* **2019**, *63*, 1232–1236. [CrossRef]
7. Rettenmaier, M.; Pietsch, M.; Schmidtler, J.; Bengler, K. Passing through the Bottleneck - The Potential of External Human-Machine Interfaces. *IEEE Intell. Veh. Symp.* **2019**, 1687–1692. [CrossRef]
8. Dietrich, A.; Willrodt, J.-H.; Wagner, K.; Bengler, K. Projection-Based External Human Machine Interfaces—Enabling Interaction between Automated Vehicles and Pedestrians. In Proceedings of the DSC 2018 Europe VR, Antibes, France, 5–7 September 2018.
9. Faas, S.M.; Mathis, L.-A.; Baumann, M. External HMI for self-driving vehicles: Which information shall be displayed? *Transp. Res. Part F* **2020**, *68*, 171–186. [CrossRef]
10. Kaß, C.; Schoch, S.; Naujoks, F.; Hergeth, S.; Keinath, A.; Neukum, A. Standardized Test Procedure for External Human–Machine Interfaces of Automated Vehicles. *Information* **2020**, *11*, 173. [CrossRef]
11. Habibovic, A.; Andersson, J.; Lundgren, V.M.; Klingegård, M.; Englund, C. External Vehicle Interfaces for Communication with Other Road Users? *Road Veh. Autom.* **2019**, *19*, 91–102. [CrossRef]

12. De Clercq, K.; Dietrich, A.; Núñez Velasco, J.P.; De Winter, J.; Happee, R. External Human-Machine Interfaces on Automated Vehicles: Effects on Pedestrian Crossing Decisions. *Hum. Factors* **2019**, *61*, 1353–1370. [CrossRef]

13. Eisma, Y.B.; Van Bergen, S.; Ter Brake, S.M.; Hensen, M.T.T.; Tempelaar, W.J.; De Winter, J.C.F. External Human–Machine Interfaces: The Effect of Display Location on Crossing Intentions and Eye Movements. *Information* **2020**, *11*, 13. [CrossRef]

14. Schieben, A.; Wilbrink, M.; Kettwich, C.; Madigan, R.; Louw, T.; Merat, N. Designing the interaction of automated vehicles with other traffic participants: Design considerations based on human needs and expectations. *Cogn. Technol. Work* **2019**, *21*, 69–85. [CrossRef]

15. *Regulation (EC) No 139/2004 Merger Procedure*; Office for Official Publications of the European Communities: Luxembourg, 2009. Available online: https://pdfs.semanticscholar.org/00f1/09017a252e7b49b2b92e1c00000ca7e9b5ba.pdf (accessed on 20 April 2020).

16. Kraftfahrt Bundesamt. Neuzulassungen von Personenkraftwagen nach Segmenten und Modellreihen im Juni 2019. 2019. Available online: https://www.kba.de/DE/Statistik/Fahrzeuge/Neuzulassungen/Segmente/2019/2019_segmente_node.html (accessed on 20 April 2020).

17. Auto Portal Angurten.de. BMW 5er Touring (G31): Abmessungen und Technische Daten. Available online: https://www.angurten.de/is/abmessungen/1681-bmw-5er-touring/1.htm#abmes-sungsbilder (accessed on 20 April 2020).

18. Andrén, B.; Brunnström, K.; Wang, K. Readability of Displays in Bright Outdoor Surroundings. *Sid Symp. Dig. Tech. Pap.* **2014**, *45*, 1100–1103. [CrossRef]

19. Luft, H. LED Leitfaden. 2016, pp. 1–13. Available online: https://www.dbz.de/download/1243759/dbz-leitfaden-2014-led.pdf (accessed on 20 April 2020).

20. German Institute for Standardization. *Ergonomics of Human-System Interaction—Part 303: Requirements for Electronic Visual Displays (ISO 9241-303:2008)*; German Registered Association: Berlin, Germany, 2009.

21. German Institute for Standardization. *Road Vehicles—Ergonomic Aspects of Transport Information and Control Systems—Specifications and Test Procedures for In-Vehicle Visual Presentation (ISO 15008:2017)*; German Registered Association: Berlin, Germany, 2017.

22. Coreman Technology Co. Products. Available online: http://www.coreman.cc/product_list.asp?bid=73 (accessed on 20 April 2020).

23. Regulation No 87 of the Economic Commission for Europe of the United Nations (UN/ECE)—Uniform Provisions Concerning the Approval of Daytime Running Lamps for Power-Driven Vehicles. In *Official Journal of the European Union*; United Nations Economic Commission for Europe: Geneva, Switzerland, 2009.

24. Model, B. (Ed.) *Raspberry Pi 4 Computer*; Raspberry Pi Trading Ltd.: Cambridge, UK, 2019. Available online: https://www.raspberrypi.org/ (accessed on 20 April 2020).

25. Zeller, H. Rpi-Rgb-Led-Matrix [Computer Software]. Available online: https://github.com/hzeller/rpi-rgb-led-matrix (accessed on 20 April 2020).

26. Zeller, H. Rpi-Rgb-Led-Matrix. Available online: https://github.com/hzeller/rpi-rgb-led-matrix/tree/master/utils (accessed on 20 April 2020).

27. Rettenmaier, M.; Bengler, K. Modeling the Interaction with Automated Vehicles in Road Bottleneck Scenarios. *Proc. Hum. Factors Ergon. Soc. Annu. Meet.* **2020**. accepted.

28. Bach, M. The Freiburg Visual Acuity Test - Automatic Measurement of Visual Acuity. *Optom. Vis. Sci.* **1996**, *73*, 49–53. [CrossRef] [PubMed]

29. Bach, M. The Freiburg Visual Acuity Test-Variability Unchanged by Post-Hoc Re-Analysis. *Graefes Arch. Clin. Exp. Ophthalmol.* **2007**, *245*, 965–971. [CrossRef]

30. Rogers, S.P.; Spiker, V.A.; Cicinelli, J. Luminance and luminance contrast requirements for legibility of self-luminous displays in aircraft cockpits. *Appl. Ergon.* **1986**, *17*, 271–277. [CrossRef]

31. JASP Team. JASP (Version 0.11.1) [Computer Software]. 2019. Available online: https://jasp-stats.org/faq/how-do-i-cite-jasp/ (accessed on 20 April 2020).

32. Cohen, J. *Statistical Power Analysis for the Behavioral Sciences*, 2nd ed.; Lawrence Erlbaum Associates: Mahwah, NJ, USA, 1988; ISBN 0-8058-0283-5.

33. Foxell, C.A.; Stevens, W.R. Measurements of visual acuity. *Br. J. Ophthalmol.* **1955**, *39*, 513–533. [CrossRef]

34. Fridman, L.; Mehler, B.; Xia, L.; Yang, Y.; Facusse, L.Y.; Reimer, B. To Walk or Not to Walk: Crowdsourced Assessment of External Vehicle-to-Pedestrian Displays. *arXiv* **2017**, arXiv:1707.02698.

35. Song, Y.E.; Lehsing, C.; Fuest, T.; Bengler, K. External HMIs and Their Effect on the Interaction Between Pedestrians and Automated Vehicles. In *Intelligent Human Systems Integration*; Karwowski, W., Ahram, T., Eds.; Springer International Publishing: Cham, Switzerland; pp. 13–18.

36. Kline, T.J.B.; Ghali, L.M.; Kline, D.W.; Brown, S. Visibility Distance of Highway Signs among Young, Middle-Aged, and Older Observers: Icons Are Better than Text. *Hum. Factors* **1990**, *32*, 609–619. [CrossRef]

37. He, Y.; Gu, Z.; Lu, W.; Zhang, L.; Okuda, T.; Fujioka, K.; Luo, H.; Yu, C.W. Atmospheric humidity and particle charging state on agglomeration of aerosol particles. *Atmos. Environ.* **2019**, *197*, 141–149. [CrossRef]

38. Zhu, Z.; Wu, J. On the standardization of VDT's proper and optimal contrast range. *Ergonomics* **1990**, *33*, 925–932. [CrossRef]

39. Lin, C.C. Effects of screen luminance combination and text color on visual performance with TFT-LCD. *Int. J. Ind. Ergon.* **2005**, *35*, 229–235. [CrossRef]

40. Woodson, W.E.; Conover, D.W. *Human Engineering Guide for equipement Designers*; University of California Press: Berkeley, CA, USA, 1964.

41. Tanaka, J.W.; Presnell, L.M. Color diagnosticity in object recognition. *Percept. Psychophys.* **1999**, *61*, 1140–1153. [CrossRef]

42. Bazilinskyy, P.; Dodou, D.; De Winter, J. Survey on eHMI concepts: The effect of text, color, and perspective. *Transp. Res. Part F* **2019**, *67*, 175–194. [CrossRef]

43. Owsley, C. Aging and vision. *Vis. Res.* **2011**, *51*, 1610–1622. [CrossRef] [PubMed]

44. German Institute for Standardization. *Light and Lighting—Lighting of Work Places—Part 1: Indoor Work Places*; German Version EN 12464-1:2011, 2011 (12464-1); German Registered Association: Berlin, Germany, 2011.

 *information*

Article

How Do eHMIs Affect Pedestrians' Crossing Behavior? A Study Using a Head-Mounted Display Combined with a Motion Suit

Lars Kooijman [1], Riender Happee [2,3] and Joost C. F. de Winter [2,*]

[1] Department of BioMechanical Engineering, Delft University of Technology, Mekelweg 2, CD 2628 Delft, The Netherlands; L.Kooijman-1@tudelft.nl
[2] Department of Cognitive Robotics, Delft University of Technology, Mekelweg 2, CD 2628 Delft, The Netherlands; R.Happee@tudelft.nl
[3] Department of Transport & Planning, Delft University of Technology, Mekelweg 2, CD 2628 Delft, The Netherlands
* Correspondence: j.c.f.dewinter@tudelft.nl

Received: 23 October 2019; Accepted: 26 November 2019; Published: 6 December 2019

Abstract: In future traffic, automated vehicles may be equipped with external human-machine interfaces (eHMIs) that can communicate with pedestrians. Previous research suggests that, during first encounters, pedestrians regard text-based eHMIs as clearer than light-based eHMIs. However, in much of the previous research, pedestrians were asked to imagine crossing the road, and unable or not allowed to do so. We investigated the effects of eHMIs on participants' crossing behavior. Twenty-four participants were immersed in a virtual urban environment using a head-mounted display coupled to a motion-tracking suit. We manipulated the approaching vehicles' behavior (yielding, nonyielding) and eHMI type (None, Text, Front Brake Lights). Participants could cross the road whenever they felt safe enough to do so. The results showed that forward walking velocities, as recorded at the pelvis, were, on average, higher when an eHMI was present compared to no eHMI if the vehicle yielded. In nonyielding conditions, participants showed a slight forward motion and refrained from crossing. An analysis of participants' thorax angle indicated rotation towards the approaching vehicles and subsequent rotation towards the crossing path. It is concluded that results obtained via a setup in which participants can cross the road are similar to results from survey studies, with eHMIs yielding a higher crossing intention compared to no eHMI. The motion suit allows investigating pedestrian behaviors related to bodily attention and hesitation.

Keywords: virtual reality; automated driving; pedestrians; decision making; crossing; eHMI

1. Introduction

Worldwide, 22% of traffic fatalities concern pedestrians [1], and over 90% of car accidents are attributable to driver error [2]. Automated vehicles (AVs) could reduce the number of accidents substantially. However, the implementation of AVs will happen gradually over several decades (e.g., [3]), which means that AVs and conventional vehicles will likely share the same roads.

In this type of mixed traffic, it may be unclear to pedestrians and other road users whether an approaching vehicle is driven manually or driving automatically. Pedestrians sometimes rely on cues from the driver, such as hand gestures and eye contact [4,5], and the absence thereof could debilitate pedestrian safety. External communication devices (i.e., external human-machine interfaces, eHMIs) on AVs might be a suitable replacement for the communication signals of drivers to pedestrians and aid pedestrians in perceiving the intention of the vehicle.

The literature on AV communication and the efficacy of eHMIs is rapidly expanding. Researchers have investigated the efficacy of eHMIs as communicative replacements for humans by means of computer surveys, lab experiments, and field tests.

Fridman et al. [6] conducted an online survey in which they presented 30 eHMI concepts and asked respondents whether they thought it was safe to cross the road based on the information provided by the eHMI. They found that text displays stating "Walk" or "Don't Walk" were considered clear, whereas green or red headlights were regarded as relatively ambiguous. Chang et al. [7] conducted a survey in which they tested five types of eHMIs, including a text display, lights, and a projection. In this study as well, participants found a text display to be the easiest to interpret. In Ackermann et al. [8], participants rated an LED light strip as highly unambiguous compared to eHMIs in the form of a display or projection that provided textual or symbolic advice. An online survey conducted by Bazilinskyy et al. [9] showed that participants tended to apply an egocentric perspective, in the sense that an eHMI with the text "Walk" or "Don't walk" was regarded as clearer and more persuasive than "Will Stop" or "Won't stop". Other online or lab-based surveys about the effectiveness of eHMIs have been conducted by Deb et al. [10], Hagenzieker et al. [11], Dey et al. [12], and Zhang et al. [13]. Typically, in surveys, the participant is shown a picture, an animation, or a recorded video from an on-road setting after which questions are asked, such as whether the participant would feel safe to cross in front of the car. An advantage of surveys is that they allow for a high number of repetitions with variations of eHMIs. A disadvantage, however, is that they require participants to imagine how they would act or feel, an approach that may have limited validity.

Others have used lab setups that immerse the participant in a traffic scenario [14–17]. De Clercq et al. [18] investigated the effect of eHMIs on the crossing intentions of pedestrians in a virtual reality environment presented via a head-mounted display (HMD). The participants stood on a curb and watched a platoon of oncoming AVs, which were devoid of or equipped with one of four eHMIs. The presence of an eHMI, indicating whether the AV would stop or not, significantly increased participants' perceived safety compared to a situation in which an eHMI was absent. A text-based eHMI was found to be the clearest overall. Using a similar setup, Ackermans [19] found that an eHMI consisting of a light animation made participants feel safer to cross compared to no eHMI. Weber et al. [20], also using an HMD, found benefits of eHMIs in terms of correct recognition rate of the vehicle's intention as well as response times. Due to their high visual field of view, HMDs offer a higher level of perceptual fidelity than survey studies. In De Clercq et al. [18] and Ackermans [19], participants were tasked to press a button when they would intend to cross or recognize the approaching vehicle's intention. In the on-road study by Walker et al. [17], participants used a physical box with a slider to indicate their willingness to cross, whereas in a simulator study by Mahadevan et al. [14], a virtual slider was used. The button and slider approach both allow performing analyses on how the crossing intention varies as a function of the distance between the pedestrian and the approaching vehicle. Although this method yields insights that questionnaires are unable to offer, the participants' behavior is measured in a subjective binary (i.e., button press/not pressed) or continuous manner (e.g., across the range of the slider), rather than an objective manner (e.g., forward gait).

Yet another method of evaluating eHMIs is to perform a field test on an actual road that is closed off from traffic. Clamann et al. [21] let participants stand on a curb while a van with eHMI was approaching. The authors extracted measures such as the moment participants turned their face to the vehicle, and the moment participants began to cross the road. The results showed no significant effect of the type and presence of the eHMI. Similar results were obtained by Palmeiro et al. [22]. They found that the presence of a sign "self-driving" did not significantly affect participants' critical gap times, that is, the last moment participants felt safe to cross. A limitation of both Clamann et al. [21] and Palmeiro et al. [22] is that participants were not permitted to step onto the road because of ethical and safety reasons; in the former study participants were physically constrained by a rope around their waist, whereas in the latter participants were asked to step *back* when they did not feel safe to cross anymore. Furthermore, field studies like these face a challenge of repeatability, as there tend to be fluctuations in the speed of the oncoming car, as well as in weather conditions [22]. Additionally, there are coding/timing challenges because pedestrian behavior is extracted from video recordings and needs to be synchronized with the AV's GPS signal.

In an overview paper, Cefkin et al. [23] described multiple research methods to examine eHMIs for highly automated vehicles used in the Renault–Nissan–Mitsubishi Alliance Innovation Lab. Among the methods were observations of stop intersections and on-road tests using a Wizard-of-Oz AV in public environments with naïve pedestrians (see also [24]). Although these approaches arguably have the highest possible level of fidelity, as pedestrians are exposed to AVs in a naturalistic manner, they too are affected by several limitations. One disadvantage is that the state of the pedestrian and AV (e.g., speed, distance) are unknown, and data need to be obtained from annotations of video recordings or interviews after the encounter. Furthermore, results vary depending on traffic conditions. Cefkin et al. [23] pointed out that in one of their studies, pedestrian traffic was limited and, therefore, only a small number of AV-pedestrian encounters could be recorded.

Summarizing, previous research on the efficacy of eHMIs has various strengths and weaknesses. One weakness is that researchers had to rely on imagined rather than actual crossing behavior. In this respect, the works of Deb et al. [10], Feldstein et al. [25], Lee et al. [26], and Schmidt et al. [16] provide promising leads. In Deb et al. [10], participants walked with a head-mounted display while encountering different types of eHMIs (e.g., an upraised hand, a colored beacon, an image of a pedestrian). Based on analyses of video recordings of the participants' walking behavior, Deb et al. [10] classified the participants' behaviors as hesitation, confusion, and stopping. A similar approach was used by Lee et al. [26] and by Schmidt et al. [16]. In Lee et al. [26], participants wore an HMD, and the dependent variable was whether participants crossed before the AV started to decelerate, during deceleration of the AV, or when the AV had stopped. However, Deb et al. [10], Lee et al. [26] and Schmidt et al. [16] did not measure the pedestrians' motion in a quantitative manner, for example, in terms of walking speed as a function of elapsed time.

We investigated whether results from survey-based and HMD-based methods replicate when using a more realistic experimental setup in which participants could cross the road. We used two eHMI concepts from De Clercq et al. [18] which corresponded to the extremities of ambiguity. That is, we selected a textual eHMI as research shows that text (i.e., "walk"/"don't walk") is generally regarded as clear and unambiguous (see, e.g., [6]). Furthermore, we selected front brake lights, a concept which is regarded as ambiguous because it may be unclear to the pedestrian whether he or she should apply an egocentric perspective (i.e., a green light on the AV means that the pedestrian can cross) or an exocentric perspective (i.e., a green light on the AV means that the vehicle will continue driving) (e.g., [9,13]).

We conducted the experiments in a virtual environment where the participants were immersed using an HMD and where a schematic representation of the participants' body (i.e., an 'avatar') was present, generated via a motion suit. This type of simulation resembles the setup of Feldstein et al. [25,27] who developed a pedestrian simulator where participants could move around freely and motions were recorded through a marker-based system. Furthermore, in Feldstein et al. [25,27], a schematic representation of the body was present in the virtual environment. We sought to examine how the eHMIs affect participants' crossing behavior as measured using the motion suit signals. We expected that the text-based eHMI would be regarded as more persuasive than the front-brake light eHMI, which, in turn, was expected to evoke higher pedestrian responsiveness than no eHMI at all. In addition to this replicative aim, we extracted information about body posture while crossing. Finally, we examined whether our setup, in which participants were able to step onto the road, would yield a higher level for self-reported realism compared to a similar setup in which participants remain static in the environment.

2. Methods

2.1. Participants

Twenty-four participants (six females, 18 males) with a mean age of 25.4 years (SD = 2.5, min = 21, max = 30) partook in the study. Participants were recruited among students, PhD candidates, and postdocs at the faculty of Mechanical, Maritime and Materials Engineering of the TU Delft. In response to the question how often the participant commuted to work or school by foot in the last 12 months, five respondents reported "never", six reported "less than once a month", four reported "once a month to once a week", four reported "1 to 3 days a week", one reported "4 to 6 days a week", and four reported "daily". Participants' nationalities were as follows: 15 Dutch, two German, two Chinese, one British and Turkish, one American, one Spanish, one Indian, and one Ukrainian. No incentive was offered and people were allowed to participate regardless of the driving side in their country of origin. All participants were living in the Netherlands. The study was approved by the Human Research Ethics Committee of the TU Delft, and each participant provided written informed consent before the start of the experiment.

2.2. Experimental Design

Participants were immersed in a virtual environment, similar to the one used in De Clercq et al. [18], via a head-mounted display. The participant was standing on a curb in front of a zebra crossing at a two-way urban road, as shown in Figure 1. In each trial, a platoon containing five cars would come driving around a corner to the far left of the participant, pass the participant, and then turn left around a corner to the right of the participant. The third car in the platoon was the stimulus vehicle. Two types of vehicles were used, namely a Smart Fortwo (small vehicle) and a Ford F150 (large vehicle). The occurrence of these types was randomized. The design of the research was within-subject, consisting of three independent variables.

Figure 1. Cars approaching a participant in the virtual environment. The starting location of the participant during the experiment was visible on the sidewalk. Cars only approached the participants from their left-hand side. Participants were instructed to cross the road up until the third zebra stripe counting from the curb. The avatar was based on the 3D measured participant motion.

The first independent variable was the type of eHMI placed on the front of the stimulus vehicle, consisting of three levels: (1) No eHMI, (2) Front Brake Lights (FBL), and (3) Text eHMI, as depicted in Table 1. The eHMIs were the same as the experiment of De Clercq et al. [18]. In our experiment, the presence of eHMIs on all other cars (i.e., the non-stimulus cars) was randomized.

Table 1. Appearance of the cars used in this experiment.

Front Brake Lights, yielding	
Front Brake Lights, nonyielding	 
Text eHMI, yielding	 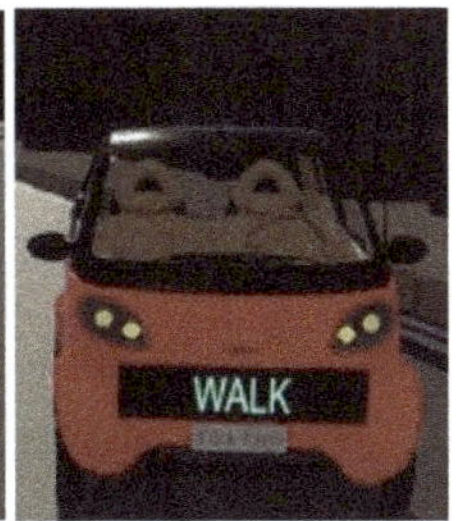
Text eHMI, nonyielding	 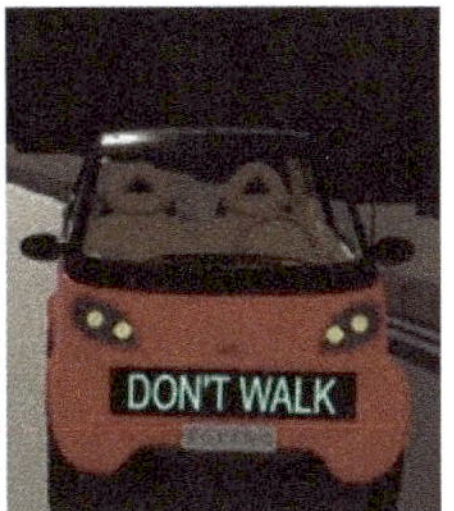
No eHMI, yielding and nonyielding	 

Note: For the No eHMI (Baseline) condition, the vehicles looked the same in their yielding and nonyielding state.

The distance between the front of the stimulus vehicle and the back of the second vehicle was the second independent variable, which was varied between 20, 30, and 40 meters. The distances between the other cars (i.e., first and second, third and fourth, fourth and fifth) were approximately 13 meters.

The yielding behavior of the cars was the third independent variable. The cars had a speed of 50 km/h while approaching the participant. In the yielding conditions, the stimulus car would start braking with a deceleration of 3.5 m/s^2 at approximately 30 meters from the zebra crossing and halt 4 meters before the zebra crossing, as can be seen in Figure 2. If an eHMI was present on the stimulus vehicle, it would change state upon braking. The fourth and the fifth car would start yielding at 40 and 50 meters from the zebra crossing, respectively, and come to a stop a few meters from one another behind the third car. After standing still for 5 seconds, the cars would pick up speed again, drive past the participant, and turn left around the corner.

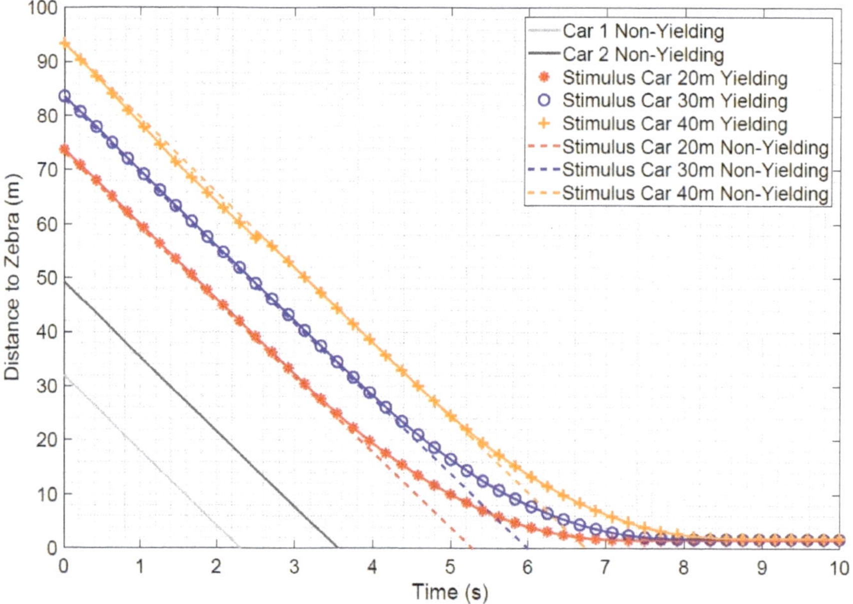

Figure 2. Distances of the front of the cars to the front edge of the zebra crossing during a trial. Time = 0 s was defined as the moment when the first vehicle in the platoon of cars was approximately 30 meters away from the zebra crossing.

Each participant completed a total of 18 trials. Each trial consisted of a unique combination of the independent variables (3 eHMI conditions × 3 gap distances × 2 yielding conditions). The order in which the combinations were presented to the participants was randomized and unique for each participant. A video of trials where cars did and did not yield is available through the link in the Supplementary Materials. The experiment lasted about 20 minutes per participant.

2.3. Participant's Task

Participants were instructed to cross the road once they felt it was safe enough to do so. They were informed that the first two cars in each of the platoons would never yield and were told not to cross in front of these two cars. Hence, the only crossing opportunity for participants was between the second car in the platoon and the stimulus vehicle or after all cars had passed. Participants were not instructed or trained about the meaning of the eHMIs. Due to space limitations of the physical environment, participants were instructed to walk to the third zebra stripe in the virtual environment. If standing on the third zebra stripe in the virtual environment, participants had one meter of unobstructed space as a safety margin around them in the physical environment.

2.4. Materials and Equipment

The experiment was run on a desktop computer with an Intel Core i7-6700 CPU (@3.4 GHz) processor, 16 GB RAM, MSI H110M Pro-D (MS-7996) motherboard, NVIDIA GeForce GTX 1070 4 GB graphics card, and a Windows 10 Pro 64-bit operating system. The participants' motion was recorded using a wireless set-up of the Xsens Link Motion Tracking Device (Enschede, The Netherlands) in combination with version 0.3b of MVN Analyze [28]. The recorded accelerations, of head, thorax, pelvis, and extremities were integrated to estimate full-body motion. The motion data were transferred from the MVN software to an avatar in the virtual environment that was built using Unity version 5.5.0f3 64-bit. The scripts and avatar used in Unity were developed by Xsens and obtained via the Unity Asset Store. The wireless transmitting device of the Xsens sent its data via an Asus RT-AC68U router to the desktop. An Oculus Rift CV1 was used to visually and audibly immerse the participant in the virtual environment. A 1-meter extension of the HDMI and USB cables of the Oculus Rift was made using a DeLOCK 1.4 HDMI and a DeLOCK USB 3.0 extension cable.

2.5. Procedure

Participants provided written informed consent before the start of the experiment. After being briefed about the goal of the experiment, namely *"to investigate whether crossing intentions of human pedestrians can be detected from body motion"*, the participants completed a questionnaire containing demographic questions and statements about pedestrians and motorists from Papadimitriou et al. [29]. Following the questionnaire, participants were familiarized with the Oculus Rift and Xsens. After putting the Xsens onto the participant and when a successful calibration was obtained, participants put on the Oculus Rift and were allowed to familiarize themselves within the virtual environment for a few minutes.

After the familiarization, the experiment was initiated. After each trial, participants verbally indicated their discomfort using the single-item misery scale (MISC). They also indicated their feeling of fear and their ability to predict the behavior of the oncoming cars on a scale from 1 (i.e., strongly disagree) to 10 (i.e., strongly agree). If participants indicated a MISC rating of 4 or higher, the experiment would be paused or aborted. Once all trials were completed, a final questionnaire was administered to measure the fidelity of the experimental environment through the use of the Virtual Reality Presence Questionnaire (VRPQ) of Witmer et al. [30].

2.6. Dependent Variables

A total of six objective and subjective measures were analyzed. The first objective dependent variable was the participants' forward (i.e., towards the zebra crossing) gait velocity as a function of elapsed time. The forward gait velocity was extracted from the pelvic sensor of the Xsens. Studies on affective body language show that stimuli of negative valence reduce gait velocity compared to when participants were confronted with stimuli of positive valence [31,32]. We used gait velocity as an index of safety perception, where we assumed that situations that were perceived by participants as ambiguous or unsafe would result in lower average gait velocity compared to situations that were perceived as unambiguous or safe.

Furthermore, using the position data from the same pelvic sensor, we computed the second dependent variable, namely the time at which the participants left the curb (Moment of Leaving Curb; MLC). In line with the above, we expected participants to leave the curb earlier when they perceived a situation that was unambiguous/safe compared to situations that were perceived to be ambiguous/unsafe.

We derived the participants' thorax angles from the T8 sensor of the Xsens as the third dependent variable. The thorax angle is expressed relative to the axis towards the zebra crossing. We expected participants to rotate their upper body earlier (i.e., to initiate forward motion) when confronted with a situation that was safer.

After each trial, the experimenter inquired the participant's wellbeing through a Misery Scale (MISC) rating. Participants reported their MISC ratings by naming an integer value between 1 and 12. The value 1 reflected that participants experienced no problems, 2 slight discomfort, 3 and 4 slight and mild nausea, and 5 and higher indicated more severe symptoms of sickness.

Furthermore, participants responded to the inquiry about whether they experienced a feeling of fear when considering crossing the road by stating an integer value between 1 and 10. The value 1 represented "strongly disagree" and 10 "strongly agree".

Lastly, participants responded to the inquiry whether it was difficult to predict the behavior of the oncoming vehicles by reporting an integer between 1 and 10. The value 1 represented "strongly disagree" and 10 "strongly agree".

2.7. Data Reduction

The motion data of the participants were recorded both in MVN Analyze and Unity. The MVN motion data were recorded at a frequency of 240 Hz, while the Unity recordings varied based on the rendering speed during the trial, and was, on average, above 40 Hz for each participant. The MVN data were filtered before they were exported to .mvnx format using the HD processor of MVN Analyze. In order to synchronize the position of the participants in the virtual environment recorded in Unity to their position recorded using MVN Analyze, the Unity motion data were interpolated to a frequency of 240 Hz. Next, the two datasets were cross-correlated to compensate for any time delay. After the cross-correlation, the MVN data were low-pass filtered with a zero-phase 10th order Butterworth filter using a cut-off frequency of 8 Hz. According to Schreven et al. [33], the optimal cut-off frequency for filtering human motion data is about 8 Hz. By means of the two filters (i.e., the one during exporting and the one in Matlab), we ensured that the signal was not contaminated with high-frequency sensor noise, while still capturing rapid limb motions of the participant.

We tested the effect of the presence of eHMIs on the participants' forward gait velocity, MLC, and their thorax angle. We performed paired-sample t-tests to compare the forward gait velocities between the eHMIs conditions at every time sample, as well as the thorax angles at every time sample. This approach has been inspired by Manhattan plots in molecular genetics research [34]. In a Manhattan plot, the x-axis shows the location on a chromosome and the y-axis depicts the common logarithm of the p-value. For the statistical tests depicted in the Manhattan plot, we used a significance level of 0.005 [35].

Despite the stringent alpha value used in the Manhattan plots, the results from t-tests per time sample should be interpreted with some caution due to a risk of false positives. Therefore, we also performed paired t-tests for a single key performance indicator: the moment participants' left the curb (MLC). Here, we applied a Bonferonni corrected significance level of $0.05/3 = 0.017$.

Additionally, we tested the effect of the eHMIs on the subjective responses of the participants per condition through paired-sample t-tests and investigated whether learning behavior occurred through linear tests of within-subject contrast. Lastly, we compared the subjective responses of our participants to the VRPQ to the responses of the participants in De Clercq et al. [18] through two-sample t-tests to investigate potential differences in subjective presence between their experimental methodology and ours.

Because the results of t-tests may be affected by outliers, we repeated the analysis using non-parametric signed-rank tests. The results for these tests, which did not alter our conclusions compared to the t-tests, can be found in the Supplementary Materials.

3. Results

3.1. Data Quality Assessment

The condition "40 meters, yielding, No eHMI" was presented to each participant as the condition "30 meters, yielding No eHMI" due to an error in our script. Subsequently, no comparison could be made for conditions "40 meters, yielding, Text eHMI" and "40 meters, yielding, FBL". The data of condition "40 meters, yielding, No eHMI" were, therefore, removed from the analysis. Furthermore, in 23 of the total of 216 trials in which the cars did not yield it was not possible to correlate the data recorded in MVN Analyze to the data recorded in Unity because the participants hardly moved during those trials.

Per yielding condition, and for each inter-vehicle distance and eHMI type, we examined whether participants left the curb too early (i.e., before the second car had passed). All participants left the curb during yielding conditions. However, if participants left the curb too early for a particular combination of inter-vehicle distance and eHMI type, their data were excluded for every eHMI of that yielding condition and inter-vehicle distance. For example, when participants crossed too early during the "20 meters, yielding, No eHMI condition", their data for "20 meters, yielding, Text" and "20 meters, yielding, and FBL" were also excluded.

3.2. Forward Gait Velocities

3.2.1. Twenty Meters Condition

Figure 3 shows the mean forward gait velocities of participants during the conditions "20 meters, yielding" and "20 meters, nonyielding". It can be seen that participants initiated forward motion already before the second car had passed and the third car started braking. If the car yielded, the presence of an eHMI on the third vehicle stimulated pedestrians to start already crossing as soon as the second car was passing, whereas in the condition where no eHMI was present they waited longer. A significant difference was observed for a duration of 1.03 seconds between the Text eHMI and No eHMI. Furthermore, a significant difference was found for 0.45 seconds, and between FBL and No eHMI. For the nonyielding trials, no significant differences were found between the eHMI conditions.

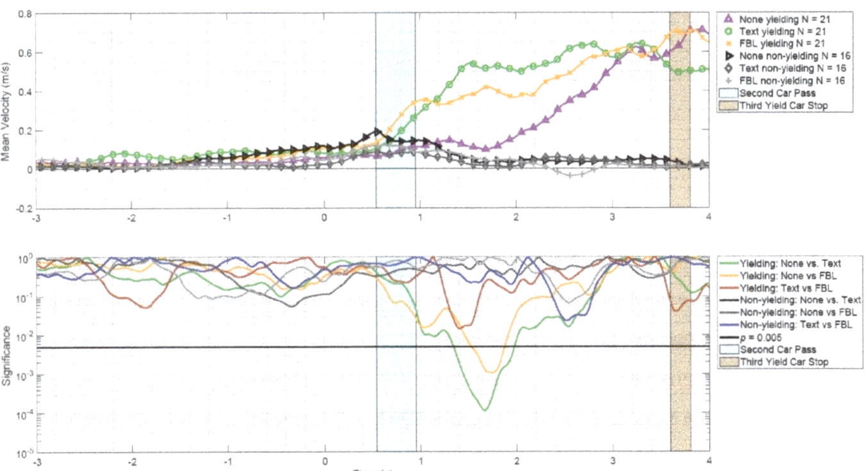

Figure 3. Top: Mean forward gait velocities during the conditions "20 meters, yielding" and "20 meters, nonyielding". Bottom: *p*-values from paired-sample *t*-tests. None = No eHMI, Text = Text eHMI, FBL = Front Brake Lights. *t* = 0 is the moment when the third vehicle in the platoon started braking in the yielding conditions or was at the same point of braking in the nonyielding conditions.

In case the cars did not yield, participants' forward velocities increased, similar to the yielding condition before the second car had passed in front of them. However, once the second car passed, the average forward velocities decreased to about zero, that is, participants halted their forward motion.

3.2.2. Thirty Meters Condition

Similar to the 20 meters condition, also in the 30-meters condition participants started walking before the second car had passed and the third car started braking, as shown in Figure 4. However, once the second car had passed and the third car advanced further, participants' mean forward velocity in the No eHMI and FBL yielding conditions did not increase further, whereas in the Text yielding condition their forward speed did increase.

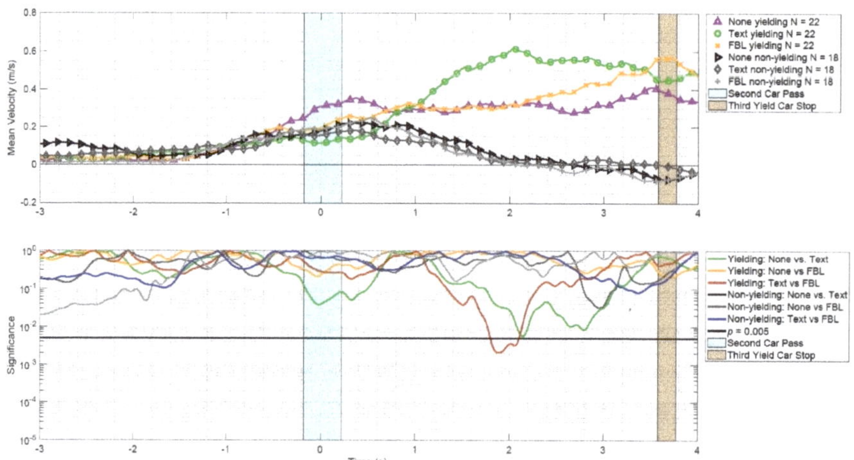

Figure 4. Top: Mean forward gait velocities during the conditions "30 meters, yielding" and "30 meters, nonyielding". Bottom: *p*-values from paired-sample *t*-tests. None = No eHMI, Text = Text eHMI, FBL = Front Brake Lights. $t = 0$ is the moment when the third vehicle in the platoon started braking in the yielding conditions or was at the same point of braking in the nonyielding conditions.

For the nonyielding conditions, participants' velocity decreased once the second car had passed, which reflects their inability to cross. For the yielding condition, significant differences were observed between the Text eHMI and FBL. For the nonyielding conditions, no significant differences were observed in forward velocity.

A depiction of separate forward velocity curves for each participant for both the 20 meters and 30 meters yielding conditions can be found in Figures S5 and S6.

3.3. Moment Leaving Curb (MLC)

In the condition "20 meters, yielding", significant differences in MLC were observed between No eHMI and when either a Text eHMI or FBL was present. For the condition "30 meters, yielding", no significant differences were found between the three eHMI conditions, as can be seen in Table 2.

Table 2. Descriptive statistics and results from paired-sample *t*-tests for the Moment of Leaving Curb (in seconds).

	None	Text	Front Brake Lights
	M (SD)	*M (SD)*	*M (SD)*
20 meters yielding	3.67 (0.99)	2.75 (1.25)	2.97 (1.21)
30 meters yielding	2.94 (1.71)	2.28 (1.02)	2.59 (1.67)
20 meters yielding	*t*-test None – Text	$t(20) = 5.79, \mathbf{p < 0.001}$	
20 meters yielding	*t*-test None – FBL	$t(20) = 3.94, \mathbf{p < 0.001}$	
20 meters yielding	*t*-test Text – FBL	$t(20) = -1.40, p = 0.175$	
30 meters yielding	*t*-test None – Text	$t(21) = 2.10, p = 0.048$	
30 meters yielding	*t*-test None – FBL	$t(21) = 1.10, p = 0.284$	
30 meters yielding	*t*-test Text – FBL	$t(21) = -1.12, p = 0.274$	

Note: None = No eHMI, Text = Text eHMI, FBL = Front Brake Lights. Significant differences ($p < 0.017$) are indicated in boldface.

3.4. Thorax Angle

The mean thorax angles relative to the *x*-axis (i.e., the crossing direction of the zebra) are depicted in Figure 5 ("20 meters, yielding" and "20 meters, nonyielding" conditions) and Figure 6 ("30 meters, yielding" and "30 meters, nonyielding" conditions). Initially, participants, on average, had their thorax rotated towards the approaching cars on the left. Participants started to rotate their upper body toward the zebra crossing before the second car had passed when an eHMI was present on the third car. In the nonyielding conditions, participants also slightly rotated their upper body before the second car had passed, but refrained from further rotation. No significant differences between the eHMI conditions were observed.

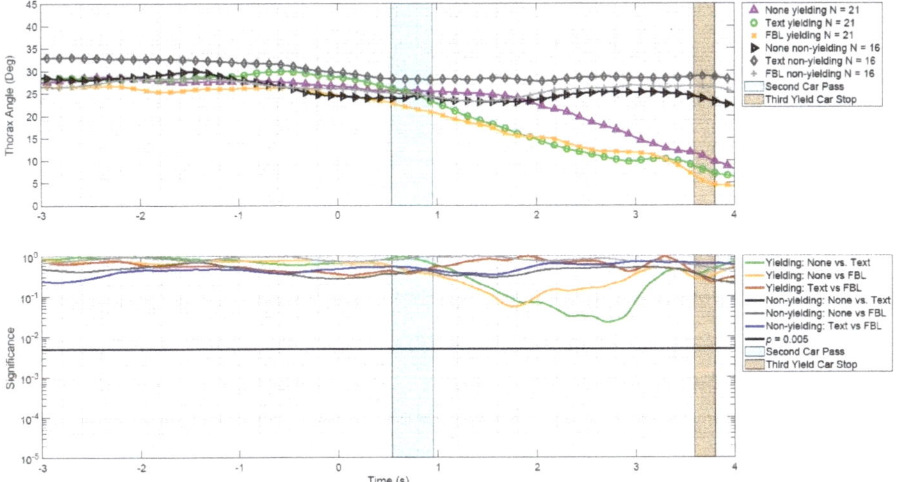

Figure 5. Top: Mean thorax angles during the conditions "20 meters, yielding" and "20 meters, nonyielding". Bottom: *p*-values over time. None = No eHMI, Text = Text eHMI, FBL = Front Brake Lights. $t = 0$ is the moment when the third vehicle in the platoon started braking in the yielding conditions or was at the same point of braking in the nonyielding conditions.

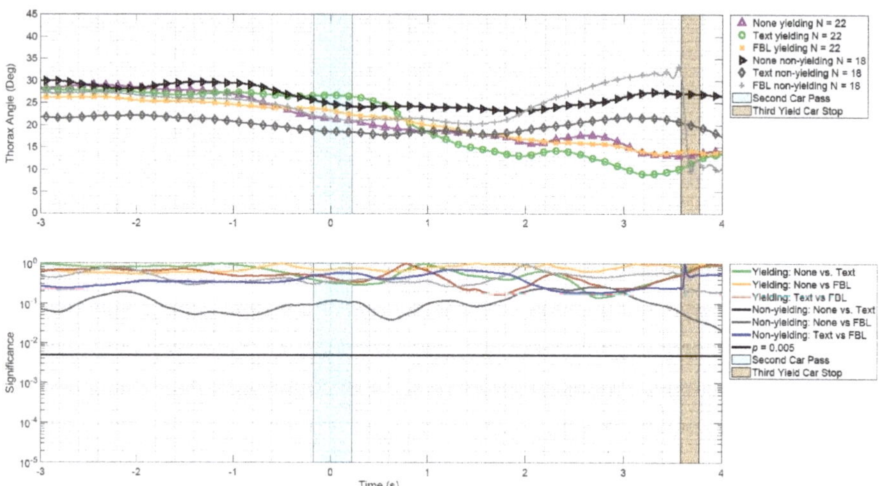

Figure 6. Top: Mean thorax angles during the conditions "30 meters, yielding" and "30 meters, nonyielding". Bottom: *p*-values over time. None = No eHMI, Text = Text eHMI, FBL = Front Brake Lights. $t = 0$ is the moment when the third vehicle in the platoon started braking in the yielding conditions or was at the same point of braking in the nonyielding conditions.

3.5. Self-Reported Predictability of Car Behavior

Participants often stated that they experienced no fear at all, either because they did not cross in the nonyielding conditions or because the car stopped in the yielding conditions. Accordingly, no meaningful comparisons between the eHMI conditions could be made for the fear responses.

Table 3 shows the means and standard deviations of the participants' difficulty to predict the behavior of the oncoming cars. Participants found it more difficult to predict the car behavior for the no eHMI condition compared to when a Text eHMI was present. Significant differences between Text and No eHMI were observed for the 20 and 30 meters yielding conditions and the 30 meters nonyielding condition. Significant differences between FBL and Text were found for the nonyielding conditions only.

Table 3. Descriptive statistics and results from paired-sample *t*-tests of the comparison between the subjective responses of participants' difficulty to predict the behavior of oncoming vehicles when an eHMI was either present or absent.

	None	Text	Front Brake Lights
	M (*SD*)	*M* (*SD*)	*M* (*SD*)
20 meters yielding	4.48 (2.18)	2.43 (1.33)	3.43 (2.23)
20 meters nonyielding	3.63 (2.42)	2.25 (1.53)	3.88 (2.09)
30 meters yielding	4.36 (2.56)	2.91 (1.97)	4.00 (2.51)
30 meters nonyielding	4.28 (2.65)	2.89 (2.22)	4.06 (2.21)
20 meters yielding	*t*-test None vs. Text	$t(20) = 4.10, \boldsymbol{p < 0.001}$	
20 meters yielding	*t*-test None vs. FBL	$t(20) = 2.02, p = 0.057$	
20 meters yielding	*t*-test Text vs. FBL	$t(20) = -1.80, p = 0.087$	
20 meters nonyielding	*t*-test None vs. Text	$t(15) = 2.59, p = 0.021$	
20 meters nonyielding	*t*-test None vs. FBL	$t(15) = -0.62, p = 0.545$	
20 meters nonyielding	*t*-test Text vs. FBL	$t(15) = -3.64, \boldsymbol{p = 0.002}$	

Table 3. Cont.

		None	Text	Front Brake Lights
		M (SD)	*M (SD)*	*M (SD)*
30 meters yielding	*t*-test None vs. Text	$t(21) = 2.71, \boldsymbol{p = 0.013}$		
30 meters yielding	*t*-test None vs. FBL	$t(21) = 0.76, p = 0.459$		
30 meters yielding	*t*-test Text vs. FBL	$t(21) = -2.11, p = 0.047$		
30 meters nonyielding	*t*-test None vs. Text	$t(17) = 2.98, \boldsymbol{p = 0.008}$		
30 meters nonyielding	*t*-test None vs. FBL	$t(17) = 0.42, p = 0.679$		
30 meters nonyielding	*t*-test Text vs. FBL	$t(17) = -2.87, \boldsymbol{p = 0.011}$		

Note: None = No eHMI, Text = Text eHMI, FBL = Front Brake Lights. Significant differences ($p < 0.017$) are indicated in boldface.

3.6. Reported User Experience

Figure 7 shows the participants' mean responses to the three statements over the course of the experiment. A downwards trend is visible for the participants' difficulty to predict the behavior of oncoming cars as well as their feeling of fear when considering crossing the road. A slight upward trend is visible for participants' self-reported MISC rating. A linear test of within-subject contrasts showed the following: participants' difficulty to predict the behavior of the oncoming vehicle: $F(1,23) = 3.79, p = 0.064$, feeling of fear: $F(1,23) = 7.34, p = 0.012$ and motion sickness: $F(1,23) = 1.53, p = 0.229$.

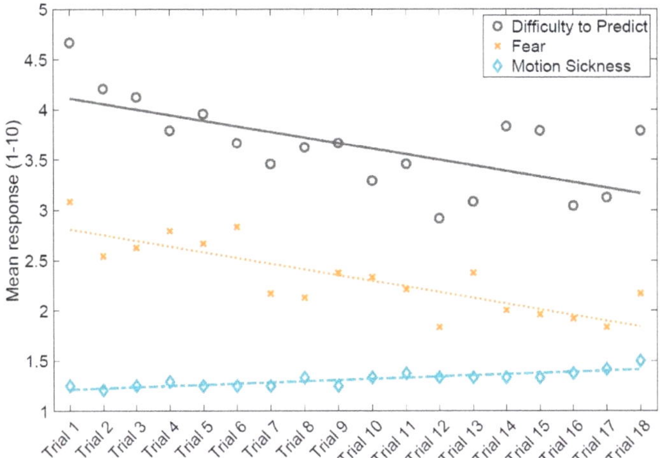

Figure 7. User experience; mean responses of the participants to the three statements after each trial, including least-squares regression lines.

Lastly, we compared our VRPQ responses to the results from [18], to investigate whether the implementation of a motion suit and virtual avatar enhanced participants' virtual immersion. Table 4 shows the mean (SD) response rates of the participants of this study and that of De Clercq et al. [18] to the factors of the VRPQ. No significant differences were found between responses in our study and that of De Clercq et al. [18] for any of the four presence factors. A detailed overview of the responses of our participants to each question of the VRPQ can be found in the Supplementary Materials (Table S3).

Table 4. Subjective evaluation of our study compared to De Clercq et al. [18]. Results from two-sample *t*-tests, including descriptive statistics.

	Responses Per Study								
	Kooijman et al.			De Clercq et al.					
	M	*SD*	*N*	*M*	*SD*	*N*	*p*	*t*	df
Involvement	5.10	0.65	24	4.94	0.62	28	0.375	0.89	50
Sensor Fidelity	4.79	1.01	24	5.07	0.72	28	0.251	−1.06	50
Adaptation/Immersion	5.56	0.64	24	5.68	0.72	28	0.546	−0.61	50
Interface Quality	3.19	0.97	24	2.99	1.03	28	0.462	0.74	50

Note: The four factors were defined according to Witmer et al. (2005). Responses are on a scale from 1 to 7.

4. Discussion

Previous literature has shown that participants consider a text-based eHMI to be less ambiguous than front brake lights (e.g., [6,9,18]). In these prior surveys and virtual reality studies, participants were not able to cross the road. Herein, we investigated the effect of eHMIs on participants' crossing behavior by visually and audibly immersing participants in a virtual environment where a schematic representation of the participant's body was present..

We hypothesized that the Text eHMI would be more persuasive than front brake lights (FBLs), and FBLs more persuasive than baseline with No eHMI. Our survey results in Table 3 are consistent with the existing literature, with No eHMI yielding the highest difficulty ratings followed by FBL and then Text. After a Bonferonni correction, the differences between eHMI conditions were significant in 5 out of 12 cases. The objective results, operationalized as forward velocities of the pedestrians, are consistent with these self-reports, with the Text eHMI yielding the highest forward velocities followed by FBL and no eHMI. Of course, these findings only apply to first-time exposures to the eHMIs. It is likely that after training or repeated exposure, participants will get to know the meaning of the front brake lights (see also [18]). The meaning of a front brake light may be confusing as participants may not know whether they should apply an egocentric or allocentric perspective [9]. An additional explanation for the efficacy of the Text eHMI is that it was larger and more salient than the FBL. A similar effect of stimulus size can be found in Ackermann et al. [8], where participants rated large street projections in front of the vehicle as more recognizable than a relatively small text display on the grill.

Significant differences in forward velocity between Text/FBL and no eHMI were found for yielding vehicles in the 20 meter gap condition. The higher *average* forward velocity for the eHMI conditions can be largely explained by the fact that participants started crossing sooner when an eHMI was present (see Table 2). This effect can also be seen from the graphs depicting the forward velocities for each participant (see Figures S5 and S6 in the Supplementary Materials). The fact that the effects were strongest for the condition with 20 meters inter-vehicle distance can be explained by the fact that crossing through a 20 meter gap was too dangerous without indication from an eHMI that the vehicle will stop. Crossing through a 30-meter gap, on the other hand, was feasible without an eHMI. Thus, the effect of an eHMI was relatively small in the 30-meter condition because some participants started to cross directly after the second car had passed. These findings indicate that a motion suit allows for extracting patterns that are not evident from self-reports.

In non-yielding conditions, no differences in forward velocity between eHMIs are to be expected because participants were unable to cross the road. Participants showed a slight forward motion in the case of nonyielding vehicles, which may point to hesitative behavior. We found clear effects on thorax motion as a function of elapsed time, with participants rotating their upper body towards the target cars, and straight ahead if they were crossing. However, we did not observe significant differences in thorax angle between the eHMI conditions. This lack of significant effect could be explained by the fact that thorax angle is subject to more inter- and intra-individual variability as compared to forward velocity.

An innovation of our setup was that participants could not only walk through the virtual environment but could also see a dynamic representation of their body while having their movements recorded. Although a similar principle has been presented by Doric et al. [36] and Feldstein et al. [25,27], no motion data were presented in those studies to investigate the effects of eHMIs on pedestrian crossing behavior. The implementation of an avatar in our study was expected to yield a compelling sense of presence. Petkova and Ehrsson [37], for example, found that some participants experienced a full-body ownership illusion in virtual reality. However, we did not find significant improvements in participants' subjective experience by utilizing a motion-tracking suit and implementing a virtual representation of the participants' bodies compared to a prior study without a motion suit by De Clercq et al. [18]. Although this comparison should be interpreted with caution, as [18] was conducted a year before the present study at a different university and used a fundamentally different measurement of willingness to cross, it does suggest that providing a person with a virtual body does not strongly enhance experienced fidelity. Although participants in our study reported high scores for being in control and able to move around, relatively low scores were obtained for quickly adjusting to the environment, visual display quality, and being able to identify sounds. Thus, it seems that the motion suit offers benefits for presence but at the same time may cause some usability issues. The lack of overall improvement in presence could be explained by the fact that the avatar was a robot-looking genderless avatar. Another factor of importance could be tactile feedback. Petkova and Ehrsson [36] found that the full-body illusion was evoked only if being stimulated by synchronous tactile-visual feedback. It is possible that, if stepping down the sidewalk onto the road could be felt by participants, this would yield a more compelling sense of presence compared to the current setup in which participants walked on the flat lab floor. Lee et al. [38] presented a cave-like simulator, argued to be the largest 4K-resolution pedestrian simulator in the world, to be used for eHMI research. Such a solution is also expected to give a high sense of presence, as participants can walk in a virtual world while being able to see their body without the need for a head-mounted display (although glasses for stereoscopic vision can be used as an option).

In conclusion, we confirmed that eHMIs influence pedestrians' actual crossing behavior compared to a baseline condition. The usage of a motion suit allows researchers to investigate subtle interaction patterns such as body angles and hesitative behaviors. Nuanced conclusions can be derived from such recordings compared to the discrete or binary information from survey studies (e.g., [6,9]) and virtual reality studies (e.g., [18,19]).

The present results allow for critical thinking about the value of high-fidelity setups for evaluating eHMIs. If a researcher's goal is merely to evaluate which type of eHMI is clearest, then an immersive virtual reality setup such as the present one may not be needed. Eisma et al. [39] found that asking people on a scale from 0 to 10 whether the eHMI is clear yielded results that correlated nearly perfectly ($r = 0.99$) with objective results measured using a response key. In turn, the present study found that pedestrians' forward velocity gave results that are similar to previous research using a response button [18]. However, if one's goal is to evaluate *how* people cross, then a high-fidelity setup such as the present one can be valuable.

For future research, we see merit in utilizing a motion-tracking suit in more complex traffic scenarios involving pedestrian-vehicle interaction. For example, it would be worthwhile to test the effectiveness of eHMIs in situations where participants have to distribute their attention, such as situations that involve bidirectional traffic flows, other pedestrians, and a mix of autonomous and conventionally driven vehicles. Additionally, the 3D recording of pedestrians in crossing situations could be beneficial to perception and modeling research (e.g., [40,41]), where the goal is to have self-driving vehicles detecting the posture of pedestrians and infer their crossing intention.

Supplementary Materials: The following are available online at http://www.mdpi.com/2078-2489/10/12/386/s1, Figures S1–S6, Tables S1–S3. A video showing two trials, raw data, and scripts can be found here: https://doi.org/10.4121/uuid:45378b74-4dab-465d-97dd-e593972d6125.

Author Contributions: Conceptualization, all authors; Methodology, all authors; Software, L.K.; Validation, J.C.F.d.W., R.H.; Formal analysis, L.K., J.C.F.d.W.; Investigation, L.K.; Resources, J.C.F.d.W., R.H.; Data curation, L.K., J.C.F.d.W.; Writing—original draft preparation, L.K., J.C.F.d.W.; Writing—review and editing, all authors; Visualization, L.K.; Supervision, J.C.F.d.W. & R.H.; Project administration, J.C.F.d.W. & R.H.; Funding acquisition, J.C.F.d.W. & R.H.

Funding: This research was supported by the research program VIDI with grant number TTW 016.Vidi.178.047 (2018–2022; "How should automated vehicles communicate with other road users?"), which is financed by the Netherlands Organisation for Scientific Research (NWO). Additionally, this research was supported by the NWO under the grant Safe Interaction with Vulnerable Road Users (SafeVRU) - TTW#14667.

Conflicts of Interest: The authors declare no conflict of interest. The funders had no role in the design of the study; in the collection, analyses, or interpretation of data; in the writing of the manuscript, or in the decision to publish the results.

References

1. WHO. *Global Status Report on Road Safety*; WHO: Geneva, Switzerland, 2015.

2. Singh, S. *Critical Reasons for Crashes Investigated in the Nation Motor Vehicle Crash Causation Survey*; National Highway Traffic Safety Administration: Washington, DC, USA, 2015.

3. Milakis, D.; Snelder, M.; Van Arem, B.; Van Wee, B.; De Almeida Correia, G.H. Development and transport implications of automated vehicles in the Netherlands: Scenarios for 2030 and 2050. *Eur. J. Transp. Infrastruct. Res.* **2017**, *17*, 63–85.

4. Matthews, M.; Chowdhary, G.; Kieson, E. Intent communication between autonomous vehicles and pedestrians. *arXiv* **2017**, arXiv:1708.07123.

5. Rasouli, A.; Kotseruba, I.; Tsotsos, J.K. Agreeing to cross: How drivers and pedestrians communicate. In Proceedings of the 2017 IEEE Intelligent Vehicles Symposium (IV), Los Angeles, CA, USA, 11–14 June 2017; pp. 264–269.

6. Fridman, L.; Mehler, B.; Xia, L.; Yang, Y.; Facusse, L.Y.; Reimer, B. To walk or not to walk: Crowdsourced assessment of external vehicle-to-pedestrian displays. *arXiv* **2017**, arXiv:1707.02698.

7. Chang, C.M.; Toda, K.; Igarashi, T.; Miyata, M.; Kobayashi, Y. A video-based study comparing communication modalities between an autonomous car and a pedestrian. In Proceedings of the 10th International Conference on Automotive User Interfaces and Interactive Vehicular Applications, Toronto, ON, Canada, 23–25 September 2018; pp. 104–109.

8. Ackermann, C.; Beggiato, M.; Schubert, S.; Krems, J.F. An experimental study to investigate design and assessment criteria: What is important for communication between pedestrians and automated vehicles? *Appl. Ergon.* **2019**, *75*, 272–282. [CrossRef] [PubMed]

9. Bazilinskyy, P.; Dodou, D.; De Winter, J.C.F. Survey on eHMI concepts: The effect of text, color, and perspective. *Transp. Res. Part F Traffic Psychol. Behav.* **2019**, *67*, 175–194. [CrossRef]

10. Deb, S.; Hudson, C.R.; Carruth, D.W.; Frey, D. Pedestrians receptivity in autonomous vehicles: Exploring a video-based assessment. *Proc. Hum. Factors Ergon. Soc. Annu. Meet.* **2018**, *62*, 2061–2065. [CrossRef]

11. Hagenzieker, M.P.; Van der Kint, S.; Vissers, L.; Van Schagen, I.N.L.G.; De Bruin, J.; Van Gent, P.; Commandeur, J.J.F. Interactions between cyclists and automated vehicles: Results of a photo experiment. *J. Transp. Saf. Secur.* **2019**. [CrossRef]

12. Dey, D.; Martens, M.; Eggen, B.; Terken, J. Pedestrian road-crossing willingness as a function of vehicle automation, external appearance, and driving behaviour. *Transp. Res. Part F Traffic Psychol. Behav.* **2019**, *65*, 191–205. [CrossRef]

13. Zhang, J.; Vinkhuyzen, E.; Cefkin, M. Evaluation of an autonomous vehicle external communication system concept: A survey study. In *International Conference on Applied Human Factors and Ergonomics*; Springer: Berlin/Heidelberg, Germany, 2017; pp. 650–661.

14. Mahadevan, K.; Sanoubari, E.; Somanath, S.; Young, J.E.; Sharlin, E. AV-Pedestrian interaction design using a pedestrian mixed traffic simulator. In Proceedings of the 2019 on Designing Interactive Systems Conference, San Diego, CA, USA, 18 June 2019; pp. 475–486.

15. Nuñez Velasco, P.; Farah, H.; Van Arem, B.; Hagenzieker, M. WEpod WElly in Delft: Pedestrians' crossing behaviour when interacting with automated vehicles using Virtual Reality. In Proceedings of the 15th International Conference on Travel Behaviour Research, Santa Barbara, CA, USA, 15–20 July 2018.

16. Schmidt, H.; Terwilliger, J.; AlAdawy, D.; Fridman, L. Hacking nonverbal communication between pedestrians and vehicles in virtual reality. *arXiv* **2016**, arXiv:1904.01931.

17. Walker, F.; Dey, D.; Martens, M.; Pfleging, B.; Eggen, B.; Terken, J. Feeling-of-safety slider: Measuring pedestrian willingness to cross roads in field interactions with vehicles. In Proceedings of the 2019 CHI Conference on Human Factors in Computing Systems, Glasgow, UK, 4–9 May 2019.

18. De Clercq, K.; Dietrich, A.; Núñez Velasco, J.P.; De Winter, J.; Happee, R. External human-machine interfaces on automated vehicles: Effects on pedestrian crossing decisions. *Hum. Factors* **2019**, *61*, 1353–1370. [CrossRef]

19. Ackermans, S.C.A. The Effects of Attitudes, Autonomous Appearance and Intention Communication in Pedestrian Interactions with Autonomous Vehicles. Master's Thesis, Eindhoven University of Technology, Eindhoven, The Netherlands, 2019.

20. Weber, F.; Chadowitz, R.; Schmidt, K.; Messerschmidt, J.; Fuest, T. Crossing the street across the globe: A study on the effects of eHMI on pedestrians in the US, Germany and China. In *HCI in Mobility, Transport, and Automotive Systems. HCII 2019*; Lecture Notes in Computer Science; Krömker, H., Ed.; Springer: Berlin/Heidelberg, Germany, 2019; Volume 11596.

21. Clamann, M.; Aubert, M.; Cummings, M.L. Evaluation of vehicle-to-pedestrian communication displays for autonomous vehicles. In Proceedings of the TRB 96th Annual Meeting Compendium of Papers 2017, Washington, DC, USA, 8–12 January 2017.

22. Palmeiro, A.R.; Van der Kint, S.; Vissers, L.; Farah, H.; De Winter, J.C.F.; Hagenzieker, M. Interaction between pedestrians and automated vehicles: A Wizard of Oz experiment. *Transp. Res. Part F Traffic Psychol. Behav.* **2018**, *58*, 1005–1020. [CrossRef]

23. Cefkin, M.; Zhang, J.; Stayton, E.; Vinkhuyzen, E. Multi-methods research to examine external HMI for highly automated vehicles. In *HCI in Mobility, Transport, and Automotive Systems. HCII 2019*; Lecture Notes in Computer Science; Krömker, H., Ed.; Springer: Berlin/Heidelberg, Germany, 2019; Volume 11596.

24. Rothenbücher, D.; Li, J.; Sirkin, D.; Mok, B.; Ju, W. Ghost driver: A field study investigating the interaction between pedestrians and driverless vehicles. In Proceedings of the 2016 25th IEEE International Symposium on Robot and Human Interactive Communication (RO-MAN), New York, NY, USA, 26–31 August 2016; pp. 795–802.

25. Feldstein, I.; Dietrich, A.; Milinkovic, S.; Bengler, K. A pedestrian simulator for urban crossing scenarios. *IFAC PapersOnLine* **2016**, *49*, 239–244. [CrossRef]

26. Le Lee, Y.M.; Uttley, J.; Solernou, A.; Giles, O.; Romano, R.; Markkula, G.; Merat, N. Investigating pedestrians' crossing behaviour during car deceleration using wireless head mounted display: An application towards the evaluation of eHMI of automated vehicles. In Proceedings of the Tenth International Driving Symposium on Human Factors in Driving Assessment, Training and Vehicle Design. 2019 Driving Assessment Conference, Santa Fe, NM, USA, 24–27 June 2019; pp. 252–258.

27. Feldstein, I.T.; Lehsing, C.; Dietrich, A.; Bengler, K. Pedestrian simulators for traffic research: State of the art and future of a motion lab. *Int.J. Human Factors Model. Simul.* **2018**, *6*, 250–265. [CrossRef]

28. MVN Analyze [Computer Software]. Motion Capture Software. Available online: https://www.xsens.com/products/mvn-analyze (accessed on 28 November 2019).

29. Papadimitriou, E.; Lassarre, S.; Yannis, G. Human factors of pedestrian walking and crossing behaviour. *Transp. Res. Procedia* **2017**, *25*, 2002–2015. [CrossRef]

30. Witmer, B.G.; Jerome, C.J.; Singer, M.J. The factor structure of the presence questionnaire. *Presence Teleoperators Virtual Environ.* **2005**, *14*, 298–312. [CrossRef]

31. Roether, C.L.; Omlor, L.; Christensen, A.; Giese, M.A. Critical features for the perception of emotion from gait. *J. Vis.* **2009**, *9*, 15. [CrossRef]

32. Crane, E.; Gross, M. Motion capture and emotion: Affect detection in whole body movement. In *International Conference on Affective Computing and Intelligent Interaction*; Springer: Berlin/Heidelberg, Germany, 2007; pp. 95–101.

33. Schreven, S.; Beek, P.J.; Smeets, J.B. Optimising filtering parameters for a 3D motion analysis system. *J. Electromyogr. Kinesiol.* **2015**, *25*, 808–814. [CrossRef]

34. Turner, S.D. qqman: An R package for visualizing GWAS results using QQ and manhattan plots. *Biorxiv* **2014**, 5165. [CrossRef]

35. Be Benjamin, D.J.; Berger, J.O.; Johannesson, M.; Nosek, B.A.; Wagenmakers, E.J.; Berk, R.; Bollen, K.A.; Brembs, B.; Brown, L.; Camerer, C.; et al. Redefine statistical significance. *Nat. Hum. Behav.* **2018**, *2*, 6–10. [CrossRef]

36. Doric, I.; Frison, A.K.; Wintersberger, P.; Riener, A.; Wittmann, S.; Zimmermann, M.; Brandmeier, T. A novel approach for researching crossing behavior and risk acceptance: The pedestrian simulator. In Proceedings of the 8th International Conference on Automotive User Interfaces and Interactive Vehicular Applications, Ann Arbor, MI, USA, 24–26 October 2016; pp. 39–44.

37. Petkova, V.I.; Ehrsson, H.H. If I were you: Perceptual illusion of body swapping. *PLoS ONE* **2018**, *3*, e3832. [CrossRef]

38. Lee, Y.M.; Madigan, R.; Markkula, G.; Pekkanen, J.; Merat, N.; Avsar, H.; Utesch, F.; Schieben, A.; Schießl, C.; Dietrich, A.; et al. interACT D.6.1. Methodologies for the Evaluation and Impact Assessment of the InterACT Solutions. Available online: https://www.interact-roadautomation.eu/wp-content/uploads/interACT_D6.1_01082019_v1.0_uploadWebsite_approved_reduced-size-1.pdf (accessed on 17 November 2019).

39. Eisma, Y.B.; Bergen, S.; Brake, S.; Hensen, M.; Tempelaar, W.; De Winter, J.C.F. External human-machine interfaces: The effect of display location on crossing intentions and eye movements. *Information* **2020**, *11*, 13.

40. Flohr, F.; Dumitru-Guzu, M.; Kooij, J.F.; Gavrila, D.M. A probabilistic framework for joint pedestrian head and body orientation estimation. *IEEE Trans. Intell. Transp. Syst.* **2015**, *16*, 1872–1882. [CrossRef]

41. Kooij, J.F.P.; Schneider, N.; Flohr, F.; Gavrila, D.M. Context-based pedestrian path prediction. In *European Conference on Computer Vision*; Springer: Berlin/Heidelberg, Germany, 2014; pp. 618–633.

Article

External Human–Machine Interfaces: The Effect of Display Location on Crossing Intentions and Eye Movements

Y. B. Eisma, S. van Bergen, S. M. ter Brake, M. T. T. Hensen, W. J. Tempelaar and J. C. F. de Winter *

Department Cognitive Robotics, Delft University of Technology, Mekelweg, 2628 CD Delft, The Netherlands;
Y.B.Eisma@tudelft.nl (Y.B.E.); S.vanBergen@student.tudelft.nl (S.v.B.); S.M.terBrake@student.tudelft.nl (S.M.t.B.);
M.T.T.Hensen@student.tudelft.nl (M.T.T.H.); W.J.Tempelaar@student.tudelft.nl (W.J.T.)
* Correspondence: j.c.f.dewinter@tudelft.nl

Received: 19 November 2019; Accepted: 17 December 2019; Published: 24 December 2019

Abstract: In the future, automated cars may feature external human–machine interfaces (eHMIs) to communicate relevant information to other road users. However, it is currently unknown where on the car the eHMI should be placed. In this study, 61 participants each viewed 36 animations of cars with eHMIs on either the roof, windscreen, grill, above the wheels, or a projection on the road. The eHMI showed 'Waiting' combined with a walking symbol 1.2 s before the car started to slow down, or 'Driving' while the car continued driving. Participants had to press and hold the spacebar when they felt it safe to cross. Results showed that, averaged over the period when the car approached and slowed down, the roof, windscreen, and grill eHMIs yielded the best performance (i.e., the highest spacebar press time). The projection and wheels eHMIs scored relatively poorly, yet still better than no eHMI. The wheels eHMI received a relatively high percentage of spacebar presses when the car appeared from a corner, a situation in which the roof, windscreen, and grill eHMIs were out of view. Eye-tracking analyses showed that the projection yielded dispersed eye movements, as participants scanned back and forth between the projection and the car. It is concluded that eHMIs should be presented on multiple sides of the car. A projection on the road is visually effortful for pedestrians, as it causes them to divide their attention between the projection and the car itself.

Keywords: eHMI; eye-tracking; attention distribution; road safety; automated driving; driverless vehicles

---

1. Introduction

In recent years, a substantial number of studies have emerged on external human–machine interfaces (eHMIs) for automated cars. In automated driving, non-verbal communication between the driver and other road users is often impossible, because the driver is not physically present in the driver seat, or because the driver is engaged in a non-driving task. One reason for employing eHMIs would be to substitute the lack of eye-contact and other types of non-verbal communication. A second reason for using eHMIs is to transmit information about the future state of the automated vehicle to other traffic participants. For example, if the path planning software of the automated driving system knows that the vehicle will slow down for an upcoming intersection, the eHMI could accordingly communicate that the vehicle is about to slow down [1]. Thus, eHMIs could communicate information that is not apparent from implicit ways of communication, for example, from the car's acceleration and deceleration.

So far, a number of different eHMIs have been designed. Bazilinskyy et al. [2] provided an overview of 22 eHMI concepts from industry, whereas Rasouli and Tsotsos [3] and Schieben et al. [4]

presented a survey of eHMIs that are studied in academic contexts. The eHMIs proposed so far come in a variety of modalities, for example as text and light strips (e.g., as in [5]), as well as in many colours (green, red, cyan; [6,7]). Research has found that text-based eHMIs are regarded as easily understood without learning [1,8], and that text has disadvantages related to legibility from a distance and cross-national interpretability [2]. A scientific consensus regarding the most efficient modality for eHMIs has not been reached so far.

A lesser studied question is where on the car the eHMI should be positioned to attain maximum compliance and decision-making efficiency. A variety of locations for eHMIs have been proposed, including:

1. The windscreen [9–12]
2. The front/grill of the car [1,12–22]
3. The roof of the car [23–26]
4. Near the wheels [27] (also proposed by Colley et al. [28])
5. A projection on the road [8,9,23,29–33]

The positioning of the eHMI is important because pedestrians (and other road users) visually sample the road environment in an intermittent matter [34]. The presented information may be critical to road safety, and should be understood early in time.

From the existing body of literature, an eHMI on the front (grill) or roof of the car seems to be the most frequently used option. These locations are justifiable because they may easily allow for mounting a communication device. An eHMI that projects a message on the road or an eHMI that is integrated with the windscreen are challenging to manufacture. However, these types of eHMIs hold promise because they can be made larger than regular screen-based eHMIs, enhancing their visibility from a distance. This notion is supported by a study using self-reports by Ackermann et al. [9]. They showed that participants found eHMIs that projected its messages on the windscreen or the ground were regarded as better recognisable than display-based eHMIs. Ackermann et al. [9] pointed out that the relatively large size of the projections was probably an underlying reason for these effects.

Even though research (e.g., [35]) shows that pedestrians and drivers do not make direct eye contact very often, an eye-tracking study by Dey et al. [36] showed that pedestrians tend to look at the windscreen when an approaching car is close by, "likely to seek the intention or information about the situational awareness of the driver" (p. 375). Accordingly, a windscreen-based eHMI may be an attractive location for presenting a message. In the same way, Bazilinskyy et al. [37] found that pedestrians often look at the wheels of parked cars; this provides motivation for using a wheel-based eHMI.

At present, it is unclear which location of the eHMI results in the best-perceived clarity and behavioural compliance among pedestrians. This lack of knowledge impedes the standardisation of eHMI designs. In the present study, we let participants view animated video clips in which automated vehicles drove with an eHMI at one of the five abovementioned locations. Participants were asked to hold the spacebar when they felt safe to cross. Consequently, we examined which type of eHMI resulted in the highest time-percentage of spacebar pressings while the automated vehicle slowed down for the participant. This is a continuous behavioural measurement method that was introduced by De Clercq et al. [1]. Additionally, we used eye-tracking to infer which type of eHMI yields the most concentrated gaze patterns.

A survey of eHMI concepts proposed by the automotive industry indicated that about 50% of the concepts contained a text message of some kind [2]. Research has also shown that the commanding text 'Walk' can be understood without particular training or prior exposure [1,2]. However, the development of commanding-text eHMIs is technologically challenging, because such design requires that the automated vehicle knows for which road user the command is meant. Another disadvantage of commanding texts concerns liability: if an automated vehicle displays 'Walk', and a pedestrian walks onto the road and collides with a third road user, the manufacturer of the automated vehicle may be at fault.

It has further been shown that a light-based eHMI can be perceived as ambiguous without learning [1,8]. For example, it may be unclear whether a green or red light signal applies to the pedestrian (egocentric perspective) or the automated vehicle (allocentric perspective; [2]).

Our eHMIs consisted of non-commanding text ('Waiting' or 'Driving') combined with an icon. The text on the eHMI was white to avoid the above-mentioned red/green dilemma. We opted for a relatively salient (i.e, large display/projection) and redundant (i.e., text combined with an icon) eHMI to ensure that participants would have no difficulty understanding what the eHMI message means. We do not aim to suggest that a text-based eHMI would be the optimal solution in real traffic. However, because the present study is concerned with examining the effect of eHMI location, we selected an eHMI design that was shown to be effective in previous research in virtual environments.

2. Methods

2.1. Participants

The participants were 51 males and 10 females. They were all aged between 19 and 27 years ($M = 23.0$, $SD = 1.8$). The participants were all students of BSc and MSc studies at the faculty of Mechanical, Maritime and Materials Engineering at the Delft University of Technology, the Netherlands. About half of the participants were recruited based on opportunity sampling within the faculty building, whereas the other half participated for course credit. All participants provided written, informed consent. The research was approved by the TU Delft Human Research Ethics Committee.

2.2. Apparatus

Eye movements were recorded at 2000 Hz using the Eyelink 1000 Plus eye-tracker v5.15 (SR-Research; Ottawa, ON, Canada). Participants were asked to place their head in the head support during the entire experiment. The stimuli were shown on a 24-inch BENQ monitor (Taipei, Taiwan) with a resolution of 1920 × 1080 pixels (531 × 298 mm). The refresh rate of the monitor was set at 60 Hz. The distance between the monitor and the head support was 95 cm. Accordingly, the monitor subtended 31 deg and 18 deg horizontal and vertical viewing angles, respectively. The experimental setup is shown in Figure 1.

Figure 1. Experimental setup. In the actual experiment, the windows were blinded with aluminium foil.

2.3. Independent Variable

The independent variable was the eHMI type. Six eHMI conditions were used: Roof, Windscreen, Grill, Projection, Wheels, and No eHMI. Figure 2 shows a car that combines all five eHMIs. In the experiment, only one eHMI condition was used at a time. The eHMI could show either 'Waiting' or 'Driving' (Figure 3). The 'Driving' message turned on when the approaching car would not stop for the pedestrian. The 'Waiting' message turned on when the approaching car would stop for the pedestrian.

This study was designed to examine participants' responses when the car was stopping and the eHMI showed 'Waiting'. The responses to the non-stopping vehicles were not analysed herein. The non-stopping vehicles were included to ensure that participants would not start to expect that all cars would stop for them. Note that stopping vehicles had a dominant effect on participants' spacebar-pressing behaviours, whereas no meaningful differences in spacebar-press behaviour between the eHMI conditions occurred for non-stopping vehicles. For example, when the stopping vehicle drove off, it became unsafe to cross, and participants released the spacebar. A non-stopping vehicle that was approaching at that time could not affect spacebar-pressing behaviour because participants already had the spacebar released. We used white text together with a symbol on a black background to achieve the highest possible contrast, because colours (e.g., red and green) already have a meaning, yet this meaning becomes ambiguous when the colour is presented on an approaching vehicle [2].

Figure 2. Car combining all five external human–machine interfaces (eHMIs). In the experiment, the car showed only one eHMI at a time. Here, the car has stopped for the pedestrian. The distance between the centre of the car and the camera (pedestrian) is 7 m longitudinal (i.e., parallel to the direction of the road) and 4.5 m lateral (i.e., perpendicular to the road). The white markings on the road were intended to create a pedestrian crossing on the road, without designated priority to the pedestrian.

Figure 3. (a) Image presented on the eHMI when the approaching car stopped for the pedestrian, (b) Image presented on the eHMI when the approaching car did not stop for the pedestrian.

2.4. Design of the Animated Video Clips

The experiment consisted of 36 non-interactive animated video clips: 6 virtual environments × 6 eHMI conditions. All cars drove at a speed of about 35 km/h unless slowing down for the pedestrian. The videos were 25 s long and played at 60 frames/s. Three environments were used: a straight road, a T-junction and an intersection, with two different preprogrammed traffic behaviours per eHMI. Accordingly, there were six videos per eHMI condition. The lane width was 3.66 m (a standard lane width, e.g., [38]). The camera perspective was from the eyes of a pedestrian waiting to cross the road at a crossing with a traffic island. The field of view of the animation was 80 deg, which ensured that a large part of the environment could be seen (e.g., cars making a right turn, cars driving straight on, and cars making a left turn). In each video, cars were driving on both lanes. The cars did not contain a driver or passenger. This was done to resemble future driverless vehicles, which may transport goods rather than people.

Within a video, all cars featured the same eHMI type. The eHMI could show one of two messages: If the approaching car passed without slowing down, the eHMI changed from blank to 'Driving' (Figure 3, right). If the approaching car did stop for the participant, the eHMI changed from blank to 'Waiting' (Figure 3, left). The change of state from blank to 'Waiting' occurred when the longitudinal distance between the center of the car and the pedestrian was 23 m. After 1.2 s, when the longitudinal distance had reduced to 11 m, the car started to decelerate to a full stop. The car came to a full stop 2.0 s after the eHMI had switched on, at a longitudinal distance of 7 m between the center of the car and the pedestrian (Figure 2). About 2 s after the car had come to a full stop, the eHMI switched to blank again. About 1.2 s later, the car drove off and passed the participant. These timing and distance parameters yielded a scenario in which cars drove by and stopped in rapid succession. The traffic was not created according to actual traffic data or models of human behaviour.

As stated above, there were six videos per eHMI condition, with each video showing a different traffic environment. The traffic environments were the same for each eHMI, except for a temporal offset (up to 10 s) of the starting moments and corresponding ending moments of the video clips. This offset was included to encourage that participants could not recognise/memorise the behaviour of the cars in the video. In each of the six traffic-environment videos for a particular eHMI condition, one or two of the approaching cars stopped and subsequently drove away. In total, across the six traffic-environment videos per eHMI condition, ten approaching cars stopped for the participant. Details about the video clips and data exclusions are available in the Supplementary Material (Figures S1–S6).

2.5. Procedure and Task

Participants first read and signed an informed consent form. Next, the eye-tracker was calibrated. Then, participants performed two 10 s training scenarios. These concerned an empty straight road, showing a single car without eHMI; this car approached, stopped and drove off. The participants' task was to press and hold the spacebar whenever they felt it was safe to cross the road. Subsequently, the participants viewed the 36 animated video clips in random order. After each scenario, the participants were asked to rate their perceived clarity with the statement: 'It was clear when I could cross' on a scale from 0 (completely disagree) to 10 (completely agree).

2.6. Dependent Variables

- We calculated the following dependent variables:
- Self-reported clarity on a scale from 0 (completely disagree) to 10 (completely agree).
- Percentage of time that the participant had the spacebar pressed since the moment the eHMI switched to 'Waiting' until 3 s after. A higher percentage indicated a better performance (i.e., indicating when it is safe to cross when it is indeed safe to cross).

- Percentage of time that the participant had the spacebar released since the moment the eHMI switched off before driving away until 3 s after. Again, a higher percentage indicates better performance (i.e., indicating that it is not safe to cross when it is indeed unsafe to cross).
- Gaze spread in pixels. We calculated, for each time sample, the distance between the participant's x and y gaze coordinates and the mean x and y gaze coordinates of all participants. The gaze spread is the average distance from the moment the eHMI switched to 'Waiting' until 3 s later.

2.7. Statistical Analyses

The effects of eHMI type on the dependent variables were assessed using a repeated-measures analysis of variance (ANOVA), after averaging the performance scores of the individual vehicle approaches per participant. Significant differences between conditions were assessed with MATLAB's *multcompare* function, using the Tukey–Kramer critical value.

3. Results

3.1. Self-Reported Clarity

Figure 4 shows the results for self-reported clarity per eHMI condition. There was a significant difference between the six eHMI conditions, $F(5,300) = 114.4$, $p < 0.001$, $\eta_p^2 = 0.66$. Pairwise comparisons showed that Roof, Windscreen, and Grill were not significantly different from each other. The mean clarity scores between the other combinations differed significantly.

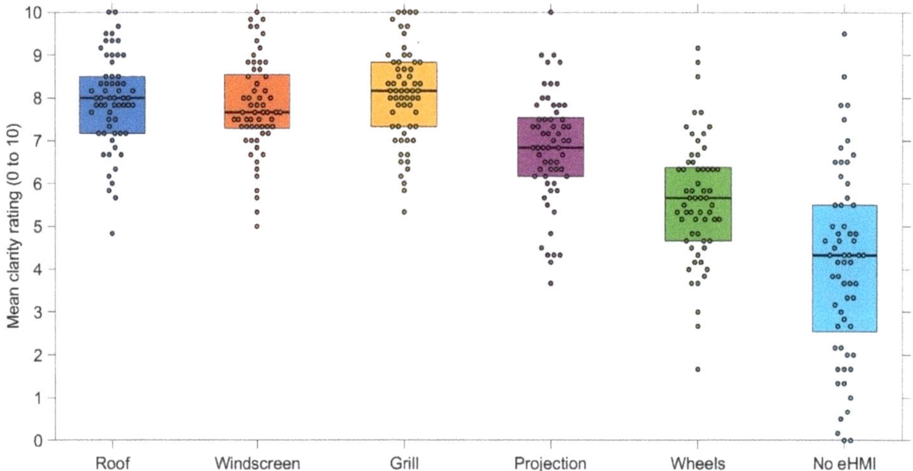

Figure 4. Mean self-reported clarity rating per participant. An average is taken of the scores of six scenarios per participant.

3.2. Performance for Approaching Cars

Figure 5 shows the performance scores, averaged for the nine approaches where the car drove straight on or made a left turn before stopping for the pedestrian. The six eHMI conditions were significantly different from each other, $F(5,300) = 130.1$, $p < 0.001$, $\eta_p^2 = 0.68$. Again, Roof, Windscreen, and Grill were not significantly different from each other, whereas all other combinations differed significantly.

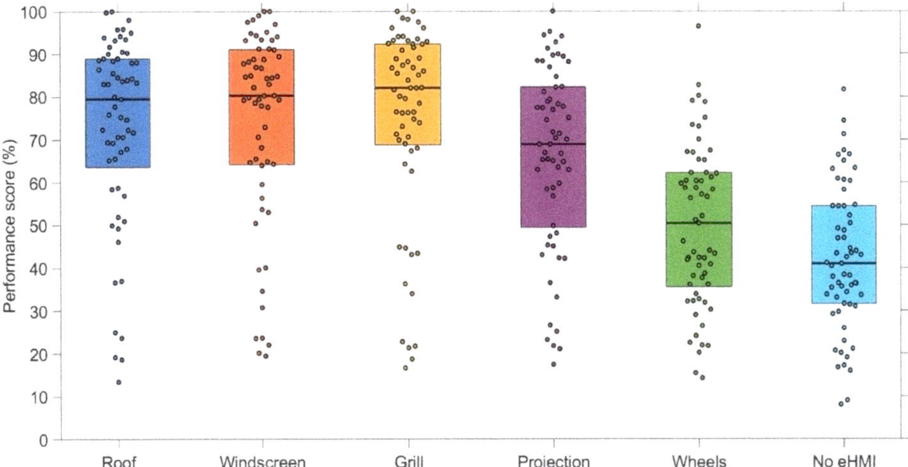

Figure 5. Mean performance score per participant for car approaches. The performance score is defined as the percentage of time that the spacebar was pressed, from the moment the eHMI turned on until 3 s later. The average is taken for the nine approaches where the car drove straight on or made a left turn before stopping for the pedestrian.

Figure 6 illustrates participants' spacebar pressing behaviour as a function of elapsed time since the moment of eHMI onset at $t = 0$ s. It can be seen that initially (between 0 and 0.5 s), the percentage of participants pressing the spacebar dropped with time, which can be explained by the fact that the approaching car kept getting closer; hence, it became less safe to cross. The Roof, Windscreen, and Grill caused participants to press the spacebar at about 0.5 s since the eHMI turned on. The Projection and especially Wheels triggered a later spacebar-press response, presumably because these eHMIs were poorly visible from a distance; see Figure 7 for an illustration. Figure 6 also shows that for No eHMI, participants only started to press the spacebar once they could detect that the car decelerated (the car decelerated between 1.2 and 2.0 s).

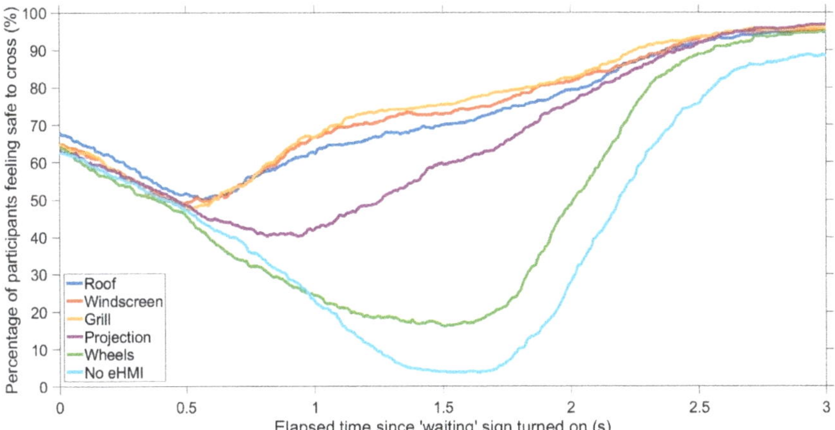

Figure 6. Percentage of participants who pressed the spacebar during car approaches. The average was taken for the nine approaches where the car drove straight on or made a left turn. $t = 0$ s: the eHMI turns on. $t = 2$ s: the car has come to a stop.

Figure 7. Screenshot of the animation in a straight approach case with the Projection eHMI. The yellow markers represent the gaze positions of all of the participants. The projection in front of the car is difficult to discern from a distance.

Figure 8 shows the performance score for one selected approach condition: a case where the approaching car made a right turn. Again, the difference in performance scores was significant, $F(5,300) = 10.6$, $p < 0.001$, $\eta_p^2 = 0.15$. All five eHMIs differed significantly from the No eHMI condition, and Wheels differed significantly from Roof and Grill. In other words, in straight and left approach cases, Wheels yielded the lowest performance (Figures 5 and 6), whereas in the right-turn case, Wheels yielded the highest performance (Figure 8).

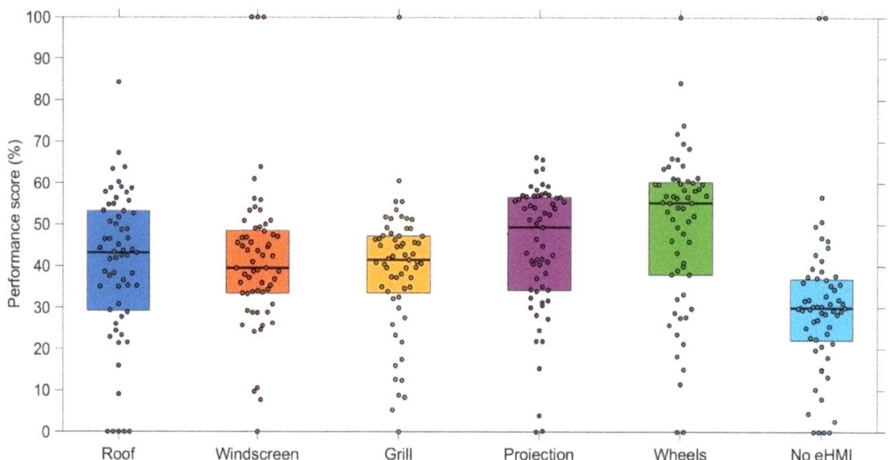

Figure 8. Mean performance score per participant for car approaches where the car made a right turn before stopping for the pedestrian. The performance score is defined as the percentage of time that the spacebar was pressed, from the moment the eHMI turned on until 3 s later.

The high performance for Wheels, and to a lesser extent for Projection, can be explained by the visibility of the sign in the right-turn case (Figure 9). The Roof, Windscreen, and Grill, however, only became visible after the car had made the turn.

Figure 9. Screenshot of the animation in the right-turn approach case with the Wheels eHMI. The yellow markers represent the gaze positions of the participants.

The results above showed similar results for self-reported clarity and objective performance. In order to describe the degree of similarity, we averaged the performance scores and clarity scores for all participants per eHMI. The results, shown in Figure 10, reveal a strong association ($r = 0.99$). In other words, in the aggregate, it appears that clarity and performance are both affected by the same mechanism, which we think is the visibility/readability of the display.

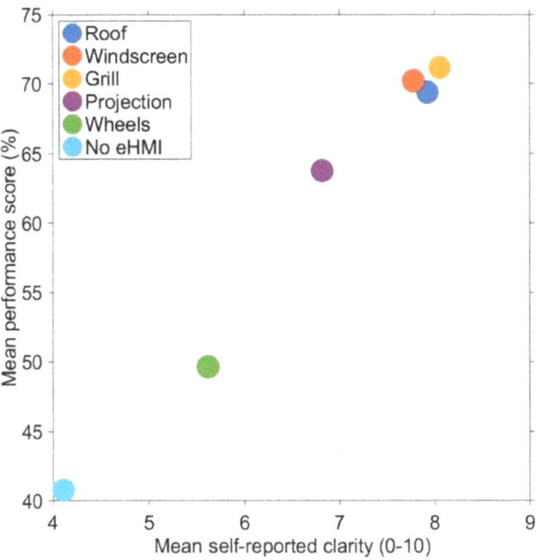

Figure 10. Overall mean self-reported clarity versus overall mean performance score during car approaches. The performance score is defined as the percentage of time that the spacebar was pressed, from the moment the eHMI turns on until 3 s later.

3.3. Eye-Movements for Approaching Cars

A visual inspection of the participants' eye movements indicated that these were often goal-directed, focusing on future interactions. For example, in Figure 11, the majority of participants looked at the approaching car even before the eHMI had turned on; participants did not necessarily look towards the nearest or more salient car. Furthermore, we found that participants' attention distribution was sometimes dispersed (e.g., when multiple cars were visible) and at other times concentrated (e.g., when a relevant car approached the participant, e.g., Figure 9). Herein, we introduce a new measure to describe the degree of gaze dispersion. We defined dispersion as the mean distance from the participants' overall mean gaze coordinate for that particular animated video clip. A dispersion score of, e.g., 200 pixels, means that participants' gaze was, on average, 200 pixels away from the mean fixation gaze position of all participants.

Figure 11. Screenshot of the animation in an intersection scenario. The yellow markers represent the gaze position of the participants.

The results of the gaze dispersion analysis (Figure 12) show that approaching cars attracted attention, as evidenced by low dispersion (<150 pixels) for the No eHMI condition while the car was approaching (0 to 2 s). The Wheels attracted attention, especially just before coming to a stop (from 1 to 2 s). The Projection, on the other hand, resulted in diversified attention, as illustrated in Figure 13. The Windscreen, on the other hand, yielded in a low gaze dispersion when the car was standing still. The eye-movement dispersion was significantly different between the six eHMI conditions, $F(5,300) = 31.4$, $p < 0.001$, $\eta_p^2 = 0.34$. The Projection yielded a significantly higher dispersion than all five other conditions. The Wheels yielded a significantly lower dispersion than all conditions, except for Windscreen. The Windscreen yielded a significantly lower dispersion than Roof and Projection.

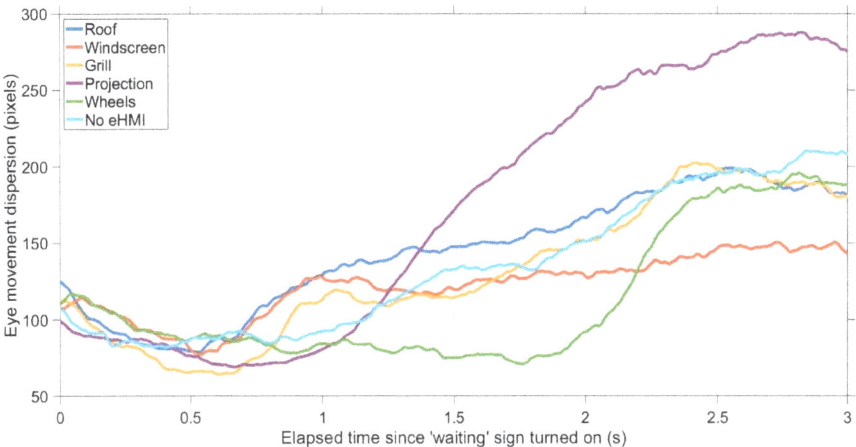

Figure 12. Eye movement dispersion score during car approaches. The average was taken of the nine approaches where the car drove straight on or made a left turn. $t = 0$ s: the eHMI turned on. $t = 2$ s: the car has come to a stop.

Figure 13. Screenshot of the animation in a straight approach scenario with the Projection eHMI. The yellow markers represent the gaze positions of the participants. The Projection results in dispersed eye gaze, with some participants looking at the eHMI on the asphalt and other participants looking at the car.

3.4. Performance for Cars Driving off

So far, we examined only the performance of eHMI for approaching cars. Another relevant aspect of eHMI evaluation is how participants respond after the eHMI switches off before the car drives away. Figure 14 shows that all eHMIs resulted in improved performance compared to No eHMI; that is, participants were more likely to release the spacebar before the car drove off. Initially (at $t = 0$ s), participants using one of the five eHMIs had the spacebar pressed, because the eHMI displayed 'Waiting' until that point. It took about 0.2 for the first participants to release the spacebar after this eHMI message disappeared. Participants in the No eHMI condition started to release the spacebar only after the car drove off (at 1.4 s), see Figure 14.

An analysis of the performance scores (Figure 15) showed a significant difference between the five eHMI conditions, $F(5,300) = 37.4$, $p < 0.001$, $\eta_p^2 = 0.38$. The No eHMI condition differed significantly from the five other eHMI conditions; there were no significant differences between Roof, Windscreen, Grill, Projection, and Wheels. In other words, participants responded similarly to the eHMI turning off, regardless of the type of eHMI.

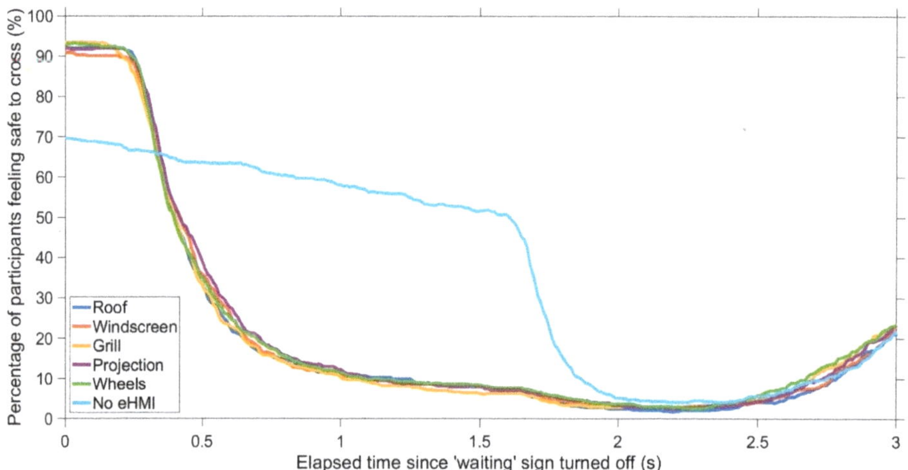

Figure 14. Eye movement dispersion score while the car was driving off. The average is taken of nine times driving off. $t = 0$ s: the eHMI turned off. $t = 1.4$ s: the car started to accelerate.

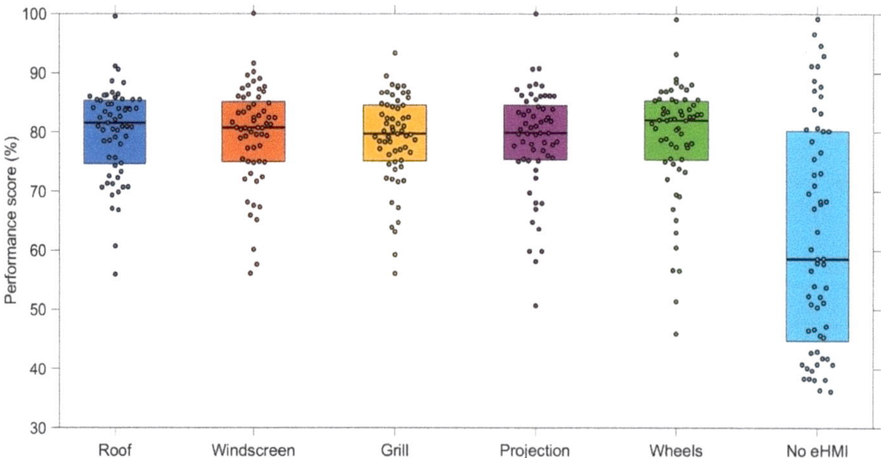

Figure 15. Mean performance score per participant for cases where the car drove off. The performance score is defined as the percentage of time that the spacebar was released, from the moment the eHMI turned off until 3 s later. For each participant, the average is taken of nine times driving off.

4. Discussion

In this study, five eHMI locations, together with a baseline No eHMI condition, were compared in a within-subjects design using a total of 61 participants. The participants viewed animated video clips and were asked to press and hold the spacebar when they thought it was safe to cross, while their eye-movements were recorded using an eye-tracker.

4.1. Performance

The results showed that the Roof, Windscreen, and Grill-based eHMIs yielded the best performance, defined in terms of the pressing time of the spacebar when it was safe to cross. However, this finding did not hold in all scenarios; the eHMI right above the wheel was found to be the best-performing eHMI when the car approached from a corner. In this specific scenario, the eHMIs on the front (Roof, Windscreen, and Grill) were not visible, and therefore failed to communicate their messages to the pedestrian. Together, our findings suggest that eHMIs should be omnidirectional if they are to be applied in traffic scenarios where cars can approach from multiple directions. Vlakveld et al. [26] showed animations of cars with an omnidirectional eHMI on the roof, whereas drive.ai [27] used multiple displays on the car's exterior. Another solution to ensure visibility from all sides is to use a light emitting diode (LED) strip as in Cefkin et al. [39], or LED patterns on the lateral surfaces of the car [40].

The Projection yielded poor spacebar-pressing performance when the car was approaching. This finding can be explained by the poor visibility of the projection at a far distance due to the shallow viewing angle. We do not mean to suggest that our results generalize to all possible projections. In a virtual reality study, Löcken et al. [31] tested different animations of eHMIs, including a projection which they dubbed F015 (after the name of the concept car presented by Mercedes–Benz USA [33]). Their results showed that the F015 yielded high ratings (5.7 on a scale from 1 to 7) on the User Experience Questionnaire. The concept of Löcken et al. [31] differed from ours, as their projection was highly salient, consisting of a bright green zebra message for the pedestrian. Our findings point to limitations in the use of projections that move with the car, as a projection may not be clear from a distance. We expect that these limitations will be more severe in real traffic. Although technologically feasible (e.g., [41]), it may require powerful lasers to ensure that a projection is visible on the road in daylight. An eHMI on a windscreen may also be technologically challenging to achieve, and may have variable contrast depending on whether or not the eHMI is mounted on a transparent windscreen or whether the windscreen is blinded (in the case of level 5 autonomous vehicles).

For the events where the car was driving away, and the eHMI switched from 'Waiting' to a blank display, all five eHMI locations were found to yield equivalent performance. These findings can be explained because the removal of the message was a salient event, which participants could detect independent of eHMI location or even message content.

Our findings indicate that it is possible to convince users to cross or not to cross before the car slows down or drives away. In other words, all eHMI locations were shown to evoke a more accurate response compared to the No eHMI condition.

4.2. Eye-Tracking

The eye-tracking results showed that the Windscreen eHMI yielded a concentrated gaze pattern, which can be explained by the fact that this eHMI is embedded in the centre of the car. This finding is in line with Dey et al. [36], who showed that pedestrians are inclined to look at the windscreen when an oncoming car gets close to the pedestrian. The Wheels eHMI also yielded a concentrated gaze pattern, but only for a brief period of about 1 s before the car came to a full stop. This finding may be explained by the fact that the Wheels eHMI was poorly visible from a distance; when the car came close to the participant, they were inclined to fixate on the eHMI to read its message.

We found that the Projection eHMI yielded a dispersed eye-movement pattern, a finding that can be attributed to the fact that participants looked at the projection and the car itself. These results are consistent with Powelleit et al. [42], who tested a projection in front of the car showing the predicted vehicle trajectory. The results of Powelleit et al. [42] showed that drivers found such a display distracting. Similarly, we see a risk that a projection on the road may result in distraction, where road users may fixate on the projection on the road at the expense of attention towards the car itself, and therefore may miss relevant implicit cues.

Such results have been found in the use of visual augmented feedback in air traffic control: Eisma et al. [43] found that augmented visual feedback helps to achieve a better task performance, but also has distraction potential.

4.3. Self-Reports

An interesting result was that, in the aggregate, self-reported clarity was strongly associated with objective performance, with a correlation of 0.99. This strong correlation may be due to a single underlying factor, such as the legibility of the display. In other words, the Projection and Wheels eHMIs were hard to read from a distance, as a result of which participants pressed the spacebar late and gave a low clarity rating. The strong correlation between subjective and objective performance is promising for those who examine eHMIs using self-reports (e.g., [8]).

4.4. Limitations and Recommendations

The present study was conducted in rather constrained conditions. We used a computer monitor that offered a physical field of view of 31 deg and a virtual field of view of 80 deg. The 36 videos followed each other in quick succession, and the cars in the videos did not behave according to a realistic traffic flow model. Furthermore, participants were given a straightforward task to press the spacebar when feeling that it was safe to cross.

It would be worthwhile to employ more ecologically-valid methods, such as a virtual reality headset combined with a motion suit [44] or a field test using a Wizard of Oz approach [39]. It remains to be investigated how participants would respond to eHMIs in real traffic, in which situations arise more naturally and in which pedestrians may be in a hurry or lack the concentration to focus on a particular eHMI. We especially recommend testing eHMIs in traffic environments that involve competing visual demands. It is possible that pedestrians in complex traffic rely on peripheral vision without sustained visual attention towards the eHMI [39,45]. Wide fields of view could be achieved using a head-mounted display or surround projections. An advantage of our setup, in which head movement was constrained, is that we were able to measure eye movements with high accuracy.

Our computer monitor had a standard resolution of 1920 × 1080 pixels. The text-based eHMIs may have been hard to read when the virtual car drove at a large distance, especially for participants that suffer from near-sightedness. As discussed above, the Projection eHMI was relatively difficult to perceive just after it has appeared. However, despite the limited display resolution, participants rated the Roof, Windscreen, and Grill eHMIs as clear, with scores of about 8 on a scale from 0 to 10, as shown in Figure 4. Furthermore, our experiment proved to be highly sensitive for detecting differences between eHMIs conditions. To illustrate, 1.5 s after the eHMI turned on, over 70% of the participants pressed the spacebar for the Roof, Windscreen, and Grill eHMIs, compared to only 4% without eHMI. The limitation of display quality also applies to other simulation environments, such as CAVE simulations and head-mounted displays (e.g., [1]). In real traffic, legibility will be affected by other types of visual factors, such as direct sunlight, rain, or smog.

Our simulation did not feature sound. In reality, pedestrians may rely on auditory information to establish the state and relative position of oncoming vehicles. Participants in the simulation were not moving through the virtual environment, and the oncoming car decelerated abruptly while not interacting with the participant. These factors should be improved in future research.

For the present experiment, we selected an eHMI consisting of a non-commanding text message combined with an icon. We do not suggest that this type of eHMI is optimal in real-life applications. Clamann et al. [14] mounted a 32-inch screen on the front of a vehicle, depicting messages that were legible from about 75 m distance. Such large screens, or even multiple screens (see [27]), may not be desirable from an aesthetics and aerodynamics point of view and will require careful system integration.

Because display clarity is an essential factor for performance, we recommend that future research examines highly salient eHMI, such as a blinking LED strip.

A final limitation is that the present experiment was conducted using young engineering students, who can be expected to have a relatively high spatial ability [46] and perceptual speed [47]. It remains to be investigated whether older people would be able to intuitively understand eHMIs, such as the ones tested in the present study.

5. Conclusions

In conclusion, eHMIs on the Grill, Windscreen, and Roof were subjectively regarded as the clearest and evoked the highest rate of compliance for approaching cars. A projection-based eHMI has limitations in the form of poor legibility and participants' visual attention distribution. Based on our results, we recommend that eHMIs should be visible from multiple directions.

Supplementary Materials: The following are available online at http://www.mdpi.com/2078-2489/11/1/13/s1, Figure S1. Percentage of participants who pressed the spacebar during the videos of Traffic environment 1; Figure S2. Percentage of participants who pressed the spacebar during the videos of Traffic environment 2; Figure S3. Percentage of participants who pressed the spacebar during the videos of Traffic environment 3; Figure S4. Percentage of participants who pressed the spacebar during the videos of Traffic environment 4; Figure S5. Percentage of participants who pressed the spacebar during the videos of Traffic environment 5; Figure S6. Percentage of participants who pressed the spacebar during the videos of Traffic environment 6. Raw data, videos, and scripts are accessible here: https://www.dropbox.com/sh/egpd8kgk9bs9yee/AABi8sbwAvfbiyVxPhKVkuota?dl=0.

Author Contributions: Conceptualization, all authors; Methodology, all authors; Software, all authors; Validation, all authors; Formal analysis, all authors; Investigation, all authors; Resources, J.C.F.d.W.; Data curation, Y.B.E. & J.C.F.d.W.; Writing—original draft preparation, Y.B.E. & J.C.F.d.W.; Writing—review and editing, Y.B.E. &, J.C.F.d.W.; Visualization, Y.B.E. &, J.C.F.d.W.; Supervision, Y.B.E. & J.C.F.d.W.; Project administration, J.C.F.d.W.; Funding acquisition, J.C.F.d.W. All authors have read and agreed to the published version of the manuscript.

Funding: This research was supported by the research program VIDI with grant number TTW 016.Vidi.178.047 (2018–2022; "How should automated vehicles communicate with other road users?"), which is financed by the Netherlands Organisation for Scientific Research (NWO).

Conflicts of Interest: The funders had no role in the design of the study; in the collection, analyses, or interpretation of data; in the writing of the manuscript, or in the decision to publish the results.

References

1. De Clercq, G.K.; Dietrich, A.; Núñez Velasco, P.; De Winter, J.C.F.; Happee, R. External human-machine interfaces on automated vehicles: Effects on pedestrian crossing decisions. *Hum. Factors* **2019**. [CrossRef] [PubMed]

2. Bazilinskyy, P.; Dodou, D.; De Winter, J.C.F. Survey on eHMI concepts: The effect of text, color, and perspective. *Transp. Res. F Traffic Psychol. Behav.* **2019**, *67*, 175–194. [CrossRef]

3. Rasouli, A.; Tsotsos, J.K. Autonomous vehicles that interact with pedestrians: A survey of theory and practice. *IEEE Trans. Intell. Transp. Syst* **2019**, in press. [CrossRef]

4. Schieben, A.; Wilbrink, M.; Kettwich, C.; Madigan, R.; Louw, T.; Merat, N. Designing the interaction of automated vehicles with other traffic participants: Design considerations based on human needs and expectations. *Cognit. Technol. Work* **2019**, *21*, 69–85. [CrossRef]

5. Benderius, O.; Berger, C.; Lundgren, V.M. The best rated human-machine interface design for autonomous vehicles in the 2016 Grand Cooperative Driving Challenge. *IEEE Trans. Intell. Transp. Syst.* **2018**, *19*, 1302–1307. [CrossRef]

6. Zhang, J.; Vinkhuyzen, E.; Cefkin, M. Evaluation of an autonomous vehicle external communication system concept: A survey study. In *Advances in Human Aspects of Transportation. AHFE 2017. Advances in Intelligent Systems and Computing*; Stanton, N., Ed.; Springer: Cham, Switzerland, 2017; Volume 597, pp. 650–661. [CrossRef]

7. Werner, A. New colours for autonomous driving: An evaluation of chromaticities for the external lighting equipment of autonomous vehicles. *Colour Turn* **2018**, *1*. [CrossRef]

8. Fridman, L.; Mehler, B.; Xia, L.; Yang, Y.; Facusse, L.Y.; Reimer, B. To walk or not to walk: Crowdsourced assessment of external vehicle-to-pedestrian displays. *arXiv* **2017**, arXiv:1707.02698.

9.   Ackermann, C.; Beggiato, M.; Schubert, S.; Krems, J.F. An experimental study to investigate design and assessment criteria: What is important for communication between pedestrians and automated vehicles? *Appl. Ergon.* **2019**, *75*, 272–282. [CrossRef] [PubMed]

10.   Nissan. IDS Concept. Available online: https://www.nissan.co.uk/experience-nissan/concept-cars/ids-concept.html (accessed on 2 December 2019).

11.   Sweeney, M.; Pilarski, T.; Ross, W.P.; Liu, C. Light Output System for a Self-Driving Vehicle. U.S. Patent No. US9902311B2, 25 December 2018.

12.   Weber, F.; Chadowitz, R.; Schmidt, K.; Messerschmidt, J.; Fuest, T. Crossing the street across the globe: A study on the effects of eHMI on pedestrians in the US, Germany and China. In *HCI in Mobility, Transport, and Automotive Systems. HCII 2019. Lecture Notes in Computer Science*; Krömker, H., Ed.; Springer: Cham, Switzerland, 2019; Volume 11596, pp. 515–530. [CrossRef]

13.   Chang, C.M.; Toda, K.; Igarashi, T.; Miyata, M.; Kobayashi, Y. A video-based study comparing communication modalities between an autonomous car and a pedestrian. In Proceedings of the Adjunct Proceedings of the 10th International Conference on Automotive User Interfaces and Interactive Vehicular Applications, Toronto, ON, Canada, 23–25 September 2018; pp. 104–109. [CrossRef]

14.   Clamann, M.; Aubert, M.; Cummings, M.L. Evaluation of vehicle-to-pedestrian communication displays for autonomous vehicles. In Proceedings of the Transportation Research Board 96th Annual Meeting, Washington, DC, USA, 8–12 January 2017.

15.   Daimler. Autonomous Concept Car Smart Vision EQ Fortwo: Welcome to the Future of Car Sharing. Available online: https://media.daimler.com/marsMediaSite/en/instance/ko.xhtml?oid=29042725 (accessed on 2 December 2019).

16.   Joisten, P.; Alexandi, E.; Drews, R.; Klassen, L.; Petersohn, P.; Pick, A.; Abendroth, B. Displaying vehicle driving mode—Effects on pedestrian behavior and perceived safety. In *International Conference on Human Systems Engineering and Design: Future Trends and Applications*; Ahram, T., Karwowski, W., Pickl, S., Taiar, R., Eds.; Springer: Cham, Switzerland, 2019; pp. 250–256. [CrossRef]

17.   Otherson, I.; Conti-Kufner, A.S.; Dietrich, A.; Maruhn, P.; Bengler, K. Designing for automated vehicle and pedestrian communication: Perspectives on eHMIs from older and younger persons. *Proceedings of the Human Factors and Ergonomics Society Europe Chapter 2018 Annual Conference*, De Waard, D., Brookhuis, K., Coelho, D., Fairclough, S., Manzey, D., Naumann, A., Onnasch, L., Röttger, S., Toffetti, A., Wiczorek, R., Eds.; 2018; 135–148.

18.   Semcon. Who Sees You When the Car Drives Itself? Available online: https://semcon.com/smilingcar (accessed on 2 December 2019).

19.   Song, Y.E.; Lehsing, C.; Fuest, T.; Bengler, K. External HMIs and their effect on the interaction between pedestrians and automated vehicles. In *International Conference on Intelligent Human Systems Integration*; Karwowski, W., Ahram, T., Eds.; Springer: Cham, Switzerland, 2018; pp. 13–18. [CrossRef]

20.   Nuñez Velasco, J.P.; Farah, H.; Van Arem, B.; Hagenzieker, M.P. Studying pedestrians' crossing behavior when interacting with automated vehicles using virtual reality. *Transp. Res. F Traffic Psychol. Behav.* **2019**, *66*, 1–14. [CrossRef]

21.   Stadler, S.; Cornet, H.; Theoto, T.N.; Frenkler, F. A tool, not a toy: Using virtual reality to evaluate the communication between autonomous vehicles and pedestrians. In *Augmented Reality and Virtual Reality*; Tom Dieck, M.C., Jung, T., Eds.; Springer: Cham, Switzerland, 2019; pp. 203–216. [CrossRef]

22.   Toyota. Concept-i. Available online: https://newsroom.toyota.eu/2018-toyota-concept-i (accessed on 2 December 2019).

23.   Deb, S.; Strawderman, L.J.; Carruth, D.W. Should I cross? Evaluating interface options for autonomous vehicle and pedestrian interaction. In Proceedings of the Road, Safety, and Simulation Conference, Iowa City, IA, USA, 14–17 October 2019.

24.   Hensch, A.C.; Neumann, I.; Beggiato, M.; Halama, J.; Krems, J.F. How should automated vehicles communicate?–Effects of a light-based communication approach in a Wizard-of-Oz study. In *International Conference on Applied Human Factors and Ergonomics*; Stanton, N., Ed.; Springer: Cham, Switzerland, 2019; pp. 79–91. [CrossRef]

25.   Mahadevan, K.; Somanath, S.; Sharlin, E. Communicating awareness and intent in autonomous vehicle-pedestrian interaction. In Proceedings of the 2018 CHI Conference on Human Factors in Computing Systems, Montreal, QC, Canada, 21–26 April 2018. [CrossRef]

26. Vlakveld, W.; Van der Kint, S.; Hagezieker, M.P. Cyclists' intentions to yield for automated cars at intersections when they have right of way: Results of an experiment using high-quality video animations. Submitted.

27. Drive.ai. The Self-Driving Car Is Here. Available online: https://web.archive.org/web/20181025194248/https://www.drive.ai/# (accessed on 2 December 2019).

28. Colley, A.; Häkkilä, J.; Pfleging, B.; Alt, F. A design space for external displays on cars. In Proceedings of the 9th International Conference on Automotive User Interfaces and Interactive Vehicular Applications Adjunct, Oldenburg, Germany, 24–27 September 2017; pp. 146–151. [CrossRef]

29. Colley, A.; Häkkilä, J.; Forsman, M.T.; Pfleging, B.; Alt, F. Car exterior surface displays: Exploration in a real-world context. In Proceedings of the 7th ACM International Symposium on Pervasive Displays, Munich, Germany, 6–8 June 2018. [CrossRef]

30. Dietrich, A.; Willrodt, J.-H.; Wagner, K.; Bengler, K. Projection-based external human-machine interfaces–Enabling interaction between automated vehicles and pedestrians. In Proceedings of the Driving Simulation Conference Europe, Antibes, France, 5–7 September 2018; pp. 43–50.

31. Löcken, A.; Wintersberger, P.; Frison, A.K.; Riener, A. Investigating user requirements for communication between automated vehicles and vulnerable road users. In Proceedings of the 2019 IEEE Intelligent Vehicles Symposium (IV'19), Paris, France, 9–12 June 2019; pp. 879–884. [CrossRef]

32. Mitsubishi Electric. Mitsubishi Electric Introduces Road-Illuminating Directional Indicators. Available online: http://www.mitsubishielectric.com/news/2015/1023.html (accessed on 2 December 2019).

33. Mercedes-Benz USA. Mercedes-Benz F 015 Luxury in Motion. Available online: https://www.youtube.com/watch?v=MaGb3570K1U (accessed on 2 December 2019).

34. Senders, J.W.; Kristofferson, A.B.; Levison, W.H.; Dietrich, C.W.; Ward, J.L. The attentional demand of automobile driving. *Highw. Res. Rec.* **1967**, *195*, 15–33.

35. AlAdawy, D.; Glazer, M.; Terwilliger, J.; Schmidt, H.; Domeyer, J.; Mehler, B.; Fridman, L. Eye contact between pedestrians and drivers. In Proceedings of the Tenth International Driving Symposium on Human Factors in Driver Assessment, Training and Vehicle Design, Santa Fe, NM, USA, 24–27 June 2019; pp. 301–307.

36. Dey, D.; Walker, F.; Martens, M.; Terken, J. Gaze patterns in pedestrian interaction with vehicles: Towards effective design of external human-machine interfaces for automated vehicles. In Proceedings of the 11th International Conference on Automotive User Interfaces and Interactive Vehicular Applications, Utrecht, The Netherlands, 22–25 September 2019; pp. 369–378. [CrossRef]

37. Bazilinskyy, P.; Wesdorp, D.; De Vlam, V.; Hopmans, B.; Visscher, J.; Dodou, D.; De Winter, J.C.F. Visual scanning behaviour on a parking lot. In preparation.

38. Liu, C.; Wang, Z. Effect of narrowing traffic lanes on pavement damage. *Int. J. Pavement Eng.* **2003**, *4*, 177–180. [CrossRef]

39. Cefkin, M.; Zhang, J.; Stayton, E.; Vinkhuyzen, E. Multi-methods research to examine external HMI for highly automated vehicles. In *International Conference on Human-Computer Interaction*; Springer: Cham, Switzerland, 2019; pp. 46–64. [CrossRef]

40. Troel-Madec, L.; Alaimo, J.; Boissieux, L.; Chatagnon, S.; Borkoswki, S.; Spalanzani, A.; Vaufreydaz, D. eHMI positioning for autonomous vehicle/pedestrians interaction. In Proceedings of the IHM 2019—31e Conférence Francophone sur l'Interaction Homme-Machine, Grenoble, France, 10–13 December 2019; pp. 1–8.

41. Ineos159challenge The Role of the Car. Available online: https://www.ineos159challenge.com/news/the-role-of-the-car/ (accessed on 2 December 2019).

42. Powelleit, M.; Winkler, S.; Vollrath, M. Cooperation through communication–Using headlight technologies to improve traffic climate. *Proceedings of the Human Factors and Ergonomics Society Europe Chapter 2018 Annual Conference*, De Waard, D., Brookhuis, K., Coelho, D., Fairclough, S., Manzey, D., Naumann, A., Onnasch, L., Röttger, S., Toffetti, A., Wiczorek, R., Eds.; 2018; 149–160.

43. Eisma, Y.B.; Borst, C.B.; Van Paassen, M.M.; De Winter, J.C.F. Augmented visual feedback: Cure or distraction? Submitted.

44. Kooijman, L.; Happee, R.; De Winter, J.C.F. How do eHMIs affect pedestrians' crossing behavior? A study using a head-mounted display combined with a motion suit. *Information* **2019**, *10*, 386. [CrossRef]

45. Moore, D.; Currano, R.; Strack, G.E.; Sirkin, D. The case for implicit external human-machine interfaces for autonomous vehicles. In Proceedings of the 11th International Conference on Automotive User Interfaces and Interactive Vehicular Applications, Utrecht, The Netherlands, 22–25 September 2019; pp. 295–307. [CrossRef]

46. Wai, J.; Lubinski, D.; Benbow, C.P. Spatial ability for STEM domains: Aligning over 50 years of cumulative psychological knowledge solidifies its importance. *J. Educ. Psychol.* **2009**, *101*, 817–835. [CrossRef]
47. Salthouse, T.A. Aging and measures of processing speed. *Biol. Psychol.* **2000**, *54*, 35–54. [CrossRef]

 *information*

Article

Efficient Paradigm to Measure Street-Crossing Onset Time of Pedestrians in Video-Based Interactions with Vehicles

Stefanie M. Faas [1,2,*], **Stefan Mattes** [1], **Andrea C. Kao** [3] **and Martin Baumann** [2]

1   Mercedes-Benz AG, Leibnizstraße 2, 71032 Böblingen, Germany; stefan.mattes@daimler.com
2   Department of Human Factors, Ulm University, Albert-Einstein-Allee 41, 89081 Ulm, Germany; martin.baumann@uni-ulm.de
3   Mercedes-Benz R&D NA, 309 N Pastoria Ave, Sunnyvale, CA 94085, USA; andrea.kao@daimler.com
*   Correspondence: stefanie.faas@daimler.com

Received: 28 May 2020; Accepted: 29 June 2020; Published: 11 July 2020

Abstract: With self-driving vehicles (SDVs), pedestrians can no longer rely on a human driver. Previous research suggests that pedestrians may benefit from an external Human–Machine Interface (eHMI) displaying information to surrounding traffic participants. This paper introduces a natural methodology to compare eHMI concepts from a pedestrian's viewpoint. To measure eHMI effects on traffic flow, previous video-based studies instructed participants to indicate their crossing decision with interfering data collection devices, such as pressing a button or slider. We developed a quantifiable concept that allows participants to naturally step off a sidewalk to cross the street. Hidden force-sensitive resistor sensors recorded their crossing onset time (COT) in response to real-life videos of approaching vehicles in an immersive crosswalk simulation environment. We validated our method with an initial study of $N = 34$ pedestrians by showing (1) that it is able to detect significant eHMI effects on COT as well as subjective measures of perceived safety and user experience. The approach is further validated by (2) replicating the findings of a test track study and (3) participants' reports that it felt natural to take a step forward to indicate their street crossing decision. We discuss the benefits and limitations of our method with regard to related approaches.

Keywords: pedestrians; self-driving vehicles; automated driving; external human-machine interface; test methods; evaluation; user studies

---

1. Introduction

Highly (SAE Level 4) and fully (SAE Level 5) automated vehicles no longer require a driver [1]. With self-driving vehicles (SDVs) and human road users sharing the road, a "mixed traffic" transition period will emerge, demanding pedestrians interact with both SDVs and conventional vehicles (CVs) [2]. The related complexity could negatively affect pedestrian safety [2]. In today's traffic, pedestrians rely on a set of elaborate communication strategies when a CV approaches to decide whether it is safe to cross, including vehicle speed [3–5], distance of the vehicle [6], and eye contact with the driver [5,7]. While pedestrians can rely on traffic lights at signalized crossings, right of way can be ambiguous at unsignalized crossings, where human drivers frequently fail to yield to pedestrians. As a consequence, pedestrians are more risk-averse and seek more eye contact with the driver at unsignalized crossings [8,9]. As a substitute to communicating with a human driver, equipping SDVs with an external Human–Machine Interface (eHMI) has been proposed, to provide information to surrounding traffic participants [10]. An eHMI may be particularly important to reduce pedestrians' uncertainty at ambiguous crossings [11]. Preceding studies showed that pedestrians feel uncomfortable when encountering a driverless vehicle [12–14]. Limiting the scope to pedestrians' crossing decisions, previous research shows that the presence of an eHMI has positive effects on perceived safety [12,13,15–18], calmness [18], trust [12], comfort [19,20], user experience [12], and crossing decisions [13,17,21–23]. It can be argued that the necessity of an eHMI is demonstrated, but the type of information and means of conveying this information need to be further examined to reach the goal of a standardized eHMI. While subjective measures such as pedestrians' perceived safety can be assessed with a questionnaire after each trial (e.g. [12,13,24]), the assessment of eHMIs' effect on traffic flow poses a challenge. Traffic flow is an objective measure that can be quantified. The sooner a pedestrian initiates street crossing, the less time s/he has to wait on the curb and the less time a quickly approaching vehicle has to remain stopped, resulting in faster traffic for both the pedestrian and the approaching vehicle. In addition, to the improved time efficiency associated with a smooth flow of traffic, there are also environmental benefits such as decreased emissions and fuel consumption [25].

In the following, we will give an overview of the preceding applied methods to measure pedestrians' street crossing decision. Then, we will explain the motivation for our method.

1.1. Previously Applied Research Methods to Capture Pedestrian Crossing Decisions

In the following, we provide an overview of methodologies applied in preceding studies to capture pedestrians' street crossing decisions, discussing their benefits and limitations.

One approach is to capture the decision-making process of street crossing in terms of a function of the distance between the pedestrian and the approaching vehicle (e.g., [26–28]). For example, in a field study by Walker et al. [26], participants were instructed to express their feeling of safety to cross the road at any moment of time between 0 ("not at all willing to cross") to 100 ("totally willing to cross") on an input device that they hold in their hand while a vehicle approaches. While this approach is promising to form a better understanding of the underlying factors influencing a street crossing decision, we believe that it is not suitable to capture traffic flow. Pedestrians' street crossing is an actual behavior that has a binary character—either a pedestrian is waiting on the curb or crosses the street. Thus, traffic flow cannot be measured on a continuous scale.

A further approach is to measure the binary crossing decision (yes/no), i.e., whether pedestrians would be willing to cross the street in front of an approaching vehicle (e.g., [15,24,29,30]). For example, Song et al. [15] conducted an online survey in which pedestrians watched videos of a vehicle approaching from an ego-perspective and had to decide after each trial whether they want to cross (pressing the space key) or let the vehicle pass (not pressing the space key). We argue that this approach does not give any indication regarding traffic flow, since it fails to produce a relationship with the point of time participants would initiate street crossing.

Another approach is to compute crossing onset time (COT) by capturing the time a pedestrian decides to cross in relation to the vehicle's action. We argue that this is the only approach that can draw conclusions about eHMI effects on traffic flow. Regarding COT, the preceding methods can be divided into unnatural approaches, requiring participants to indicate their decision to cross in an explicit manner via pressing a button [11,17,22,23] or raising their hand [31], and natural approaches which allow participants to indicate their decision to cross with the actual behavior of taking a step forward [12,21,32,33]. We believe that methods requiring participants to imagine how they would act or feel make their decision explicit, which might limit their validity. We argue that, in terms of ecological validity, the natural behavior of stepping forward constitutes the best approach to measure COT. For example, in a test track study by Faas et al. [12], pedestrians watched an approaching vehicle coming to a stop at an intersection and had to cross the street as soon as they felt safe to do so. The vehicle encounters were video recorded for later analysis to estimate the time gap between the vehicle coming to a stop and the pedestrians' COT. Street crossing can be seen as an unreflective skillful action [34]. When crossing a street, pedestrians often act adequately, yet without deliberation. Street-crossing decisions are not guided by explicit reasoning, but constitute a form of embodied intelligence or cognition. Bodily processes or so-called "gut-feelings" might be of enormous importance for street crossing decision making [35]. It can be argued that pedestrians make their decision to cross unconsciously as soon as they feel that it is safe to cross, which is usually as soon as they are sure that the vehicle intends to yield for them. Their embodied nature makes individuals' street crossing decisions sensitive to aspects of the situation [34], such as the presence of a visible driver or an eHMI. However, to date, only a few test track studies [12,33] and VR studies [21,32] have assessed COT by allowing pedestrians to take a step forward.

1.2. Proposed Concept to Capture Street Crossing Onset Time (COT)

In this paper, we propose a parsimonious, safe, and reproducible paradigm for video-based lab studies that can capture COT in a natural way to test the efficacy of eHMI concepts for SDV and pedestrian interaction. We present a method in which participants indicate their COT by actually stepping off a "sidewalk" onto a "crosswalk". We conducted the experiment in a lab environment where participants were immersed using two large TV screens for a panoramic street view. With adhesive tape, we sketched a sidewalk and a crosswalk onto the floor. Under the sidewalk, we hid two force-sensitive sensors to capture COT. When the participant stepped onto the sidewalk, the videos were triggered and the COT timer was started. The COT was recorded when the participant stepped off the sidewalk to enter the crosswalk, with force-sensitive resistor sensors making data analysis time-efficient.

For the experiment, we contrasted three eHMI variants (no eHMI, status eHMI, status+intent eHMI) to address the research question of which information an eHMI should communicate. We used two light-based eHMI concepts adapted from Faas et al. [12]. The status eHMI is a steady blue-green light indicating the automated driving mode, as recommended by the SAE [36]. For the status+intent eHMI, an additional slowly flashing blue-green light (adapted from [37]) indicated the SDV's intent to yield as soon as the vehicle was braking, thus resembling the frontal brake light concept of previous eHMI studies [13,18,24,38]. We put the encounters with a driverless SDV in relation to encounters with a CV steered by a driver. We conducted three measuring points to study the stability of eHMI effects. The results of the study are published in Faas et al. [39]. The study showed that pedestrians benefit from an eHMI communicating SDVs' status, and that additionally communicating SDVs' intent adds further value. These eHMI effects last (acceptance, user experience) or even increase (COT, perceived safety, trust, learnability, reliance) with time.

The present paper focuses on the description and validation of the applied research method. For the present paper, we specifically re-evaluated the data of the first measuring time of the longitudinal study of Faas et al. [39], since we argue that our method is able to compare the efficacy of eHMI variants with one measuring time only. Furthermore, the present paper includes additional procedures that were not reported in Faas et al. [39] to validate the applied research method. To this end, we compared participants' responses in the lab study of Faas et al. [39] with participants' responses in the test track study of Faas et al. [12] to investigate potential differences attributed to the applied experimental methodology. Additionally, we analyzed participants' self-reported naturalism in the study setup. In this paper, we provide a detailed description of our method to allow others to adopt it. We validate our method by showing that it is able to detect significant eHMI effects on COT (and thus traffic flow) and subjective measures of perceived safety and user experience. Our approach is further validated by replicating findings of a test track study. Finally, participants reported that it felt natural to take a step forward to indicate that they would cross the street. We conclude that our paradigm allows relative comparisons of eHMI variants.

2. Materials and Methods

2.1. Participants

Thirty-four pedestrians (19 male, 15 female) in the age range of 22 to 69 years ($M = 41.5$, $SD = 15.8$ years) took part in the study. A third-party agency recruited the participants. For screening, potential participants specified which modes of transportation they use during a typical work week by distributing the percentage out of 100% among driving, public transit, biking, walking, and other.

Those, who distributed at least 20% to walking, received an invitation to participate in the study. All participants were living in the San Francisco Bay Area, CA, USA. All subjects gave their informed consent for inclusion before they participated in the study. The study was conducted in accordance with the Declaration of Helsinki, and the protocol was approved by the RD Ethical Clearing Committee of Daimler AG.

2.2. Independent Variable

Figure 1 gives an overview of the study procedure, including the independent variable, that is, vehicle type.

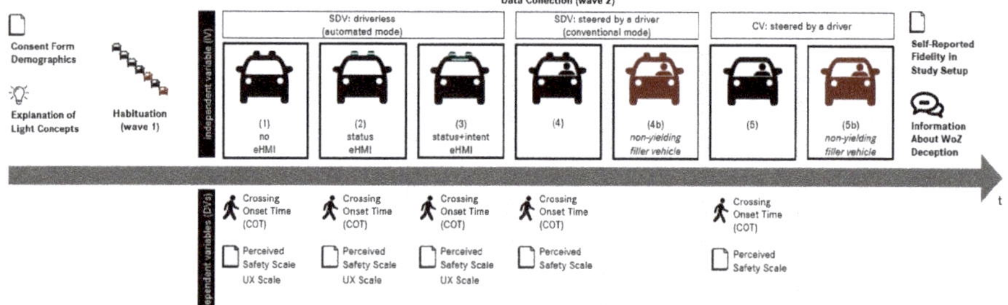

Figure 1. Procedure. The top row represents the study flow and the independent variable (IV). The driverless self-driving vehicle (SDV) (automated mode) is equipped with no eHMI, status eHMI or status+intent eHMI (test conditions 1, 2, 3). Both the SDV steered by a driver (conventional mode) and the conventional vehicle (CV) are either yielding (test conditions 4, 5) or non-yielding (filler test conditions 4b, 5b). In a randomized order, each participant experienced all seven test conditions once for habituation (wave 1) and once for data collection (wave 2). The bottom row represents the dependent variables (DVs) assessed in wave 2. The crossing onset time (COT) data were recorded by an Arduino Uno through the logs of two force-sensitive resistor sensors. For the subjective measures, participants filled in questionnaires after each trial. While perceived safety was measured for all yielding vehicle trials, we applied the user experience scales only after trials with a driverless SDV. Reproduced with permission from Faas, Kao and Baumann, A longitudinal video study on communicating status and intent for self-driving vehicle–pedestrian interaction, *Proceedings of the SIGCHI Conference on Human Factors in Computing Systems (CHI '20)*; published by ACM, 2020, doi: 10.1145/3313831.3376484.

Three eHMI test conditions were contrasted without a driver, i.e., self-driving (Figure 2):

1. Driverless SDV without eHMI: there is no indication whether the vehicle is in automated mode, i.e., self-driving, or conventional mode, i.e., steered by a driver;
2. Driverless SDV with status eHMI: steadily emitting blue-green lights on each fake Lidar sensor indicates that the vehicle is in automated mode. The design follows the recommended practice of the SAE [36];
3. Driverless SDV with status+intent eHMI: additionally to the "status" message, the "intent" signal was turned on when the approaching car started to brake, thus resembling the frontal brake light concept of previous eHMI studies [13,18,24,38]. To communicate the SDV's intent to yield, a light above the windshield flashed with a frequency at 0.5 Hz and a sinus cycle from 30% to 100% light intensity. The design follows the recommendation of Faas et al. [37]. The video of the status+intent eHMI test condition is available through the link: https://dl.acm.org/doi/fullHtml/10.1145/3313831.3376484.

The driverless SDV was shown to always yield to pedestrians. We chose a driverless setup to resemble a future automated vehicle on its way to pick up a passenger. Furthermore, to realize a mixed traffic environment, we incorporated encounters with vehicles steered by a visible driver.

The human-driven vehicles (Figure 3) were either yielding (test conditions 4, 5) or non-yielding (filler test conditions 4b, 5b):

4   SDV steered by a driver: yielding;
4b. SDV steered by a driver: non-yielding (filler test condition);
5.   CV steered by a driver: yielding;
5b. CV steered by a driver: non-yielding (filler test condition).

Figure 2. The study compared three eHMI test conditions within a driverless self-driving vehicle (SDV). This figure shows the status+intent eHMI (test condition 3). Two steady lights at the fake sensors indicate the automated status; a slowly flashing light at the windshield indicates its intent to yield to the pedestrian. For the status eHMI (test condition 2), the two steady lights were engaged. Without an eHMI (test condition 1), no lights were engaged. Reproduced with permission from Faas, Kao and Baumann, A longitudinal video study on communicating status and intent for self-driving vehicle–pedestrian interaction, *Proceedings of the SIGCHI Conference on Human Factors in Computing Systems (CHI '20)*; published by ACM, 2020, doi: 10.1145/3313831.3376484.

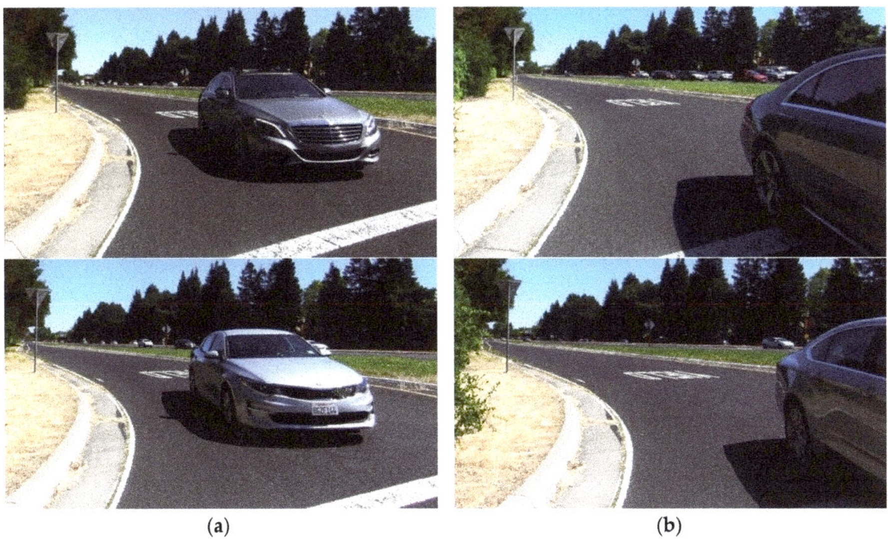

(a)	(b)

Figure 3. To provide a mixed-traffic environment, a visible driver steered the self-driving vehicle (SDV; top row) respectively a conventional vehicle (CV; bottom row). They were either (**a**) yielding to let the pedestrian cross first (test condition 4, 5); or (**b**) non-yielding (test condition 4b, 5b) so that the pedestrian has to wait for the vehicle to go first and crosses the empty street afterwards safely.

This study was designed to examine participants' responses when the car was yielding. Thus, the responses to the non-yielding vehicles were not analyzed (test conditions 4b, 5b). The non-yielding vehicles were included to ensure that participants would not habituate to all cars stopping for them, which might lower their attention and COT. We deliberately chose not to include any non-yielding driverless SDV encounters. While it can be argued that human drivers differ in their driving style, vehicle automation is programmed to adhere to traffic laws, thus always yielding at a pedestrian crossing.

2.3. Materials and Equipment

The experiment took place at the lab facilities of Mercedes-Benz Research and Development North America in Sunnyvale, CA, USA. We immersed participants with two large TV screens (25.5 inches (width) by 44 inches (length)) displaying the real-life video clips. The TV screens were set up at an angle of 60 degree to create a panoramic view. With adhesive tape and a mat, we sketched a "sidewalk" and a "crosswalk" onto the floor (Figure 4). Under the "sidewalk", we fixated two force-sensitive resistor sensors with the dimensions 44.45 × 44.45 nm (1.75 × 1.75 in). On the "sidewalk", we sketched two footprints at the same level as the force-sensitive resistor sensors. An Arduino Uno analog-to-digital converter was used to read the variable resistance of the force-sensitive resistor sensors. A 1k resistor was used to create a voltage divider. The software Arduino IDE (version 1.8.9) was used to code the data. A timer was added to display the elapsed time. When participants stepped onto the footprints (respectively putting force on each sensor), the COT timer started and the video clips were triggered starting with a three-second countdown. To provoke natural behavior, the participants' task was to cross the street when they felt safe to do so by entering the "crosswalk". When participants stepped off the "sidewalk" (respectively removing force on either sensor), the COT timer stopped.

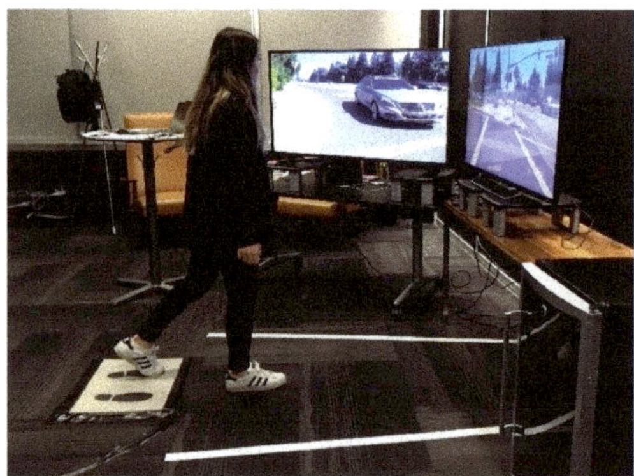

Figure 4. Study setup. Reproduced with permission from Faas, Kao and Baumann, A longitudinal video study on communicating status and intent for self-driving vehicle–pedestrian interaction, *Proceedings of the SIGCHI Conference on Human Factors in Computing Systems (CHI '20)*; published by ACM, 2020, doi: 10.1145/3313831.3376484.

For the real-life videos with the SDV, we created a Wizard-of-Oz setup [40]. On the roof of a silver Mercedes-Benz S-Class (Series W222), we mounted fake Lidar sensors similar to those of SDVs currently test-driving on public roads (e.g., [41,42]) as a reminder of the vehicle's ability to drive automated (see [43]). On the fake sensors, we attached LED light stripes to simulate the eHMIs. To create the deception of a driverless vehicle (test conditions 1, 2, 3), the driver controlling the vehicle wore a seat costume (adapted from [14]). For the videos in conventional driving mode (test conditions 4, 4b), the driver steering the vehicle was visible. For the videos with the CV and a visible driver (test conditions 5, 5b), we used three silver sedan models, namely a Chevrolet Impala, a Dodge Charger, and a Kia Optima. The occurrence of these models was randomized. All videos were cropped to a length of 15 s. Five observers who were not associated with the study checked the videos to ensure that they all displayed the same driving behavior.

2.4. Real-World Video Clips

2.4.1. Real-World Crossing Scenario

For the traffic scenario, we chose an intersection that requires pedestrians to cross an expressway exit lane while a vehicle approaches. The crossing has no traffic lights, but the request "YIELD" is written onto the street. In a preceding workshop, this traffic scenario was identified to be ambiguous for pedestrians. Workshop participants reported that, while the law states designated priority to pedestrians, the norm is that some approaching vehicles do not stop. In ambiguous traffic scenarios, communication strategies with the driver become especially prominent [9].

The video clips were recorded on a sunny day on a public highway. The camera perspective was from the viewpoint of a pedestrian standing on the sidewalk waiting to cross the road (see Figures 2 and 3). Specifically, the approaching vehicle was exiting Central Expressway to enter North Mary Avenue in Sunnyvale, CA, USA. Figure 5 shows the traffic scenario from a bird's eye-view.

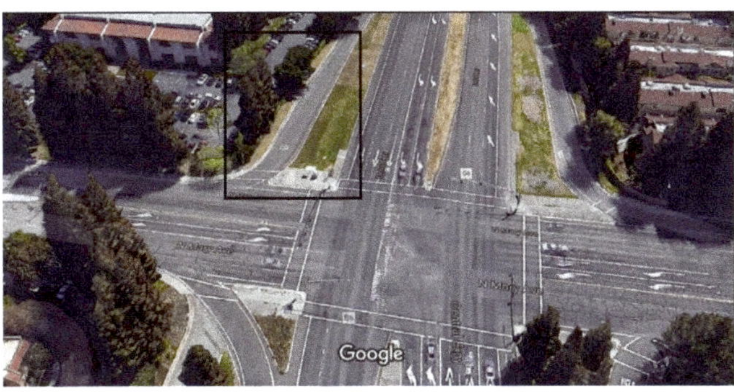

Figure 5. Traffic scenario from a bird's-eye view. The crossing scenario at the exit lane is framed in black. Copyright: Imagery ©2020 Google, Map data ©2020.

2.4.2. Video Flow

The experiment employed seven test conditions with yielding (test conditions 1, 2, 3, 4, 5) and non-yielding (test conditions 4b, 5b) vehicles in a within-subjects test design. Test conditions were randomized according to a Latin Square. Table 1 shows an overview of the video flow. The left TV screen showed the street with the approaching vehicles and the right screen showed the crosswalk. To allow time for participants to focus their attention back to the TV screens, each test condition started with a 3 s countdown on the left screen. Then, the video of the corresponding test condition was triggered. In each video, a vehicle approaches with a constant speed of 25 mph.

Table 1. Participants' task and video flow.

Participants' Task	Left Screen	Right Screen
Participant is ready for the next trial and asked to step on the "sidewalk".	*When you are ready* **Step onto the sidewalk with <u>both feet</u> to start the scenario**	*When you are ready* **Step onto the sidewalk with <u>both feet</u> to start the scenario**
Participant steps on the footprint . . .		
. . . which triggers the 3s countdown . . .	**3**	
. . . followed by the video of the approaching vehicle.		
To indicate her/his crossing decision, the participant steps off the "sidewalk" to enter the "crosswalk" . . .		
. . . which is a safe decision for yielding videos (test conditions 1, 2, 3, 4, 5), triggering a crossing video.		
. . . which is a safe decision if letting the vehicle go first for non-yielding videos (test conditions 4b, 5b), triggering a crossing video.		
. . . which is an unsafe decision if the vehicle is still approaching for non-yielding videos (test conditions 4b, 5b), triggering a visual warning and a passing car video.	*Not safe to cross!*	

For the yielding videos (test conditions 1, 2, 3, 4, 5), the vehicle approached with a constant speed for 3 s, decelerated to come to a stop at the intersection for 8 s, and waited for the pedestrian to cross for 4 s. After participants stepped off the "sidewalk", we provided visual feedback on their crossing decision through a street crossing video from an ego perspective on the right screen. On the left screen, the vehicle was waiting for the pedestrian to cross.

For the non-yielding videos (test conditions 4b, 5b), the vehicle slightly decelerated to make a right turn, but did not yield to the pedestrian. If participants succeeded by waiting for the car to pass to enter the street, a video was triggered on the right screen showing a street crossing from an ego perspective, while on the left screen the road was empty. If participants entered the crosswalk while the vehicle was still approaching, a red screen with the message "not safe to cross!" (left screen) and a video of a passing car (right screen) was triggered. In this case, the test condition was repeated.

2.5. Procedure and Participants' Task

Prior to the experiment, participants provided written informed consent. Participants completed a demographic questionnaire. Then, participants were introduced with the definition of high driving automation (SAE Level 4). Participants were told that the SDV they will encounter *"has both an automated and a manual driving mode. The vehicle can thus either be self-driving or be controlled manually by a driver."* Next, the three eHMI concepts were explained to participants. Subsequently, participants' understanding of the eHMI concepts was tested by asking them *"What does the light signal indicate?"*

Following, participants were familiarized with the study setup by the experimenter going through the participants' task. Participants were shown an example scenario with the status+intent eHMI (test condition 3). First, they were asked to imagine that the mat is a "sidewalk".

Then, namely "The next slide lets you know that at this time, you can step on the sidewalk to begin the scenario. When you step on the sidewalk, please make sure your feet are aligned with the footprints. Once both feet are on these footprints, the scenario will begin." Participants were told that in each scenario a vehicle will be approaching, but not all vehicles are going to yield. The participants' task was "to safely cross the road at an intersection as a pedestrian while different vehicles approach. As soon as you feel safe to cross, please do so. You must cross for all scenarios. To cross, just step off the sidewalk as if you're going to enter the crosswalk." Thus, with each trial, participants indicated their COT by stepping of the "sidewalk" to enter the "crosswalk" (see Table 1). The field of view was panoramic in the way that pedestrian had to bend their head to the left to observe the approaching vehicle and step forward to initiate street crossing.

Subsequently, the room's light was dimmed to allow a better contrast for the participant to see the contents of the TV screens clearly. Participants encountered two waves consisting of seven trials, each with vehicles that yielded to the pedestrian in five trials (test conditions 1, 2, 3, 4, 5) and non-yielding vehicles in two trials (test conditions 4b, 5b). Participants experienced one wave for habituation. After habituation, the second wave followed for data acquisition. We assessed participants' COTs and subjective measures for all yielding vehicle trials. The crossing onset data were recorded by an Arduino Uno. After each trial, participants filled in a questionnaire to indicate subjective measures of perceived safety and user experience (see Figure 1).

After all trials, participants were asked to rate the naturalism of our paradigm. We informed participants that the encountered vehicle had not been driving automated at any time. Total testing time was about 30 min per participant.

2.6. Dependent Variables

In this paper, we report the following objective measure:

- Crossing Onset Time (COT): After each yielding vehicle trial (test conditions 1, 2, 3, 4, 5), we determined COT. COT indicates the time in seconds between the vehicle yielding and the pedestrian stepping off the "sidewalk". Hence, to calculate the COT, we have subtracted the time between the pedestrian entering the "sidewalk" and the vehicle yielding (3s countdown + 3s vehicle approaching at constant speed). We used COT as an index of traffic flow. Shorter times indicate an earlier crossing decision. The earlier pedestrians cross when it is safe to do so, the more efficient the traffic flows. We excluded extreme cases from data analysis, defined as more than three times the interquartile range (IQR) greater than the upper or lower quartile (2 values of $N = 1$ participant excluded).

Furthermore, we report the following subjective measures, all measured on a scale from −3 (very negative) to +3 (very positive):

- Perceived Safety: After each yielding vehicle trial (test conditions 1, 2, 3, 4, 5), participants reported their perceived safety with four items (based on [44]) with semantic differentials answered on a 7-point scale ranging from −3 to +3 ("anxious–relaxed", "agitated–calm", "unsafe–safe", "timid–confident"). Reliability was excellent, with Cronbach's α = 0.90 to 0.96;
- User Experience (UX) Qualities: After each driverless SDV trial (test condition 1, 2, 3), participants completed the short version of the User Experience Questionnaire (UEQ-S) [45]. The scale consists of two dimensions: pragmatic quality (PQ) and hedonic quality (HQ). Participants reported their user experience with semantic differentials ranging from −3 (negative) to +3 (positive). The reliability of all subscales was good to excellent, with Cronbach's α = 0.80 to 0.94;
- Naturalism: In the post-experiment interview, participants rated the items "How immersive was the study setup?" and "How natural was it to take a step forward to indicate that you would cross the street?" (based on [33]) on a scale from −3 ("not at all") to +3 ("extremely").

2.7. Data Analysis

We used repeated measures ANOVAs to test the effect of vehicle type (test conditions 1, 2, 3, 4, 5) on COT and perceived safety. As an additional analysis, we performed cluster analyses to categorize the participating pedestrians into groups according to their COT obtained for each yielding test condition. To classify pedestrians into groups, we used Ward's method in combination with squared Euclidean distances (see [46,47]). As a hierarchical procedure, the Ward's method successively merges cases into clusters such that the variance within a cluster is associated with the smallest possible increase (see [46,47]).

Next, we used repeated measures ANOVAs to test the effects of eHMI type (test conditions 1, 2, 3) on UX qualities (HQ and PQ).

Finally, we compared the subjective responses to the PQ scale of our participants and the participants in the test track study of Faas et al. [12] to investigate potential differences attributed to the applied experimental methodology. For this purpose, we used the data of the no eHMI, status eHMI, and status+intent eHMI test conditions that were assessed with N = 30 participants at an intersection traffic scenario on a test track in Immendingen, Germany. We believe that this comparison is valuable, although the experiments differ regarding participants' nationality (U.S. vs. German) and traffic scenario (exit lane vs. four-way intersection). The study participants of this lab study and the test track study did not differ regarding age, $t(57)$ = −0.37, p = 0.714, or gender, $\chi^2(1)$ = 0.04, p = 0.838. We chose the PQ scale for the following comparison, since it is the only standardized questionnaire that has been applied in both studies. We used two-sample t-tests to investigate whether pedestrians' subjective PQ ratings of the three eHMI variants (no eHMI, status eHMI, status+intent eHMI) differ among experimental methodology (lab study vs. test track study).

For all ANOVAs, the data were checked for sphericity using Mauchly's test, and, where violated, Greenhouse–Geisser and Hyunh–Feldt corrections were applied (as recommended by [48]). Where needed, we used Bonferroni-corrected post-hoc t-tests.

3. Results

3.1. Crossing Onset Time (COT)

On COT, the one-way repeated measures ANOVA revealed a significant effect of vehicle, $F(4, 128)$ = 12.47, $p < 0.001$, η_p^2 = 0.28. Figure 6 shows the mean values. Post-hoc t-tests revealed that participants started crossing earlier if the driverless SDV (automated mode) was equipped with a status+intent eHMI (M = 6.74, SD = 2.17) compared to no eHMI (M = 7.86, SD = 1.40), p = 0.003, 95% CI [0.30–1.95]. However, there was no improvement from no-eHMI to the status eHMI (M = 7.46, SD = 1.35), p = 0.439, or from the status eHMI to the status+intent eHMI, p = 0.117. Regarding human-driven vehicles, there is no difference in COT between an SDV steered by a driver ($M = 8.14$, $SD = 1.34$) and a CV steered by a driver ($M = 8.08$, $SD = 1.35$), $p = 1.000$. When comparing driverless vehicles and vehicles steered by a driver, participants initiated street crossing at the same time for encounters with a driverless SDV without eHMI and an SDV steered by a driver, $p = 1.000$, or a CV steered by a driver, $p = 1.000$. However, if the driverless SDV is equipped with a status eHMI, $p = 0.005$, 95% CI [0.15–1.21], or a status+intent eHMI, $p < 0.001$, 95% CI [0.62–2.19], participants initiated crossing earlier than if encountering a SDV steered by a driver. Analogously, if the driverless SDV is equipped with a status eHMI, $p = 0.068$, 95% CI [−0.03–1.27], or a status+intent eHMI, $p < 0.001$, 95% CI [0.51–2.18], participants (tended to) initiate crossing earlier than if encountering a CV steered by a driver.

To account for pedestrians' individual crossing strategies [12], we performed cluster analyses, classifying pedestrians into groups according to their COT obtained for each yielding test condition. A dendrogram graphically illustrates the formation of clusters at the individual fusion stages (Figure 7a). To determine the number of clusters into which pedestrians can be meaningfully clustered, we computed a structogram (Figure 7b). The stuctogram graphically illustrates that the fourth cluster contributes significantly less to the variance than the first three clusters. Because of the considerable drop in the Sum of Squared Errors (ΔSSE), it seems reasonable to assume a solution with three clusters. Figure 8 shows the individual COT for each participant sorted by the three derived clusters from cluster analyses. Visual inspection suggests the following description of the three clusters: The first cluster ($N = 7$) includes early crossers who cross before the vehicle comes to a stop and are strongly influenced by the test conditions, particularly by the presence of a status+intent eHMI. The second cluster ($N = 20$) describes intermediate crossers who initiate crossing at about the same time as the vehicle comes to a stop. They are slightly influenced by the test conditions and constitute the biggest cluster. The third cluster ($N = 7$) includes late crossers who wait for the vehicle to come to a stop before crossing the street. These late crossers are slightly influenced by the test conditions.

In summary, pedestrians initiated street-crossing the soonest with a status+intent eHMI. Compared to a CV or SDV steered by a driver, pedestrians initiated crossing at the same time if the driverless SDV was not equipped with an eHMI and sooner if it was equipped with an eHMI displaying the SDV's status and intent (see also: Faas et al. [39]). The significant effect of status+intent eHMI seems to be carried by a cluster of pedestrians, who are likewise characterized by a tendency to cross the street early, also with human-driven vehicles.

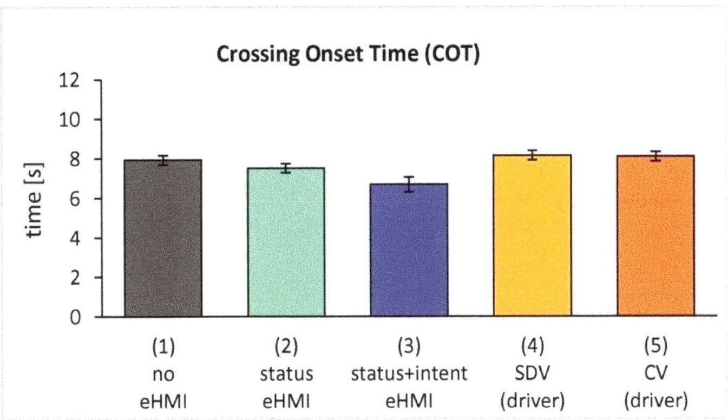

Figure 6. Mean crossing onset time (COT) for all yielding test conditions. Error bars: ±1 SE.

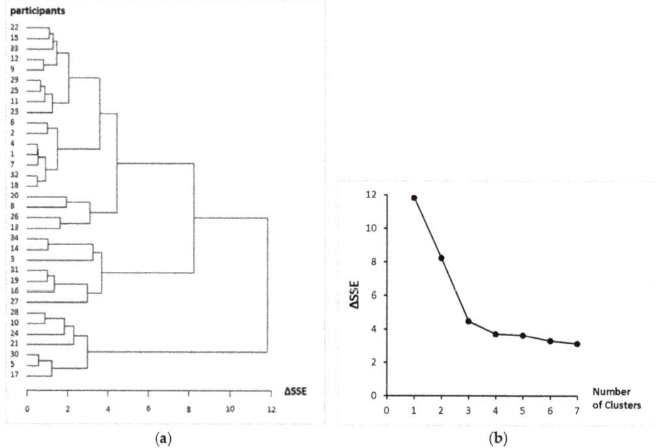

Figure 7. Cluster analyses. (**a**) Dendrogram. On the x-axis, the Sum of Squared Errors (ΔSSE) are plotted. (**b**) Structogram for the number of clusters ranging from 1 to 7. The structogram indicates that by splitting the 3 clusters into 4 clusters, the ΔSSE values do not decrease much. Thus, three clusters seem appropriate for this scheme.

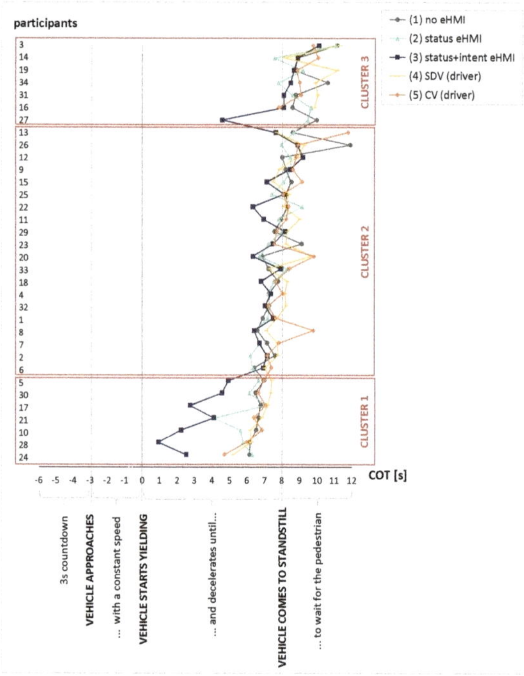

Figure 8. Individual crossing onset time (COT) per participant for all yielding test conditions. Participants were sorted according to the three clusters derived from cluster analyses: (1) early crossers who are strongly influenced by the presence of the eHMIs, particularly the status+intent eHMI ($N = 7$) as well as (2) intermediate ($N = 20$) and (3) late crossers ($N = 7$), who are both slightly influenced by the eHMIs. Within each cluster, participants were sorted according to their average COT over all test conditions (e.g., within cluster 1, SP24 crosses the earliest, SP5 the latest over all yielding test conditions).

3.2. Perceived Safety

On perceived safety, the one-way repeated measures ANOVA found a significant effect of vehicle, $F(2.59, 85.56) = 8.65$, $p < 0.001$, $\eta_p^2 = 0.21$. Figure 9 shows the results. Pedestrians feel significantly safer if the driverless SDV (automated mode) is equipped with a status eHMI ($M = 0.31$, $SD = 1.67$) than if it is without eHMI ($M = -0.43$, $SD = 1.73$), $p = 0.011$, 95% CI [0.12–1.37]. With a status+intent eHMI ($M = 1.17$, $SD = 1.32$), pedestrians feel safer than with a status eHMI, $p = 0.026$, 95% CI [0.07–1.65], and, thus, also safer than without eHMI, $p < 0.001$, 95% CI [0.69–2.51], drawing the following pattern: status+intent eHMI > status eHMI > no eHMI. Regarding human-driven vehicles, participants feel equally safe with an SDV steered by a driver ($M = 1.06$, $SD = 1.46$) and a CV steered by a driver ($M = 1.06$, $SD = 1.51$), $p = 1.000$. Compared to an SDV steered by a driver or a CV steered by a driver, participants felt less safe encountering a driverless SDV without eHMI, all $ps < 0.01$. However, if the driverless SDV was equipped with a status eHMI or a status+intent eHMI, participants felt as safe as with a human driven vehicle, all $ps > 0.05$.

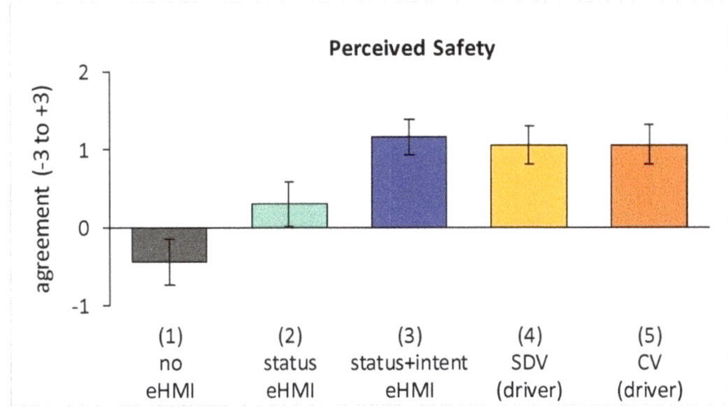

Figure 9. Mean perceived safety scores for all yielding test conditions. Error bars: ±1 SE.

In summary, pedestrians felt safest with a status+intent eHMI. With any eHMI, pedestrians felt as safe as with human-driven vehicles. However, if the driverless SDV is not equipped with an eHMI, pedestrians felt less safe than with human-driven vehicles (see also: Faas et al. [39]).

3.3. User Experience

The one-way repeated measures ANOVAs found a significant effect of eHMI on PQ, $F(2, 66) = 54.27$, $p < 0.001$, $\eta_p^2 = 0.62$, and HQ, $F(1.60, 52.84) = 22.20$, $p < 0.001$, $\eta_p^2 = 0.40$. Figure 10 shows the results.

Pedestrians rate PQ significantly higher if the driverless SDV (automated mode) is equipped with the status eHMI ($M = 1.03$, $SD = 1.37$) than without eHMI ($M = -0.49$, $SD = 1.30$), $p < 0.001$, 95% CI [0.98–2.06]. With a status+intent eHMI ($M = 1.93$, $SD = 0.86$), pedestrians rate PQ higher than with a status eHMI, $p = 0.001$, 95% CI [0.32–1.48], and, thus, higher than without eHMI, $p < 0.001$, 95% CI [1.77–3.07], revealing the following pattern: status+intent eHMI > status eHMI > no eHMI.

Accordingly, pedestrians rate HQ significantly higher if the driverless SDV (automated mode) is equipped with the status eHMI ($M = 1.43$, $SD = 1.34$) than without eHMI ($M = 0.76$, $SD = 1.67$), $p = 0.003$, 95% CI [0.21–1.13]. With a status+intent eHMI ($M = 2.11$, $SD = 0.86$), pedestrians rate HQ higher than with a status eHMI, $p = 0.001$, 95% CI [0.27–1.09], and, thus, also higher than without eHMI, $p < 0.001$, 95% CI [0.71–1.98], leading to the same pattern: status+intent eHMI > status eHMI > no eHMI.

Based on Hinderks et al. [49], the UX scores can be interpreted as bad (PQ) and below average (HQ) for no eHMI, below average (PQ) and good (HQ) for the status eHMI and excellent (PQ, HQ) for the status+intent eHMI (see also: Faas et al. [39]).

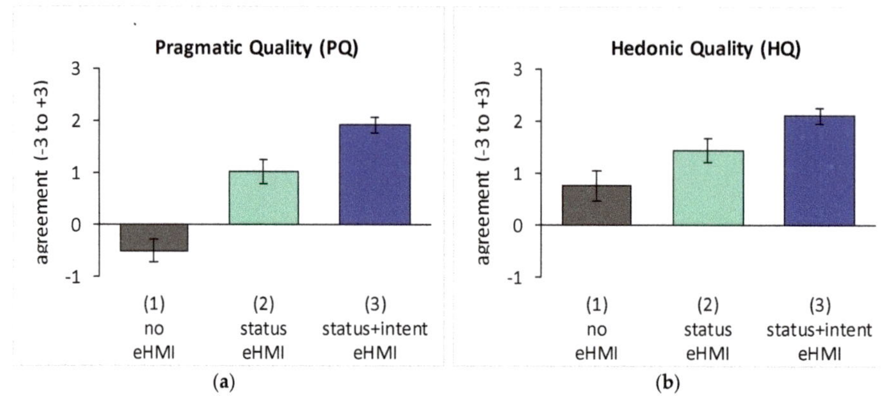

Figure 10. Mean UX scores for all driverless self-driving vehicle (SDV; automated mode) test conditions, as shown by the subscales: (**a**) Pragmatic Quality (PQ) and (**b**) Hedonic Quality (HQ). Error bars: ±1 SE.

3.4. Comparison of Participants' PQ Ratings in This Lab Study and a Test Track Study

We compared the PQ responses of this lab study to the PQ results of the test track study of Faas et al. [12] to investigate whether the different experimental methodologies lead to different results. We used two-sample t-tests to investigate whether pedestrians' PQ ratings of the three eHMI variants (no eHMI, status eHMI, status+intent eHMI) differ among experimental methodology (this lab study vs. test track study of Faas et al. [12]). Levene's test for equality of variances was not violated for any t-test. Table 2 and Figure 11 show the results. For no eHMI, participants' PQ ratings were significantly lower in this lab study compared to the test track study of Faas et al. [12], $t(62) = -2.10, p = 0.40, r = 0.26$. However, both mean scores lead to the same interpretation of a bad user experience according to the benchmarks of Hinderks et al. [49]. Accordingly, for the status eHMI there was a trend implicating that participants' PQ ratings were lower in this lab study compared to the test track study of Faas et al. [12], $t(62) = -1.71, p = 0.92, r = 0.21$. For the status+intent eHMI, we found no significant difference between the studies, $p = 0.822$.

Furthermore, both studies revealed the same results regarding participants' PQ ratings of the three eHMI conditions: status+intent eHMI > status eHMI > no eHMI, leading to similar conclusions (see also: Faas et al. [12], Faas et al. [39]).

Table 2. Two-sample t-tests.

Test Condition	This Lab Study [1]		Test Track Study [2]			t-Tests		
	M	**SD**	**M**	**SD**	**df**	t-**Value**	p-**Value**	r
(1) no eHMI	−0.49	1.30	0.31	1.74	62	−2.10	$p = 0.040$ *	0.26
(2) status eHMI	1.03	1.37	1.56	1.05	62	−1.71	$p = 0.092$	0.21
(3) status+intent eHMI	1.93	0.86	1.98	0.85	62	−0.23	$p = 0.822$	

[1] in this lab study, $N = 34$ participants experienced the three eHMIs within-subject though real-world video clips; [2] in the test track study of Faas et al. [12], $N = 30$ participants experienced the three eHMIs within-subject at an intersection with a real vehicle. * $p < 0.05$.

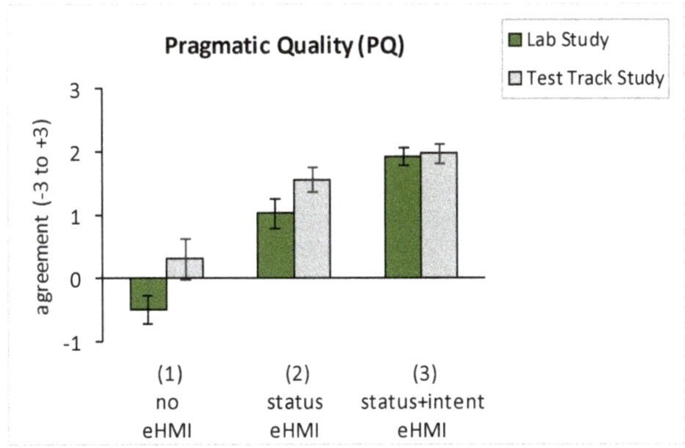

Figure 11. Pedestrians' Pragmatic Quality (PQ) ratings of the three eHMI variants (test conditions 1, 2, 3) in this lab study and in the test track study of Faas et al. [12]. Error bars: ±1 SE.

3.5. Self-Reported Naturalism

After all trials, participants rated the naturalism of the experiment on a scale from −3 ("not at all") to +3 ("extremely"). The mean score to the question "How immersive was the study setup?" was $M = 0.62$ ($SD = 1.37$), suggesting a fair immersion. The mean score to the question "How natural was it to take a step forward to indicate that you would cross the street?" was $M = 1.82$ ($SD = 1.03$), suggesting good validity.

4. Discussion

This paper presents an innovative method to study SDV–pedestrian interactions in a safe, reproducible, and a natural manner for video-based eHMI studies. We developed a cost-efficient concept that allows participants to show natural behavior (i.e., entering a street). Participants make an actual street-crossing decision; that is, they are instructed to take a step off a sketched "sidewalk" to enter a sketched "crosswalk" to measure COT as a means to assess traffic flow. In the following, we discuss how the eHMI effects, which have been brought to light by our approach, validate its application. Furthermore, we discuss our method with regard to related approaches as well as the limitations and further improvements of our methodology.

4.1. Validation

We showed that our method is able to detect statistically significant eHMI effects that are comparable to a real-life study on a test track, and further displays a good level of self-reported naturalism.

The results of the eHMI study, yielding significant and meaningful results, validate the use of our approach. We found that, compared to human-driven vehicles, pedestrians feel less safe encountering a driverless SDV if it has no eHMI. However, pedestrians feel as safe if the driverless SDV is equipped with an eHMI displaying its status and, eventually, intent. When comparing the eHMI variants, all subjective measures (perceived safety, HQ, PQ) revealed the same pattern: status+intent eHMI > status eHMI > no eHMI. On COT, we found that pedestrians make earlier (thus more efficient) crossing decisions with a status+intent eHMI than with no eHMI. The significant effect of status+intent eHMI seems to be carried by a cluster of participants, suggesting individual crossing strategies among pedestrians (comparable to different lane changing strategies among drivers, see for example, [50]). Thus, providing pedestrians with information on SDVs' automated status and imminent intent supports a feeling of safety and HQ. Pedestrians perceive an eHMI to be useful information (PQ), supporting them in their decision to cross the road as observed in earlier COTs (for a textual discussion, see Faas et al. [39]).

The approach is further validated by the fact that the study outcomes confirm previous research showing eHMI effects on perceived safety [12,13,15–18] and crossing onset [13,17,21–23], suggesting that our method is as suitable as other approaches to detect eHMI effects. This becomes particularly clear as our method replicates the findings of a test track study by Faas et al. [12]. Both studies compared the effect of light-based eHMI concepts on PQ at an ambiguous crossing traffic scenario. Both studies revealed the same significant pattern regarding pedestrians' rating of PQ: status+intent eHMI > status eHMI > no eHMI. Thus, both studies showed that communicating an SDV's intent adds further benefit for pedestrians over just displaying the automated status. However, in the current lab study (Faas et al. [39]) pedestrians rated the no-eHMI test conditions as significantly worse, and the status eHMI test condition as slightly worse, than participants of the test track study (Faas et al. [12]). We believe that the worse ratings emerged because, in the lab study, a vehicle without an eHMI could mean a real disadvantage, potentially representing a non-yielding vehicle. On the contrary, in the test track study (Faas et al. [12]) all vehicles yielded, so the participants' safety was guaranteed. Further, a lab study is more controlled than a test track study. Thus, while showing the same pattern of eHMI ratings (status+intent eHMI > status eHMI > no eHMI), the lab study produced more variance in participants' ratings, leading to a more differentiated evaluation of the eHMIs variants.

Finally, participants reported that it felt natural to take a step forward to indicate their street-crossing decisions ($M = 1.82$ on a scale from -3 to $+3$), suggesting a good validity.

4.2. Benefits with Regard to Related Approaches

The benefits of our method are its natural approach to assess COT in a parsimonious, reproducible, and safe manner.

Most previous approaches assessed crossing decisions in an unnatural manner, instructing participants to indicate their decision via pressing a button [13,15,17,22,23,29,30], a slider [26–28], or raising their hand [31]. Those approaches make the participants' crossing decisions explicit, creating an intermediary step that may affect their behavior. Participants have to transfer their implicit crossing decision to an explicit motor decision with their hand. Furthermore, participants may have to look at the button or slider, so they cannot observe the approaching vehicle at all times. For example, in the study of Walker et al. [26], 29% of the participants reported that they were not able to use the slider naturally, thus not able to indicate their feeling of safety valid. Since street-crossing can be seen as an unreflective skillful action, which is a form of embodied intelligence or cognition [34,35], we argue that COT should be measured in a natural way, by actually stepping off a sidewalk onto a crosswalk. Our approach allows participants to show natural street-crossing behavior (i.e., entering a street) if they feel safe to cross. Thus, with our method, participants are closer to the processes that take place in real-world traffic situations, which improves ecological validity.

Only a few test track studies [12,33] and VR studies [21,32] allowed participants to indicate their decision to cross in a natural manner via the actual behavior of making a step forward. However, test track and VR studies require high-priced apparatus and materials as well as time-consuming data analysis. For example, the required resources for an eHMI study on a test track include a test track location, a real vehicle, a light setup (e.g., LED stripes), and a driver steering the vehicle, possibly in a seat costume. These resources are required for several days. For later analysis, videos of each vehicle encounter need to be visually analyzed to extract the crossing onset measure (e.g., [12,33,37]). Similarly, to conduct and analyze VR studies, researchers need technologically advanced software and hardware (for an overview, see [51]). Participants might suffer simulation sickness [52]. Compared to previous studies on a test track or in VR, our approach requires only a few materials. Video-based studies are cost-efficient in comparison. The material required for our approach include two TV screens, adhesive tape, two force-sensitive resistor sensors, an Arduino Uno analog-to-digital converter, and a laptop with the software Arduino IDE. For our real-world eHMI video clips, we needed a vehicle, fake

Lidar sensors with LED light stripes, and a seat costume to create the illusion of a driverless vehicle. If researchers do not have access to those materials, future studies could use animated videos instead, just as VR studies do (e.g., [17,21,32]). An advantage of animated videos is that they allow researchers to have absolute control of any variable they might want to manipulate. However, their physical accuracy is lower than real-life videos [53]. Data analysis of our approach is as time-efficient as the Arduino Uno records COT in real-time.

Furthermore, video-based studies allow for flexibility and variety in eHMI test conditions. Researchers need to conduct only one video of an approaching vehicle and can use animations to create eHMI variants. The study is reproducible.

Lastly, one advantage of video studies is the possibility of incorporating non-yielding vehicle encounters while ensuring participants' safety. In contrast, test track studies need to meet high ethical standards and safety provisions, limiting their representativeness for complex urban traffic scenarios. For example, to guarantee participants' safety, non-yielding vehicle encounters should not be incorporated. Our approach allows participants to experience safety critical situations without actually endangering them. Although non-yielding vehicle encounters are not of research interest, they prevent participants from habituating to all cars stopping for them, which might lower their attention and, thus, the validity of the study.

4.3. Limitations and Recommendations

While our approach is promising, we acknowledge that there are limitations that require further attention. The first one refers to the absence of a real safety risk. The fact that participants cannot be harmed ensures participants' safety, but it also limits the realism of our approach. Since pedestrians do not have to fear any real risks from non-yielding vehicles, they might behave in a riskier manner than in normal traffic. The second limitation refers to participants' fair evaluation of the approach's immersiveness ($M = 0.62$ on a scale from -3 to $+3$), which might be rooted in the participants' constrained field of view. While real-life videos from the perspective of a pedestrian exhibit a high level of physical accuracy, their operationalization is not as good as experiencing a traffic situation in a real environment [53]. Thus, our method is suitable for relative comparisons (i.e., detecting differences between eHMI concepts) but not to establish the true value of COT for a certain eHMI concept. However, this limitation applies to all research studies that use simulation. To make the setup more realistic, future studies could setup the "sidewalk" with a real curb so that participants need to take a step down onto the "crosswalk" compared to the current setup with a flat lab floor (suggestion made by Koojman et al. [21]). Moreover, the use of VR glasses instead of TV screens may increase the participants' degree of immersion. However, despite these limitations, our approach proved its sensitivity to detect eHMI effects on pedestrians' COT, perceived safety, and user experience.

5. Conclusions

This paper introduces a novel paradigm to study SDV–pedestrian interaction that is relatively easy to implement and can find a balance between a natural and parsimonious study setup. We propose the use of two TV screens and a simulated sidewalk with hidden force-sensitive resistor sensors as the input device. We believe that street crossing behavior should be grasped by the actual action of stepping off a sidewalk onto a street. We propose that the study design shows clear advantages, as opposed to an artificial design with participants watching videos on a screen in a sitting position and/or indicating their crossing decision with a button or slider. We believe that this experimental design can be valuable and effective for future video studies examining vehicle-pedestrian interaction.

Within the presented approach, it was possible to demonstrate the need for an eHMI for the communication between SDVs and pedestrians in an ambiguous traffic scenario. The eHMI concepts revealed significant differences in terms of COT, perceived safety, and User Experience (for a textual discussion, see Faas et al. [39]). Further, we validated our method's efficacy by showing that its results are not only comparable, but more differentiated than the results produced by a test track approach.

Furthermore, our method displays a good level of self-reported naturalism. Thus, the presented method is validated as a suitable tool to make relative comparisons between eHMI concepts. We conclude that the method can be applied in future studies comparing eHMI concepts from a pedestrians' point of view.

Author Contributions: Conceptualization, S.M.F., S.M., A.C.K. and M.B.; Data curation, S.M.F. and A.C.K.; Formal analysis, S.M.F.; Investigation, S.M.F. and A.C.K.; Methodology, S.M.F., S.M., A.C.K. and M.B.; Project administration, S.M.F. and A.C.K.; Resources, S.M.F. and A.C.K.; Software, A.C.K.; Supervision, S.M.F.; Validation, S.M.F. and A.C.K.; Visualization, S.M.F.; Writing—original draft, S.M.F.; Writing—review and editing, S.M.F., S.M., A.C.K. and M.B. All authors have read and agreed to the published version of the manuscript.

Funding: This research received no external funding.

Acknowledgments: We would like to thank Juergen Arnold, Jeff Bertalotto, Michael Boehringer, Sean Cannone, Katarina Carlos, Edwin Danner, Kevin Gee, Peter Goedecke, Ulrich Hipp, Ralf Krause, Eric Larsen, Laura Neiswander, Frank Ruff, and Ellen Tyler for their help with the study, as well as all study participants for their time and feedback.

Conflicts of Interest: The authors declare no conflict of interest.

References

1. SAE International. *Taxonomy and Definitions for Terms Related to Driving Automation Systems for On-Road Motor Vehicles (J3016)*; SAE International: Warrendale, PA, USA, 2018.
2. Sivak, M.; Schöttle, B. Road Safety with Self-Driving Vehicles: General Limitations and Road Sharing with Conventional Vehicles (Report No. UMTRI-2015-2). 2015. Available online: https://deepblue.lib.umich.edu/bitstream/handle/2027.42/111735/103187.pdf?sequence=1&isAllowed=y (accessed on 25 July 2019).
3. Ackermann, C.; Beggiato, M.; Bluhm, L.-F.; Löw, A.; Krems, J.F. Deceleration parameters and their applicability as informal communication signal between pedestrians and automated vehicles. *Transp. Res. Part F Traffic Psychol. Behav.* **2019**, *62*, 757–768. [CrossRef]
4. Petzoldt, T.; Schleinitz, K.; Banse, R. The potential safety effects of a frontal brake light for motor vehicles. *IET Intell. Transp. Syst.* **2018**, *12*, 449–453. [CrossRef]
5. Šucha, M.; Dostal, D.; Risser, R. Pedestrian-driver communication and decision strategies at marked crossings. *Accid. Anal. Prev.* **2017**, *102*, 41–50. [CrossRef] [PubMed]
6. Liu, Y.-C.; Tung, Y.-C. Risk analysis of pedestrians' road-crossing decisions: Effects of age, time gap, time of day, and vehicle speed. *Saf. Sci.* **2014**, *63*, 77–82. [CrossRef]
7. Rodríguez, P. Safety of Pedestrians and Cyclists When Interacting with Automated Vehicles: A Case Study of the Wepods. 2017. Available online: https://www.raddelft.nl/wp-content/uploads/2017/06/Paola-Rodriguez-Safety-of-Pedestrians-and-Cyclists-when-Interacting-with...pdf (accessed on 19 April 2019).
8. Li, Y.; Dikmen, M.; Hussein, T.; Wang, Y.; Burns, C. To Cross or Not to Cross: Urgency-Based External Warning Displays on Autonomous Vehicles to Improve Pedestrian Crossing Safety. In Proceedings of the 10th International Conference on Automotive User Interfaces and Interactive Vehicular Applications Pages, Toronto, ON, Canada, 23–25 September 2018; pp. 188–197.
9. Färber, B. Communication and Communication Problems between Autonomous Vehicles and Human Drivers. In *Autonomous Driving*, 1st ed.; Maurer, M., Gerdes, J.C., Lenz, B., Winner, H., Eds.; Springer: Berlin/Heidelberg, Germany, 2016; pp. 125–144.
10. Schieben, A.; Wilbrink, M.; Kettwich, C.; Madigan, R.; Louw, T.; Merat, N. Designing the interaction of automated vehicles with other traffic participants: Design considerations based on human needs and expectations. *Cogn. Technol. Work* **2019**, *2019*, 69–85. [CrossRef]
11. Jayaraman, S.K.; Creech, C.; Tilbury, D.M.; Yang, X.J.; Pradhan, A.K.; Tsui, K.M.; Robert, L.P. Pedestrian trust in automated vehicles: Role of traffic signal and AV driving behavior. *Front. Robot. AI* **2019**, *6*, 14. [CrossRef]
12. Faas, S.M.; Mathis, L.-A.; Baumann, M. External HMI for self-driving vehicles: Which information shall be displayed? *Transp. Res. Part F Traffic Psychol. Behav.* **2020**, *68*, 171–186. [CrossRef]
13. De Clercq, K.; Dietrich, A.; Núñez Velasco, J.P.; de Winter, J.; Happee, R. External human-machine interfaces on automated vehicles: Effects on pedestrian crossing decisions. *Hum. Factors* **2019**, *61*, 1353–1370. [CrossRef] [PubMed]

14. Rothenbücher, D.; Li, J.; Sirkin, D.; Mok, B.; Ju, W. Ghost Driver: A Field Study Investigating the Interaction between Pedestrians and Driverless Vehicles. In Proceedings of the 25th IEEE International Symposium on Robot and Human Interactive Communication (IEEE Ro-Man '16), New York, NY, USA, 26–31 August 2016; pp. 795–802.

15. Song, Y.E.; Lehsing, C.; Fuest, T.; Bengler, K. External HMIs and Their Effect on the Interaction between Pedestrians and Automated Vehicles. In Proceedings of the 1st International Conference on Intelligent Human Systems Integration (IHSI '18), Dubai, United Arab Emirates, 7–9 January 2018; pp. 13–18.

16. Böckle, M.-P.; Brenden, A.P.; Klingegård, M.; Habibovic, A.; Bout, M. SAV2P: Exploring the impact of an interface for shared automated vehicles on pedestrians' experience. In Proceedings of the 9th International Conference on Automotive User Interfaces and Interactive Vehicular Applications (AutomotiveUI '17), New York, NY, USA, 24–27 September 2017; pp. 136–140.

17. Chang, C.-M.; Toda, K.; Sakamoto, D.; Igarashi, T. Eyes on a Car: An Interface Design for Communication between an Autonomous Car and a Pedestrian. In Proceedings of the 9th International Conference on Automotive User Interfaces and Interactive Vehicular Applications, New York, NY, USA, 24–27 September 2017; pp. 65–73.

18. Habibovic, A.; Andersson, J.; Malmsten Lundgren, V.; Klingegård, M.; Englund, C.; Larsson, S. External Vehicle Interfaces for Communication with Other Road Users? In *Road Vehicle Automation 5*; Meyer, G., Beiker, S., Eds.; Springer: Cham, Germany, 2019; pp. 91–102.

19. Hudson, C.R.; Deb, S.; Carruth, D.W.; McGinley, J.; Frey, D. Pedestrian Perception of Autonomous Vehicles with External Interacting Features. In Proceedings of the 9th International Conference on Applied Human Factors and Ergonomics (AHFE '18), Orlando, FL, USA, 21–25 July 2018; pp. 33–39.

20. Ackermann, C.; Beggiato, M.; Schubert, S.; Krems, J.F. An experimental study to investigate design and assessment criteria: What is important for communication between pedestrians and automated vehicles? *Appl. Ergon.* **2019**, *75*, 272–282. [CrossRef]

21. Kooijman, L.; Happee, R.; de Winter, J.C.F. How do eHMIs affect pedestrians' crossing behavior? A study using a head-mounted display combined with a motion suit. *Information* **2019**, *10*, 386. [CrossRef]

22. Mahadevan, K.; Sanoubari, E.; Somanath, S.; Young, J.E.; Sharlin, E. AV-Pedestrian Interaction Design Using a Pedestrian Mixed Traffic Simulator. In Proceedings of the 2019 on Designing Interactive Systems Conference (DIS '19), San Diego, CA, USA, 23–28 June 2019; pp. 475–486.

23. Eisma, Y.B.; van Bergen, S.; ter Brake, S.M.; Hensen, M.T.T.; Tempelaar, W.J.; de Winter, J.C.F. External human–machine interfaces: The effect of display location on crossing intentions and eye movements. *Information* **2020**, *11*, 13. [CrossRef]

24. Lagström, T.; Lundgren, V.M. Automated Vehicle's Interaction with Pedestrians. 2015. Available online: http://publications.lib.chalmers.se/records/fulltext/238401/238401.pdf (accessed on 20 April 2019).

25. Texas A&M Transportation Institute. Variable Speed Limits. 2018. Available online: https://mobility.tamu.edu/mip/strategies-pdfs/active-traffic/technical-summary/Variable-Speed-Limit-4-Pg.pdf (accessed on 10 February 2020).

26. Walker, F.; Dey, D.; Martens, M.; Pfleging, B.; Eggen, B.; Terken, J. Feeling-of-Safety Slider: Measuring Pedestrian Willingness to Cross Roads in Field Interactions with Vehicles. In Proceedings of the Extended Abstracts of the 2019 CHI Conference on Human Factors in Computing Systems, Scotland, UK, 4–9 May 2019.

27. Dey, D.; Walker, F.; Martens, M.; Terken, J. Gaze Patterns in Pedestrian Interaction with Vehicles. In Proceedings of the 11th International Conference on Automotive User Interfaces and Interactive Vehicular Applications (AutomotiveUI '19), Utrecht, The Netherlands, 22–25 September 2019; pp. 369–378.

28. Dey, D.; Martens, M.; Eggen, B.; Terken, J. Pedestrian road-crossing willingness as a function of vehicle automation, external appearance, and driving behaviour. *Transp. Res. Part F Traffic Psychol. Behav.* **2019**, *65*, 191–205. [CrossRef]

29. Bazilinskyy, P.; Dodou, D.; de Winter, J. Survey on eHMI concepts: The effect of text, color, and perspective. *Transp. Res. Part F Traffic Psychol. Behav.* **2019**, *67*, 175–194. [CrossRef]

30. Fridman, L.; Mehler, B.; Xia, L.; Yang, Y.; Facusse, L.Y.; Reimer, B. To Walk or not to walk: Crowdsourced assessment of external vehicle-to-pedestrian displays. *arXiv* **2017**, arXiv:1707.02698.

31. Fuest, T.; Michalowski, L.; Träris, L.; Bellem, H.; Bengler, K. Using the Driving Behavior of an Automated Vehicle to Communicate Intentions: A Wizard of Oz Study. In Proceedings of the 21st International Conference on Intelligent Transportation Systems (ITSC '18), Maui, HI, USA, 4–7 November 2018; pp. 3596–3601.

32. Lee, Y.M.; Uttley, J.; Solernou, A.; Giles, O.; Romano, R.; Markkula, G.; Merat, N. Investigating Pedestrians' Crossing Behaviour During Car Deceleration Using Wireless Head Mounted Display: An Application Towards the Evaluation of eHMI of Automated Vehicles. In Proceedings of the Tenth International Driving Symposium on Human Factors in Driver Assessment, Training and Vehicle Design, Santa Fe, NM, USA, 24–27 June 2019; pp. 252–258.

33. Palmeiro, A.R.; van der Kint, S.; Vissers, L.; Farah, H.; de Winter, J.C.F.; Hagenzieker, M. Interaction between pedestrians and automated vehicles: A Wizard of Oz experiment. *Transp. Res. Part F Traffic Psychol. Behav.* **2018**, *58*, 1005–1020. [CrossRef]

34. Rietveld, E. Situated normativity: The normative aspect of embodied cognition in unreflective action. *Mind* **2008**, *117*, 973–1001. [CrossRef]

35. Herbert, B.M.; Pollatos, O. The body in the mind. On the relationship between interoception and embodiment. *Top. Cogn. Sci.* **2012**, *4*, 692–704. [CrossRef] [PubMed]

36. SAE International. *Automated Driving System (ADS) Marker Lamp (J3134)*; SAE International: Warrendale, PA, USA, 2019.

37. Faas, S.M.; Baumann, M. Yielding Light Signal Evaluation for Self-Driving Vehicle and Pedestrian Interaction. In Proceedings of the 2nd International Conference on Human Systems Engineering and Design: Future Trends and Applications (IHSED '19), Munich, Germany, 16–18 September 2019; pp. 189–194.

38. Mahadevan, K.; Somanath, S.; Sharlin, E. Communicating Awareness and Intent in Autonomous Vehicle-Pedestrian Interaction. In Proceedings of the SIGCHI Conference on Human Factors in Computing Systems (CHI '18), Montreal, QC, Canada, 21–27 April 2018; pp. 1–12.

39. Faas, S.M.; Kao, A.C.; Baumann, M. A longitudinal Video Study on Communicating Status and Intent for Self-Driving Vehicle—Pedestrian Interaction. In Proceedings of the SIGCHI Conference on Human Factors in Computing Systems (CHI '20), Oahu, HI, USA, 25–30 April 2020.

40. Dahlbäck, N.; Jönsson, A.; Ahrenberg, L. Wizard of Oz Studies: Why and How. In Proceedings of the 1st International Conference on Intelligent User Interfaces (IUI '93), Orlando, FL, USA, 4–7 January 1993; pp. 193–200.

41. Garsten, E. Mercedes-Benz, Bosch Launch Robocar Ride-Hailing Pilot in San Jose. 2019. Available online: https://www.forbes.com/sites/edgarsten/2019/12/09/mercedes-benz-bosch-launch-robocar-ride-hailing-pilot-in-san-jose/#441deb7e3c5b (accessed on 21 June 2020).

42. Randazzo, R. Waymo's Driverless Cars on the Road: Cautious, Clunky, Impressive. 2019. Available online: https://eu.azcentral.com/story/money/business/tech/2018/12/05/phoenix-waymo-vans-how-self-driving-cars-operate-roads/2082664002/ (accessed on 21 June 2020).

43. Ackermans, S.; Dey, D.; Ruijten, P.; Cuijpers, R.H.; Pfleging, B. The Effects of Explicit Intention Communication, Conspicuous Sensors, and Pedestrian Attitude in Interactions with Automated Vehicles. In Proceedings of the SIGCHI Conference on Human Factors in Computing Systems (CHI '20), Honolulu, HI, USA, 25–30 April 2020.

44. Bartneck, C.; Kulić, D.; Croft, E.; Zoghbi, S. Measurement instruments for the anthropomorphism, animacy, likeability, perceived intelligence, and perceived safety of robots. *Int. J. Soc. Robot.* **2009**, 71–81. [CrossRef]

45. Schrepp, M.; Hinderks, A.; Thomaschewski, J. Design and evaluation of a short version of the user experience questionnaire (UEQ-S). *IJIMAI* **2017**, 103–108. [CrossRef]

46. Bortz, J.; Schuster, C. *Statistik für Human- und Sozialwissenschaftler*, 7th ed.; Springer: Berlin/Heidelberg, Germany, 2010.

47. Field, A. Discovering Statistics: Cluster Analysis. 2017. Available online: https://www.discoveringstatistics.com/2017/01/13/cluster-analysis/ (accessed on 21 June 2020).

48. Field, A. *Discovering Statistics Using IBM SPSS Statistics*, 5th ed.; SAGE Publications Ltd.: London, UK, 2018.

49. Hinderks, A.; Schrepp, M.; Thomaschewski, J. UEQ Data Analysis Tool. 2019. Available online: https://www.ueq-online.org/Material/Short_UEQ_Data_Analysis_Tool.xlsx (accessed on 22 June 2019).

50. Sun, D.J.; Elefteriadou, L. Lane-Changing Behavior on Urban Streets: An "In-Vehicle" Field Experiment-Based Study. *Comput.-Aided Civ. Infrastruct. Eng.* **2012**, *27*, 525–542. [CrossRef]

51. Feldstein, I.T.; Lehsing, C.; Dietrich, A.; Bengler, K. Pedestrian simulators for traffic research: State of the art and future of a motion lab. *Int. J. Hum. Factors Model. Simul.* **2018**, *6*, 250–265. [CrossRef]

52. Hettinger, L.J.; Riccio, G.E. Visually induced motion sickness in virtual environments. *Presence Teleoperators Virtual Environ.* **1992**, *3*, 306–310. [CrossRef]
53. Weiß, T.; Petzoldt, T.; Bannert, M.; Krems, J.F. Einsatz von computergestuetzten Medien und Fahrsimulatoren in Fahrausbildung, Fahrerweiterbildung und Fahrerlaubnispruefung. *Ber. Bundesanst. Straßenwesen Reihe M (Mensch Sicherh.)* **2009**, *202*. Available online: https://bast.opus.hbz-nrw.de/opus45-bast/frontdoor/deliver/index/docId/1/file/BASt_Schlussbericht_November_2007.pdf (accessed on 10 July 2020).

Review

Usability Evaluation—Advances in Experimental Design in the Context of Automated Driving Human–Machine Interfaces

Deike Albers [1,*], **Jonas Radlmayr** [1], **Alexandra Loew** [1], **Sebastian Hergeth** [2], **Frederik Naujoks** [2], **Andreas Keinath** [2] **and Klaus Bengler** [1]

1   Chair of Ergonomics, Technical University of Munich, Boltzmannstraße 15, 85748 Garching, Germany; jonas.radlmayr@gmail.com (J.R.); alexandra.loew@tum.de (A.L.); bengler@tum.de (K.B.)
2   BMW Group, Knorrstraße 147, 80937 Munich, Germany; Sebastian.Hergeth@bmw.de (S.H.); Frederik.Naujoks@bmw.de (F.N.); andreas.keinath@bmw.de (A.K.)
*   Correspondence: deike.albers@tum.de; Tel.: +49-89-289-15420

Received: 27 March 2020; Accepted: 23 April 2020; Published: 28 April 2020

Abstract: The projected introduction of conditional automated driving systems to the market has sparked multifaceted research on human–machine interfaces (HMIs) for such systems. By moderating the roles of the human driver and the driving automation system, the HMI is indispensable in avoiding side effects of automation such as mode confusion, misuse, and disuse. In addition to safety aspects, the usability of HMIs plays a vital role in improving the trust and acceptance of the automated driving system. This paper aggregates common research methods and findings based on an extensive literature review. Empirical studies, frameworks, and review articles are included. Findings and conclusions are presented with a focus on study characteristics such as test cases, dependent variables, testing environments, or participant samples. These methods and findings are discussed critically, taking into consideration requirements for usability assessments of HMIs in the context of conditional automated driving. The paper concludes with a derivation of recommended study characteristics framing best practice advice for the design of experiments. The advised selection of scenarios and metrics will be applied in a future validation study series comprising a driving simulator experiment and three real driving experiments on test tracks in Germany, the USA, and Japan.

Keywords: conditionally automated driving; human–machine interface; usability; validity; method development

---

1. Introduction

The introduction of conditionally automated driving (CAD) vehicles drastically alters the role of the human in the car. Based on the definition of the Society of Automotive Engineers (SAE), CAD or Level 3 automated driving means that the automated driving system (ADS) is responsible for the entire driving task, while the human operator is ready to respond as necessary to ADS-issued requests to intervene and to system failures by resuming the driving task [1]. The transition of the human driver from the role of operator to the passenger role implies a paradigm change relative to the Level 2 or partially automated systems that are available today [1,2]. This paradigm change, including transitions back and forth to lower levels of automated driving, affects the human–machine interface. CAD implies that the human must take back control of the driving task in cases where the system reaches a system boundary and in doing so, to resume manual driving. The resulting transition of the driving task from the automation system to the human requires an appropriate communication strategy as well as a human–machine interface (HMI) that supports the interaction between the two parties in general. New challenges in both HMI design for automated driving and CAD in particular are addressed in this review paper.

This paper gives an overview of the status quo for usability assessments for automated driving HMIs. Current practice is presented by summarizing the methodological approaches of study articles. Additionally, theoretical articles such as literature reviews are included. Both are considered in the derivation of best practice advice for experimental design. This best practice advice will be applied in an international validation study for assessing the usability of CAD HMIs comprising four experiments in three countries and two testing environments. In Germany, a driving simulator experiment and a test track experiment are planned. Two further test track experiments are planned for Japan and the USA. All four experiments will apply the same study design, ensuring the comparability of the results. The articles in this paper have been aggregated using a predefined set of six categories. These categories were identified in the research phase of the validation project and represent differences in the methodological approaches.

Basing on the existing literature, this paper aims to derive a feasible practical and theoretical experimental design that will be validated in the study series described above. The developed experimental design serves as best practice for future studies in which the aim is to assess the usability of CAD HMIs.

2. Paper Selection and Aggregation

This paper reviews 16 scientific articles that cover the usability assessment of CAD HMIs. The selection includes study articles and theoretical articles. The selection process and the aggregated data are presented in the following sections.

2.1. Paper Selection

Literature searches have been conducted in the literature manager Mendeley and Researchgate, resulting in seven articles.

Additionally, a systematic review has been conducted via the search engine for scientific literature Google Scholar. The process followed the guideline Preferred Reporting Items for Systematic Reviews and Meta-Analyses (PRISMA) and is visualized in Figure 1 [3]. This guideline enhances the transparency of the selection process by describing the stepwise narrowing of the chosen articles for the review. For the identification of potential articles, different combinations of keywords such as "Usability", "Human–Machine Interface", and "Conditionally Automated Driving" are applied. The first step in the process resulted in 553 articles. The next step included the articles identified in other libraries or databases, respectively. In total, 188 duplicates were removed. A first screening of the titles and the abstracts lead to the exclusion of 346 further articles. After reading all articles, 10 more articles were excluded due to a lack of relevance for this review. In the resulting selection of 16 articles, the usability assessment of ADS HMIs for CAD is described.

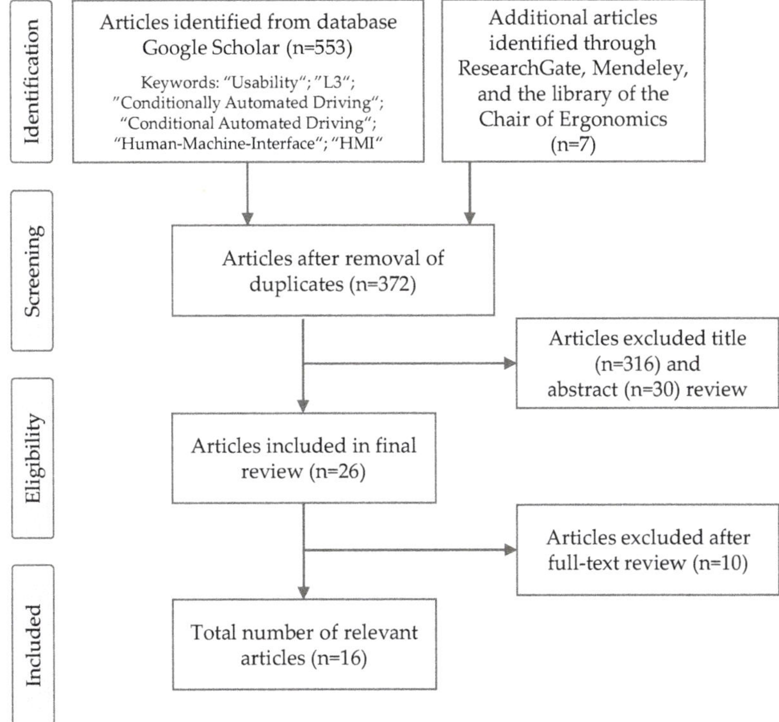

Figure 1. Process of the literature review based on the PRISMA guideline [3].

2.2. Aggregation

The final selection includes 16 articles. Nine articles present experiments, which are hereafter referred to as study articles [4–12]. The seven other articles are of theoretical nature and are therefore referred to as theoretical articles [13–19]. There are several characteristics that define a study design. By taking into account both common practice and theoretical considerations, this review paper aims to derive best practice advice for researchers interested in the usability of CAD HMIs.

Six experiment characteristics were chosen to meet the challenge of assessing usability in the development process. The literature search yielded different approaches for the usability testing of ADS HMIs. The differences identified in the first research phase resulted in the selection of six categories. These provide the structure of this paper, including the study characteristics' dependent variables, and the testing environment. The definitions of the term usability applied in each of the selected articles are used to understand the research focus of each article. Furthermore, the sample characteristics, the test cases, and the conditions of use, i.e., initial versus repeat contact (see below), are considered. The characteristics listed below provide an insight into the methodological approaches of the nine empirical study articles and the discussed and recommended methodologies of the seven theoretical articles:

- Definition of Usability
- Testing Environment
- Sample Characteristics
- Test Cases
- Dependent Variables
- Conditions of Use

The characteristics listed or applied in the 16 articles are summarized in the first paragraph of the following subsections and the respective tables. Every subsection closes with a critical discussion of the findings resulting in a recommendation of an experimental procedure or method. These recommendations form the best practice advice for usability assessments of CAD HMIs.

2.2.1. Definition of Usability

The understanding of the term usability has a considerable influence on the experimental design that researchers choose. Different definitions and operationalizations may result in a different study design. To reflect these potential differences in design, the information on usability given in the selected articles is compared in this subsection. Table 1 shows 12 of the 16 articles. Four articles do not define or operationalize the term usability [4,5,8,15]. Five of the remaining articles [9,11,13,18,19] give an insight into the authors' understanding of the construct usability by the chosen dependent variable(s), e.g., the acceptance, or metrics, e.g., the System Usability Scale (SUS) [20], the Post-Study System Usability Questionnaire (PSSUQ) [21], or the acceptance scale of Van der Laan (VDL) [22]. Four articles [6,7,12,17] cite ISO Standard 9241 with its definition of usability as the "extent to which a system, product or service can be used by specified users to achieve specified goals with effectiveness, efficiency and satisfaction in a specified context of use" [23] (p. 2). However, the complete definition is used only once [7], whereas three articles focus on the effectiveness and efficiency while leaving out the construct satisfaction [6,12,17]. Ref. [12] adds the term "usefulness" to the constructs effectiveness and efficiency. Two other articles cite the minimum requirements of the National Highway Traffic Safety Administration (NHTSA) [16,17]. These requirements impose that the user of an ADS HMI must be able to understand if the ADS is "(1) functioning properly; (2) currently engaged in ADS mode; (3) currently 'unavailable' for use; (4) experiencing a malfunction; and/or (5) requesting control transition from the ADS to the operator" [24] (p. 10). Ref. [17] applies the NHTSA minimum requirements to the two constructs effectiveness and efficiency. The remaining two articles [10,14] cite Nielsen [25] who builds usability from five constructs: learnability, efficiency, memorability, errors, and satisfaction.

Table 1. Aggregation of the definitions of usability.

Article	ISO Standard 9241 [23]	Nielsen [25]	NHTSA Minimum Requirements [24]	Operationalization Through Dependent Variables
Forster et al. (2019c) [6]	Effectiveness and efficiency			
Forster et al. (2019d) [7]	x			
Kettwich et al. (2016) [9]				Satisfaction and usefulness (VDL [22]), expectations, suggestions
Morgan et al. (2018) [10]		x		
Naujoks et al. (2017) [11]				Comprehensibility, SUS [20]
Richardson et al. (2018) [12]	Efficiency, effectiveness, and usefulness			
Forster et al. (2018) [1] [13]				SUS [20]/PSSUQ [21]
François et al. (2016) [1] [14]		x		
Naujoks et al. (2018) [1] [16]			Usability and safety	
Naujoks et al. (2019a) [1] [17]	Effectiveness and efficiency		x	
Naujoks et al. (2019b) [1] [18]				20-item guideline
Pauzie and Orfila (2016) [1] [19]				Acceptability, acceptance, trust, situation awareness, workload

[1] Theoretical article.

In addition to the implicit operationalization through dependent variables, only three sources are cited for the 16 selected articles. These are the ISO Standard 9241 [23], the NHTSA minimum requirements [24], and the Nielsen model for usability [25]. During examination of the articles, both theoretical articles and study articles posed difficulties in working out the authors' understanding of usability. Considering that usability forms the focus of the research question, the underlying definition or at least the operationalization should be communicated to the readers. We strongly advice applying ISO Standard 9241 that comprises the constructs effectiveness, efficiency, and satisfaction [23]. Since the ISO Standard does not elaborate on the detailed testing procedure, further operationalizations are recommended, e.g., whether the effectiveness is tested in a setting with novice or experienced users. When citing the NHTSA requirements for usability tests, researchers choose a different approach to defining the term usability that considers the context of automated driving. Moreover, the usability is rated according to the comprehension of the user that the ADS is "(1) functioning properly, (2) currently engaged in ADS mode, … " [24] (p. 10). This narrows down the practical realization of the usability assessment. A combination of this approach and the definition of the general term usability based on ISO Standard 9241 seems to be most applicable.

2.2.2. Testing Environment

Four of the theoretical articles provide no information on the testing environment in which the usability assessment should be conducted [13–16]. Of the remaining 12 articles (shown in Table 2), the use of an instrumented car is recommended twice [18,19], while the additional use of a high-fidelity driving simulator is recommended by [18]. Ten of the 12 articles recommend or use a driving simulator [4–11,17,18]. The details of the simulator are specified in most of the study articles. A fix-base simulator is used in four articles [4,7,9,10], a moving-base simulator is used in two cases [5,6], and in one other case, a low-fidelity simulator is described [11]. Ref. [12] does not use an instrumented car or driving simulator; rather, desktop methods are applied where paper and video prototypes are evaluated.

Table 2. Aggregation of the testing environments.

Article	Driving Simulator	Instrumented Car	Desktop Methods
Forster et al. (2019a) [4]	Fix-base		
Forster et al. (2019b) [5]	Moving-base		
Forster et al. (2019c) [6]	Moving-base		
Forster et al. (2019d) [7]	Fix-base		
Guo et al. (2019) [8]	x		
Kettwich et al. (2016) [9]	Fix-base		
Morgan et al. (2018) [10]	Fix-base		
Naujoks et al. (2017) [11]	Low-fidelity		
Richardson et al. (2018) [12]			Workshop
Naujoks et al. (2019a) [1] [17]	x		
Naujoks et al. (2019b) [1] [18]	High-fidelity	x	
Pauzie and Orfila (2016) [1] [19]		x	

[1] Theoretical article.

Driving simulators are the prevalent testing environment in the field of usability assessments of ADS HMIs for CAD. Only two of the theoretical articles stress the need for real driving experiments, e.g., with instrumented cars. Driving simulators provide efficient and risk-free testing environments that provide valuable results [26]. For some research questions, they may even be the only realizable testing environment, e.g., for testing critical situations in automated driving such as near crashes and system failures in high-speed conditions. As the name implies, driving simulators do not equate with reality. High-fidelity driving simulators increase the match with reality and are to be preferred over low-fidelity simulators or desktop methods. The validity of driving simulators is assessed in several studies [27]. For research results obtained in driving simulators used to assess the usability of CAD

HMIs, the validity is yet to be verified. For practical reasons, driving simulators constitute the best testing environment. However, a validation check for the research results is needed.

2.2.3. Sample Characteristics

This subsection aggregates the sample characteristics. Usability tests can be conducted with experts or potential users [28,29]. Information on the participant group is provided by 14 articles of this review [4–14,16–18]. Three theoretical articles recommend including both sample groups, i.e., experts and participants in the development process of an ADS HMI [13,16,18]. Two other theoretical articles list users as participants [14,17]. Ref. [17] recommends a diverse age distribution as advised in [30]. Moreover, the authors emphasize that participants should not be affiliated with the tested system. Of the nine study articles, two conducted the usability assessment with 6 or 5–9 experts, respectively [11,12]. In these articles, experts were described as working in the "field of cognitive ergonomics" or "field of ergonomics, HMI, and function development from university and industry". The seven other study articles conducted their usability tests with potential users [4–10]. The reported sample size varies between 12 and 57. The age distribution ranges between 20 and 62, except for [10], where older adults between 47 and 88 years old were tested. Attention should be drawn to the fact that of the seven experiments with potential users, five experiments were conducted with employees of a car maker [4–8]. Table 3 shows an overview of the sample characteristics.

Table 3. Aggregation of the sample characteristics.

Article	Users	Experts
Forster et al. (2019a) [4]	n = 24; age 20–62; BMW employees	
Forster et al. (2019b) [5]	n = 52; age 20–62; BMW employees	
Forster et al. (2019c) [6]	n = 55; age 20–62; BMW employees	
Forster et al. (2019d) [7]	n = 57; age 25–60; BMW employees	
Guo et al. (2019) [8]	n = 22; age 24–61; Renault or IRT System X employees	
Kettwich et al. (2016) [9]	n = 12; age 23–49	
Morgan et al. (2018) [10]	n = 31; age 47–88	
Naujoks et al. (2017) [11]		n = 6; field of cognitive ergonomics
Richardson et al. (2018) [12]		$n_1 = 5$, $n_2 = 9$; field of ergonomics, HMI, driver assistance systems; from university and industry
Forster et al. (2018) [1] [13]	x	x
François et al. (2016) [1] [14]	x	
Naujoks et al. (2018) [1] [16]	x	x
Naujoks et al. (2019a) [1] [17]	n > 20; diverse age distribution [30]; potential users, comparable prior experience, not affiliated with tester's company	
Naujoks et al. (2019b) [1] [18]	x	n > 4

[1] Theoretical article.

Conducting tests with potential users is the predominant method in the articles of this review. Using experts as participants represents an efficient approach for identifying major usability issues early in the development process. At advanced stages, tests with potential users are indispensable. The participants should be selected with high demands to the representativeness. The population of potential users of ADS has a high level of variability in its characteristics, e.g., prior experience or physical and cognitive abilities. User testing is most valid and productive when a sample representing potential users is being tested. Research using subpopulations could lead to biased results [31]. Therefore, when testing the usability of CAD HMIs, efforts should be made to keep the number of participants with affiliations to technical or automotive domains to a minimum. Further characteristics such as age or gender should be selected according to the represented user group. The sample size varies greatly in the selected articles. The decision on sample size should be defined by the statistical procedure used to identify potential effects of interest.

2.2.4. Test Cases

The test cases in an experiment are strongly dependent on the research question. As the research questions in the selected articles of this review all focus on the usability assessment of ADS HMIs for CAD, the test cases are comparable. However, no details are considered; Table 4 shows only test case categories. Ten of the 13 articles that provide information on test cases list transition scenarios [4–7,11,12,15–18]. Downward transitions are found in each of these 10 articles. A more detailed view shows that seven of these articles describe transitions to manual driving [6,7,11,12,16–18]. Eight articles [4–7,12,16–18] list test cases with upward transitions, e.g., SAE Level 0 (L0) to SAE Level 3 (L3) [1]. The system mode as well as the availability of automated driving modes are listed as dedicated test cases in four articles [12,16–18]. Likewise, three experiments include test cases with information on planned maneuvers, e.g., lane changes [7,11,12]. Two articles include test cases that represent different traffic scenarios, e.g., traffic density [8,9]. Use of the navigation function is the focus of [10].

Table 4. Aggregation of the test cases.

Article	Upward Transitions [2]	Downward Transitions [2]	System Mode/Availability [2]	Others
Forster et al. (2019a) [4]	L0 → L2 L0 → L3 L2 → L3	L3 → L2		
Forster et al. (2019b) [5]	L0 → L2 (driver) L0 → L3 (driver) L2 → L3 (driver)	L3 → L2 (driver)		
Forster et al. (2019c) [6]	L0 → L2 L0 → L3 L2 → L3	L3 → L0 L3 → L2 L2 → L0		
Forster et al. (2019d) [7]	L0 → Lx (initial) L0 → Lx (re-activation) L0 → Lx (re-activation)	Lx → L0 (driver) Lx → L0 (system; TOR) Lx → L0 (driver; TOR)		Maneuver (lane change, speed adaptation)
Guo et al. (2019) [8]				Highway entry section with different traffic conditions
Kettwich et al. (2016) [9]				Environment (traffic light)
Morgan et al. (2018) [10]				Operating a navigation system
Naujoks et al. (2017) [11]		Lx → L0		Maneuver and environment (splitting lanes, curvature, speed limit)
Richardson et al. (2018) [12]	L0 → Lx	Lx → L0	x	
Gold et al. (2017) [1] [15]		x		
Naujoks et al. (2018) [1] [16]	84 TC	84 TC	14 TC	
Naujoks et al. (2019a) [1] [17]	L2 → L3	L3 → L2 (driver) L3 → L2 (system) L3 → L1 (system) L3 → L0 (system)	L2 steady state L3 steady state L3 degraded L3 unavailable	
Naujoks et al. (2019b) [1] [18]	L0 → Lx	Lx → L0	x	

[1] Theoretical article. [2] [1].

In the articles considered in this review, most of the test cases comprise transitions between or the availability of different automation modes, which are mostly referred to as SAE levels [1]. Successful transitions and the operator's understanding of the automated driving modes are important for the safe and efficient handling of the ADS. If the usability is tested and the human operator fails to understand the information communicated by the HMI, improvement measures for the HMI are inevitable. Therefore, the interaction of the operator with the ADS should be tested regarding these functions. In addition to test cases directly related to automation modes, another type of test case can be applied when assessing the usability. These are test cases where usability evaluations refer to the handling of additional systems such as navigation systems or the radio. Non-driving-related activities (NDRA) are of high importance for usability evaluations where the human operator is involved in the driving task [2]. With the introduction of CAD, the focus of usability assessments is on transitions and the automation modes themselves. Additionally, this review concludes with a recommendation for testing non-critical scenarios. Critical situations are important for assessing safety aspects. These situations have a low probability of occurring. In particular, situations with high criticality are not suitable for usability assessments, e.g., tests that determine the range of reaction times with a crash rate of 100%. For a thorough evaluation of usability, comprising constructs such as satisfaction of the ISO Standard 9241 [23], recurring non-critical situations are more appropriate.

2.2.5. Dependent Variables

Three of the theoretical articles do not provide information on dependent variables [14–16]. The dependent variables stated in the theoretical articles or applied in the study articles of the remaining 13 articles are shown in Table 5. The dependent variables are categorized in constructs, while information on the specific metrics is added in the respective cells. More generally, the variables can be categorized into observational and subjective data. Six articles recommend or report the use of observational data [4,5,8,11,13,19]. Ref. [13] recommends collecting both data types; the interaction performance with a system or secondary task, as well as the visual behavior. Two other articles name visual behavior (e.g., the number of gaze switches) as a suitable metric [5,19]. The interaction performance is assessed either directly based on the reaction time or the number of operating steps/errors or indirectly by expert assessments. In total, four articles list this type of a dependent variable [4,8,11,13]. The SUS [20] is widely used and belongs to the subjective measures. The questionnaire is listed by six of the 13 articles [6,7,10–13]. Two other dedicated usability questionnaires are utilized in one article each; the Post-Study System Usability Questionnaire [21] by [13] and the standardized ISO 9241 Questionnaire [32], as cited by [12]. Other constructs that interrelate with usability such as acceptance, which correlates with the construct satisfaction of ISO 9241 [23], are tested by several articles in this review. These constructs report further questionnaires. Questionnaires on acceptance are used three times [7,9,19], e.g., the VDL [22] or the Unified Theory of Acceptance and Use of Technology (UTAUT) [33]. Questionnaires on trust such as the General Trust Scale (GTS) [34] or the Universal Trust in Automation scale (UTA) [35] are reported three times [7,10,19]. Constructs such as workload (cited by [10,19]), measured, for example, using the metric NASA Task Load Index (NASA-TLX) [36], situation awareness (cited by [10,19]), measured, for example, using the metric Situation Awareness Global Assessment Technique (SAGAT) [37], or the mental model of drivers (cited by [4,5]), measured, for example, using the mental model questionnaire by Beggiato [38], are each listed twice. Additional questionnaires that are reported only once can be found in Table 5. In addition to questionnaires, methods such as the Thinking Aloud Technique [39], applied by [8,9,11], or heuristic evaluations [40], applied by [12,17,18], are commonly used, especially for expert studies. Furthermore, interviews, expert evaluations, and spaces for suggestions and comments are often used to gain insights that standardized methods cannot provide [8,9,11,19].

Table 5. Aggregation of the dependent variables. NDRA: non-driving-related activities. UEQ: User Experience Questionnaire. meCUE: modular evaluation of key Components of User Experience. SART: Situation Awareness Rating Technique. ATCQ: Attitudes Towards Computers Questionnaire. DALI: Driving Activity Load Index.

Article	Observational Metrics (Visual Behavior, Interaction and NDRA Performance, etc.)	Usability Questionnaire	Other Constructs (Questionnaires) and Methods
Forster et al. (2019a) [4]	Experimenter rating		Mental model [38]
Forster et al. (2019b) [5]	Visual behavior (no. of gaze switches)		Mental model [38]
Forster et al. (2019c) [6]		SUS [20]	
Forster et al. (2019d) [7]		SUS [20]	Acceptance (VDL [22], UTAUT [33]); trust (Trust in Automated Systems [41], UTA [35]); user experience (AttrakDiff [42], UEQ [43], meCUE [44])
Guo et al. (2019) [8]	Time & frequency of button use		Interview; Thinking Aloud Method [39]
Kettwich et al. (2016) [9]			Acceptance (VDL [22]); interview thinking aloud method [39]
Morgan et al. (2018) [10]		SUS [20]	Workload (NASA-TLX [36]); Trust (ATS [41], GTS [34]); Situation Awareness (SART [45]); Technical Affiliation (ATCQ [46])
Naujoks et al. (2017) [11]	Take-Over Performance No. of unnecessary system deactivations	SUS [20]	Interview; Expert Evaluation
Richardson et al. (2018) [12]		SUS [20], ISO 9241 [32] as cited by [12]	Desirable HMI Aspects [47]; Thinking Aloud Method [39]; Heuristic Evaluation [40]
Forster et al. (2018) [1] [13]	Visual Behavior; Reaction Times; Interaction and NDRA Performance; Expert Assessment	SUS [20], PSSUQ [21]	
Naujoks et al. (2019a) [1] [17]			Heuristic Evaluation [40]
Naujoks et al. (2019b) [1] [18]			Heuristic Evaluation [40]
Pauzie, & Orfila (2016) [1] [19]	Visual Behavior		Acceptance; Workload (DALI [48]); Trust; Situation Awareness (SAGAT [37], SART [45]); Interview

[1] Theoretical article.

Summarizing the listed dependent variables, usability appears as a well-defined construct ([23]) that can be assessed via multifaceted metrics. Depending on the research questions, different dependent variables seem more applicable than others. Nevertheless, patterns can be detected. A combination of observational and subjective data is used by 6 of the 13 articles that provide information on dependent variables [4,5,8,11,13,19]. The SUS [20] is widely used by the researchers cited in this review. Where individual research questions are concerned, further questionnaires can be used to evaluate constructs such as trust, acceptance, or workload. If information for specific research interests cannot be extracted via standardized methods, interviews, the Thinking Aloud Technique or heuristic evaluations can be applied. When combining these dependent variables, mutual impacts should be considered. For example, applying the Thinking Aloud Technique is not suitable in combination with interaction performance measurements such as reaction times. For tests with potential users, this review recommends a combination of observational metrics that measure the behavior and subjective metrics that gather the operator's personal impressions. For observational data, analysis of the visual behavior based on ISO 15007 [49] and the interaction performance using the ADS HMI seem most applicable. Possible metrics are the number of operating errors, the reaction time for a button press or the percent time on an area of interest, e.g., the instrument cluster. The SUS is recommended as a valid and widely used usability questionnaire. Supplementary questionnaires should be selected with regard to the specific research question. If usability is not the only construct of interest in an experiment, the link between the dependent variables and the constructs should be clearly stated. Standardized metrics should be used to enable comparisons between experiments and create transparency with other researchers. Short interviews provide valuable insights that can be tailored to the specific research question. Interviews should be conducted after test trials and questionnaires to avoid distorted results.

2.2.6. Conditions of Use

When the usability of a system is tested, the duration of use and the prior experience need to be considered. The conditions can range between the first contact between a novice user and the system and everyday use by an experienced user. The first contact can be tested with prior experience, e.g., after reading the manual, after conducting an interactive tutorial, or after being instructed by an advisor. Prolonged use can be interpreted as a series of repeat contacts between the user and the operator within a few hours or in the scope of a long-term study. The articles analyzed in this review generally do not provide detailed information on the conditions of use that is of research interest. Table 6 shows an overview with aggregated information on the nature of the usability testing provided by 14 of the articles in this review. In all 14 articles, first contact is tested or, in the case of the theoretical articles, it is recommended to be tested [4–14,16–18]. In four of these cases, the testing circumstances are specified as testing intuitive use without having first being given detailed instructions [5,6,8,16]. Another article investigates the influence of different tutorials and therefore tests both tutorials and intuitive use [4]. Of the 14 articles that test first contact, seven also assess repeat contact with the system [5,6,10,12–14,17].

Table 6. Aggregation of the conditions of use.

Article	First Contact	Repeat Use
Forster et al. (2019a) [4]	Intuitive use, manual, and interactive tutorial	
Forster et al. (2019b) [5]	Intuitive use	x
Forster et al. (2019c) [6]	Intuitive use	x
Forster et al. (2019d) [7]	x	
Guo et al. (2019) [8]	Intuitive use	
Kettwich et al. (2016) [9]	x	
Morgan et al. (2018) [10]	x	x
Naujoks et al. (2017) [11]	x	
Richardson et al. (2018) [12]	x	x
Forster et al. (2018) [1] [13]	x	x
François et al. (2016) [1] [14]	x	x
Naujoks et al. (2018) [1] [16]	Intuitive use	
Naujoks et al. (2019a) [1] [17]	x	x
Naujoks et al. (2019b) [1] [18]	x	

[1] Theoretical article.

Testing the first contact when assessing the usability of an ADS HMI appears to be the predominant method. Only a few of the selected articles tested repeat contacts when assessing usability. Prolonged use in the form of a long-term study testing everyday use is not considered in the articles selected for this review. Both first contact and prolonged use are important aspects to consider when evaluating usability. A successful first contact is highly important from the point of view of safety. This means that the handling of a system is intuitively understandable without consulting the manual—similar, for example, to a human driver using a rental car without first familiarizing themselves with the car's handling. For research effects such as disuse, misuse, or even abuse of a system, the consideration of prolonged use in everyday situations is critical [50]. As it poses a different type of research question, it requires a different kind of experiment. In alignment with most of the articles, this review concludes by recommending first contact tests. The NHTSA minimum requirements state that an HMI must be designed in such a way that a user understands if the ADS is "(1) functioning properly; (2) currently engaged in ADS mode; (3) currently "unavailable" for use; (4) experiencing a malfunction; and/or (5) requesting control transition from the ADS to the operator" [24] (p. 10). The fulfillment of these requirements can be checked by assessing usability in a first contact situation. This requires that participants are not given detailed instructions, such as pictures of the HMI requesting a control transition, prior to the first contact. Instead, participants should only receive instructions with general information on ADS.

3. Discussions

In this review paper, 16 selected articles focusing on usability assessments for ADS HMIs for CAD are analyzed. Information on methodological approaches, study characteristics, as well as the understanding of the term usability has been aggregated. The insights gained are used to draw conclusions on best practice for researchers investigating the usability of CAD HMIs. In this section, the recommendations are discussed and incorporated in more general advice on usability testing.

Three different sources are cited for the understanding of the term usability [23–25]. Yet, many articles do not provide information on the definition used. The definitions of the three sources result in different study designs than those that would have been derived had a different definition been selected. In order to assess the usability of CAD HMIs, we advise applying a combination of ISO 9241 and the NHTSA minimum requirements [23,24]. However, other definitions, e.g., [25], might be better suited for specific research questions. In general, it is important to provide an operationalization of the term usability when conducting assessments, especially where standards are not applied.

For practical reasons, the review concludes with the recommendation of high-fidelity driving simulators. Depending on the development stage, other testing environments may prove more applicable. For early prototypes, desktop methods provide valuable insights with minimal resource input. Real driving tests can help in the refinement process of preproduction products.

This review recommends that usability tests should be conducted with potential end users. These tests are indispensable for final usability evaluations. Other participants, such as experts or users that represent only a segment of the user population, e.g., students or participants with affiliations to technical or automotive domains, can provide valuable insights at earlier stages of the development process.

The test cases listed in the best practice advice of this review focus on transitions between, and the availability of, different automation levels in non-critical situations. These test cases are recommended for general usability assessments of ADS HMIs for CAD. Other test cases in this review cover usability assessments of HMIs displaying information on more complex scenarios, such as maneuvers, navigation systems, or dense traffic. These test cases are relevant for specific research questions, e.g., the design of integrated functions in the CAD HMI.

A set of metrics for testing the usability of CAD HMIs is listed in this paper. Depending on the study design and the research question, further metrics might prove suitable in obtaining valuable research findings. Researchers should clearly indicate the link between dependent variables and the respective definition or construct of interest.

This review recommends that the usability tests be performed in first contact situations without in-depth instructions on how to use the system having been provided prior to the testing situation. Where research questions not focusing on the NHTSA minimum requirements are concerned, the use of manuals or tutorials might be applicable in order to equalize the knowledge and experience level of the participants. In addition to testing first contacts, the everyday use of ADS is of great interest, especially in the context of CAD. The transition of the human driver from operator to passenger could generate side effects such as disuse, misuse, or abuse of the ADS, which might impair safety. The assessment of these effects poses an interesting and important topic for future research.

4. Conclusions

This paper reviews 16 articles, comprising both study and theoretical articles. These articles are analyzed in respect of six study characteristics. The insights into common practice and theoretical considerations lead to a derivation of best practice advice. This advice is aimed at helping researchers who are interested in usability assessments of CAD HMIs in the planning phase of a study. Furthermore, the comparability of studies in this field increases with the application of similar experimental designs. Table 7 summarizes the key statements of the derived best practice.

Table 7. Best practice advice for testing the usability of conditionally automated driving (CAD) human–machine interfaces (HMIs). ADS: automated driving system.

Study Characteristic	Best Practice Advice
Definition of Usability	General Definition: "extent to which a system, product or service can be used by specified users to achieve specified goals with effectiveness, efficiency and satisfaction in a specified context of use" [23] (p. 2) Practical Realization: the user understands that the ADS is "(1) functioning properly; (2) currently engaged in ADS mode; (3) currently "unavailable" for use; (4) experiencing a malfunction; and/or (5) requesting control transition from the ADS to the operator" [24] (p. 10)
Testing Environment	Driving Simulator
Sample Characteristics	Sample Group: represents the potential user population (age, gender, prior experience, affiliation with technical devices, etc.) Sample Size: determined by the statistical procedure
Test Cases	Scenarios: (1) transitions between different automation modes and (2) availability of different automation modes Criticality: non-critical situations
Dependent Variables	General: Combination of observational and subjective metrics Observational metrics: (1) visual behavior according to [49] (e.g., percent on Area of Interest) and (2) the interaction performance with CAD HMI (e.g., operating errors or reaction time for a button press) Subjective Metrics: (1) System Usability Scale [20], (2) short interviews after test trials and questionnaires, and (3) supplementary standardized questionnaires
Conditions of Use	First contact between user and ADS Instructions contain only general information on the ADS

5. Outlook

In this review, driving simulators are identified as the prevalent testing environment in the field of usability assessments of ADS HMIs for CAD. As an efficient and risk-free alternative to real driving experiments, simulators offer a convenient and valuable testing environment. Since the validity of driving simulators has not yet been assessed, the transferability of results to the real world is not assured. A thorough validation study comparing a simulator and a test track experiment is advisable. This forms the foundation for a future validation study series comprising a driving simulator experiment and three real driving experiments on test tracks in Germany, the USA, and Japan.

Author Contributions: Conceptualization, D.A., J.R., S.H., F.N., A.K., and K.B.; Methodology, D.A., J.R., A.L., S.H., and F.N.; Formal Analysis, D.A., J.R., A.L., and S.H.; Writing—Original Draft, D.A.; Writing—Review and Editing, D.A., J.R., A.L., S.H., F.N., A.K., and K.B.; Supervision, A.K., and K.B. All authors have read and agreed to the published version of the manuscript.

Funding: This research was funded by the BMW Group.

Conflicts of Interest: The authors declare no conflict of interest. The funders had no role in the collection, analyses, or interpretation of the data.

References

1. SAE. *Taxonomy and Definitions for Terms Related to Driving Automation Systems for On-Road Motor Vehicles*; Society of Automotive Engineers: Warrendale, PA, USA, 2018; Volume 3016, pp. 1–16.
2. Lorenz, L.; Kerschbaum, P.; Hergeth, S.; Gold, C.; Radlmayr, J. Der Fahrer im Hochautomatisierten Fahrzeug. Vom Dual-Task zum Sequential-Task Paradigma. In Proceedings of the 7. Tagung Fahrerassistenz, München, Germany, 25–26 June 2015.

3. Moher, D.; Liberati, A.; Tetzlaff, J.; Altman, D.G.; The PRISMA Group. Preferred Reporting Items for Systematic Reviews and Meta-Analyses: The PRISMA Statement. *PLoS Med.* **2009**, *6*. [CrossRef]

4. Forster, Y.; Hergeth, S.; Naujoks, F.; Krems, J.; Keinath, A. User Education in Automated Driving: Owner's Manual and Interactive Tutorial Support Mental Model Formation and Human-Automation Interaction. *Information* **2019**, *10*, 143. [CrossRef]

5. Forster, Y.; Hergeth, S.; Naujoks, F.; Beggiato, M.; Krems, J.F.; Keinath, A. Learning and Development of mental models during interactions with driving automation: A simulator study. In Proceedings of the Tenth International Driving Symposium on Human Factors in Driver Assessment, Training and Vehicle Design, Santa Fe, NM, USA, 24–27 June 2019.

6. Forster, Y.; Hergeth, S.; Naujoks, F.; Beggiato, M.; Krems, J.F.; Keinath, A. Learning to use automation: Behavioral changes in interaction with automated driving systems. *Transp. Res. Part F Traffic Psychol. Behav.* **2019**, *62*, 599–614. [CrossRef]

7. Forster, Y.; Hergeth, S.; Naujoks, F.; Krems, J.F.; Keinath, A. Self-report measures for the assessment of human–machine interfaces in automated driving. *Cogn. Technol. Work* **2019**. [CrossRef]

8. Guo, C.; Sentouh, C.; Popieul, J.-C.; Haué, J.-B.; Langlois, S.; Loeillet, J.-J.; Soualmi, B.; Nguyen That, T. Cooperation between driver and automated driving system: Implementation and evaluation. *Transp. Res. Part F Traffic Psychol. Behav.* **2019**, *61*, 314–325. [CrossRef]

9. Kettwich, C.; Haus, R.; Temme, G.; Schieben, A. Validation of a HMI Concept Indicating the Status of the Traffic Light Signal in the Context of Automated Driving in Urban Environment. In Proceedings of the 2016 IEEE Intelligent Vehicles Symposium (IV), Gothenburg, Sweden, 19–22 June 2016.

10. Morgan, P.L.; Voinescu, A.; Alford, C.; Caleb-Solly, P. Exploring the Usability of a Connected Autonomous Vehicle Human Machine Interface Designed for Older Adults. In Proceedings of the International Conference on Applied Human Factors and Ergonomics, Orlando, FL, USA, 21–25 July 2018; pp. 591–603. [CrossRef]

11. Naujoks, F.; Forster, Y.; Wiedemann, K.; Neukum, A. A Human-Machine Interface for Cooperative Highly Automated Driving. In *Advances in Human Aspects of Transportation*; Stanton, N.A., Landry, S., Di Bucchianico, G., Vallicelli, A., Eds.; Springer International Publishing: Cham, Switzerland, 2017; pp. 585–595. ISBN 978-3-319-41681-6.

12. Richardson, N.T.; Lehmer, C.; Lienkamp, M.; Michel, B. Conceptual Design and Evaluation of a Human Machine Interface for Highly Automated Truck Driving. In Proceedings of the 2018 IEEE Intelligent Vehicles Symposium (IV), Changshu, China, 26–30 June 2018; pp. 2072–2077.

13. Forster, Y.; Hergeth, S.; Naujoks, F.; Krems, J.F. How Usability Can Save the Day—Methodological Considerations for Making Automated Driving a Success Story. In Proceedings of the 10th International ACM Conference on Automotive User Interfaces and Interactive Vehicular Applications (AutomotiveUI '18), Toronto, ON, Canada, 23–25 September 2018; pp. 278–290. [CrossRef]

14. François, M.; Osiurak, F.; Fort, A.; Crave, P.; Navarro, J. Automotive HMI design and participatory user involvement: Review and perspectives. *Ergonomics* **2016**, *60*, 541–552. [CrossRef] [PubMed]

15. Gold, C.; Naujoks, F.; Radlmayr, J.; Bellem, H.; Jarosch, O. Testing Scenarios for Human Factors Research in Level 3 Automated Vehicles. In Proceedings of the International Conference on Applied Human Factors and Ergonomics, Los Angeles, CA, USA, 17–21 July 2017; pp. 551–559. [CrossRef]

16. Naujoks, F.; Hergeth, S.; Wiedemann, K.; Schömig, N.; Keinath, A. Use Cases for Assessing, Testing, and Validating the Human Machine Interface of Automated Driving Systems. In Proceedings of the Human Factors and Ergonomics Society Annual Meeting, Philadelphia, PA, USA, 1–5 October 2018; Volume 62, pp. 1873–1877. [CrossRef]

17. Naujoks, F.; Hergeth, S.; Wiedemann, K.; Schömig, N.; Forster, Y.; Keinath, A. Test procedure for evaluating the human-machine interface of vehicles with automated driving systems. *Traffic Inj. Prev.* **2019**, *20*, S146–S151. [CrossRef] [PubMed]

18. Naujoks, F.; Wiedemann, K.; Schömig, N.; Hergeth, S.; Keinath, A. Towards guidelines and verification methods for automated vehicle HMIs. *Transp. Res. Part F Traffic Psychol. Behav.* **2019**, *60*, 121–136. [CrossRef]

19. Pauzie, A.; Orfila, O. Methodologies to assess usability and safety of ADAS and automated vehicle. *IFAC PapersOnLine* **2016**, 72–77. [CrossRef]

20. Brooke, J. SUS: A 'Quick and Dirty' Usability Scale. In *Usability Evaluation in Industry*; Jordan Patrick, W., Thomas, B., Weerdmeester, B.A., McClelland, I.L., Eds.; Taylor & Francis: London, UK, 1996; pp. 189–194.

21. Lewis, J.R. Psychometric Evaluation of the PSSUQ Using Data from Five Years of Usability Studies. *Int. J. Hum. Comput. Interact.* **2002**, *14*, 463–488. [CrossRef]

22. Van der Laan, J.D.; Heino, A.; De Waard, D. A Simple Procedure for the Assessment of Acceptance of Advanced Transport Telematics. *Transp. Res. Part C Emerg. Technol.* **1997**, *5*, 1–10. [CrossRef]

23. ISO 9241-11: 2018. *Ergonomics of Human-System Interaction. Part. 11: Usability: Definitions and Concepts*; International Organization of Standardization: Geneva, Switzerland, 2018.

24. NHTSA. *Automated Driving Systems 2.0: A Vision for Safety*; NHTSA: Washington, DC, USA, 2017.

25. Nielsen, J. *Usability Engineering*; Elsevier: Amsterdam, The Netherlands, 1993; ISBN 0125184069.

26. Caird, J.K.; Horrey, W.J. Twelve Practical and Twelve Practical and Useful Questions About Driving Simulation. In *Handbook of Driving Simulation for Engineering, Medicine, and Psychology*; Fisher, D.L., Rizzo, M., Caird, J.K., Lee, J.D., Eds.; CRC Press: Boca Raton, FL, USA, 2011; ISBN 978-1-4200-6101-7.

27. Mullen, N.; Charlton, J.; Devlin, A.; Bédard, M. Simulator Validity: Behaviors Observed on the Simulator and on the Road. In *Handbook of Driving Simulation for Engineering, Medicine, and Psychology*; Fisher, D.L., Rizzo, M., Caird, J.K., Lee, J.D., Eds.; CRC Press: Boca Raton, FL, USA, 2011; ISBN 978-1-4200-6101-7.

28. Nielsen, J. Usability Inspection Methods. In *Conference Companion on Human Factors in Computing Systems*; Association for Computing Machinery: New York, NY, USA, 1994; pp. 413–414.

29. Tan, W.-S.; Liu, D.; Bishu, R. Web Evaluation: Heuristic Evaluation vs. User Testing. *Int. J. Ind. Ergon.* **2009**, *39*, 621–627. [CrossRef]

30. NHTSA. *Visual–Manual NHTSA Driver Distraction Guidelines for In-Vehicle Electronic Devices*; NHTSA: Washington, DC, USA, 2014.

31. Henrich, J.; Heine, S.J.; Norenzayan, A. The weirdest people in the world? *Behav. Brain Sci.* **2010**, *33*, 61–83; Discussion 83–135. [CrossRef] [PubMed]

32. Prümper, J.; Anft, M. ISONORM 9241/110 (Langfassung): Beurteilung von Software auf Grundlage der Internationalen Ergonomie-Norm DIN EN ISO 9241-110. Available online: http://people.f3.htw-berlin.de/Professoren/Pruemper/instrumente/ISONORM%209241-110-L.pdf (accessed on 2 November 2017).

33. Ventakesh, V.; Morris, M.G.; Davis, G.B.; Davis, F.D. User Acceptance of Information Technology: Toward a Unified View. *MIS Q.* **2003**, *27*, 425–478.

34. Mcknight, D.H.; Carter, M.; Thatcher, J.B.; Clay, P.F. Trust in a specific Technology. *ACM Trans. Manag. Inf. Syst.* **2011**, *2*, 1–25. [CrossRef]

35. Chien, S.-Y.; Semnani-Azad, Z.; Lewis, M.; Sycara, K. Towards the Development of An Inter-Cultural Scale to measure Trust in Automation. In *Cross-Cultural Design*; Rau, P.L.R., Ed.; Springer International Publishing: Cham, Switzerland, 2014; ISBN 978-3-319-07307-1.

36. Hart, S.G.; Staveland, L.E. Development of NASA-TLX (Task Load Index): Results of Empirical and Theoretical Research. *Adv. Psychol.* **1988**, *52*, 139–183.

37. Endsley, M.R. Measurement of Situation Awareness in Dynamic Systems. *Hum. Factors* **1995**, *37*, 65–84. [CrossRef]

38. Beggiato, M.; Pereira, M.; Petzoldt, T.; Krems, J. Learning and Development of Trust, Acceptance and the Mental Model of ACC. A longitudinal on-road Study. *Transp. Res. Part F Traffic Psychol. Behav.* **2015**, *35*, 75–84. [CrossRef]

39. Boren, T.; Ramey, J. Thinking Aloud: Reconciling Theory and Practice. *IEEE Trans. Prof. Commun.* **2000**, *43*, 261–278. [CrossRef]

40. Nielsen, J.; Molich, R. Heuristic Evaluation of User Interfaces. In Proceedings of the SIGCHI Conference on Human Factors in Computing Systems, Seattle, WA, USA, 1–5 April 1990; pp. 249–256.

41. Jian, J.-Y.; Bisantz, A.M.; Drury, C.G.; Llinas, J. *Foundations for an Empirically Determined Scale of Trust in Automated Systems*; Air Force Research laboratory: Buffalo, NY, USA, 2000.

42. Hassenzahl, M.; Burmester, M.; Koller, F. AttrakDiff: Ein Fragebogen zur Messung wahrgenommener hedonischer und pragmatischer Qualität. In *Mensch & Computer 2003: Interaktion in Bewegung*; Szwillus, G., Ziegler, J., Eds.; B. G. Teubner: Stuttgart, Germany, 2003; pp. 187–196.

43. Laugwitz, B.; Held, T.; Schrepp, M. Construction and Evaluation of a User Experience Questionnaire. *HCI Usability Educ. Work* **2008**, 63–76. [CrossRef]

44. Minge, M.; Thüring, M.; Wagner, I.; Kuhr, C.V. The meCUE Questionnaire. A Modular Tool for Measuring User Experience. In *Advances in Ergonomics Modeling, Usability & Special Populations, Proceedings of the 7th Applied Human Factors and Ergonomics Society Conference, Orlando, FL, USA, 27–31 July 2016*; Soares, M., Falcão, C., Ahram, T.Z., Eds.; Springer International Press: Cham, Switzerland, 2016; pp. 115–128.
45. Taylor, R.M. Situational Awareness Rating Technique (SART): The Development of a Tool for Aircrew Systems Design. In "Situation Awareness in Aerospace Operations". In *Situational Awareness*; Salas, E., Dietz, A.S., Eds.; Routledge: New York, NY, USA, 2016; pp. 111–128. ISBN 9780754629733.
46. Jay, G.M.; Willis, S.L. Influence of Direct Computer Experience on Older Adults' Attitudes toward Computers. *J. Gerontol. Psychol. Sci.* **1992**, *47*, 250–257. [CrossRef] [PubMed]
47. Van den Beukel, A.P.; Van der Voort, M.C. Design Considerations on User-Interaction for Semi-Automated Driving. In Proceedings of the FISITA 2014 World Automotive Congress, Maastricht, The Netherlands, 2–6 June 2014; pp. 1–8.
48. Pauzié, A. A Method to assess the Driver Mental Workload: The Driving Activity Load Index (DALI). *IET Intell. Transp. Syst.* **2008**, *2*, 315–322. [CrossRef]
49. ISO 15007-2. *Road Vehicles—Measurement of Driver Visual Behaviour with Respect to Transport Information and Control Systems. Part. 2: Equipment and Procedures*; International Organization for Standardization: Geneva, Switzerland, 2013.
50. Parasuraman, R.; Riley, V. Humans and Automation: Use, Misuse, Disuse, Abuse. *Hum. Factors* **1997**, *39*, 230–253. [CrossRef]

 *information*

Discussion

Checklist for Expert Evaluation of HMIs of Automated Vehicles—Discussions on Its Value and Adaptions of the Method within an Expert Workshop

Nadja Schömig [1,*], Katharina Wiedemann [1], Sebastian Hergeth [2], Yannick Forster [2], Jeffrey Muttart [3], Alexander Eriksson [4], David Mitropoulos-Rundus [5], Kevin Grove [6], Josef Krems [7], Andreas Keinath [2], Alexandra Neukum [1] and Frederik Naujoks [2]

[1]   Würzburg Institute for Traffic Sciences GmbH (WIVW), D-97209 Veitshöchheim, Germany; wiedemann@wivw.de (K.W.); neukum@wivw.de (A.N.)

[2]   BMW Group, D-80937 München, Germany; Sebastian.Hergeth@bmw.de (S.H.); Yannick.Forster@bmw.de (Y.F.); Andreas.Keinath@bmw.de (A.K.); Frederik.Naujoks@bmw.de (F.N.)

[3]   Crash Safety Research Center, LLC, 201 W High Street, B8 East Hampton, New York, NY 06424, USA; muttartj@gmail.com

[4]   Volvo Car Corporation, Torslanda, SE-405 31 Göteborg, Sweden; alexander.eriksson.2@volvocars.com

[5]   Hyundai American Technical Center, Superior Township, MI 48198, USA; DMRUNDUS@HATCI.COM

[6]   Virginia Tech Transportation Institute, 3500 Transportation Research Plaza, Blacksburg, VA 24061, USA; kgrove@vtti.vt.edu

[7]   Department of Behavioural and Social Sciences, Technische Universität Chemnitz, 09111 Chemnitz, Germany; josef.krems@psychologie.tu-chemnitz.de

*   Correspondence: schoemig@wivw.de; Tel.: +49-931-78009-208

Received: 28 February 2020; Accepted: 20 April 2020; Published: 24 April 2020

Abstract: Within a workshop on evaluation methods for automated vehicles (AVs) at the Driving Assessment 2019 symposium in Santa Fe; New Mexico, a heuristic evaluation methodology that aims at supporting the development of human–machine interfaces (HMIs) for AVs was presented. The goal of the workshop was to bring together members of the human factors community to discuss the method and to further promote the development of HMI guidelines and assessment methods for the design of HMIs of automated driving systems (ADSs). The workshop included hands-on experience of rented series production partially automated vehicles, the application of the heuristic assessment method using a checklist, and intensive discussions about possible revisions of the checklist and the method itself. The aim of the paper is to summarize the results of the workshop, which will be used to further improve the checklist method and make the process available to the scientific community. The participants all had previous experience in HMI design of driver assistance systems, as well as development and evaluation methods. They brought valuable ideas into the discussion with regard to the overall value of the tool against the background of the intended application, concrete improvements of the checklist (e.g., categorization of items; checklist items that are currently perceived as missing or redundant in the checklist), when in the design process the tool should be applied, and improvements for the usability of the checklist.

Keywords: automated vehicles; automated driving systems; HMI; guidelines; heuristic evaluation; checklist; expert evaluation

---

1. Background

With the Federal Automated Vehicles Policy, the U.S. National Highway Traffic Safety Administration (NHTSA) has provided an outline that can be used to guide the development and validation of automated driving systems (ADS).

With regard to the human–machine interface (HMI), the policy proposes that an automated vehicle (AV) HMI at minimum shall inform the user that the system is either of the following [1]:

1. Functioning properly;
2. Engaged in automated driving mode;
3. Currently 'unavailable' for use;
4. Experiencing a malfunction; and/or
5. Requesting a control transition from ADS to the operator.

A suitable design of mode indicators should effectively support the driver in using an ADS and prevent a false understanding of the current driving mode. NHTSA encourages implementing and documenting a process for the testing, assessment, and validation of each element [1]. However, details on how entities can assess and validate if a specific HMI meets these requirements are not proposed. Therefore, a test procedure was developed that serves to evaluate the conformity of SAE level 3 (conditional automation) ADS HMIs with the requirements outlined in NHTSA's Automated Vehicles policy (for an overview, see [2]). Before this publication, no standardized tools for the assessment of the usability and safety of ADS HMIs existed.

The proposed evaluation protocol includes (1) a method to identify relevant use cases for testing on the basis of all theoretically possible system states and mode transitions of a given ADS (see [3]); (2) an expert-based heuristic assessment to evaluate whether the HMI complies with applicable norms, standards, and best practices (the topic of the present paper); and (3) an empirical evaluation of ADS HMIs using a standardized design for user studies and performance metrics [2]. An overview of the complete test procedure can be seen in Figure 1 (for further information, see [2]).

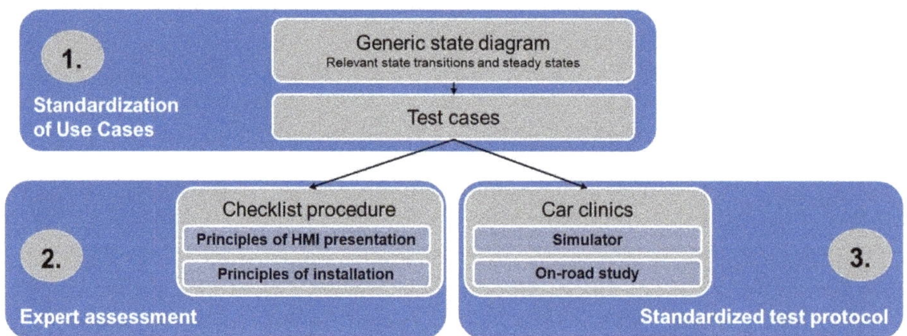

Figure 1. Overview of the test procedure for the evaluation of automated driving system (ADS) human–machine interfaces (HMIs) based on U.S. National Highway Traffic Safety Administration (NHTSA) requirements.

The present paper deals with the reviewing of the heuristic evaluation method that can be used by human factors and usability experts to evaluate and document whether an HMI meets the above-mentioned minimum requirements. In usability engineering, such heuristic assessment methods are commonly applied during the product development cycle and can be used as a quick and efficient tool to identify potential usability issues associated with the HMI [4].

The heuristic assessment method consists of a set of ADS HMI guidelines together with a checklist that can be used as a systematic HMI inspection and a problem reporting sheet. This version of the checklist and the considered HMI principles are reported in [5] and [6].

In comparison with existing approaches that test the usability via user studies/car clinics, the heuristic evaluation can be applied through rapid iteration early in the product cycle, and is thus able to correct identified issues and reduce late-stage design changes. Using experts has the advantage that inadequate mental models that might influence evaluations of naïve users can be better controlled. Furthermore, experts are trained to concentrate on single HMI aspects separately from each other in their evaluations. In addition, by means of the checklist, experts can evaluate an HMI in absolute values independently from a comparison with other HMIs. However, both heuristic evaluation and car clinics are recommended to be used as complementary methods in the evaluation protocol (see Figure 1).

The paper at hand has the goal to disseminate the already published work on the developed test procedure to a scientific community and to further adapt the checklist based on the results of the expert workshop. Suggestions for improvement from human factors experts and practitioners are discussed against the background of feasibility (keeping it an easy-to-use tool) and appropriateness for use in a checklist compared with other methods.

2. Content and Usage of the Checklist

2.1. Checklist Items

The aim of the assessment is to evaluate whether a set of pre-defined HMI principles (the "heuristics") are met. Thus, the checklist consists of 20 items summarizing the most important design recommendations for visual-auditory and visual-vibrotactile HMIs derived from existing norms, applicable standards, design guidelines, and empirical research, pertaining to in-vehicle interfaces. The complete list of items is presented in Table 1. The derivations of these items from the literature are elaborately described in [5].

Table 1. List of heuristics (see also [5]). HMI, human–machine interface.

#	Item
1	Unintentional activation and deactivation should be prevented
2	The system mode should be displayed continuously
3	Mode changes should be effectively communicated
4	Visual interfaces used to communicate system states should be mounted to a suitable position and distance. High-priority information should be presented close to the driver's expected line of sight
5	HMI elements should be grouped together according to their function to support the perception of mode indicators
6	Time-critical interactions with the system should not afford continuous attention
7	The visual interface should have a sufficient contrast in luminance and/or color between foreground and background
8	Texts (e.g., font types and size of characters) and symbols should be easily readable from the permitted seating position
9	Commonly accepted or standardized symbols should be used to communicate the automation mode. Use of non-standard symbols should be supplemented by additional text explanations
10	The semantic of a message should be in accordance with its urgency
11	Messages should be conveyed using the language of the users (e.g., national language, avoidance of technical language, use of common syntax)
12	Text messages should be as short as possible
13	Not more than five colors should be consistently used to code system states (excluding white and black)
14	The colors used to communicate system states should be in accordance with common conventions and stereotypes
15	Design for color-blindness by redundant coding and avoidance of red/green and blue/yellow combinations
16	Auditory output should raise the attention of the driver without startling her/him or causing pain
17	Auditory and vibrotactile output should be adapted to the urgency of the message
18	High-priority messages should be multimodal
19	Warning messages should orient the user towards the source of danger
20	In case of sensor failures, their consequences and required operator steps should be displayed

2.2. Method Description

The method should be conducted by a pair of HMI experts. Preferably, experts should have received formal training in human factors and usability engineering or have demonstrable practical experience in the assessment and evaluation of automotive HMIs. However, the evaluators should have no prior experience with the vehicle and features to be tested. The most suitable testing environment depends on the maturity of the product. In the very early development stages, where there is only a prototype available and series production is far away, it is recommended to use a driving simulator. For series production vehicles or high-fidelity prototypes, it is advised to conduct the study on-road as this provides the most realistic conditions for testing. Each of the two evaluators completes a fixed set of use cases, observes the visual, auditory and haptic HMI output, and records potential usability issues arising from the non-compliance with the checklist items (see Figure 2 for an example). The use case set consists of the various system states and the transitions between them (e.g., activating the system, deactivating the system, switching between system modes, required control transition from the system to the operator) and depends on the specific design of the ADS with respect to the available levels of automation (e.g., whether only manual or conditional automation are available, or if partial automation (level 2) is also available within the same vehicle). While one of the evaluators is the driver, the other one is seated in the passenger seat, providing step-wise instructions about the desired system state to the driver at appropriate times during the drive. To ensure that both observers are able to experience each use case and resulting system and user reactions and responses in a comparable way, they switch position after one driving session and repeat the drive. The aim of the heuristic assessment is twofold:

1. For the minimum HMI requirements to be fulfilled, each of the use cases should be reflected in a mode indicator or the change of a mode indicator that must be present in the in-vehicle HMI. At minimum, a persistent mode indicator should be presented visually. In addition, auditory, tactile, or kinaesthetic cues for mode transitions are recommended.
2. The design of the respective mode indicator should be in accordance with the common HMI standards and best practices that are the basis of the checklist.

Checklist compliance and identified usability issues should be initially documented independently by each of the evaluators. Each of the checklist items should be answered using the following rating categories:

- "major concerns": non-compliance with guideline;
- "minor concerns": partial fulfilment of guideline, but some aspects of the HMI are non-compliant;
- "no concerns": compliance of all HMI aspects with guideline;
- "measurement necessary"/"subject to verification": no definite conclusion can be given on the basis of the checklist and empirical testing is needed. This category should be chosen if highly innovative designs are used that are not covered by current standards and best practices. An example would be the use of other communication channels than the above-mentioned (e.g., olfactory cues);
- "not applicable": respective design recommendation not applicable to the system under investigation (e.g., HMI without vibrotactile output).

The reasons for "major" and "minor" concerns should be documented in a separate reporting sheet. After the individual assessment, the results should be discussed between the evaluators to come to a unanimous rating decision for each item, which should also be documented. Figure 2 shows a simplified flow chart of the test procedure.

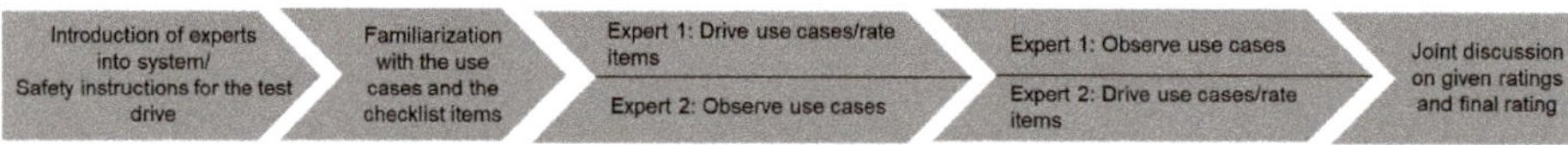

Figure 2. Simplified flow chart of the test procedure.

Figure 3 shows an example of the format/appearance of the checklist and an item notionally judged by an evaluator. Each checklist item contains the requirement. Additionally, positive and/or negative examples for a good/insufficient HMI solution of the requirement are given below the heuristic. Please note that the handwritten notes in Figure 2 do not refer to one of the systems investigated within the workshop, but do serve as exemplary problems that could potentially be identified during the heuristic evaluation. The complete checklist can be found in the Appendix A. It was used in a slightly adapted version in the workshop.

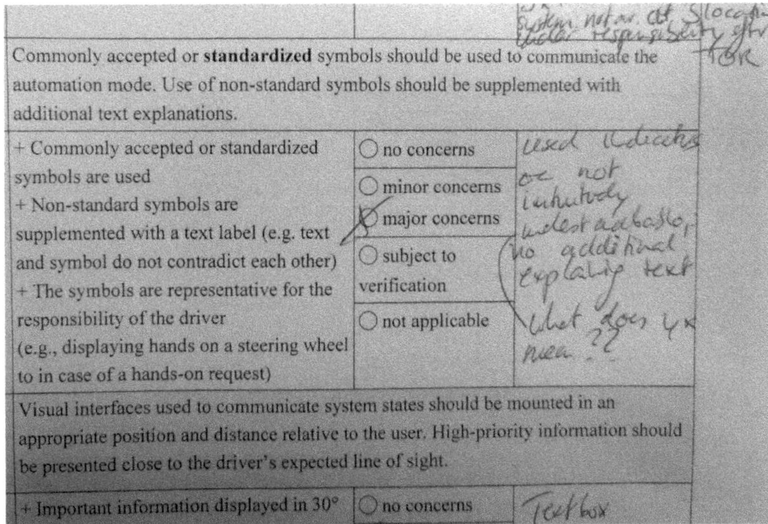

Figure 3. Example of the format of the checklist and an evaluated item.

2.3. Application Domain of the Method

In its current version, the checklist should be viewed as a living document that can be modified to account for gaps where research in the field of automated vehicles is still emerging. The checklist expresses one set of guidelines and is intended as a first step towards guideline development and vehicle-verification methods (the first version of the checklist and three validation studies were published in [6]). In addition, it should be noted that it was developed with the intention to evaluate ADS of level 3 systems. However, it may also be applied to L2 automation as most of the heuristics refer to general design guidelines that should be met by all types of automated system HMIs for ensuring proper usage. Users could misunderstand or misuse the capabilities of an L2 system, treating it like an L3 system when the automated driving mode is not clearly indicated, as demanded by the checklist items. Vehicles equipped with L3 systems may also be usable in an L2 operational state. Understanding user interactions in lower modes of automation may inform best practices in higher modes. This justifies the application towards L2 systems, as was done in the conducted workshop.

For the sake of practicability and efficiency, the list of guidelines was kept as short as possible; therefore, it is likely that it will not cover every aspect of the HMI for ADS. At this point, it is important to emphasize that new and innovative HMI designs may rely on other HMI elements than the ones covered by the sources used to compile the checklist. In this case, the evaluators are encouraged to give a positive assessment when they judge the respective guideline to be fulfilled during the on-road test and their level of expertise on this topic allows it, even if this cannot be based on the design recommendations. In the case of sufficient inter-evaluator agreement, verification of the guideline can be assumed. Otherwise, they are suggested to rate the item as "measurement necessary".

It should further be noted that the fulfilment of the HMI guidelines should facilitate regular and safe usage of the ADS, such as switching the automation on, checking the status of the ADS, or taking back manual control from the automation (either driver-initiated or as a result of a take-over request).

A more comprehensive evaluation of the ADS will most likely need to incorporate (a) usability testing with real users in instrumented vehicles or high-fidelity driving simulators and (b) investigating other domains such as the assessment of controllability of system limits or failures and foreseeable misuse of the ADS as defined in the RESPONSE Code of Practice [7] and ISO 26262 [8]. The aim of the proposed method is thus not to replace usability testing with participant samples, but to complement empirical approaches with an economical heuristic evaluation tool. It may serve as a guide or sieve to identify and improve poorly performing HMIs before going into further user studies. Further limitations of the current approach are also discussed in the corresponding sections of this paper.

3. Intention of the Workshop

The intention of the conducted workshop at the Driving Assessment 2019 symposium in Santa Fe, New Mexico was to present the developed method and its application to human factors experts in the scientific community of automated vehicle research in order to further improve the method. A higher level goal was to stimulate the discussion within a larger scientific and technical community towards future standards that may be appropriate for guiding the development of automated vehicle HMIs. Another goal is to facilitate a more rapid convergence towards an agreed-upon set of robust guidelines and verification methods that can serve the industry in the important evolution of automation within vehicles.

4. Evaluation Procedure during the Workshop

The workshop was organized by WIVW GmbH (a company providing research services in the field of human factors, especially from the automotive sector) and was held following the closing session of the conference, on 27 June 2019 at the El Dorado Hotel in Santa Fe. In total, 14 participants took part in the workshop. The workshop was announced via the conference website and was open for application to all conference participants. The workshop attendees were selected based on their scientific background either as practitioners or academics in the automotive domain with previous experience in the HMI design, development, or evaluation methods of driver assistance systems. The resulting workshop group consisted of agents from the automotive industry, scientific institutes, and national agencies. In preparation of the workshop, the participants received publications that described the background of the checklist method [3] and its application [6].

Regarding the agenda, the workshop started with a short introduction of the organizers, who gave an outline of the workshop. Afterwards, the method and its background were presented. Then, the 14 participants were split into small groups of 3–4 people who would apply the method while riding together in two vehicles equipped with L2 driving automation systems: a Tesla Model 3 with the Autopilot system (AP) and a Cadillac with the GM Supercruise (SC) system.

These two systems were chosen as their system architecture, system operation principles, as well as their system performance (e.g., regarding the threshold required for overriding the system by steering input) are different from each other. The Tesla had a large center-stack touch screen for the HMI display, while the Cadillac used a classical instrument cluster and a light bar at the steering wheel for communicating the current system mode. The Cadillac had many of the automation controlling buttons on the steering wheel, while the Tesla system is operated by a lever behind the steering wheel. The two systems also offered different warning and alerts, different both in the type of alerts and when they occurred. Systems also differed in their approaches for driver monitoring; while the Tesla used a hands-on detection system, the SC determines whether the driver has enough visual attention on the road via a camera system. Furthermore, the Cadillac had more constraints to use than the Tesla L2 system (for further descriptions of the system, see Chapter 5).

Before starting the drive, each participant received a copy of the checklist and familiarized themselves with the items and the rating procedure. It was emphasized that the rating of the systems itself was not the relevant outcome of the evaluation process and that methodological issues emerging when applying the method were of greater importance.

The participants experienced both systems as passengers in a 30 min drive on the interstate 25 (Denver–Albuquerque). The first system was experienced while being driven in one direction on the interstate. After a stop at an exit, the groups switched vehicles and experienced the second system on the way back to the interstate exit. The drive from the conference site to the interstate took about 10 min.

Owing to safety and insurance reasons, the workshop organizers drove the vehicles. Therefore, the evaluation process during the workshop did not follow exactly the one proposed in the heuristic evaluation procedure, where the evaluators should really drive the vehicle (see [6]). On the way to the interstate, the drivers briefly explained the control elements (button, lever, and so on) used for operating the system and which HMI elements had to be observed by the evaluators. The test itself started when the vehicle reached the interstate. Up to here, the Tesla theoretically permitted to use the Autopilot, while the SC could not be used before entering the interstate. The drivers conducted several use cases in which the system could be experienced:

- The activation of the system;
- Driving with active L2 for a longer time interval (i.e., 4–5 min);
- Experiencing the driver monitoring system, which required a take-over in case the driver did not react to it;
- The deactivation of the system;
- Short-term standby modes, for example, in the case of non-detection of lane markings or lane changes;
- Planned system limits when exiting the interstate.

After getting back to the workshop room, participants were asked to fill out the checklist based on the second system they experienced. After all four groups had experienced the two vehicles, all workshop participants jointly discussed methodological issues they noticed during the application of the method.

5. Description of the Evaluated Systems

Although the rating of the systems themselves was not in the scope of the evaluation, a short description of both systems is inserted here to better understand the resulting discussions.

5.1. The GM Supercruise System

The GM Supercruise system is operated via buttons on the steering wheel. The system mode is indicated by presenting graphics, icons, and text messages at the instrument cluster (see Figure 4). In addition, there is a light bar on the steering wheel that is also used to indicate the current system mode by different colours and pulsation (static vs. flashing lights). The system is geofenced, meaning that it is only available on certain roads, such as interstates. Other preconditions for activation are that the driver has to drive in the center of the lane and that adaptive cruise control (ACC) is active. If the driver tries to activate the system outside these conditions, they receive a text message at the right side of the instrument cluster. The system state is indicated by a specific area on the left side of the instrument cluster as well as by telltales in the centre of the cluster. In order to activate Supercruise, first, ACC has to be shifted into standby mode (separate button) and activated by setting the speed. After that, Supercruise can be activated by a separate button. The activity of Supercruise is indicated by a green steering wheel together with green horizontal bars for ACC shown in the cluster. A short-term degraded standby mode (meaning lateral control is not active) is indicated by a blue steering wheel and the steering wheel light bar in blue. Lateral control is automatically resumed if the conditions are fulfilled once again. The Supercruise system does not require the driver to keep their hands permanently at the steering wheel, as UN-CE-R-79 [9] regulating this matter does not apply for the United States. However, if the driver tries to deactivate the system without having their hands at the wheel, a take-over request is triggered by a text message and red indicators. The driver monitoring system consists of a camera on the top of the steering wheel that determines whether the driver is looking towards the road. If the time not looking at the road exceeds a certain threshold, the steering wheel first flashes green before it turns red and lateral control is deactivated.

Figure 4. HMI elements of the GM Supercruise system.

5.2. The Tesla Autopilot

The Autopilot function by Tesla can theoretically be activated on all roads without any restriction (other than laws). In order to use the additional lane change assistance function, the navigation system has to be active. The system state is exclusively indicated on the left area of the touch display in the center stack console, which is used for all driving-related and non-driving-related information (replacing the instrument cluster; see Figure 5 left). The active L2 system is indicated by a blue trajectory on ego lane. The dynamic display additionally shows adjacent lanes and other vehicles surrounding the ego-vehicle. The system is activated by pulling the gear switch twice towards the driver (see Figure 5 right). After each activation process, the driver is requested to keep the hands on the steering wheel by a text message. In the case in which the system detects no steering interventions of the driver for a longer time interval, it requests the driver to exert a slight force on the steering wheel with the display flashing in blue and a symbol indicating a steering wheel with hands on it. If the driver does not react to such a hands-on request, the system will be switched off completely and can no longer be used for the remaining drive. Lateral control can be easily deactivated by a steering wheel intervention of the driver. There is no standby mode, meaning that lateral control has to be activated by the driver each time after it has been deactivated.

Figure 5. HMI elements of Tesla Autopilot.

6. Methodological Issues Discussed during the Workshop

After having experienced the two vehicles and after having applied the checklist, several methodological aspects arose from the discussions between the workshop participants that were grouped into the following topics.

6.1. Design Issues of the Checklist

With regard to better usability of the checklist, it was proposed to reorganize its design. For a better overview, some items could be grouped together into higher-level categories (e.g., with regard to color usage). Another suggestion for grouping the items was to categorize them with regard to use cases, for example, group all items with regard to change of system mode together. However, with the intention of the checklist to only test the minimum requirements set by NHTSA policy, this idea would prove as impractical as this would mean repetitions of some items that are valid for several use cases, which would unnecessarily stretch the checklist. The idea of shifting the positive and negative design examples to an appendix as added material was judged as not appropriate as the raters could profit from the current position of the examples, while there is a chance that an appendix tends to be overlooked during the rating process.

In addition to the current rating categories, it was proposed to add a category for "suggested improvements", not only in the final reporting sheet, but also on the item level to encourage experts to think about better solutions instead of simply marking "concerns".

6.2. Missing and Redundant Items

One concern regarding the selection of the checklist items was that some of them should not be evaluated subjectively by experts, but must be better objectively measured by technicians. These items comprise the following:

- Displaying HMI elements close to the line of sight (part of item 4);
- Checking color contrast (item 7);
- Checking text size (character height and stroke width, item 8);
- Design of auditory or vibrotactile feedback with regard to length, loudness, and frequency (item 17).

It is agreed that, for later stages of the HMI development, objective measurement by a technician is necessary, while in an early stage, a heuristic assessment of these items might be acceptable. Therefore, methods used for early stage and later stage assessment of the HMI might differ. It was discussed whether only those items should remain in the checklist, which must be subjectively assessed by experts. Other objectively measurable items could be deleted from the checklist and inserted into a separate technical checklist. Finally, there was some discussion about items that should be added to the checklist as they seem to cover aspects that are currently not adequately addressed.

Regarding extensions of the checklist, the greatest benefit, but also the greatest challenge, would be to rate the overall complexity of the system/HMI. The (perceived) complexity in using a system will heavily influence acceptance and trust in the system (e.g., [10]). The term includes two types of complexity. First, the system complexity, meaning the logic behind the various system modes and its transitions (e.g., are lateral and longitudinal control separate sub-functions that can be used in combination as well as independently of each other? Are standby modes included?). The system complexity will likely influence the complexity of system operation (e.g., sequence of operational steps to be performed or number of possible operation steps in order to reach a certain system state) or the demands on the distinctiveness of the several system modes (how many different indicators are necessary and how are they designed in order to clearly identify the current system mode). The latter is linked to the second type of complexity, the display complexity, which can be described by the arrangement of the information elements on the display, for example, in terms of display layout, number of display elements relative to display size (so-called visual clutter), spatial proximity of elements (e.g., in terms of overlapping), and so on.

Another possibility to operationalize the term complexity would be the categorization into different types of demands that are put on the operator of the system and that result in a certain level of perceived complexity in system usage. One possibility would be to define dimensions according to typical categorizations of workload based on Wicken's multiple resource model [11]. This model categorizes workload based on the following:

- Visual demands that result from the design of the visual displays, including content and arrangement;
- Motoric demands that result from the number and arrangement of the operational devices;
- Cognitive demands that result from the system logic and the difficulty in understanding the various system modes and the conditions for the transitions between them;
- In addition, temporal demands, meaning the requirements on the reaction to hazards play an important role and are influenced by the design of warning messages and the take-over request.

To sum up this issue, in the general discussion about the aim of the checklist, workshop participants tended to agree that complexity can be reasonably assessed by experts. Therefore, it was proposed to consider the system complexity in future iterations of the checklist. It is recommended to reflect the multidimensionality of this item in the checklist. One way would be to define multiple items addressing the defined sub-dimensions from the chapter above and group them together in the more global category of "complexity". The other option (which might be more appropriate as concrete standards for the assessment of complexity are missing) would be to formulate one single more generic item with the defined sub-items as positive/negative examples.

With regard to the evaluation of system operation, currently, there is only one item included that deals with the avoidance of unintentional activation and deactivation of the system (see also, for example, currently ongoing work by UNECE on the ACSF regulation (automatically commanded steering function) [12]). Thé reason for the limitation to only this item is that, at the creation of the checklist, there were no clear design guidelines or recommendations on how system operations should be designed. At the moment, there are some concrete specifications on activation, deactivation, and driver input principles under ongoing consideration in the UNECE ACSF document (e.g., the system should be deactivated when the driver overrides the system by steering, braking, or accelerating while holding the steering control). However, issues concerning the system logic are outside the scope of the checklist. However, the design of operational devices might be an extension of the checklist when more research and valid guidelines on this topic are available.

Highly correlated to the term complexity is the learnability of the system's logical operation and the HMI. Learnability is said to be one major attribute of usability (beside effectiveness, error tolerance, satisfaction, and memorization; for example, see [4]) and will be influenced by interface design (e.g., visibility of successful operations, feedback, continuity of task sequences, design conventions, information presentation, user assistance, error prevention) and conformity to users' expectations to the car manufacturer's philosophy (differences in functionality, differences in interaction style, concept clarity, and completeness of information [13]). Beside an intuitive first contact with a system, the concept of learnability should also include the aspect of re-learning the use of the system again after a longer interval of non-usage and the resources involved. However, it seems difficult for experts to provide a meaningful rating regarding learnability as an expert involved in system design, and assessment is likely biased when it comes to learnability owing to their experience. The same will be true for in-house testers who have extensive knowledge about currently developed products. This aspect should thus better be tested with naïve users. A small sample may be enough, and may include people not involved in ADS design. According to Nielsen [4], most of the usability problems can be identified by a number of five experts.

Another issue that could be considered by the checklist is the evaluation of other display elements beside the conventional ones, such as instrument cluster and head-up displays. This should contain not only the mere presence of peripheral displays, which are considered as an example in item 4 ("peripheral displays supporting noticing of mode changes, e.g., by movement or size of displays"), but also more concrete items referring to the design of those displays (e.g., steering wheel light bars). However, up to now, there are no concrete design guidelines, but a few empirical studies exist on the positive effects of ambient displays on mode awareness and take-over performance (e.g., [14]). Beside concrete design aspects, it can be requested that such elements should be congruent with the ones displayed in the instrument cluster, as otherwise, this might be problematic for understanding of system modes.

For the design of warnings, it was discussed whether to consider additional aspects beside the ones that are already included dealing with the communication channels to be used (in multiple modalities; item 18) and the desired effect of not distracting the driver (item 19). Such aspects are, for example, nomenclature choices and linguistic complexity (i.e., fault messages based on engineering nomenclature vs. easily comprehensible names of system modes). In addition, the content of the warning could be defined more explicitly. It was proposed to positively evaluate if the potential consequences of a system limit are displayed (e.g., what would happen if the user does not intervene and how can the user recover or reactivate the system, for example, in the case of repeated hand-off warnings).

American National Standards Institute (ANSI) suggests that a safety warning should include the following (ANSI Z353 [15]):

1. Identification of the hazard;
2. Identification of a means to avoid the hazard;
3. The consequences of not avoiding the hazard.

While there is not a need to adhere closely to the ANSI warning standard, such a standard could be considered as a guideline. In the context of automated driving, warnings can occur owing to less time critical hazards such as sensor failures that do not inevitably require an immediate action like a forward collision warning. However, reaching system limits can be interpreted as imminent hazards that require the driver to immediately take over the driving task to avoid an accident. Signal words that can be used for describing the identification of the hazard are "Danger", "Warning", "Caution", or "Notice". Then, a notice of what to do next to avoid a hazard should be given.

In the case of an urgent take-over request, these first two aspects are probably the most critical points. Typically, the HMI addresses them by displaying a short text such as "Take Over" together with a warning sign. The third aspect, conveying the consequence of inaction, seems to be the most problematic point in the case of take-over requests, as it is not always clear what happens in the case in which the system is deactivated without the immediate reaction of the driver. This information might be better explained by a user manual of the system instead of by the HMI in the imminent situation.

Finally, there were some suggestions for new items to be included in the checklist. One example was about the usage of a dynamic environment display that shows the surrounding traffic. Such a display is currently included in the Tesla Autopilot HMI. However, up to the current state-of-art, the benefit of such a display has not yet been established, and it is thus not clear how such an additional display should be evaluated. Does it have a positive effect on situation awareness or might it distract the driver from extracting relevant status information from the display and promote over trust in the system? It is possible that displays utilizing motion/animation may blunt the driver's response to warnings as motion is a powerful attention grabber, and thus a driver may start filtering out display content that is more relevant in the respective situation (i.e., notices, state changes). Owing to the lack of empirical evidence on the potential effects, the formulation of a checklist item regarding such a display is problematic, and will thus be postponed until more research is available.

Another example is a potential item for the ease of overriding the system. This item would address the controllability of a system, which was not the initial scope of the checklist. Currently, it is not clear whether it is good or bad if a system can be easily overridden by the driver. In addition, it is argued that such highly complex interactions between various factors (e.g., the degree of lateral control will play an important factor on this issue) seem to be better assessed in user studies. Therefore, no item regarding this issue will be included for now.

6.3. Human Factors Aspects to be Considered by Other Methods

All workshop participants jointly agreed that other aspects such as differences in system behaviour, system logic, system operation logic (beside some describable specific aspects in the item of complexity), and their effects on system usability require additional evaluation methods, as the correlation between these different factors is not yet clear in order to formulate concrete design recommendations. Moreover, their effects on system acceptance and system trust must be assessed by user studies. User experience (UX)-related aspects such as the hedonistic quality of the system and the HMI are also recommended to be evaluated by user studies. In this way, real users can experience the system and report their emotions and attitudes towards the system.

High agreement was also achieved for the fact that, especially on the type of control, user studies are needed (e.g., operation via steering wheel buttons vs. touch screen) with regard to performance times or distraction potential before checklist items can be deduced.

6.4. Test Procedure

Regarding the test procedure, the workshop participants recommended to put more emphasis on the fact that the experts should take the perspective of naive users. A naive user can be defined as "a person that tests the ADAS under evaluation with no more experience and prior knowledge of the system than a later customer would have" ([7], p. 7). This should allow that the requirements are valid for the average population. The inclusion of certain items also makes it possible to address the needs of certain specific driver groups, for example, drivers with colour-blindness.

Nevertheless, it should be kept in mind that this method should not replace, but rather supplement other approaches like user studies that allow for eye tracking, reaction time testing, and performance measurement on tasks dealing with the handling of the system. Both methods are proposed to be conducted within the complete evaluation protocol (see Figure 1).

The proposed test procedure (a team of two experts rates the system after having once experienced the use cases themselves as a driver and having watched the other evaluator driving) was rated as a reasonable approach. This approach has the advantage that both experts do not merely observe someone interacting with the system, but really experience that interaction themselves. In addition, the fact that one evaluator can directly document their first impression while the other evaluator is driving (compared with a retrospective documentation) avoids negative effects such as memory decay. For later reference, it is suggested to capture video of the driving experience (by scenario and system response) using small video cameras mounted in locations that capture cluster, head unit, and other displays and controls, while not covering the rater's view on these elements.

It was further proposed to conduct the evaluation with a larger group of experts if no time and resource constraints object to this approach. However, as this might complicate the process, to reach an agreement in a joint discussion, we would recommend only consulting a third external evaluator if no such agreement can be made between the evaluators even after a longer discussion.

Owing to the variety of systems that can be evaluated and the fact that new and innovative HMI designs are currently not covered by the checklist items, there will be situations in which adaptations need to be made by the experts to accommodate for specific circumstances. In this case, we suggest that experts follow the following approach.

1. Search for common published standards;
2. If 1 does not apply, evaluate the system by extrapolating those concepts from the checklist that seems to be transferable to the innovative HMI;
3. If 1 and 2 do not apply, conduct empirical testing.

6.5. General Value of the Checklist

It was jointly agreed that the developed method is a useful tool in the design process of AV HMIs. It is primarily intended to facilitate the assessment of system usability. It is able to check whether the minimum requirements proposed in the NHTSA policy are fulfilled. It is also reasonable as the current rapid evolvement of automated systems makes it extremely difficult to identify the "best design" of such a system and its HMI. This method serves as a tool to guide and make quick changes during the development process, that is, testing several concepts and narrowing down options, as well as ensuring a "basic" compliance throughout the design loops.

As said above, for a global evaluation, the assessment of aspects such as different system logics, different concepts for system operation (e.g., longitudinal and lateral control as separate systems, L2 as add on to L1, stepwise activation of L1, then L2, and so on), and different design philosophies should be considered, which are better answered by user studies.

7. Conclusions and Outlook

On the basis of the discussions with the workshop participants, the following adaptions of the checklist were decided:

- The structure of the checklist will be revised in order to achieve a better usability for the experts. This will mainly refer to a re-arrangement of the items into more global categories and underlying subcategories.
- Items regarding measurable aspects such as text sizes, line of sight, or colour contrasts will remain in the checklist to be assessed subjectively by the experts (as confirmation from a user perspective based on technical tests that will be conducted later in the design process). Absolute measurable numbers will be removed from the examples list.
- A new category of perceived complexity will be included in the checklist. This category will comprise several items/examples, which still have to be defined. Issues that should be considered are

  - the visual demands of the HMI in general;
  - the cognitive demands resulting from the complexity of the system's logic;
  - the motoric demands resulting from the number, positioning, and arrangement of operational devices;
  - the ease of learning the interaction with the system.

- The following new items will be included in the checklist

  - An item on the appropriate design of other display elements;
  - An item on the content of a warning/take-over request.

- The test procedure itself will remain in the proposed manner with two experts experiencing the system to be evaluated within a defined set of use cases in real drives, first separately filling out the checklist, and finally give a global rating based on a joint discussion.

It is planned to transfer the checklist into a computer-application that can be used, for example, on tablets in order to support the experts in the documentation of the tests, the discussion of its output, and the recommendations for system improvements.

Author Contributions: Workshop organization: K.W., N.S., Workshop participation: N.S., K.W., Y.F., F.N., J.M., A.N., D.M.-R., K.G., J.K.; Writing – review and editing: All authors. All authors have read and agreed to the published version of the manuscript.

Funding: This research received no external funding.

Acknowledgments: The development of the method was funded by BMW Group. The workshop was organized by WIVW GmbH and hosted by the Driving Assessment Conference. Thanks to all workshop participants for their valuable inputs to the methodological discussions during the workshop. Special thanks to all of those workshop participants who additionally contributed to the present publication with input and comments (Jeffrey Muttart, Alexander Eriksson, Kevin Grove, David Mitropoulos-Rundus, Josef Krems, Sebastian Hergeth, Frederik Naujoks, and Yannick Forster).

Conflicts of Interest: The authors declare no conflict of interest.

Appendix A

Table A1. Extended checklist items (see [5]). NHTSA, U.S. National Highway Traffic Safety Administration. Abbreviations in the table: NDRT = Non-driving-related task; DDT = Dynamic Driving Task.

1	Unintentional activation and deactivation should be prevented.	Assessment
	+ System design ensures driver readiness before transfer of control (e.g., pushing of two buttons simultaneously, need to have both hands on the steering wheel, need to have eyes on the road) - Surprising or inexplicable driver-initiated activation/deactivation during regular use	o major concerns o minor concerns o no concerns o measurement necessary
2	The system mode should be displayed continuously.	
	+ Minimum set of mode indicators present (1) functioning properly (2) currently engaged in an automated driving mode (3) currently unavailable for automated driving (4) experiencing a malfunction (5) requesting a control transition from the automated driving system to the operator - Indicators missing - Indicators not distinguishable from each other - Indicators only displayed for short periods of time - Mode indication discontinued (e.g., through pop-ups)	o major concerns o minor concerns o no concerns o measurement necessary
3	System state changes should be effectively communicated.	
	+ Recognizable change of pictorial indicator + Auditory/haptic feedback + Communication of responsibility (e.g., by disclaimer) + Pop-up messages + Error messages are provided in case (e.g., failed activation) + Delayed reaction to control input displayed in HMI	o major concerns o minor concerns o no concerns o measurement necessary
4	Visual interfaces used to communicate system states should be mounted to a suitable position and distance. High-priority information should be presented close to the driver's expected line of sight.	
	+ Important information displayed in 30° cone about normal line of sight + Safety-critical information displayed in 20° cone about normal line of sight + Peripheral displays support noticing of mode changes (e.g., movement or size of displays) + Status information mirrored on NDRT device	o major concerns o minor concerns o no concerns o measurement necessary
5	HMI elements should be grouped together according to their function to support the perception of mode indicators.	
	- Unnecessary glances to retrieve information from display (e.g., to interpret a symbol and perceive accompanying text) + Indicators pertaining to the automation are grouped together + High priority messages are easily distinguished from low-priority messages	o major concerns o minor concerns o no concerns o measurement necessary
6	Time-critical interactions with the system should not afford continuous attention.	
	- Important information is displayed too shortly (e.g., only for a few seconds) - While the driver is responsible for the DDT, sustained attention (longer than 1.5 s) is needed to accomplish an interaction	o major concerns o minor concerns o no concerns o measurement necessary
7	The visual interface should have a sufficient contrast in luminance and/or color between foreground and background.	
	+ Sufficient color and/or luminance contrast to identify different automation modes	o major concerns o minor concerns o no concerns o measurement necessary
8	Texts (e.g., font types and size of characters) and symbols should be easily readable from the permitted seating position.	
	+ Displayed text and symbols are big enough to be easily readable + Display resolution is good enough to be easily readable + Character width and stroke width appear to be appropriate + Text-fonts are easily readable	o major concerns o minor concerns o no concerns o measurement necessary
9	Commonly accepted or standardized symbols should be used to communicate the automation mode. Use of non-standard symbols should be supplemented by additional text explanations.	
	+ Commonly accepted or standardized symbols are used + Non-standard symbols are supplemented with a text label + The symbols are representative for the responsibility of the driver (e.g., displaying hands on a steering wheel to in case of a hands-on request)	o major concerns o minor concerns o no concerns o measurement necessary

Table A1. *Cont.*

10	The semantic of a message should be in accordance with its urgency.	
	+ Use of notification-style to present non-critical information + Use of command-style to present critical information + Wording in accordance with criticality of the situation (e.g., "caution", "danger", "warning")	o major concerns o minor concerns o no concerns o measurement necessary
11	Messages should be conveyed using the language of the users (e.g., national language, avoidance of technical language, use of common syntax).	
	+ Use of national language + Use of simple language + Avoidance of abbreviations + Displaying functionality rather than SAE/NHTSA/BASt-level	o major concerns o minor concerns o no concerns o measurement necessary
12	Text messages should be as short as possible.	
	+ Messages are as short as possible + Not more than four chunks of information are displayed	o major concerns o minor concerns o no concerns o measurement necessary
13	Not more than five colors should be consistently used to code system states (excluding white and black).	
	+ Colors are used consistently throughout an automated driving mode + Not more than five colors are used	o major concerns o minor concerns o no concerns o measurement necessary
14	The colors used to communicate system states should be in accordance with common conventions and stereotypes.	
	+ Colors are in accordance with common stereotypes of the user population + Red = imminent danger, yellow/amber = caution, green = hazard-free operating state	o major concerns o minor concerns o no concerns o measurement necessary
15	Design for color-blindness by redundant coding and avoidance of red/green and blue/yellow combinations.	
	+ green/red and yellow/blue combinations are avoided + system states are redundantly coded in a suitable way	o major concerns o minor concerns o no concerns o measurement necessary
16	Auditory output should raise the attention of the driver without startling her/him or causing pain.	
	Generic auditory output + suitable length (100 ms–500 ms) + suitable loudness (50 dB–90 dB, should be 15 dB above background noise) + frequencies between 500 and 4000 Hz	o major concerns o minor concerns o no concerns o measurement necessary
	Vibrotactile output + suitable length (50 ms–200 ms) + comfortable stimuli 15–20 dB above threshold + frequencies between 150 and 300 Hz	o major concerns o minor concerns o no concerns o measurement necessary
17	Auditory and vibrotactile output should be adapted to the urgency of the message.	
	Generic auditory output + Auditory output of varying urgency is distinguishably different by pulse rate, frequency, or loudness + Low-priority information is either unobtrusive or without auditory output	o major concerns o minor concerns o no concerns o measurement necessary
	Vibrotactile output + urgency is coded through a variation of location and timing, not frequency and amplitude	o major concerns o minor concerns o no concerns o measurement necessary
18	High-priority messages should be multimodal.	
	+ high priority information is presented in more than one modality + auditory or vibrotactile stimuli are also visually presented	o major concerns o minor concerns o no concerns o measurement necessary
19	Warning messages should orient the user towards the source of danger.	
	+ warning messages lead to an orienting response to the source of danger, causing the driver to look in the direction of the hazard + warning messages to not focus the driver's attention to a display	o major concerns o minor concerns o no concerns o measurement necessary
20	In case of sensor failures, their consequences and required operator steps should be displayed.	
	+ unavailability of sub-systems because of sensor degradation is displayed + consequences of sensor degradation are displayed + required operator behavior is displayed	o major concerns o minor concerns o no concerns o measurement necessary

References

1. National Highway Traffic Safety Administration. *Automated Driving Systems 2.0: A Vision for Safety*; NHTSA: Washington, DC, USA, 2017.
2. Naujoks, F.; Hergeth, S.; Wiedemann, K.; Schömig, N.; Forster, Y.; Keinath, A. Test procedure for evaluating the human–machine interface of vehicles with automated driving systems. *Traffic Inj. Prev.* **2019**, *20*, 146–151. [CrossRef] [PubMed]
3. Naujoks, F.; Hergeth, S.; Wiedemann, K.; Schömig, N.; Keinath, A. Use cases for assessing, testing, and validating the human machine interface of automated driving systems. In Proceedings of the Human Factors and Ergonomics Society Annual Meeting, Philadelphia, PA, USA, 1–5 October 2018.
4. Nielsen, J. *Usability Engineering*; Academic Press: Boston, MA, USA, 1993.
5. Naujoks, F.; Wiedemann, K.; Schömig, N.; Hergeth, S.; Keinath, A. Towards guidelines and verification methods for automated vehicle HMIs. *Transp. Res. Part F Traffic Psychol. Behav.* **2019**, *60*, 121–136. [CrossRef]
6. Naujoks, F.; Hergeth, S.; Keinath, A.; Wiedemann, K.; Schömig, N. Development and application of an expert based assessment for evaluating the usability of SAE Level 3 ADS HMIs. In Proceedings of the ESV Conference 2019, Eindhoven, The Netherlands, 10–13 June 2019.
7. RESPONSE Consortium. Code of Practice for the Design and Evaluation of ADAS; RESPONSE 3: A PReVENT Project; 2006. Available online: https://www.acea.be/uploads/publications/20090831_Code_of_Practice_ADAS.pdf (accessed on 21 April 2020).
8. ISO 26262. *Road Vehicles Functional Safety*; International Organization for Standardization: Geneva, Switzerland, 2008.
9. UN ECE R79: Uniform Provisions Concerning the Approval of Vehicles with Regard to Steering Equipment. 2017. Available online: https://www.unece.org/fileadmin/DAM/trans/main/wp29/wp29regs/2017/R079r3e.pdf (accessed on 21 April 2020).
10. Hoff, K.A.; Bashir, M. Trust in Automation: Integrating Empirical Evidence on Factors That Influence Trust. *Hum. Factors* **2015**, *57*, 407–434. [CrossRef] [PubMed]
11. Wickens, C.D. Processing resources in attention. In *Varieties of Attention*; Parasuraman, R., Davies, R., Eds.; Academic Press: New York, NY, USA, 1984; pp. 63–101.
12. UNECE. Uniform Provisions Concerning the Approval of Vehicles with Regard to Automated Lane Keeping System. Informal Document ACSF-25-03. Available online: https://wiki.unece.org/download/attachments/92013066/ACSF-25-23%20%28Chairs%29%20Draft%20UN%20Regulation%20for%20ALKS%20for%20GRVA.pdf?api=v2 (accessed on 18 February 2020).
13. Linja-aho, M. Creating a framework for improving the learnability of a complex system. *Hum. Technol.* **2006**, *2*, 202–224. [CrossRef]
14. Borojeni, S.S.; Chuang, L.; Heuten, W.; Boll, S. Assisting Drivers with Ambient Take-Over Requests in Highly Automated Driving. In *Automotive' UI: Proceedings of the 8th International Conference on Automotive User Interfaces and Interactive Vehicular Applications*; Association for Computing Machinery: New York, NY, USA, 2016; pp. 237–244.
15. ANSI Z535.4-2011 (R2017). American National Standard for Product Safety Signs and Labels. Available online: https://webstore.ansi.org/preview-pages/NEMA/preview_ANSI+Z535.4-2011+(R2017).pdf (accessed on 18 February 2020).

 *information*

Article

Human–Vehicle Integration in the Code of Practice for Automated Driving

Stefan Wolter [1],*, Giancarlo Caccia Dominioni [2], Sebastian Hergeth [3], Fabio Tango [4], Stuart Whitehouse [5] and Frederik Naujoks [3]

1 Ford Werke GmbH, 52072 Aachen, Germany
2 Toyota Motor Europe NV/SA, 1930 Zaventem, Belgium; Giancarlo.Caccia.Dominioni@toyota-europe.com
3 BMW Group, 80937 Munich, Germany; sebastian.hergeth@bmw.de (S.H.); frederik.naujoks@bmw.de (F.N.)
4 Centro Ricerche Fiat SCpA, 10043 Orbassano (Turino), Italy; fabio.tango@crf.it
5 Jaguar Land Rover, Coventry CV34LF, UK; swhiteh6@jaguarlandrover.com
* Correspondence: swolter3@ford.com

Received: 21 April 2020; Accepted: 22 May 2020; Published: 27 May 2020

Abstract: The advancement of SAE Level 3 automated driving systems requires best practices to guide the development process. In the past, the Code of Practice for the Design and Evaluation of ADAS served this role for SAE Level 1 and 2 systems. The challenges of Level 3 automation make it necessary to create a new Code of Practice for automated driving (CoP-AD) as part of the public-funded European project L3Pilot. It provides the developer with a comprehensive guideline on how to design and test automated driving functions, with a focus on highway driving and parking. A variety of areas such as Functional Safety, Cybersecurity, Ethics, and finally the Human–Vehicle Integration are part of it. This paper focuses on the latter, the Human Factors aspects addressed in the CoP-AD. The process of gathering the topics for this category is outlined in the body of the paper. Thorough literature reviews and workshops were part of it. A summary is given on the draft content of the CoP-AD Human–Vehicle Integration topics. This includes general Human Factors related guidelines as well as Mode Awareness, Trust, and Misuse. Driver Monitoring is highlighted as well, together with the topic of Controllability and the execution of Customer Clinics. Furthermore, the Training and Variability of Users is included. Finally, the application of the CoP-AD in the development process for Human-Vehicle Integration is illustrated.

Keywords: automated driving; human factors; human machine interface; controllability; L3Pilot

1. Introduction

The European research project L3Pilot focuses on different activities with regard to automated driving. Split into seven subprojects, the main objective of the L3Pilot subproject 2 is to define a Code of Best Practice for Automated Driving (CoP-AD). The CoP-AD shall provide comprehensive guidelines for supporting the automotive industry and relevant stakeholders in the development of the automated driving technology. Thus, the CoP is meant to provide best practice guidance that can be used by designers and engineers throughout the lifecycle of automated driving systems. The guidelines are derived from knowledge gained in the industry as well as best practices collected on this topic.

Previously, for systems up to and including SAE level 2 [1], the Code of Practice for advanced driver assistance systems, derived from the Response project [2], served as a guideline for the development of such functions. With the advent of SAE level 3 systems and above, its application is no longer appropriate. Nonetheless, the existing code of practice was analyzed in order to apply the lessons learnt and to make use of the aspects, which remain appropriate for SAE level 3.

In order to define the scope of the document, a framework for the Code of Practice for Automated Driving was defined at the beginning of this project. It serves as a baseline for the work to be done for creating the CoP-AD. In the second section of this paper, the development process is outlined, culminating in the definition of the topics to be addressed, which were classified into four different categories. It also includes the applicable development phases and furthermore, the geographical regions, operational design domains, and SAE levels affected. The template on how to phrase and execute the questions that will form the checklist of aspects to consider when developing an Automated Driving Function (ADF) is also outlined and explained.

The third section shows the draft content of the Human–Vehicle Integration (HVI) category. This is one of the four main categories in the CoP-AD. It focuses on the topics related to the interaction between the vehicle and the user. This ranges across a broad area covering human factors, user experience, usability, and cognitive ergonomics. The section is divided into the areas of Guidelines for HVI, Mode Awareness, Trust and Misuse, Driver Monitoring, Controllability and Customer Clinics, and finally Driver Training and Variability of Users. The topics are explained, and examples are given on how to apply them as part of the CoP-AD.

In the final section, some general conclusions have been drawn, and further conclusions are highlighted with a focus on the HVI category. This paper is based on the L3Pilot deliverable D2.2 [3], which is a draft of the CoP-AD used to gather feedback from external partners outside of the L3Pilot consortium.

2. Development Process

At the beginning of the L3Pilot project, a survey was distributed to all L3Pilot partners in order to collect the requirements of all key stakeholders for the CoP-AD. This includes experts from both industry and research institutes. The relevant topics to be covered in best practices were derived using this feedback. The topics collected as part of the survey were selected based on predefined criteria during a subsequent workshop. The criteria for inclusion of a topic are listed in Table 1.

Table 1. Criteria for inclusion of topics into the Code of Practice for automated driving (CoP-AD).

The topic or process poses a common challenge in the development process that requires cooperation.
A wrongly applied approach for the topic or process would lead to serious consequences (e.g., malfunctions in certain traffic situations leading to non-release of the function).
A frequent misapplication of an approach for a topic or process is highly likely.
The topic/process has already been identified as relevant by others.
The topic or process can be described in a general way that does not lead to unreasonable limitations in the development process (company independent).
And the optional criteria: the topic or process is of relevance for L3Pilot prototype vehicles and can be evaluated in this project.

With regard to the actual process of applying the CoP-AD, the decision was made to use the existing Code of Practice for Advanced Driver Assistance Systems as a baseline. Figure 1 shows the selected development phases for the CoP-AD. Compared to the Code of Practice for Advanced Driver Assistance Systems, the number of phases was reduced from six to four during the actual development. The second and fourth phase originally consisted of two separate stages, but these were condensed into the Concept Selection Phase and the Validation and Verification Phase for greater simplicity. An additional phase for the time post start of production was added to cover the entire lifecycle of the ADF. The conceptual stage consists of the Definition Phase and Concept Selection Phase, while the Design Phase and the Validation and Verification Phase constitute the series development stage. During the Definition Phase, the basic requirements are defined and based on this, the best concept is chosen in the Concept Selection Phase. The Design Phase requires the detailed design of the system. Then, it is validated and verified in the final phase before the start of production. Post start of production, further data can be gathered and improvements can be applied. This process is not necessarily linear; iterative improvements with repetitions of important steps might be possible. The process has been designed to remain abstract on purpose, so that the CoP-AD can be applied to the many different development processes in place in the industry at various companies.

Figure 1. Development phases.

In order to clearly summarize the topics that were collected, a number of categories were defined to cluster them. Table 2 shows the categories finally chosen with the pertaining topics. They are based on extensive expert discussions, clustering all the available topics in a meaningful way. The last row on Human–Vehicle Integration is the key focus of this paper.

Table 2. Categories and topics. HMI: Human–Machine Interface, ODD: Operational Design Domain.

Category	Topics
Overall Guidelines and Recommendations	Minimal Risk Manoeuver Documentation Existing Standards
ODD Vehicle Level	Requirements Scenarios and Limitations Performance Criteria and Customer Expectations Architecture Testing (incl. Simulation)
ODD Traffic System and Behavioral Design	Automated Driving Risks and Coverage of Interaction with Mixed Traffic V2X Interaction Traffic Simulation Ethics and Other Traffic-Related Aspects
Safeguarding Automation	Functional Safety Cybersecurity Implementation of Updates Safety of the Intended Functionality (SOTIF) Data Recording, Privacy and Protection
Human-Vehicle Integration	Guidelines for HMI Mode Awareness, Trust, and Misuse Driver Monitoring Controllability and Customer Clinics Driver Training and Variability of Users

The first category is quite generic and focusses on overall guidelines and recommendations, such as a minimal risk manoeuver. The Operational Design Domain (ODD) on the Vehicle Level offers a description of the function and scenarios at the level of the vehicle. The category ODD on the Traffic System Level, including Behavioral Design, offers a description of the function on the level of the overall environment and a description of the behavior of other road users. Safeguarding Automation is about how to ensure a safe operation of the function, primarily the functional safety, but also the cybersecurity and data privacy aspects. Human–Vehicle Integration is the interaction between the driver and the vehicle's displays and control elements.

The topics within each of the categories were distributed along the development process phases in a workshop. In order to better address the topics derived from previously held expert sessions, a thorough literature review was done to back up the topics with research results and existing best practices. Based on this, the questions for the CoP-AD checklist were phrased. These questions underwent a rigorous iterative improvement process, improving overall quality and reducing the overall number of available questions to the most important ones. This enabled the deliverable D2.2 [3] to be written, which is a draft used to gather feedback from external partners outside the L3Pilot consortium. This will culminate in the deliverable D2.3, the final CoP-AD, to be presented in 2021.

In order to apply the CoP-AD appropriately, a template was defined for all questions; this can be seen in Table 3. The reference number for each question can be found in the top left cell of the table, and the development phases associated with the question have been marked in the top right. In the body of the table, the main question is on the left, supported where applicable by sub-questions on the right. Only the main question needs to be answered directly with yes or no. Ideally, independent evaluators (e.g., individuals from other departments or external sources such as research institutes) who have formal training or experience in the subject matter of the topics are also involved in the application of the CoP-AD. For example, for the Human–Vehicle Integration topic, the evaluator should have experience in human factors, usability engineering, or cognitive ergonomics.

Table 3. Template for questions.

Question X-Y-Z	Relevant Phase(s)	DF	CO	DS	VV	PS
Main question ( ) Yes/( ) No			• Sub-Question 1 • Sub-Question 2 • Sub-Question 3			

Following the CoP means that all of the questions should be answered positively, or, that the issue raised by the items has been solved in another way. The sub-questions serve as an elaboration. The main question is phrased in a way that an answer with yes always means that the question has been addressed sufficiently. However, even in case a no is given as an answer, this may still be appropriate, as there might be good reasons why something could not be done or answered, or is simply not applicable in a given case, as long as the underlying problem is solved and documented. For some of the items, accepted pass/fail criteria are available (such as the number of participants that need to pass a controllability confirmation test), others are relying on norms (e.g., legibility of displays) or expert assessments if these kinds of thresholds are not available. In a further step, the questions may be transferred to an Excel file or another software tool for easy application and editing.

The CoP-AD was scoped to cover motorway and parking scenarios for SAE level 3 and level 4 functions. Although only EU markets are currently in scope, it is assumed that the CoP-AD may also be applied to non-EU regions, as well as urban or rural traffic scenarios, and even driverless robot taxis. This needs to be investigated in further research.

In the third section of this paper, the HVI category is explained in detail. This also includes examples of the questions asked.

3. Draft Content Human–Vehicle Integration

The HVI category comprises all factors related to the interaction between the vehicle and the user. This ranges across a broad area covering human factors, user experience, usability, and cognitive ergonomics. The introduction of automated driving systems that allow fallback-ready users to disengage from driving and engage in non-driving-related tasks introduces a range of potential human factors problems that must be considered in the development process. First, the transitions from automated driving to manual driving must be supported so that users are capable of taking over the driving task in a safe way in case of system limits and malfunctions. Furthermore, the possibility of different automated driving modes being available within the same vehicle, each requiring different levels of responsibility from the user, creates the need to communicate the active driving mode unambiguously. Thus, the design of the Human–Machine Interface (HMI) is a central element in the design process to ensure proper mode awareness and controllable transitions to manual driving. Secondly, the "availability" of the driver to react to requests to intervene needs to be ensured, which is mainly a function of non-driving-related tasks carried out during the automated ride. Thus, the design of the ADF should be made with foreseeable non-driving-related tasks that might likely be carried out by users during the automated ride. Thirdly, whether the ADF will be used in accordance with the intended usage, or whether users will misuse it (possibly because of over trust in the ADF) will depend on the training and information users receive.

Display and control concepts, i.e., the HMI, must be developed in a way that they are easily and safely operated by the user of an ADF. The HVI is about the harmonious interaction between the user and the vehicle in a broader sense, whereas the HMI is more specifically about the hardware and software interface between them. In order to streamline the various aspects related to HVI, this category is divided into five different topics. The first topic covers the general guidelines on how to design the HMI. This includes the acceptance of the ADF, as well as the usability and the user experience-related aspects. The Mode Awareness, Trust, and Misuse topic is primarily about the driver's awareness of the ADF's current driving mode. This also relates to the users' trust in the ADF and their potential for misuse. Driver monitoring is about assessing the user's state when operating an ADF, which is a topic closely related to the users' mental models and their workload. An important aspect of this is the impact of non-driving-related tasks (in the following referred to as secondary tasks) carried out while driving with a highly automated function. The Controllability and Customer Clinics topic refers to the question of an ADF's controllability from the user's perspective on the one hand and how to conduct a study with participants to test the controllability and other properties of the ADF on the other. Driver Training and Variability of Users is the final topic. It covers the area of user training required for an ADF. Furthermore, it also relates to the variability of users to be taken into account. Together, these topics, comprising 39 main questions, form a comprehensive overview on the overall category of HVI. All the main questions from this and all other categories are available in [3].

3.1. Guidelines for Human–Machine Interface

Guidelines for the ADF's HMI are prominently addressed as a topic in the CoP-AD. Following appropriate guidelines is key to producing a well-executed user experience and usability, which in turn will create a much higher level of underlying safety in the ADF [4]. On a generic level, this topic is about using HMI design guidelines to define, assess, and validate an HMI concept. They should be followed during the whole development process of the HMI for an ADF. There are various HMI guidelines available (e.g., [5,6]), and the guidelines used during the ADF development should be selected carefully to ensure they are suitable for the SAE level 3 systems. Guidelines adapted to HMIs for conditionally automated vehicles were presented by Naujoks et al. [7] and validated in empirical studies [8,9]. The HMI should be standardized where possible following industry standards that are consistent with the user's mental models [10,11]. This will minimize the time required to familiarize oneself with the HMI, therefore improving the experience of first-time users. Still, guidelines may differ for certain demographics, as different groups of people may prefer different communication methods such as symbols or color coding.

Table 4 shows an example question from the Guidelines for HMI topic. The question aims to determine whether unintentional activations and deactivations of the ADF are prevented or not. Unintentional deactivation of an ADF by the user is an event that needs to be avoided. The driver may be focusing on a secondary task and will not be ready to take over control of the driving task if necessary. The HMI concept should be designed so that it is not possible for the driver to inadvertently initiate a transfer of control. At the same time, it is important to prevent unintentional activations of the ADF. Unexpected longitudinal or lateral input from the ADF may have a detrimental effect on the user's trust in the ADF.

Table 4. Example question Human–Vehicle Integration (HVI) guidelines. ADF: Automated Driving Function.

Question 4-1-2	Relevant Phase(s)	CO
Are unintentional activations and deactivations of the ADF prevented? ( ) Yes/( ) No		

Furthermore, the visual interface shall be designed to be easy to read and interpret [12]. This item focuses on the importance of having a clear strategy for the visual HMI. Guidelines and standards need to be followed to ensure that the visual feedback is easy and intuitive to understand. Icons can be designed to be interpreted quickly if standard symbols and colors are used where possible. Where icons cannot be used, text messages shall be applied. However, it is important that the text can be understood in short glances, so that the driver is not forced to remove the eyes from the road for extended periods of time [6,13,14]. Finally, it is important to cluster relevant HMI elements in similar locations so that the driver can intuitively understand where an HMI should appear [5,14,15].

The HMI shall be designed to portray the urgency of the message to be conveyed [11,12,16]. During the use of an ADF, the user may be subject to many types of HMI feedback with various levels of urgency. It is important that the driver understands which HMI elements are of high priority and are conveying urgent feedback to the driver [17]. Equally, it is important that the driver understands that other messages are provided primarily for informational purposes and therefore do not require immediate action. Assessing the user acceptance is also a key point. Customer clinics, heuristic expert assessments, and various other user trials can be carried out to gain both subjective and objective data on user acceptance.

3.2. Mode Awareness, Trust, and Misuse

This topic addresses the correct understanding of the role shared between the driver and the ADF, concerning the active mode, as well as the correct usage of and the trust in the ADF.

An example question is given in Table 5. This is about ensuring the drivers fully understand their responsibilities and the function's capabilities during each of the defined ADF modes. They may be informed by several means, such as in-product advertisements and written explanations in the owner's manual [18]. Drivers may get explicit information from the in-vehicle HMI, before, during, and after activation of the ADF itself. They may of course also learn by experience [19]. Additionally, a simple and intuitive HMI can improve the driver's situational awareness and help them to take the correct actions when necessary.

Table 5. Example question Mode Awareness, Trust, and Misuse.

Question 4-2-9	Relevant Phase(s)	CO	DS	VV
Is the communication to the driver, of the driver's responsibilities in each defined automated driving mode(s) investigated and confirmed? ( ) Yes/( ) No	<ul><li>Is a method implemented to clearly inform the user of their responsibilities and of vehicle capabilities and possibly of the result of not acting within these capabilities?</li><li>Is the communication to the user, of the ADF's capabilities in each defined automated driving mode(s) investigated and confirmed?</li><li>Is there clear information in the user's manual, about the ADF's boundaries, and has this been confirmed?</li><li>Is additional training material to communicate the ADF's boundaries and the user's responsibilities considered?</li><li>Is a process defined on how the user will be informed about any new potential functionality of the ADF based on software updates?</li></ul>			

All possible automated driving modes shall be explicitly defined in terms of how the driver should acknowledge them. The goal of this item is to ensure that the possible ADF modes are clearly defined from a user's perspective. It is important that a user is aware of the possible automated driving modes of the ADF to avoid any misunderstanding.

It is key to know whether the HMI modalities to communicate the relevant active (automated) driving modes are described. This item focuses on how the active automated driving modes are communicated to both the driver and the other road users, in terms of modalities (visual, auditory, haptic, etc.).

All reasonably foreseeable mistakes and misuse cases of the ADF in relation to the HMI shall be described. The purpose of this question is to ensure that possible driver mistakes, failures and misuses have been addressed in the best possible way, in order to be able to define countermeasures for them [2,20].

Communicating the automated driving modes to the driver in an appropriate and clear way shall be investigated and confirmed. For an ADF, a clear communication of the mode is crucial. This question focuses on the HMI to communicate the ADF modes, the consideration of a permanent display of the modes, how to communicate the mode changes, and how well these HMI elements are recognized by both the driver and other road users. A test procedure to assess whether basic mode indicators are capable of informing the driver about relevant modes and transitions has been proposed by Naujoks et al. [21]. Additional information regarding this topic is provided by JAMA [22], Albers et al. [23], and Schömig et al. [24].

A multimodal HMI to improve driver alertness and minimize the time to get back in the loop should be investigated. However, it should also be ensured that the HMI is no more intrusive than necessary. Therefore, it is necessary to find a balance between the effectiveness of the HMI and the level of annoyance that it may cause the users [25]. Speech is another possibility to communicate a take-over request. The impact of the HMI on relevant driver indicators such as eyes-on-road time should be investigated [26].

Information shall be provided to the driver about an ADF-initiated minimum risk manoeuver [27]. A minimum risk manoeuver typically happens if the driver fails to appropriately take over the controls, or if the function does not have enough time to make a proper take-over request (for example, due to a sudden unexpected situation). This item aims to consider how to inform the driver in the event that the function has initiated the minimum risk manoeuver in order to provide the driver with the necessary information, such as what is going on, why, and what action the driver should take.

The communication to the driver, of the driver's responsibilities in each defined automated driving mode should be investigated and confirmed. It shall be considered how and to what extent the operational design domain information will be displayed to the driver. The driver awareness of automated driving modes shall be investigated as well.

Driver expectations regarding the ADF's features need to be considered. It is crucial to confirm whether user expectations are met. This is a broad subject that would need to be narrowed down to precise specifications, and this question is there to make sure that this process will be considered. For example, in terms of HVI, the balance between the amount of information and its conciseness or simplicity should be investigated.

The driver's trust in the ADF is an important aspect to consider [28]. It is necessary that the users trust the function, in order for them to feel comfortable using it. On the other hand, it is necessary to avoid over-trust, as this may lead to unintended misuse of the function [29]. Again, a good balance should be targeted in order to ensure the correct amount of trust. The appropriate usage of the ADF should be assessed and confirmed, encouraging the intended use and preventing misuse.

Long-term effects of the ADF on the users shall be investigated. Typically, the main risks of long-term effects are skill degradation and building over-trust in the function [30]. The impact of the HMI on driver workload and other aspects over long journeys shall be investigated as well.

3.3. Driver Monitoring

This topic addresses the correct application of driver monitoring, specifically the identification and classification of the driver's status and the recognition of the actions made inside the vehicle. Monitoring a driver's attention is a crucial topic, especially when discussing automated driving [31]. Since driving is a complex phenomenon, involving the performance of various tasks (including simultaneous quick and accurate decision making), fatigue, workload, and distraction drastically increase human response time, which may result in an inability to drive correctly or to respond properly to a take-over request [32].

Table 6 shows an example item for this topic. The question is assessing whether all relevant secondary tasks are considered when defining the driver monitoring requirements. This item addresses which secondary tasks are allowed during automated driving. The idea is to consider what is currently available and what will become available in the future. In addition, one sub-question focuses on metrics that shall be taken into account when a driver monitoring function is present within the vehicle. Moreover, the possibility to add additional apps or secondary tasks to the HMI in the future shall be considered as well.

Table 6. Example question on Driver Monitoring.

Question 4-3-1	Relevant Phase(s)	DF
Are all relevant secondary tasks considered? ( ) Yes/( ) No		• Are plausible secondary tasks possible today and in the near future taken into account? • Which secondary tasks are legal, and in what timeframe will they become legal? • Which metrics shall be measured via a driver monitoring function? • Are the metrics appropriate for the automated driving function defined? • Which apps/secondary tasks can be integrated into the vehicle HMI?

A further important question is whether the HMI is connected with the driver monitoring function. It is essential to provide crucial information on driver's state directly to the driver, as an impairment may compromise the safety of the situation. Thus, unsafe driver states such as drowsiness need to be communicated effectively [33].

Furthermore, it should be taken into account whether it is possible to mirror the user's devices on the HMI [34,35]. If it is legally allowed, then it is important to consider how to prompt the driver to take back control of the vehicle while their device is being mirrored. For example, this could be done by overlaying a take-over request on the user's device. This way, the driver can be taken back into the control loop in an effective manner. Device-pairing offers further benefits; for instance, the larger in-vehicle screens may be used as opposed to the relatively small smartphone screens. Due to the use of dedicated controls and displays, driver distraction is also minimized. The impact of typical secondary tasks on take-over time and quality should be identified as well. It is useful to measure the impact of secondary tasks on the take-over request.

After the start of production, data may be gathered to assess the types of secondary tasks, the amount of time users spend doing them, and their impact on driving behavior, traffic safety, etc. This is related to measuring the long-term effects of secondary tasks on driver behavior.

3.4. Controllability and Customer Clinics

SAE level 3 automated driving will still require the driver to take over the driving task in case of system failures and malfunctions. Thus, it has to be ensured that drivers are able to control transitions to manual or assisted driving and avoid safety critical consequences. Driver-initiated transitions should also be considered from this perspective. This topic is one of the key elements in the existing Code of Practice for Advanced Driver Assistance Systems [2].

Table 7 shows an example question for this topic. It is about the suitability of testing environments for controllability. In the verification phase, controllability assessments should be carried out in suitable test environments, ranging from laboratory to test tracks, etc. When these controllability assessments are carried out on test tracks or on public roads, precautions regarding the safety of participants and other road users should be taken.

Table 7. Example question for Controllability and Customer Clinics.

Question 4-4-7	Relevant Phase(s)	VV
Are the testing environments for controllability confirmation tests suitable? ( ) Yes/( ) No		• Are the venues for the customer clinics adequate (laboratory, test track, etc.)? • Are adequate precautions taken for real world testing, especially with naive participants?

During the definition phase, it shall be ensured that user needs regarding controllability are taken into account. For example, the design of the HMI should consider the transition from automated driving to lower levels of automation with respect to function failures and system limits as well as driver-initiated transitions. Relevant and applicable guidelines for the design of the HMI should be considered in the design phase in order to ensure that they are in line with generally accepted standards and best practices in view of the targeted user population [7,36,37].

Limitations of the human driver should be taken into account. Careful consideration of the driver's sensory and motor limitations (e.g., inability to move freely) need to be applied. The concept selection should thus consider topics such as color-blindness, general vision, sensory-motor, and hearing impairments.

The development should also account for a clear and understandable description of the ADF and its limits. Most importantly, if the driver is informed about function limits, that will trigger requests to intervene [38]. These should be described in the user manual and other available multimedia-based information, together with a description of the expected reaction. It also comprises the selection of a transition-of-control concept. Furthermore, it shall be tested if the vehicle is controllable in the case of a malfunction or by overruling or switching off the function.

The behavior of the ADF should not lead to uncontrollable situations from the perspective of other road users. The design should also consider the limitations and perception of other traffic participants that are not equipped with an ADF. The automated vehicle's behavior shall be designed in a way that it is controllable for these traffic participants and does not exceed the motion ranges of drivers who are driving manually in non-emergency situations.

Even in the early design phase, a preliminary assessment of the controllability can be carried out, which is normally based on expert assessments. A suitable prototype should be used that allows for an assessment of function limits and failures, but also normal driver-initiated transitions [39,40]. The final controllability verification can be based on different evaluation methods such as expert assessments, controllability verification tests, or customer clinics [40].

A suitable post-production evaluation strategy should be implemented that assesses the impact of the ADF on possible negative behavioral adaptations such as skill degradation and misuse. This way, the ADF is adequately evaluated from a human factors perspective after the start of production.

3.5. Driver Training and Variability of Users

This topic covers the training required for ADF users and the variability of these users, which needs to be considered. The training aspect is about the issue of providing users with the appropriate knowledge and skills to operate an ADF. As there is a huge variability of users, different age groups, gender, cultural backgrounds, and different levels of previous experience need to be considered. Both topics are combined here, as they share various aspects.

Table 8 shows an example question for this topic, asking if the information that the user needs to operate the ADF is available to create a training course. Creating the user training for the ADF requires a specification of the ADF's operation to serve as a baseline. Due to the complexity of ADFs generally, a user training course may be required or at least recommended. Ideally, this is unnecessary due to a well-executed intuitive system design. The training methods shall be defined in more detail to produce a course that could use one of many of the following mediums: a training course provided by the dealer, user manuals integrated within the vehicle, online material for home training, or the use of digital assistants. A reasonable combination of training methods shall be considered taking individual learning preferences into account [20,41,42].

Table 8. Example question Driver Training and Variability of Users.

Question 4-5-2	Relevant Phase(s)	CO	DS
Is the information that the user needs to operate the ADF available to create a training course? ( ) Yes/( ) No		• Is there a training course needed for test drivers? • Is there a driver training course for ordinary users planned? • Is a process to train users of an ADF established? • Are the possible training methods for the user defined (e.g., dealer training, online material for home training, material in car, manual, use of virtual reality, digital assistants, etc.)?	

There may be huge differences between user groups. The questions in the CoP-AD target the difference between countries and geographical regions. Infrastructural differences with regard to roads and traffic control functions as well as driver behavior in general have a huge impact on the design of ADFs and so these differences need to be handled appropriately. An ADF designed for only a specific country or geographical region without taking into account the local infrastructure and the requirements of their user groups must be avoided. Another factor to be taken into account are elderly drivers. Due to their degrading physical abilities, driving becomes more cumbersome. Therefore, during the definition of ADFs, the physical impairments of elderly drivers should be addressed. There is also a significant variability in users' physical dimensions and anthropometry. Size and strength differences between genders can play a role, and so the ADF shall be designed to be operated by a variety of different users, including those with non-age-related disabilities.

There shall also be a representative test sample for user studies. Depending on the exact user study to be conducted, this may range from age, gender, and socio-cultural background to test candidates with previous experience with ADFs or technology in general. The test participants in a sample should be selected accordingly.

A solid mix of customer education and information shall be made available to the users post start of production. Developers need to ensure that there is enough information available for the users of an ADF to properly operate it. There should be sufficient training material available inside the vehicle to provide users with the required knowledge to operate the ADF safely on the road. To reduce the likelihood of people over-estimating the possibilities offered by the ADF, the marketing shall support user information and training with realistic information regarding its abilities.

4. Conclusions

The introductory part gave an overview on the development process applied to finalize the draft of the CoP-AD. This comprises all the different main categories such as the ODD Vehicle Level, the ODD Traffic System and Behavioral Design as well as Safeguarding Automation. The draft results of the CoP-AD presented here with a focus on the HVI category offer the first insight on how the interaction between the driver of the vehicle and the automated driving system shall be part of a standardized development process. Whereas the first category focuses on available guidelines in general, the other topics concentrate more specifically on topics of interest for designing an appropriate interaction between the driver and the vehicle equipped with an automated driving system. Mode awareness, including the aspects of trust and misuse is a cornerstone on how to make people aware of the automated system's abilities, improving trust and at the same time preventing misuse. Driver monitoring plays a major role when taking into account the state of the driver and its importance for a safe operation of the automated driving function. Controllability and customer clinics actually focus on two distinct but interrelated topics. Ensuring the controllability of a system is key, especially in case of minimum risk manoeuvers. This shall be tested in user studies, which in turn serve as a primary method to test many of the guidelines and assumptions mentioned in this text. Driver training again emphasizes the importance of giving drivers the education they need, and in a medium that they can consume and learn from most effectively. In addition, the variability of users is taken into account, including the cultural and infrastructural differences between different cultures and geographical regions.

It must be emphasized that the proposed CoP-AD is based on current best practices, research, and applicable norms. Many of the published studies have been conducted using driving simulators or proving grounds; however, as automated vehicles have not been deployed, final proof that the proposed CoP-AD will be able to eliminate all possible design issues is not yet possible. The current publication is meant to stimulate the ongoing discussion in the technical and scientific community to further improve and converge current research and evaluation practice. It should also be noted that the current paper lays out a draft version of the CoP-AD that will be further refined based on available feedback. This does not only include the HVI, but also the other categories mentioned in this paper. The final CoP-AD needs to be available in an easy-to-use way, preferably as some kind of software application, either Excel-based or standalone. During the development process of an ADF, the questions presented here as examples, and those being part of the final document will guide the engineers from the concept phase up to the time post start of production.

The scope of this document is currently on highway driving and parking, primarily on SAE Level 3 and to a certain extent on SAE Level 4, for the European regions. Further work is required to see if it may be applied to other regions outside of the EU as well. Of particular interest are the USA and China. Automated driving systems that operate within the city or in rural areas shall also be applicable to the CoP-AD. Otherwise, future iterations will have to be adapted to be also applicable to other areas. This is also true for applications regarding robot taxis, reaching from geo-fenced SAE Level 4 up to SAE Level 5 systems. Until then, the CoP-AD will serve as an important guideline for the development of automated driving functions.

Author Contributions: For research Conceptualization, S.W. (Stefan Wolter), G.C.D., S.H., F.T., S.W. (Stuart Whitehouse) and F.N.; Supervision, S.W. (Stefan Wolter); Writing—original draft, S.W. (Stefan Wolter); Writing—review and editing, G.C.D., S.H., F.T., S.W. (Stuart Whitehouse) and F.N. All authors have read and agreed to the published version of the manuscript.

Funding: This paper results from the L3Pilot project. This project has received funding from the European Union's Horizon 2020 research and innovation programme under grant agreement No 723051. The sole responsibility of this publication lies with the authors. The authors would like to thank all partners within L3Pilot for their cooperation and valuable contribution.

Conflicts of Interest: The authors declare no conflict of interest.

References

1. SAE International. *Taxonomy and Definition for Terms Related to Driving Automation Systems for On-Road Motor Vehicles (J3016)*; SAE International: Warrendale, PA, USA, 2018.
2. Knapp, A.; Neumann, M.; Brockmann, M.; Walz, R.; Winkle, T. Code of Practice for the Design and Evaluation of ADAS, Deliverable of PReVent-Preventive and Active Safety Applications Integrated Project, Version 5.0. 2009. Available online: https://www.acea.be/uploads/publications/20090831_Code_of_Practice_ADAS.pdf (accessed on 15 May 2020).
3. Fahrenkrog, F.; Schneider, M.; Naujoks, F.; Tango, F.; Knapp, A.; Wolter, S.; Cao, Y.; Griffon, T.; Demirtzis, E.; Lorente Mallada, J.; et al. L3Pilot Deliverable D2.2. In *Draft and Results from Pilot Application of Draft CoP*; p. 2020.

4. Campbell, J.L.; Brown, J.L.; Graving, J.S.; Richard, C.M.; Lichty, M.G.; Bacon, L.P.; Morgan, J.F.; Li, H.; Williams, D.N.; Sanquist, T. *Human Factors Design Guidance for Level 2 and Level 3 Automated Driving Concepts (Report No. DOT HS 812 555)*; National Highway Traffic Safety Administration: Washington, DC, USA, 2018.

5. Transport Research Laboratory. *A Checklist for the Assessment of In-Vehicle Information Systems (IVIS)*; TRL: Wokingham, UK, 2011.

6. Campbell, J.L.; Carney, C.; Kantowitz, J.L. *Human Factors Design Guidelines for advanced Traveler Information Systems (ATIS) and Commercial Vehicle Operations (CVO)*; Report No. FHWA-RD-98-057; Federal Highway Administration: Washington, DC, USA, 1997.

7. Naujoks, F.; Wiedemann, K.; Schömig, N.; Hergeth, S. Towards guidelines and verification methods for automated vehicle HMIs. *Transp. Res. Part F Traffic Psychol. Behav.* **2019**, *60*, 121–136. [CrossRef]

8. Forster, Y.; Hergeth, S.; Naujoks, F.; Krems, J.F.; Keinath, A. Empirical Validation of a Checklist for Heuristic Evaluation of Automated Vehicle HMIs. In Proceedings of the International Conference on Applied Human Factors and Ergonomics, Washington, DC, USA, 24–28 July 2019; Springer: Cham, Switzerland, 2019; pp. 3–14.

9. Naujoks, F.; Hergeth, S.; Keinath, A.; Wiedemann, K.; Schömig, N. Development and Application of an expert assessment method for evaluating the usability of SAE L3 ADS HMIs. In Proceedings of the ESV Conference Proceedings, Eindhoven, The Netherlands, 10–13 June 2019.

10. HARDIE Consortium. *HARDIE Design Guidelines Handbook*; HARDIE Project; Commission of the EC: Brussels, Belgium, 1996.

11. Transport Research Laboratory. *Design Guidelines for Safety of In-Vehicle Information Systems*; TRL: Wokingham, UK, 2002.

12. ISO 15008. *Road Vehicles–Ergonomic Aspects of Transport Information and Control Systems–Specifications and Test Procedures for In-Vehicle Visual Presentation*; International Organization for Standardization: Geneva, Switzerland, 2007.

13. SAE International. *Development of Design and Engineering Recommendations for In-Vehicle Alphanumeric Messages (J2831)*; SAE International: Warrendale, PA, USA, 2012.

14. Kelsch, J.; Dziennus, M.; Schieben, A.; Schömig, N.; Wiedemann, K.; Merat, N.; Louw, T.; Madigan, R.; Kountouriotis, G.; Aust, M.L.; et al. Final Functional Human Factors Recommendations. AdaptIVe Deliverable D3.3. 2017. Available online: http://www.adaptive-ip.eu/index.php/AdaptIVe-SP3-v23-DL-D3.3-Final%20Functional%20Human%20Factors%20Recommendations_Core-file=files-adaptive-content-downloads-Deliverables%20&%20papers-AdaptIVe-SP3-v23-DL-D3.3-Final%20Functional%20Human%20Factors%20Recommendations_Core.pdf (accessed on 15 May 2020).

15. NASA-STD-3001. *NASA Space Flight Human-System Standard Volume 2: Human Factors, Habitability and Environmental Health*; NASA: Washington, DC, USA, 2011.

16. ISO 15623. *Intelligent Transport Systems–Forward Vehicle Collision Warning Systems–Performance Requirements and Test Procedures*; International Organization for Standardization: Geneva, Switzerland, 2013.

17. Naujoks, F.; Mai, C.; Neukum, A. The effect of urgency of take-over requests during highly automated driving under distraction conditions. *Adv. Hum. Asp. Transp.* **2014**, *7 Pt I*, 431.

18. Forster, Y.; Hergeth, S.; Naujoks, F.; Krems, J.F.; Keinath, A. What and how to tell beforehand: The effect of user education on understanding, interaction and satisfaction with driving automation. *Transp. Res. Part F Traffic Psychol. Behav.* **2020**, *68*, 316–335. [CrossRef]

19. Forster, Y.; Hergeth, S.; Naujoks, F.; Beggiato, M.; Krems, J.F.; Keinath, A. Learning to use automation: Behavioral changes in interaction with automated driving systems. *Transp. Res. Part F Traffic Psychol. Behav.* **2019**, *62*, 599–614. [CrossRef]

20. Campbell, J.; Brown, J.; Graving, J.; Richard, C.; Lichty, M.; Sanquist, T.; Bacon, P.; Woods, R.; Li, H.; Williams, D.; et al. *Human Factors Design Guidance for Driver-Vehicle Interfaces*; NHTSA Report DOT HS 812 360; NHTSA: Washington, DC, USA, 2016.

21. Naujoks, F.; Hergeth, S.; Wiedemann, K.; Schömig, N.; Forster, Y.; Keinath, A. Test procedure for evaluating the human–machine interface of vehicles with automated driving systems. *Traffic Injury Prev.* **2019**, *20* (Suppl. S1), 146–151. [CrossRef] [PubMed]

22. Japan Automobile Manufacturers Association (JAMA). *Guidelines for In-Vehicle Display Systems—Version 3.0*; Japan Automobile Manufacturers Association (JAMA): Tokio, Japan, 2004.

23. Albers, D.; Radlmayr, J.; Loew, A.; Hergeth, S.; Naujoks, F.; Keinath, A.; Bengler, K. Usability Evaluation—Advances in Experimental Design in the Context of Automated Driving Human–Machine Interfaces. *Information* **2020**, *11*, 240. [CrossRef]

24. Schömig, N.; Wiedemann, K.; Hergeth, S.; Forster, Y.; Muttart, J.; Eriksson, A.; Mitropoulos-Rundus, D.; Grove, K.; Krems, J.; Keinath, A. Checklist for Expert Evaluation of HMIs of Automated Vehicles—Discussions on Its Value and Adaptions of the Method within an Expert Workshop. *Information* **2020**, *11*, 233. [CrossRef]

25. Naujoks, F.; Forster, Y.; Wiedemann, K.; Neukum, A. Improving usefulness of automated driving by lowering primary task interference through HMI design. *J. Adv. Transp.* **2017**, 6105087. [CrossRef]

26. Yang, Y.; Götz, M.; Laqua, A.; Caccia Dominioni, G.; Kawabe, K.; Bengler, K. *A Method to Improve Driver's Situation Awareness in Automated Driving*; HFES Europe Chapter: Groningen, The Netherlands, 2017.

27. National Highway Traffic Safety Administration. *Automated Driving Systems 2.0: A Vision for Safety*; NHTSA: Washington, DC, USA, 2016.

28. Hergeth, S.; Lorenz, L.; Vilimek, R.; Krems, J.F. Keep your scanners peeled: Gaze behavior as a measure of automation trust during highly automated driving. *Hum. Factors* **2016**, *58*, 509–519. [CrossRef] [PubMed]

29. Hergeth, S. Automation Trust in Conditional Automated Driving Systems: Approaches to Operationalization and Design (Doctoral dissertation). 2016. Available online: https://www.qucosa.de/api/qucosa%3A20560/attachment/ATT-0/ (accessed on 15 May 2020).

30. Saffarian, M.; De Winter, J.C.; Happee, R. Automated driving: Human-factors issues and design solutions. *Proc. Hum. Factors Ergon. Soc. Annu. Meet.* **2012**, *56*, 2296–2300. [CrossRef]

31. Cunningham, M.L.; Regan, M.A. Driver distraction and inattention in the realm of automated driving. *IET Intell. Transp. Syst.* **2018**, *12*, 407–413. [CrossRef]

32. Naujoks, F.; Höfling, S.; Purucker, C.; Zeeb, K. From partial and high automation to manual driving: Relationship between non-driving related tasks, drowsiness and take-over performance. *Accid. Anal. Prev.* **2018**, *121*, 28–42. [CrossRef]

33. Sato, T. *Driver Distraction and Inattention in the Realm of Automated Driving*; SIP-Adus Workshop: Tokyo, Japan, 2017.

34. Wandtner, B.; Schömig, N.; Schmidt, G. Effects of non-driving related task modalities on takeover performance in highly automated driving. *Hum. Factors* **2018**, *60*, 870–881. [CrossRef]

35. Kim, A.; Choi, W.; Park, J.; Kim, K.; Lee, U. Interrupting Drivers for Interactions: Predicting Opportune Moments for In-vehicle Proactive Auditory-verbal Tasks. *Proc. ACM Interact. Mob. Wearable Ubiquitous Technol.* **2018**, *2*, 1–28. [CrossRef]

36. Naujoks, F.; Hergeth, S.; Wiedemann, K.; Schömig, N.; Keinath, A. Use cases for assessing, testing, and validating the human machine interface of automated driving systems. *Proc. Hum. Factors Ergon. Soc. Annu. Meet.* **2018**, *62*, 1873–1877. [CrossRef]

37. Gold, C.; Naujoks, F.; Radlmayr, J.; Bellem, H.; Jarosch, O. Testing scenarios for human factors research in level 3 automated vehicles. In *International Conference on Applied Human Factors and Ergonomics*; Springer: Cham, Switzerland, 2017; pp. 551–559.

38. Kraus, J.; Scholz, D.; Stiegemeier, D.; Baumann, M. The more you know: Trust dynamics and calibration in highly automated driving and the effects of take-overs, system malfunction, and system transparency. *Hum. Factors* **2019**. [CrossRef] [PubMed]

39. Bengler, K.; Drüke, J.; Hoffmann, S.; Manstetten, D.; Neukum, A. UR: BAN Human Factors in Traffic. In *Approaches for Safe, Efficient and Stress-free Urban Traffic*; Springer: Wiesbaden, Germany, 2018.

40. Naujoks, F.; Wiedemann, K.; Schömig, N.; Jarosch, O.; Gold, C. Expert-based controllability assessment of control transitions from automated to manual driving. *MethodsX* **2018**, *5*, 579–592. [CrossRef] [PubMed]

41. Brusque, C.; Bruyas, M.P.; Carvalhais, J.; Cozzolino, M.; Gelau, C.H.; Kaufmann, L.; Macku, I.; Pereira, M.; Rehnova, V.; Risser, R.; et al. *Effects of System Information on Drivers' Behaviour*; INRETS Synthesis No. 54; INRETS: Arcueil, France, 2007.

42. SIP-Adus. *SIP-Adus Workshop 2017 Summary Report*; Conference report; SIP-Adus Workshop: Tokyo, Japan, 2017.

 *information*

Article

Sleep Inertia Countermeasures in Automated Driving: A Concept of Cognitive Stimulation

Johanna Wörle [1,*], Ramona Kenntner-Mabiala [1], Barbara Metz [1], Samantha Fritzsch [1], Christian Purucker [1], Dennis Befelein [1] and Andy Prill [2]

[1]   Würzburg Institute for Traffic Sciences; 97209 Veitshöchheim, Germany; kenntner@wivw.de (R.K.-M.); metz@wivw.de (B.M.); fritzsch@wivw.de (S.F.); purucker@wivw.de (C.P.); befelein@wivw.de (D.B.)

[2]   Hyundai Motor Europe Technical Center GmbH; 60547 Rüsselsheim, Germany; amprill@hyundai-europe.com

*   Correspondence: woerle@wivw.de; Tel.: +49-931-78009-124

Received: 26 May 2020; Accepted: 28 June 2020; Published: 30 June 2020

Abstract: When highly automated driving is realized, the role of the driver will change dramatically. Drivers will even be able to sleep during the drive. However, when awaking from sleep, drivers often experience sleep inertia, meaning they are feeling groggy and are impaired in their driving performance—which can be an issue with the concept of dual-mode vehicles that allow both manual and automated driving. Proactive methods to avoid sleep inertia like the widely applied 'NASA nap' are not immediately practicable in automated driving. Therefore, a reactive countermeasure, the sleep inertia counter-procedure for drivers (SICD), has been developed with the aim to activate and motivate the driver as well as to measure the driver's alertness level. The SICD is evaluated in a study with N = 21 drivers in a level highly automation driving simulator. The SICD was able to activate the driver after sleep and was perceived as "assisting" by the drivers. It was not capable of measuring the driver's alertness level. The interpretation of the findings is limited due to a lack of a comparative baseline condition. Future research is needed on direct comparisons of different countermeasures to sleep inertia that are effective and accepted by drivers.

Keywords: highly automated driving; sleep; sleep inertia; HMI design

---

1. Introduction

Highly automated driving systems (ADS) are about to be introduced to the market and they have the potential to change the way we travel fundamentally. Technologies such as Internet of Things, Big Data and Connected Vehicles further promote the progress in the development of ADS [1]. Surveys that were conducted on user requirements with regard to ADS reveal how potential users want to spend the gained time that they do not have to spend on controlling the vehicle. Among the most desired activities are phoning, mailing, interacting with passengers, eating and drinking, watching movies and resting [2]. In a more recent survey in five countries, "sleeping and relaxing" was stated as the preferred way to spend an automated drive [3]. This desire can be explained by the requirements of the modern lifestyle with long working hours and extended time spent on commuting. Some 30% of U.S. employees report sleeping less than six hours per night [4]. Thus, the option to use the commute to work for a nap appears highly promising for drivers.

Those ADS that are currently on the market (such as Tesla Autopilot, General Motors Super Cruise, or Mercedes-Benz Distronic Plus) do not offer the option for the driver to sleep. They are partially automated (level 2 according to the taxonomy of the Society of Automobile Engineers, SAE [5]) and therefore have to be supervised by the driver at all times. However, the ADS technology is advancing fast, and when reaching the level of high automation (SAE level 4), sleep can be implemented as a use case during the automated drive. The concept of "dual mode vehicles" in level 4 automation includes the option for a driver to engage in manual driving or the ability of a driver to take over in situations the system cannot handle [5]. Hence, questions arise on the impact of driver state after sleep. After awakening from sleep, humans experience a "period of transitory hypovigilance, confusion, disorientation of behavior and impaired cognitive and sensory-motor performance" ([6], p. 834), called "sleep inertia".

Sleep inertia is widely recognized and well-regulated in operational domains. In aviation, e.g., where pilots are allowed to take a nap during a flight, standardized nap protocols are in place to avoid performance impairment due to sleep inertia. Pilots on long-haul flights are allowed and even advised to sleep to restore their alertness throughout the flight. In order to avoid performance decrements after sleep, and thus potential safety risks, a procedure called the "NASA nap" is implemented. The NASA nap is a standardized rest period of 40 min with the opportunity to sleep followed by a 20-min period of wakefulness to overcome sleep inertia before returning to duty [7]. The duration of sleep is restricted to avoid deep sleep which produces the highest magnitude of sleep inertia. The NASA nap is recommended with slight differences in various aviation operator guidelines [8,9].

In the AD domain, there are no guidelines and no common understanding on how to deal with a sleeping driver. The first driving simulator studies on human performance after sleep indicate that after sleep, drivers are impaired in their ability to engage in vehicle control and their driving performance is worsened [10,11]. Adverse driver states are a major safety issue in conventional driving to date. In automated driving, the safety impact of an adverse driver state is especially crucial in take-over situations, i.e., in the period after taking back vehicle control from automated to manual driving. EuroNCAP, the European car safety assessment program, introduces reliable driver state monitoring and effective action when an adverse driver state is detected as a primary safety measure [12]. This could mean that, in the case of the driver monitoring system detecting a driver getting too drowsy, it will warn the driver or even initiate a safety manoeuver. Drivers who awaken from sleep experience critical impairments in their take-over and driving performance [10]. The duration of sleep inertia depends on various factors such as the duration of prior sleep or the sleep stage the driver is awakened from [13]. It is thus critical to assess the driver's readiness to engage in vehicle control after sleep to avoid safety-critical situations. The second approach is to actively counter sleep inertia and thus performance impairment and reduce the severity and duration of impaired performance after awakening.

The aim of the paper is to make a proposal of a reactive strategy to deal with sleep inertia in AD and start a discussion on sleep inertia countermeasures in this new field of application.

1.1. Sleep and Sleep Inertia

Sleep is broadly defined as a "reversible behavioral state of perceptual disengagement from and unresponsiveness to the environment" [14] (p. 15). Sleep itself is not a constant state but rather characterized by an alternation of different sleep stages. The sleep stages according to the American Academy of Sleep Medicine (AASM) standard [15] are:

- W: Wakefulness
- N1: Light sleep or dozing
- N2: Stable sleep
- N3: Deep sleep or slow wave sleep
- R: REM-sleep or dream sleep

The transitional phase from sleep to wakefulness is also a distinct state characterized by "hypovigilance, confusion, disorientation of behavior and impaired cognitive and sensory-motor performance" ([6], p. 834), called "sleep inertia". Physiologically, the state of sleep inertia is characterized by a decreased cerebral blood flow [16]. Spectral analyses of the EEG show higher power in the delta-theta and alpha frequency range and a lower power in the beta frequency range which indicates low general alertness [17,18]. Hilditch and McHill [19] suggest that the function of sleep inertia might be for the organism to promote sleep upon awakening so that sleep can be maintained when the awakening is undesired.

In the post-awakening period, performance impairment is evident in a wide range of tasks. Most laboratory studies investigate human performance after sleep with highly standardized tasks: the Psychomotor Vigilance Task (PVT) is widely used in studies on fatigue, but also on sleep inertia [20–22]. The PVT is a standardized test that measures alertness. Subjects have to respond to a visual stimulus as quickly as possible. One of the advantages of the PVT is that it has no learning curve. After sleep, subjects react slower to the stimuli [21,22] and they have more lapses [20]. Other studies assessed the working memory upon awakening: In the n-back task, a subject is presented with a sequence of stimuli and they have to react when the current stimulus matches the one from n steps earlier. Groeger and colleagues [23] applied a 1-, 2- and 3-back task to investigate impairments in working memory on tasks of rising difficulty after 90-min naps. They found stronger performance decrements on tasks which highly rely on executive functions. The Digit Symbol Substitution Test (DSST) assesses working memory and processing speed by presenting digit-symbol pairs followed by a list of digits. Subjects have to assign the correct symbol as fast as possible. The number of correct responses was lower after sleep than before [22,24].

The magnitude and duration of sleep inertia is shaped by many factors. A circadian influence seems apparent with sleep inertia being stronger in the circadian low i.e., during the biological night [21,25,26]. An important factor that influences the magnitude of sleep inertia is the sleep stage prior to awakening. Deep sleep (or slow wave sleep, SWS) produces the highest impairments due to sleep inertia [25,27]. For other sleep stages, results are ambiguous: Cavallero and Versace [28] found a higher impairment of performance on a reaction time task after N2 sleep than after REM sleep. Reaction times were prolonged after N2 sleep compared to N1 sleep [29]. Scheer and colleagues [21] found no differences in the performance of an addition task between subjects awakening from N2, deep sleep or REM sleep. The duration of sleep inertia ranges from 1 min up to 4 h depending on the study design. However, without major sleep deprivation, a duration of more than 30 min is unlikely [13].

1.2. Countering Sleep Inertia

Sleep inertia can be a serious safety issue especially in settings where optimal human performance is crucial under adverse conditions. Such conditions can be extended working hours, working during the circadian trough or traveling through different time zones. In those operational domains, Fatigue Risk Management Systems (FMRS) are in place to avoid safety risks due to impaired alertness, e.g., fatigue or sleep inertia. Different strategies can be distinguished to counter sleep inertia: proactive strategies are commonly recommended in work guidelines that regulate, e.g., sleep schedules or nap durations to avoid sleep inertia. Reactive countermeasures are implemented after awakening when sleep inertia is already present.

Proactive strategies to mitigate sleep inertia are well-established in operator guidelines or shift schedules, e.g., in hospitals or the transportation industry. They include recommendations on sleep schedules, avoid awakening during the circadian low and strategic naps. A common proactive strategy to minimize sleep inertia is the 'NASA nap'. The total sleep duration is restricted to avoid deep sleep, and after awakening, operators have to wait for 20 min to return to duty to overcome sleep inertia. It has to be mentioned, however, that the 40-min rest opportunity of the NASA nap does not fully avoid deep sleep. Even in the original study on the NASA nap, some pilots entered deep sleep within this short period [7]. Since deep sleep occurs in cycles throughout the sleep period, awakening after 80−100 min could be an alternative. After this time the first whole sleep cycle with the first deep sleep period is normally finished [13]. It was found, for instance, that sleep inertia magnitude was greater after a 40 min nap than after a 60 min nap [30]. Ferrara and De Gennaro [6] suggest that awakenings after extended periods of sleep deprivation and during the circadian trough (i.e., during the night) should be avoided. The implementation of proactive strategies is that they require a planned sleep opportunity and a planned wake-up time.

Reactive countermeasures are not commonly implemented and empirical evidence on their effectiveness is incomplete. Examples for applications of reactive countermeasures are light alarms that claim to wake the user more gently and thus minimize sleep inertia. Hilditch, Dorrian and Banks [31] give an overview of the literature on reactive sleep inertia countermeasures. The review includes studies on caffeine, light (postwaking), light (prewaking), sound, temperature, self-awakening and face-washing. Studies included in the review assess the impact of these countermeasures on either subjective alertness or objective alertness (i.e., physiology or performance) or both. One main conclusion of the authors is that there is a gap in the evidence-base of research on sleep inertia countermeasures. Caffeine administered before sleep is suggested as the most effective reactive sleep inertia countermeasure. Empirical evidence on the effectiveness of light or temperature is not sufficient to draw conclusions at this point.

1.3. Implications of Sleep Inertia in Automated Driving

Driving automation has not yet progressed to a level that allows drivers to sleep during the drive. Current ADS require the driver to supervise the ADS at all times. Despite that, videos are making the headlines that show drivers sleeping behind the wheel of their automated vehicles [32,33]. At the current stage, sleep is a clear misuse and has to be avoided at any cost. However, with progress in the development of ADS, the systems will be able to execute all parts of the driving task reliably within the system boundaries. Fully automated driving is not realized throughout all road sections of a trip. That is why at some point of the drive (e.g., at a motorway exit, or when boarding a ferry) the user will be required to execute the driving task manually. The concept of dual-mode vehicles outlined by the SAE [5] explicitly refers to this design option where the user of a highly ADS has the option to request manual driving if she or he wants. The user can therefore switch actively between a user state and a driver state. Users therefore will be allowed to sleep but when they take back the driving task as a driver, it has to be ensured that they are fit to drive after awakening. An exemplary use case could be a business trip where a saleswoman starts her drive with a highly automated vehicle early in the morning. The trip consists of two hours of motorway driving and after leaving the motorway, a rural road leads to her destination. The ADS only supports driving on the motorway, but not on rural roads. The saleswoman uses the motorway section to get some more sleep and is alerted by the ADS before the motorway exit. This way she is able to take back vehicle control before entering the rural road. The ADS ensures that after awakening, the driver's manual driving ability is not impaired due to sleep inertia.

When humans awake from sleep, they experience sleep inertia and are therefore impaired in their ability to drive [10,11]. Sleep inertia as a driver state has barely been an issue in road transport research so far. A study of take-over performance after sleep yielded clearly impaired performance when drivers take control back from the ADS [10]. Drivers' take-over reactions (i.e., glance at the road, glance at the mirrors, hands on the wheel) were all delayed by a few seconds after sleep compared to an awake baseline. It also seemed that the ADS's HMI display was a more important source of information. After sleep, drivers first checked the information on the HMI display before taking over. In contrast to that, the drivers ignored it when taking over after wakefulness. Most importantly, the drivers' overall performance in the take-over situation was worse after sleep. Drivers' lane keeping was clearly impaired and they performed fewer safety glances when changing lanes. Drivers subjectively perceived the take-over situations as more critical after sleep. In another driving simulator study, the focus was on the drivers' manual driving behavior during the first 10 min after sleep [11]. After being awakened by a request to intervene, drivers had to drive manually on a monotonous motorway for about 10 min. Lane-keeping performance was clearly impaired after sleep. This effect was mainly evident in the first two minutes of the manual drive. After sleep, drivers drove at a reduced speed and they had problems keeping to a constant speed. The speed-keeping performance did not improve significantly in the course of the 10-min drive.

The presented findings from previous studies emphasize the necessity for a framework to minimize sleep inertia and associated safety risks in automated driving. Established strategies from other operational domains might only be partly transferable to AD. One approach could be—similar to the NASA nap—to limit rest periods of the driver during the drive, in order to avoid deep sleep. However, this might hardly be acceptable for the driver. If, for example, a two hour period of uninterrupted AD is available, it might not be communicable to the driver that they are only allowed to rest for 40 min. Another strategy could be to awaken the driver early enough to let sleep inertia dissipate before they re-engage in driving. In this case however, the ADS has to ensure that the driver does not go back to sleep in the meantime. Proactive approaches to deal with sleep inertia are usually designed for professionals like pilots or hospital workers that are trained in alertness management. It cannot be expected from regular drivers to stick to such protocols. Therefore, we argue that technical solutions are to be preferred in AD.

Instead of proactive strategies to avoid sleep inertia, reactive strategies that counteract it seem highly promising in AD. While proactive strategies are thoroughly investigated and implemented, e.g., in industry guidelines, there is not much research on strategies to minimize sleep inertia after awakening. Some authors [6] suggest that sleep inertia can be reduced by stimulating or activating the individual after sleep. Everyday strategies like washing one's face with cold water or intensive stretching is not applicable in the vehicle cabin (although one could think of physical activities that could be performed while seated, similar to exercises suggested by some airlines for on-board fitness).

Due to the mainly cognitive requirements of the task of vehicle control, a more promising approach is to cognitively stimulate the driver after sleep. A very popular approach in daily life is a task-based mobile alarm app for smartphones. The basic principle is, that after awaking from sleep, one has to complete a task on their smartphone to ensure that they wake up reliably. Examples for tasks are taking a picture or solving math problems [34]. The activation through cognitive stimulation can promote cerebral activity on the one hand and be motivating because of its playful character on the other. However, there is barely any empirical evidence for the effectiveness of such task-based alarms.

Besides cognitive stimulation and motivation of the driver, our concept had a third aim: similar to established measurements of alertness such as the PVT, we aimed at assessing the alertness level of the driver after waking to assess the driver's fitness to drive.

A prototype sleep inertia countermeasure was developed and is tested in a driving simulator setup to evaluate its effectiveness in terms of:

- cognitively and physiologically activating the driver after sleep.
- motivating the driver after sleep.
- assessing the driver's alertness level to determine their readiness to drive.

2. Materials and Methods

2.1. The Concept of the Sleep-Inertia Counter-Procedure for Drivers

Two expert workshops were conducted with the aim to work out prototype wake-up concepts and a framework for a concept to counter sleep inertia in automated driving. N = 8 and N = 9 experts with backgrounds in human factors, traffic psychology and HMI design participated in the workshops. After several iterations, a concept for a sleep inertia counter-procedure for drivers (SICD) was developed and implemented as a tablet application in the driving simulation.

The basic idea of the SICD was to minimize sleep inertia by activating the driver after sleep as it is suggested by [6]. Suggestions such as washing one's face with cold water or physical exercise were discussed but rejected since they were not practicable in a vehicle cabin. Other reliable methods to counter sleep inertia such as caffeine administration [20] were rejected because they were judged to be too intrusive. One approach that was assessed to be feasible as an HMI solution was to cognitively activate the driver with a challenging task. Another advantage of a cognitive task is that it can also be used as a diagnostic tool to measure the alertness level of drivers after sleep similar to, e.g., the PVT. The SICD was designed with a gamification approach so that it was perceived as motivating and created a positive feeling.

The SICD was implemented as a gaming application on a tablet similar to a classical choice-reaction task, see Figure 1. Purple and turquoise dots appeared on a tablet screen every 1–2 s at random positions of the play area. Both the position of appearance and the time point were defined by random number generators. Frequency and position varied to avoid predictability and thus repetitive behavior and boredom. Drivers had to hit all target stimuli (purple dots) and avoid distractor stimuli (turquoise dots). To promote drivers' motivation during the task they received motivating messages such as "You are doing great". The duration of the SICD was 10 min.

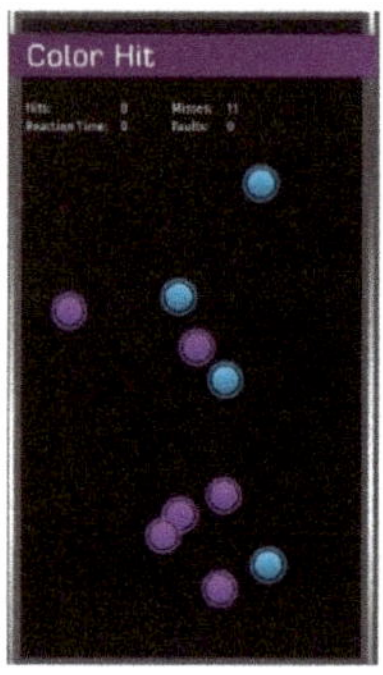

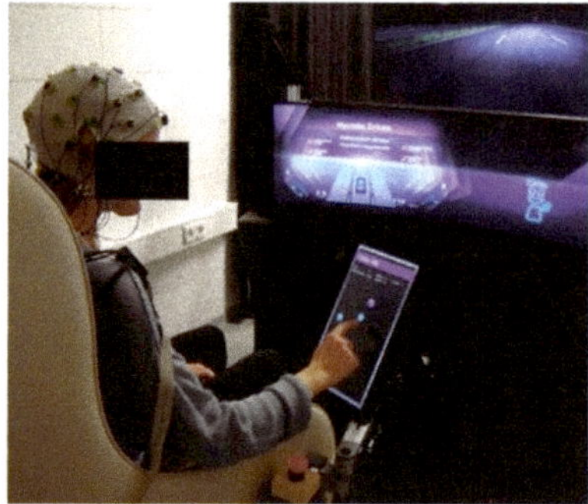

Figure 1. Screenshot of the sleep inertia counter-procedure for drivers (SICD, **left**) and participant equipped with EEG electrodes performing the SICD (**right**).

If participants hit a purple target dot, the counter registered a "hit". For all hits, reaction times were recorded. If participants hit a turquoise dot, a "fail" was counted. If a target dot was not hit, a "miss" was counted. Reaction times were calculated starting with the appearance of the dot on the screen until it was hit. The four parameters hits, fails, misses and reaction times were supposed to serve as measures for alertness, similar to established alertness measurements, such as the PVT.

2.2. An L4 Concept Driving Simulator to Investigate Sleep

The evaluation study was conducted in a driving simulator using the simulation software SILAB. The simulator was specifically designed for evaluating HMI concepts for automated driving. The main components were a dashboard with a steering wheel and a large diagonal display. Accelerator and brake pedal were available in manual driving mode. The driver was seated in a comfortable seat with a central infotainment touch display. All relevant components of the driving simulator were equipped with electric linear actuators and could be controlled via a computer. It was possible to move the seat to a lying position and to retract the steering wheel and pedals so that the driver had more space. Therefore, the cabin concept in the "manual driving mode" was different with the seat in the upright position and steering wheel and pedals extended while in the "automated driving mode" the seat was moved backwards and steering wheel and pedals were retracted and in "sleep mode" the seat was moved to a horizontal position.

Two different wake-up procedures were developed and implemented. The first wake-up procedure focused on a reliable awakening with a loud and sharp sound and flashing lights. The second wake-up procedure focused on a comfortable and pleasant wake up with soft music and a warm yellow light concept. The two wake-up procedures were tested between-subjects in the driving simulator study. However, no results will be presented on the acceptance and effectiveness of the wake-up procedures since this is outside the scope of this paper.

2.3. Study Design

The study was conducted at the premises of the Würzburg Institute for Traffic Sciences (WIVW). N = 21 test participants (10 female, mean$_{age}$ = 33, sd = 8) completed two driving sessions in an L4 driving simulator using a highly automated driving system. All participants were recruited from the WIVW driver panel. Session 1 was scheduled during the daytime and session 2 was scheduled at 6 a.m. after a night of partial sleep deprivation, i.e., drivers were allowed to sleep no more than 4 h. The aim was to get the drivers to fall asleep in the driving simulator. Each session started with a prequestionnaire and ended with a postquestionnaire.

In session 1, drivers first gave their informed consent, filled in the prequestionnaire and were then familiarized with the driving simulator, i.e., they learned the system handling and drove manually for 10 min. Then they practiced the SICD. After the familiarization, the test drive started. For a graphical representation of the test drive, see Figure 2. The test drive started on a parking lot and drivers entered the highway. On the highway they activated the automated mode, then the vehicle drove automatically and the vehicle cabin also changed: the steering wheel folded back, the wide screen moved in near to the participant and the driving seat moved backwards and tilted back slightly so that the driver was in a more comfortable position. Then, the system offered the sleep mode that the driver confirmed with a button press. The screen turned darker and the driver's seat tilted to a lying position. Then, drivers were instructed to close their eyes and relax but not to sleep. After two minutes, drivers were alerted with either of two "wake-up procedures". Then they were asked to rate their subjective arousal and their subjective well-being on a slightly adapted version of the 9-point Self-Assessment Manikin (SAM) scale [35]. The SAM-scale is a "non-verbal pictorial assessment technique" ([35], p. 49). For our purposes, the valence scale showed five manikins displaying a scale ranging from an unhappy face expression to a happy face expression. Participants were asked "How good do you feel?" On the arousal scale the manikins ranged from a relaxed looking manikin with closed eyes to a very active manikin. Participants were asked "How activated are you?"

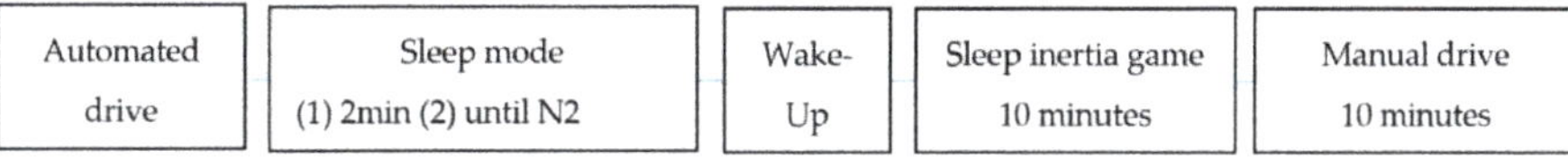

Figure 2. Schematic sequence of the test drive.

After the rating, the SICD was offered on the tablet screen and was started by the drivers via button press. The SICD was executed for 10 min and when it was finished, drivers were asked again to rate their well-being and their arousal. Then they performed a 10-min manual drive on a 3-lane highway with low traffic volume. After the manual drive, drivers rated their subjective well-being and arousal.

Session 2 had a similar procedure with the only differences that it took place at 6 a.m. and drivers were sleep deprived. Participants arrived at the test facilities by taxi and after filling in the pre-questionnaire, they were equipped with the EEG. Electrodes were placed according to the International 10–20 system [36]. The procedure of the test drive was the same as in session 1, but drivers were awakened when a sleep expert confirmed sleep stage N2 via EEG evaluation. Sleep stage N2 was chosen since it is the "deepest" stage that is considered appropriate during a nap in most operational guidelines. After awakening, drivers engaged in the SICD for 10 min and then drove manually for 10 min. Then, the AD was available again and drivers tried to sleep again. If sleep stage N2 was confirmed a second time, drivers were awakened and the procedure with first the SICD and then the manual drive was triggered again. During both driving sessions, heart rate was measured with the Polar T34 chest belt as a measure for physiological activation.

Alertness is either measured with self-report measures such as a visual analogue scale or the Karolinska Sleepiness Scale, measures of cognitive performance—of which the PVT is arguably the most common—or physiologic measures [37]. All of them have a high inter-correlation and it is advised to use a combination of different measures. We therefore chose to use a combination of different alertness measures with the SAM-Scale as a self-report measure, the cardiac parameters as physiologic measures and the performance parameters of the SICD as measures of cognitive performance.

2.4. Data Analysis

For all indicators of arousal, repeated measures ANOVAs were conducted with the factors state and time. For state, three manifestations of the driver state were compared: after wakefulness (wakefulness, session 1), after drivers were asleep for the first time (after Sleep 1) and after drivers were asleep for the second time (after Sleep 2, both session 2). Furthermore, a factor time was analyzed which showed the change of the indicators over time.

For the subjective state, changes in arousal and well-being were compared for three points in time, directly after being awakened, after the SICD and after the manual drive. Mean and standard deviation of heart rate were analyzed as objective indicators for physiological arousal. Starting from the beginning of the SICD, parameters for heart rate were calculated for time segments of one minute duration, starting two minutes prior to the start of the SICD, including the 10 min of the SICD and 8 min of successive manual driving. In a similar approach, indicators measuring the performance in the SICD were calculated for segments with one minute of duration and analyzed over time. The proportion of hits and reaction times were calculated.

3. Results

3.1. Subjective Arousal and Well-Being

For the subjective arousal, there was a significant main effect of driver state [$F(2, 36) = 9.898$, $p < 0.000$], a significant main effect of time [$F(2, 36) = 17.069$, $p < 0.000$] and a significant interaction effect driver state*time [$F(4, 72) = 6.499$, $p < 0.000$]. Tukey post hoc test revealed that before the SICD (time point wake-up), the arousal after wakefulness was higher than after sleep. During the SICD, arousal increased for all three states. After the SICD, the differences between wakefulness and after sleep were no longer significant and were reduced further until after the drive. The development of the subjective arousal over time did not differ between after Sleep 1 and after Sleep 2.

For the subjective well-being, there was a main effect of driver state [$F(2, 36) = 9.537$, $p < 0.000$], no effect of time [$F(2, 36) = 0.159$, $p = 0.853$] and an interaction effect driver state*time [$F(4, 7) = 3.719$, $p = 0.008$]. As can be seen in Figure 3, the only effect that could be interpreted is the interaction. After wake-up, subjective well-being was significantly lower after sleep than after wakefulness. Then, after sleep, there was a slight increase of well-being over time. On the contrary, subjective well-being decreased in the awake baseline condition. After the drive, subjective well-being after sleep and after wakefulness were on a similar level. Again, there was no difference in subjective state between Sleep 1 and Sleep 2.

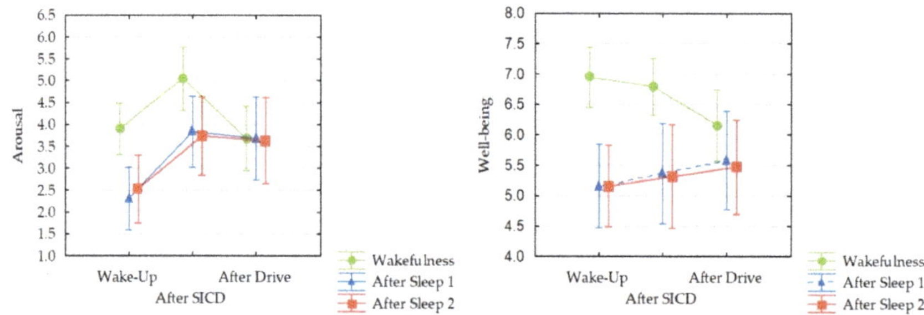

Figure 3. Subjective arousal (**left**) and well-being (**right**) after wake-up, after the SICD and after the manual drive for drivers after wakefulness and twice after sleep. The graph shows means and 95% confidence intervals.

3.2. Physiological Activation

For the mean heartrate, there was a significant effect of time [$F(19, 361) = 20.7$, $p < 0.001$] and a signficant interaction effect [$F(38, 722) = 1.6177$, $p = 0.012$]. For all states, there was an increase in heart rate with the beginning of the SICD which was more pronounced after sleep. During the SICD, the mean heart rate stayed on a constant level. After wakefulness, the increase of mean heart rate during the SICD was followed by a decrease during the manual drive. This decrease could not be found after sleep. For the standard deviation of heart rate, there was a significant effect of time [$F(19,361) = 31.22$, $p < 0.001$], of state [$F(2, 38) = 4.42$, $p = 0.019$] and a significant interaction [$F(38, 722) = 10.505$, $p < 0.000$]. All effects were based on a strong increase of heart rate variability during the process of waking up and starting with SICD after sleep. After wakefulness, heart rate variability stayed on a constant level throughout the analysed time frame. Means and standard deviations of heart rate during the SICD as well as during the manual drive are shown in Figure 4.

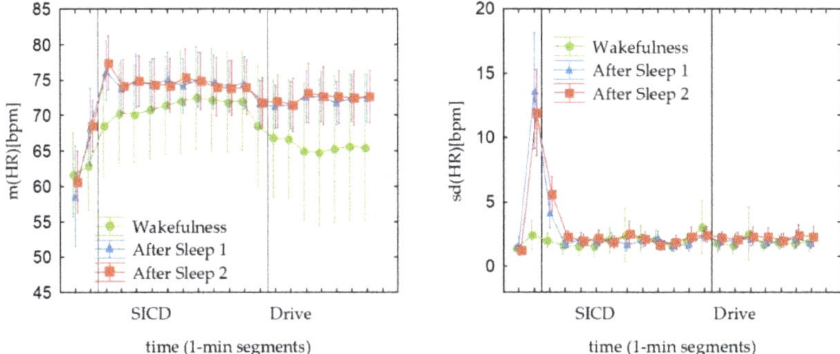

Figure 4. Mean (**left**) and standard deviation (**right**) of heart rate in time segments of 1 min before the SICD, during the SICD and during the successive manual drive.

Figure 5 shows an example of one driver's heart rate in the course of the drive. It illustrated how the heart rate was low during the automated drive due to a low arousal level of the driver. When the driver was awakened by the ADS, there was a sharp increase in heart rate and an overall higher level during the SICD and the manual drive. The arousal lowered as soon as the ADS was activated again.

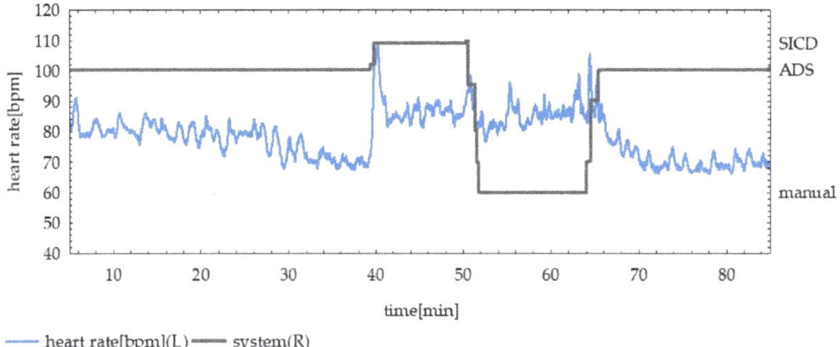

Figure 5. Example for change of heart rate for automated drive (ADS), during the SICD and during the manual drive and back to ADS. The driver was asleep during the automated drive (ADS).

3.3. Subjective Evaluation of the SICD

For assessing the acceptance of the SICD by users, the drivers were asked to rate the SICD on the 9-point acceptance scale after the drive in the second session. One sample *t*-tests were calculated against the scale mean (0). The SICD was perceived as assisting (M = −0.57, SD = 1.07, p = 0.024) and marginally as good (M = 0.38, SD = 0.8, p = 0.057). All other scales did not differ from the scale mean. Drivers' acceptance of the SICD is depicted in Figure 6.

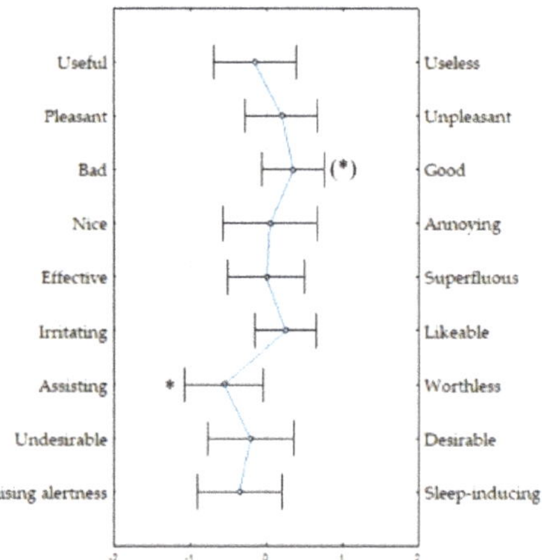

Figure 6. Means and 95% confidence intervals for the acceptance of the SICD. Significant differences from the scale mean are marked with *, nearly significant differences with (*).

In the postquestionnaire, drivers were asked to judge their driving behaviour after sleep with an open question. Some answers revealed information about the evaluation of the SICD. Those answers were, e.g., "Directly at the beginning, I felt more awake, but this effect quickly changed after the game was finished", "When I first played the game, it was refreshing and activating, but when I had to play it again, it was rather sleep-inducing.", "The game was sleep-inducing, because it is too long and not varied enough", "At the beginning of the game, it is arousing and raises alertness, but after a while it becomes annoying and monotonous".

3.4. Performance on the SICD

For the parameters mean reaction time and standard deviation of reaction time there were no significant effects. For the proportion of hits there was a significant effect of time [$F(9, 162)$ = 2.1521, p = 0.028], which was based on an increase during the first three minutes after the start of SICD. Figure 7 shows two of the performance parameters, as examples, of the SICD, the percentage of hits and the mean reaction time in the course of 10 min.

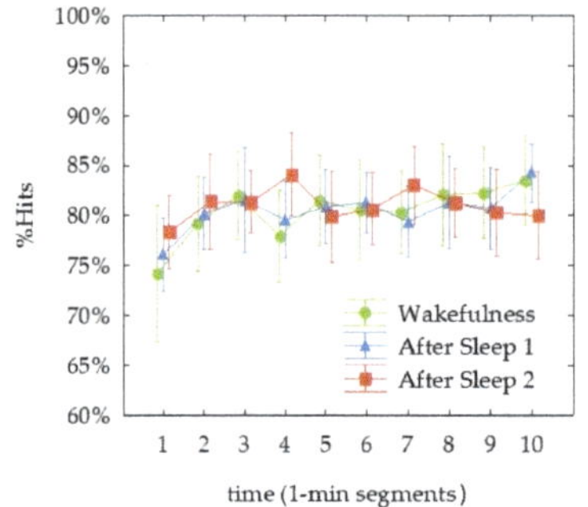

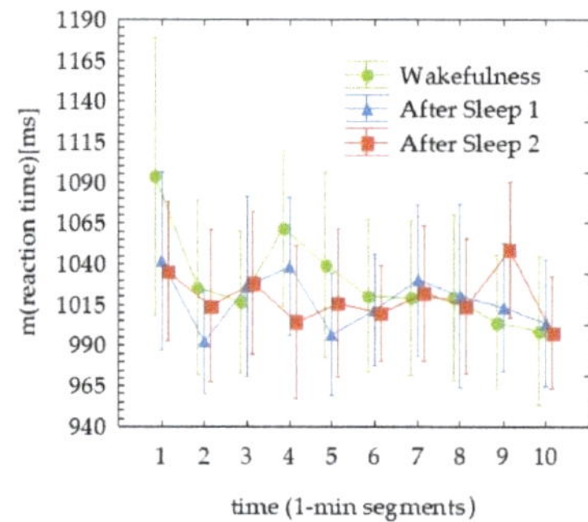

Figure 7. Proportion of hits of the SICD (**left**) and mean reaction times (**right**) for drivers after wakefulness, and twice after sleep.

4. Discussion

Sleeping drivers in automated vehicles are already an issue today at a level of automation where the driver is clearly required to stay alert [30,31]. With the progression of automated driving technology and the development of dual-mode vehicles, sleep will arise as a use case and thus as a new driver state to be considered, e.g., in safety research and vehicle design. After sleep, human performance is impaired due to sleep inertia [6]. First studies on sleep in automated driving show that performance impairments are evident after awakening. There are more errors in take-over performance and manual driving performance is impaired [10,11]. The aim of the presented study was to develop a first implementation of a countermeasure to sleep inertia for drivers who are awakened during an automated drive. A sleep inertia counter-procedure for drivers (SICD) was developed. The purpose of the SICD was threefold: First, to activate the driver after awakening, second, to improve the driver's mood and motivation and third, to measure the driver's alertness.

The effectiveness and acceptance of the SICD was evaluated with N = 21 drivers who completed the SICD (a) after wakefulness and (b) after sleep. In both sessions we assessed physiological activation, subjective arousal and well-being, subjective evaluation of the SICD, as well as performance on the SICD.

4.1. Activation of the Driver

The SICD was designed especially to be activating. Drivers had to react quickly to achieve as many hits as possible and had to avoid distractor cues. Drivers report a higher subjective arousal after the SICD than before. This was the case after drivers had slept but also when they were awake. The activating effect of the SICD lasted until the end of the successive manual drive for drivers who had slept before but not for drivers who had been awake. The subjective arousal is also reflected by the physiological activation of the drivers. Drivers' activation, measured by the mean heart rate, is higher during the SICD than during the rather monotonous manual drive for drivers who had not slept. However, when drivers were awakened from sleep, the awakening process was highly activating and the heart rate was on a rather high level throughout the SICD but also during the manual drive. It seems that when drivers were asleep prior to the SICD, the SICD was able to physiologically activate the driver. This effect even lasted until after the SICD was finished and drivers drove manually. When drivers had not slept before, the SICD had a similar activating effect. However, after being awake, this effect did not last after the SICD was finished. The effects found for subjective arousal and for mean heart rate are similar: There was an increase of activation during the SICD which was followed again by a decrease during the manual drive after wakefulness. After sleep, the arousal level reached during SICD remained quite stable during the manual drive. Both the subjective and the objective activation effects after sleep were very stable and occurred in a similar way after both awakenings during the drive. Therefore, the aim of activating the driver with the SICD can be confirmed.

4.2. Driver Mood and Attitude towards the SICD

The drivers' subjective well-being was generally higher in the "awake" condition than after sleep which can be explained by the affective component of sleep inertia which is described as "grogginess". The SICD did not improve the drivers' well-being after sleep. The subjective evaluation of the SICD by drivers on the acceptance scale [38] was neutral. However, the drivers perceived the SICD as "assisting". The drivers stated that the SICD was too long and monotonous. The SICD could thus be improved by shortening the duration or by adding features that help to reduce monotony.

4.3. Measuring Driver Alertness with the SICD

One basic idea of the SICD was to use it as a diagnostic tool that reveals whether the driver's alertness has improved enough to consider her or him "ready to drive". Therefore, similar to classical measures of alertness like the PVT, performance parameters "hits", "misses", "fails" and "reaction times" were measured. Unfortunately, there is no effect of the drivers' state on the parameters of the SICD. In the current implementation it can therefore not be considered a valid diagnostic tool to assess drivers' readiness to drive. A simpler task is indicated where no learning effects can be expected. However, this goal might be challenging to be combined with the aim to implement a less monotonous task.

4.4. Conclusions

While the SICD proved to be subjectively and physiologically activating, the trade-off between motivating appeal and diagnostic capacity—similar to classic alertness tasks—turned out to be the essential challenge. Standardized and validated tasks like the PVT have the advantage that, due to the simplicity of the design, there are no learning effects and the subject's alertness can directly be derived from the performance parameters. However, this task could clearly not be considered motivating and rather, was annoying to the subject. We therefore tried to design our SICD such that it was more varied and added motivating messages. The drivers considered the SICD to be of assistance and the subjective arousal scales as well as the heart rate show that it was also activating. On the other hand, it was not capable of measuring alertness and thus its diagnostic properties could not be confirmed. The SICD was accepted by drivers and at least did not worsen the driver's well-being. On the other hand, driving behavior is impaired [11] and therefore, before handing the vehicle control over to the driver, some kind of performance check of the driver is indicated. The SICD did not reveal information about the driver's alertness. In summary, the SICD was able to fulfill two of the aims: to activate and to motivate the driver. The third aim, to measure the driver's alertness, was not accomplished.

4.5. Limitations

The main limitation for the interpretation of the results is the chosen study design. The two experimental conditions compared the full SICD in the state of sleep inertia to wakefulness as a baseline condition. However, to draw clear conclusions about the effectiveness of the SICD, the treatment (SICD) should be compared to a baseline (no SICD). The obtained physiological, subjective and performance data can only be evaluated in a timely perspective. The variance in the data might not only depend on the SICD but also on time effects and consequently a clear interpretation of the data is difficult. Future studies should directly compare different approaches to deal with sleep inertia. One approach could be to test the presented cognitive stimulation approach to the NASA nap paradigm, thus a reactive approach to a proactive approach. As other approaches, physical exercise or a combined physical-cognitive activation task (e.g., a cue-search task incorporating the whole vehicle cockpit) could be introduced. Despite the limitations posed by the study design, we can conclude that the approach of cognitive stimulation is a promising framework for activating the driver after sleep. However, direct comparison of different approaches is indicated to assess their effectiveness.

4.6. Directions for Future Research

The approach of cognitive stimulation proved to be effective to physiologically and subjectively activate the driver. However, the SICD was not capable of assessing the driver's readiness to drive. This could either be done by a sophisticated driver monitoring system that detects the driver state [37] or by a performance check as it was conceptualized in the SICD. The driver's performance capabilities are clearly reduced after sleep [9]. It is crucial for driving safety to detect the driver's readiness to drive [12]. If the driver is detected as being not ready to drive, appropriate actions have to be taken. When the driver of a dual-mode vehicle is awakened from sleep, the ADS has to ensure that the driver is cognitively alert to engage in vehicle control. The SICD was developed similar to the PVT, as a validated measure of alertness. The difference was only that it was conceptualized not as a single reaction task but as a choice-reaction task to make it more varied and therefore more appealing and motivating. Measuring alertness with cognitive tasks is an established approach, however, our task was not able to measure alertness reliably. Future task designs should be more similar to established tasks, e.g., simpler single-choice tasks where no learning effects can be expected.

Another promising approach is physiological activation instead or in combination with cognitive activation [39]. Cerebral blood flow is decreased in the sleep inertia period which delays the reinstatement of alertness. Physical exercises have the potential to increase the overall blood circulation and therefore counter physiological sleep inertia. The implementation of physical exercises in the vehicle cabin is restricted. However, it is imaginable to instruct the driver to do stretching exercises. Physiological activation on the other hand, does not ensure that vehicle control—a primarily cognitive task—can be safely executed. Therefore, a combination of physiological and cognitive stimulation seems promising. Future research is needed in order to compare the effectiveness of different SICD approaches and to develop a method that is capable of successfully activating the driver, of measuring alertness and is motivating at the same time.

It is clearly critical to establish a framework to avoid sleep inertia from becoming a safety issue in automated driving. In other operational areas, e.g., aviation, standardized guidelines are in place to avoid sleep inertia. In automated driving, there is no such framework. Our proposed approach of cognitive stimulation has the potential to activate the driver. However, a sleep inertia countermeasure can only be considered effective when the driver's alertness and thus readiness to drive can be determined reliably.

Author Contributions: Conceptualization, J.W., R.K.-M.; C.P., D.B., S.F. and A.P.; methodology, J.W. and R.K.-M.; software, S.F.; investigation, J.W. and R.K.-M.; resources, A.P.; data curation, R.K.-M., J.W. and B.M.; writing—original draft preparation, J.W.; writing—review and editing, B.M., C.P. and R.K.-M.; project management, C.P. and A.P.; project acquisition: C.P. and A.P.; funding acquisition, A.P. All authors have read and agreed to the published version of the manuscript.

Funding: This research was funded by Hyundai Motor Europe Technical Center GmbH.

Acknowledgments: The authors would like to thank the co-workers of the WIVW who supported the research team with providing a fruitful and creative research environment and with conceptualizing and building the driving simulator hardware and the SILAB driving simulation implementation used in the study: Alexandra Neukum, Mathias Gold, Michael Hanig, Stefan Ludwig, and Markus Tomzig.

Conflicts of Interest: The authors declare no conflict of interest.

References

1. Azmat, M.; Kummer, S.; Moura, L.T.; Gennaro, F.D.; Moser, R. Future Outlook of Highway Operations with Implementation of Innovative Technologies Like AV, CV, IoT and Big Data. *Logistics* **2019**, *3*, 15. [CrossRef]

2. Kyriakidis, M.; Happee, R.; de Winter, J.C. Public opinion on automated driving: Results of an international questionnaire among 5000 respondents. *Transp. Res. Part F Traffic Psychol. Behav.* **2015**, *32*, 127–140. [CrossRef]

3. Becker, T.; Herrmann, F.; Duwe, D.; Stegmüller, S.; Röckle, F.; Niko, U. *Enabling the Value of Time*; Fraunhofer Institute for Industrial Engineering IAO: Stuttgart, Germany, 2018; pp. 1–27.

4. Upender, R.P. Sleep Medicine, Public Policy, and Public Health. In *Principles and Practices of Sleep Medicine*; Kryger, M., Roth, T., Eds.; Elsevier: Philadelphia, PA, USA, 2017; Volume 6, pp. 638–645.

5. SAE. *Taxonomy and Definitions for Terms Related to Driving Automation Systems for On-Road Motor Vehicles*; SAE: Warrendale, PA, USA, 2018; Volume J3016.

6. Ferrara, M.; De Gennaro, L. The sleep inertia phenomenon during the sleep-wake transition: Theoretical and operational issues. *Aviat. SpaceEnviron. Med.* **2000**, *71*, 843–848.

7. Rosekind, M.R.; Smith, R.M.; Miller, D.L.; Co, E.L.; Gregory, K.B.; Webbon, L.L.; Gander, P.H.; Lebacqz, J.V. Alertness management: Strategic naps in operational settings. *J. Sleep Res.* **1995**, *4*, 62–66. [CrossRef]

8. EASA. Commission Regulation (EU) 965/2012 on air operations. Amendment 16. In *Acceptable Means of Compliance (AMC) and Guidance Material (GM) to Annex IV: Commercial Air Transport Operations [Part-CAT]*; EASA: Cologne, Germany, 2019.

9. CASA. Safety Behaviours. In *Human Factors Resource Guide for Engineers*; Civil Aviation Safety Authority: Canberra, Australia, 2013.

10. Wörle, J.; Metz, B.; Othersen, I.; Baumann, M. Sleep in Highly Automated Driving: Take-over Performance after Waking Up. *Accid. Anal. Prev.* **2020**, *144*. [CrossRef]

11. Wörle, J.; Metz, B.; Baumann, M. Investigating sleep inertia in automated driving: Methodological considerations and results from a driving simulator study. *Accid. Anal. Prev.* **2020**. Under review.

12. EuroNCAP. *Euro NCAP 2025 Roadmap*; EuroNCAP: Leuven, Belgium, 2017; pp. 1–17.

13. Tassi, P.; Muzet, A. Sleep inertia. *Sleep Med. Rev.* **2000**, *4*, 341–353. [CrossRef]

14. Carskadon, M.A.; Dement, W.C. Normal human sleep: An overview. In *Principles and Practice of Sleep Medicine*, 6th ed.; Kryger, M., Roth, T., Eds.; Elsevier: Philadelphia, PA, USA, 2017; Volume 4, pp. 15–24.

15. AASM. *The AASM Manual for the Scoring of Sleep and Associated Events: Rules, Terminology and Technical Specifications*; American Academy of Sleep Medicine: Darien, IL, USA, 2017.

16. Balkin, T.J.; Braun, A.R.; Wesensten, N.J.; Jeffries, K.; Varga, M.; Baldwin, P.; Belenky, G.; Herscovitch, P. The process of awakening: A PET study of regional brain activity patterns mediating the re-establishment of alertness and consciousness. *Brain* **2002**, *125*, 2308–2319. [CrossRef]

17. Ferrara, M.; Curcio, G.; Fratello, F.; Moroni, F.; Marzano, C.; Pellicciari, M.C.; De Gennaro, L. The electroencephalographic substratum of the awakening. *Behav. Brain Res.* **2006**, *167*, 237–244. [CrossRef]

18. Marzano, C.; Ferrara, M.; Moroni, F.; De Gennaro, L. Electroencephalographic sleep inertia of the awakening brain. *Neuroscience* **2011**, *176*, 308–317. [CrossRef]

19. Hilditch, C.J.; McHill, A.W. Sleep inertia: Current insights. *Nat. Sci. Sleep* **2019**, *11*, 155–165. [CrossRef] [PubMed]

20. Van Dongen, H.P.; Price, N.J.; Mullington, J.M.; Szuba, M.P.; Kapoor, S.C.; Dinges, D.F. Caffeine eliminates psychomotor vigilance deficits from sleep inertia. *Sleep* **2001**, *24*, 813–819. [CrossRef] [PubMed]

21. Scheer, F.A.; Shea, T.J.; Hilton, M.F.; Shea, S.A. An endogenous circadian rhythm in sleep inertia results in greatest cognitive impairment upon awakening during the biological night. *J. Biol. Rhythm.* **2008**, *23*, 353–361. [CrossRef] [PubMed]

22. Hilditch, C.J.; Centofanti, S.A.; Dorrian, J.; Banks, S. A 30-minute, but not a 10-minute nighttime nap is associated with sleep inertia. *Sleep* **2016**, *39*, 675–685. [CrossRef]

23. Groeger, J.A.; Lo, J.C.; Burns, C.G.; Dijk, D.-J. Effects of sleep inertia after daytime naps vary with executive load and time of day. *Behav. Neurosci.* **2011**, *125*, 252. [CrossRef] [PubMed]

24. McHill, A.W.; Hull, J.T.; Cohen, D.A.; Wang, W.; Czeisler, C.A.; Klerman, E.B. Chronic sleep restriction greatly magnifies performance decrements immediately after awakening. *Sleep* **2019**, *42*, zsz032. [CrossRef]

25. Dinges, D.F.; Orne, M.T.; Orne, E.C. Assessing performance upon abrupt awakening from naps during quasi-continuous operations. *Behav. Res. MethodsInstrum. Comput.* **1985**, *17*, 37–45. [CrossRef]

26. Silva, E.J.; Duffy, J.F. Sleep inertia varies with circadian phase and sleep stage in older adults. *Behav. Neurosci.* **2008**, *122*, 928. [CrossRef]

27. Bruck, D.; Pisani, D.L. The effects of sleep inertia on decision-making performance. *J. Sleep Res.* **1999**, *8*, 95–103. [CrossRef]

28. Cavallero, C.; Versace, F. Stage at awakening, sleep inertia and performance. *Sleep Res. Online* **2003**, *5*, 89–97.

29. Hayashi, M.; Motoyoshi, N.; Hori, T. Recuperative power of a short daytime nap with or without stage 2 sleep. *Sleep* **2005**, *28*, 829–836.

30. Signal, T.L.; Gander, P.; van den Berg, M.; O'Keeffe, K. *Magnitude and Time Course of Sleep Inertia*; Sleep/Wake Research Center: Wellington, New Zealand, 2008.

31. Hilditch, C.J.; Dorrian, J.; Banks, S. Time to wake up: Reactive countermeasures to sleep inertia. *Ind. Health* **2016**, *54*, 528–541. [CrossRef] [PubMed]

32. Solon, O. Who's driving? Autonomous cars may be entering the most dangerous phase. *Guardian* **2018**, *9*, 2019.

33. Guarino, B. *Man Appears to Snooze at the Wheel of His Tesla while the Car Drives Itself on L.A. Highway*; The Washington Post: Washington, DC, USA, 2016.

34. Oh, K.T.; Shin, J.; Kim, J.; Ko, M. Analysis of a Wake-Up Task-Based Mobile Alarm App. *Appl. Sci.* **2020**, *10*, 3993. [CrossRef]

35. Bradley, M.M.; Lang, P.J. Measuring emotion: The self-assessment manikin and the semantic differential. *J. Behav. Ther. Exp. Psychiatry* **1994**, *25*, 49–59. [CrossRef]

36. Jasper, H. Report of the committee on methods of clinical examination in electroencephalography. *Electroencephalogr. Clin. Neurophysiol.* **1958**, *10*, 370–375.

37. Gabehart, R.J.; van Dongen, H.P. Circadian Rhythms in Sleepiness, Alertness, and Performance. In *Principles and Practice of Sleep Medicine*; Kryger, M.H., Roth, T., Eds.; Elsevier: Philadelphia, PA, USA, 2017; Volume 5, pp. 388–394.
38. Van Der Laan, J.D.; Heino, A.; De Waard, D. A simple procedure for the assessment of acceptance of advanced transport telematics. *Transp. Res. Part C Emerg. Technol.* **1997**, *5*, 1–10. [CrossRef]
39. Kovac, K.; Ferguson, S.A.; Paterson, J.L.; Aisbett, B.; Hilditch, C.J.; Reynolds, A.C.; Vincent, G.E. Exercising caution upon waking–can exercise reduce sleep inertia? *Front. Physiol.* **2020**, *11*, 254. [CrossRef] [PubMed]

Article

Methodological Approach towards Evaluating the Effects of Non-Driving Related Tasks during Partially Automated Driving

Cornelia Hollander [1,*], Nadine Rauh [1], Frederik Naujoks [2], Sebastian Hergeth [2], Josef F. Krems [1] and Andreas Keinath [2]

[1] Research Group Cognitive and Engineering Psychology, Chemnitz University of Technology, 09120 Chemnitz, Germany; Nadine.Rauh@psychologie.tu-chemnitz.de (N.R.); Josef.Krems@psychologie.tu-chemnitz.de (J.F.K.)

[2] BMW Group, 80937 Munich, Germany; Frederik.Naujoks@bmw.de (F.N.); Sebastian.Hergeth@bmw.de (S.H.); Andreas.Keinath@bmw.de (A.K.)

* Correspondence: cornelia.hollander@psychologie.tu-chemnitz.de; Tel.: +49-371-531-32044

Received: 27 May 2020; Accepted: 26 June 2020; Published: 30 June 2020

Abstract: Partially automated driving (PAD, Society of Automotive Engineers (SAE) level 2) features provide steering and brake/acceleration support, while the driver must constantly supervise the support feature and intervene if needed to maintain safety. PAD could potentially increase comfort, road safety, and traffic efficiency. As during manual driving, users might engage in non-driving related tasks (NDRTs). However, studies systematically examining NDRT execution during PAD are rare and most importantly, no established methodologies to systematically evaluate driver distraction during PAD currently exist. The current project's goal was to take the initial steps towards developing a test protocol for systematically evaluating NDRT's effects during PAD. The methodologies used for manual driving were extended to PAD. Two generic take-over situations addressing system limits of a given PAD regarding longitudinal and lateral control were implemented to evaluate drivers' supervisory and take-over capabilities while engaging in different NDRTs (e.g., manual radio tuning task). The test protocol was evaluated and refined across the three studies (two simulator and one test track). The results indicate that the methodology could sensitively detect differences between the NDRTs' influences on drivers' take-over and especially supervisory capabilities. Recommendations were formulated regarding the test protocol's use in future studies examining the effects of NDRTs during PAD.

Keywords: partially automated driving; non-driving related tasks; take-over situations; test protocol development; user studies (simulator; closed circuit)

---

1. Introduction

1.1. Theoretical Background

In recent years, researchers and practitioners alike have been increasingly motivated to enhance driving assistance and automation resulting in different vehicle automation levels [1], with the overarching goal to improve driving comfort, traffic safety and to reduce traffic congestion [2].

The Society of Automotive Engineers (SAE) [3] defines six automation levels ranging from no automation (level 0) to full automation (level 5). However, only SAE level 1 and level 2 systems are currently available to consumers. Of those, partially automated driving (PAD, SAE level 2) provides continuous steering as well as brake and acceleration support to the driver; however, the driver must constantly supervise these support features and be prepared to steer, brake, or accelerate as needed to maintain safety [3].

Even though the driver is partially relieved from the driving task and its demands [4], this does not automatically improve driving safety. New problems may arise during PAD when changing the drivers' role from an active to a passive system supervisor [5,6]. For instance, the (partial) relief of driving demands and decreased driving task engagement during automated driving can reduce the drivers' workload and cause cognitive underload [4]. In combination with lengthy supervision periods that are likely to result in both boredom and monotony [7], it can lead to fatigue and therefore inattention towards safety critical aspects of the driving task [8]. Drivers' inattention can be divided into (1) driver restricted attention, which is reflected by the driver's mind wandering for instance [7] and (2) driver diverted attention, emerging through the drivers focusing attention towards driving unrelated tasks (i.e., non-driving related tasks (NDRTs)) [9].

To a certain degree, engaging in NDRTs while driving might not considerably impair driving performance due to drivers' spare attentional capacity. For the manual driving task, which is a predominantly visual task [10], literature proposes that 20–25% [11] or, depending on the driving environment's complexity, even up to 50% [12] of drivers' visual attention is focused on objects unrelated to the driving task (e.g., advertisements, scenery), reflecting drivers' spare attentional capacity during manual driving. This suggests that drivers could execute certain NDRTs during manual driving without performance losses in their gaze and driving behavior as the resources are obtained from the spare attentional capacity [13]. Secondary tasks, identified as suitable for manual driving according to the National Highway Traffic Safety Administration (NHTSA) [14], are likely executed based on these resources. An on-road driving study showed that participants tapped into this spare capacity to increase visual attention towards a newly provided in-vehicle-display during manual driving, indicated by fewer glances towards the speedometer or periphery [15]. Glances towards the street ahead or the mirrors were not affected. Nevertheless, the driving environment's complexity and the workload of the driving task itself e.g., [16,17] might influence how much spare capacity drivers have (e.g., the more complex driving environment, the less free visual attentional capacity available). For PAD, the driving demands are partially reduced since the automated system controls longitudinal and lateral position [4]. Hence, it could be assumed that more spare attentional capacity would be available, for instance, to execute NDRTs without performance losses. However, drivers constantly need to supervise the driving environment as well as the automated system to take over the driving task if necessary, which demands a crucial amount of visual attention. Hence, if the amount of attentional resources needed exceeds the available spare capacity, NDRT execution during PAD will likely negatively impact drivers' supervisory and take-over performance.

Several simulator and real-world environment studies have revealed that drivers tended to engage in NDRTs (e.g., watching a DVD, reading or interacting with a smartphone) to reduce boredom, monotony, and cognitive underload associated with automated driving e.g., [18–20]. However, this behavior might overly distract drivers from vigilantly supervising the system and driving scene, thus reducing their situation awareness if the resources needed to execute the NDRTs are taken from beyond the drivers' spare capacity.

In general, cognitive underload, driver inattention and the associated decrease in system supervision are all related to slower and poorer reactions [7], complete failures to react and poor decision making in the event of a system failure [20,21], which could counteract the potential safety benefits of automated driving [22] and exemplify potential new challenges connected with PAD. Hence, the drivers' new passive role leading to the new problems mentioned, together with the fact that these intermediate automation levels are not consistently reliable yet [23] underline the importance of reasonable usage and implementation of automated driving [14].

1.2. Examining the Effects of NDRTs during PAD

Thus far, many studies have examined the effects of NDRT execution during automated driving. However, most of these studies have focused on automation levels other than PAD. For instance, Carsten et al. [20] observed voluntary NDRT execution (e.g., eating or watching a DVD) during semiautomated driving, which they defined as automated lateral or longitudinal control, or highly automated driving in a simulator. In contrast, other studies have focused on driver NDRT execution during higher automation levels, such as conditional automation (i.e., SAE level 3) e.g., [18,24].

In addition, many of the PAD studies only included NDRTs as a secondary aspect, while focusing on other main aspects. For example, one simulator study [25] mainly focused on how anticipatory information affected drivers' supervisory behavior during PAD while executing NDRTs (e.g., reading or interacting with a smartphone) on a voluntary basis. A different simulator study mainly centered on how participants' self-regulation during secondary task engagement would affect their supervisory behavior [26]. In contrast, other studies have required participants to execute the NDRTs during PAD instead of leaving it optional. However, the core focus still remained on aspects other than the NDRT execution itself. For instance, Large et al. [27] compared behavioral cues of distraction during NDRT execution (i.e., reading task) across three automation levels: manual driving, PAD, and highly automated driving. Another simulator study examined whether a NDRT could reduce fatigue during PAD [28]. However, concentrating on the systematic evaluation of NDRT execution during PAD is highly important since drivers are likely to engage in these tasks due to their spare attentional capacity available and to reduce the monotony and boredom of the supervisory task. Moreover, it is essential because these tasks might have similar negative and distractive effects on the driver during PAD as they have during manual driving.

In addition to the fact that NDRT execution is often only a secondary aspect, PAD studies often differ considerably regarding their applied methods as well as PAD specifications. For instance, some PAD studies involved take-over requests to redirect the participants' attention towards the driving task e.g., [25,27], whereas others have examined the drivers' ability to detect automation failures during NDRT execution without warning [29]. Moreover, several studies did not include any situations or automation failures requiring participants to regain vehicle control e.g., [20]. Additionally, some studies employed PAD to assist with navigating traffic congestions and managing speeds under 50 km/h e.g., [25], while other studies employed PAD for managing higher speeds, such as 130 km/h e.g., [30].

Hence, even though these studies often applied similar "[...] paradigms when participants are instructed to undertake a period of automated driving, and additionally given the option to (and are free to when/if comfortable) engage in a range of secondary activities available to them while sitting in the driver seat" [22] (p. 3), the varying methods and specifications these studies used complicate the generalization of the findings. Moreover, the different studies yielded varying results regarding the effects of NDRT execution during PAD. For instance, one simulator study revealed that reaction time during hazardous situations clearly increased when driving with NDRT execution compared to without NDRT execution [25]. Another simulator study permitted participants to freely engage in various smartphone activities during PAD and highly automated driving [30]. On the one hand, results showed that drowsiness and highly motivational NDRTs negatively affected driving performance during PAD in terms of slower reactions. On the other hand, NDRTs with low to moderate visual and mental workloads improved driving performance in a hazardous situation [30].

This brief overview underlines the general need to examine the effects of NDRT execution on drivers during PAD by systematically manipulating various NDRTs. Moreover, the different methods and specifications employed in the studies emphasize the importance of incorporating a standardized methodology that is comparable to, for example, the methodology used by the NHTSA to examine NDRTs during manual driving [14]. Previous efforts in developing standardized methods either focused on higher automation levels, such as the overview of current research questions and relevant methical approaches in the conditional automated driving field (SAE level 3) [31], or on evaluating

PAD system and human-machine-interface (HMI) designs e.g., [32,33]. This clearly emphasizes the need to fill this gap and develop a standardized method to enhance the comparability, reproducibility, and generalization of these studies and their results. The standardization supports continuing examination of NDRT execution effects on drivers' supervisory and take-over capabilities to reach the important, overarching goal of safe PAD usage.

1.3. Developing a Standardized Methodology to Evaluate NDRT Execution during PAD

For manual driving, well-established methodologies and guidelines exist that detail how the effects of NDRTs on driving performance and gaze behavior can be evaluated. For example, NHTSA's well-established methodology focuses on examining different visual-manual NDRTs [14]. In addition to the standardized methodology, NHTSA provides guidelines and cut-off values that clearly regulate whether a NDRT is suitable for execution during manual driving [14]. For instance, to be acceptable, the single gaze durations towards a NDRT should not exceed two seconds. Further, neither driving performance nor gaze behavior during NDRT execution should be poorer than during execution of the manual radio tuning task [34], which is NHTSA's [14] recommended reference task. Employing the methodology and guidelines standardizes the evaluation of visual-manual NDRTs and enhances the comparability and reproducibility between the studies incorporating them.

However, the driving task during PAD is considerably different than manual driving. Hence, the evaluation specifications for manual driving performance (e.g., lane maintenance, speed or distance to another vehicle) no longer apply given that the automated system takes over these tasks in PAD. Instead, the drivers' ability to vigilantly supervise the system during a prolonged PAD period and to take over vehicle control immediately and in a safe manner, if necessary, become more important during PAD. Moreover, since the driving demands during PAD are lower than those of manual driving allowing for more available cognitive resources, the question arises whether the cut-off values for manual driving proposed by the NHTSA [14] are still applicable for PAD.

Therefore, to be useful when examining NDRT execution during PAD, existing methods need to be adapted and fulfill several additional requirements. Firstly, the new method needs to capture and sensitively evaluate drivers' capabilities to perform the new tasks that are important during PAD (i.e., vigilant supervision and taking-over the driving task if needed). To fulfill these requirements, the PAD periods must be interrupted by critical situations in which drivers must recognize the need and be able to take-over vehicle control due to a system failure or limit based on their vigilant system supervision. Secondly, the methodology must be sensitive to different NDRTs with varying distractive potentials and to other aspects relevant to the automobile context (e.g., different (in-vehicle) display locations). Lastly, the methodology must enable the establishment of cut-off values, comparable to those for manual driving, based on an adequate number of testing. A further beneficial characteristic of the method would be the ability to adapt it based on the research questions of interest.

Due to safety considerations, any new methodology for testing PAD should initially be applied in a driving simulator. However, the effects of NDRT execution in a real-world environment are potentially more safety critical than in a driving simulator. Therefore, it is also necessary to examine the external validity of such methods. Accordingly, a high external validity would greatly enhance the methods' generalizability.

1.4. Objectives of the Present Research Project

Since the methodology assessing manual driving is not applicable for PAD, the current project's overarching goal was to fill this methodological gap and take the initial steps towards developing a test protocol and providing recommendations for the systematic evaluation of drivers' supervisory and take-over capabilities during PAD while engaging in different NDRTs. To achieve this, the well-established methodology for manual driving [14] was extended for PAD based on the formulated requirements (see Section 1.3). The new test protocol was developed, validated and adapted through the course of three studies.

- The first two studies took place in a driving simulator to determine the potential for a new test protocol to assess the effects of NDRT execution during PAD.
- In the second simulator study, the new test protocol was also extended to other relevant aspects, such as (in-vehicle) display locations.
- The third study was conducted in a partially automated vehicle to validate the test protocol in a real vehicle on a closed test track. The main goal was to determine whether the test protocol was applicable to a real driving environment.

The following research questions were addressed within the three studies:

- Research question 1 (RQ1): Can the test protocol sensitively detect differences, as they are expected based on the literature, in the drivers' supervisory and take-over capabilities during PAD depending on various influencing factors?

  - Research question 1a (RQ1a): Is the new test protocol sensitive to the effects of visual-manual NDRTs with varying distractive potentials on the drivers' supervisory and take-over capabilities during PAD?
  - Research question 1b (RQ1b): Is the new test protocol sensitive to the effects of (in-vehicle) display locations with varying proximity to the driving scene on the drivers' supervisory and take-over capabilities during PAD?

- Research question 2 (RQ2): What parameters are minimally necessary and sufficient to sensitively capture and evaluate the take-over and supervisory capabilities of the drivers in light of these aspects?

The focus of the current article and the corresponding research questions solely lie with the examination of the proposed test protocols' suitability to sensitively evaluate the effects of NDRTs on the drivers during PAD. The evaluation of the NDRTs' particular effects on the drivers' take-over and supervisory capabilities using the provided test protocol were not the focus of the current manuscript. The presentation of the specific results will be described in more detail in separate papers.

2. Test Protocol Development

The following chapters will describe the relevant aspects of the test protocol's development based on the literature and existing research and specific implementation by the authors within the three studies of the current project, beginning with a presentation of the driving scenario and the implemented take-over situations (Section 2.1), followed by a description of the evaluated independent variables (Section 2.2) and the assessed dependent variables (Section 2.3). In the subsequent chapters, the equipment and materials used are presented (Section 2.4), as well as the detailed experimental design and procedure (Section 2.5). This is proceeded by a presentation of the data preparation and analysis (Section 2.6). The final chapter (Section 2.7) describes the participants of all three studies. In addition, two subchapters are integrated into each chapter, providing unique details of the driving simulator studies (Section 2.x.1 Driving Simulator Implementation) and the test track study (Section 2.x.2 Test track implementation).

2.1. Driving Scenario and Take-Over Situations

This section will describe the development and the specific implementation of the driving scenario and the take-over situations. To test how visual-manual NDRTs affect drivers during manual driving, the NHTSA methodology recommends incorporating a car-following scenario on a highway road [14]. Effects are evaluated by judging the drivers' gaze behavior and driving performance while executing the NDRTs. More precisely, drivers are evaluated on their ability to maintain distance to the lead vehicle, speed, proper lane maintenance during the car-following scenario, as well as how long, in terms of single and total glance durations, the drivers are glancing towards the NDRTs [14].

To standardize examination, NHTSA [14] prescribes several specifications for the test track and driving scenario. Firstly, NHTSA recommends a car-following scenario where drivers attempt to maintain a certain speed (80 km/h) and distance to the lead vehicle (70 m), which allows the examination of the drivers' ability to fulfill this task during NDRT execution [14]. Moreover, NHTSA advises using a straight highway route with two lanes per direction and a predefined lane width. This reflects a realistic setting and enables examination of the drivers' ability to stay within the lane for instance [14]. Accordingly, straight road segments should be used to examine the drivers' gaze and driving behavior, although curved segments can be included occasionally [14]. Lastly, NHTSA recommends using a generic driving environment that excludes any external cues (i.e., trees, houses) [14], though they allow occasional (oncoming) traffic during the car-following task.

Due to the changed driving task during PAD, parameters for manual driving are not applicable anymore and it is necessary to evaluate drivers' supervisory and take-over capabilities when a system limit is reached. Therefore, generic take-over situations had to be implemented that simulate such a system limit. The current studies included two take-over situations: (a) *lead vehicle deceleration* and (b) *drifting of the participant's partially automated vehicle (i.e., ego vehicle)*. Both situations represented system limitations directly corresponding to the main driving tasks taken over by the partially automated system (lateral and longitudinal vehicle control).

Although these two take-over situations were based on earlier research see [35], several adaptations were made to match the NHTSA scenario specifications more closely. During the take-over situation with *lead vehicle deceleration* (addressing longitudinal vehicle control) the lead vehicle slowed down without brake lights. To mimic realistic braking movement, the vehicle slowed down based on a predefined value. Without any driver intervention (i.e., braking), a collision with the lead vehicle would occur. During the second take-over situation involving *ego vehicle drifting* (addressing lateral vehicle control), the vehicle drifted to the left or right see [35]. To prevent a guardrail collision, the participants had to notice the drifting and steer in the opposite direction. A collision would occur without any driver intervention. To ensure comparability between the two distinct situations, the outcome and time to collision (TTC) were identical: without any driver intervention, a collision (outcome) with the lead vehicle or guardrails would occur after the same predefined TTC.

Following Signal Detection Theory (SDT), the most critical situations are missed warnings, in which errors or events occur without any warnings to the system supervisor [36]. When a system limitation is reached during PAD, the automated system neither gives a warning nor issues any take-over request for the drivers, thus the drivers must vigilantly supervise any system changes [3]. Therefore, any system warning or take-over request for the two take-over situations were excluded. All environmental (i.e., trees or houses) and vehicle (i.e., lead vehicle's brake lights or steering wheel movement) visual cues were excluded to reduce the predictability of the take-over situations.

2.1.1. Driving Simulator Implementation

The driving scenario and take-over situations were implemented as follows in the two driving simulator studies. Based on the NHTSA methodology, an identical car-following scenario on a straight, four-lane highway route with two lanes in each direction was included. Moreover, the same specifications for speed (80 km/h) and distance to the lead vehicle (70 m) were applied. The participants drove a partially automated vehicle that controlled the longitudinal and lateral position. The two take-over situations (*lead vehicle deceleration* and *ego vehicle drifting)* were implemented with the general specifications discussed in Section 2.1. The specific braking speed of the decelerating lead vehicle was 2.3 m/s^2, which corresponded to an electric vehicle with a regenerative braking movement. Without any driver intervention, a collision with the lead vehicle would have occurred after seven seconds. A collision with the guardrails would have occurred after seven seconds in the ego vehicle drifting scenario if the driver did not intervene in time. The participants were introduced to react by braking or steering, respectively, to regain control from the partially automated system.

The test track was 11 km long in the first and 9 km long in the second simulator study, which was programmed using the Silab 5.0 simulation environment. To reduce the predictability of the take-over situations, the driving environment was as generic as possible, excluding any visual cues (i.e., trees). In contrast to the NHTSA guidelines, the simulation did not include any traffic other than the lead and ego vehicle. This allowed for a controlled execution of the take-over situations without needing to, for instance, check for rear traffic before braking. Further, the aim was to reduce any potential distractions, especially during the reference trial, in which boredom might have encouraged drivers to gaze towards irrelevant vehicles instead of focusing on the system and lead vehicle. Although this aspect is also important in terms of situation awareness, it was not the focus of our studies.

2.1.2. Test Track Implementation

To ensure participant safety as well as a standardized data collection free of any interference, the third study occurred on a closed test track in a parking lot. The limited space restricted the precise application of the specifications used in the driving simulator, resulting in several adaptations. These adaptations resulted in differences between the simulator and test track studies regarding, for instance, the execution of the driving scenario and take-over situations. These differences potentially reduced the comparability between the results of the two study types (see Table 1 for a comparative overview).

Firstly, compared to the simulator studies, the driving scenario was downscaled for the test track study in terms of the driving environment (with landmarks), test track (one lane, with curves), speeds (max. 25 km/h), the distance between the two vehicles (speed of ego vehicle/2 + 7 m) and the particular execution of the two take-over situations. Nonetheless, the goal was to mimic the scenario as much as possible by finding a test track with as few curves as possible and with at least one long, straight segment for the *lead vehicle deceleration* take-over situation. The driving scenario and take-over situations relied heavily on non-automated, human execution (i.e., lead vehicle or lateral ego vehicle control maintained via Wizard-of-Oz). Therefore, the take-over situations were always executed on the same track segment to enhance reproducibility and comparability as well as to reduce the chance for human error. Figure 1 (top row, left) shows the final test track with the two segments chosen for the two take-over situations, the execution of lateral vehicle control (top row, right) and the experimental setup within the ego vehicle (bottom row, left) as well as the two vehicles involved in the driving scenario (bottom row, right).

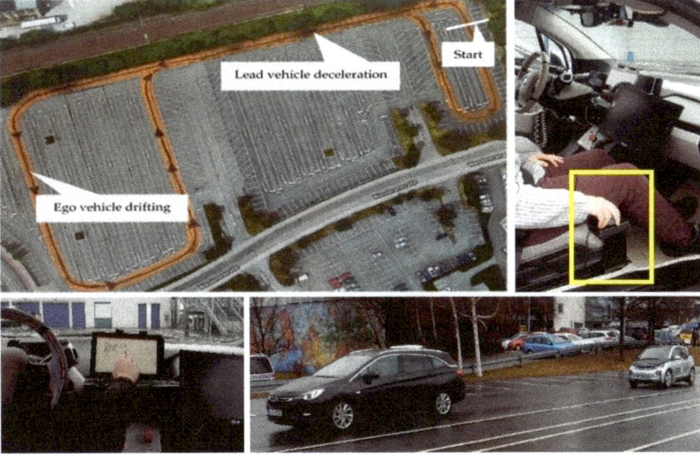

Figure 1. Test track depicting the two take-over situation locations (**top row, left**), the execution of lateral vehicle control via Wizard-of-Oz (**top row, right**), experimental setup within the ego vehicle (**bottom row, left**), and the two vehicles involved in the driving scenario (**bottom row, right**).

Table 1. Differences and similarities between the simulator and test track.

	Simulator	Test Track	Relevant Differences
Study Environment	Simulator	Test track	Yes, since the experienced risk and situation criticality likely differ between the two study environments. For instance, they are likely higher in a real vehicle (on a test track) than in a simulator, where collisions pose no risk to the participants' safety.
Driving Scenario			
Car-following scenario	Yes	Yes	/
Duration	Approx. 8–9 min	Approx. 12 min	No, since the difference is only minor and therefore unlikely to elicit different levels of monotony or fatigue.
Route	Straight highway	Test track on parking lot with many curves	Yes, because a test track with several curves potentially results in different supervisory behavior (e.g., enhanced supervision of the street) compared to a straight test track. Hence, the supervisory behavior on the simulated straight highway might be underestimated compared to the more realistic curved test track.
Surrounding	Without landmarks	With many landmarks	No, since the landmarks (i.e., trees) possessed low levels of visual attraction and were static. Hence, they were less distractive than, for example, dynamic landmarks.
Take-over Situations			
Types	Lead vehicle deceleration and Ego vehicle drifting	Lead vehicle deceleration and Ego vehicle drifting	/
Number of situations per trial	4	4	/
Collision risk	Yes	No	No, since participants were aware that the theoretical consequence of failing to react in a real vehicle would be a collision, a certain level of situation criticality still existed.
Duration (e.g., time to collision)	7 s	Not precisely realizable	No, since the time to collision of the take-over situations was comparably short in all studies (i.e., only several seconds) and participants needed to react as soon as they noted the take-over situations. The differences in distance and speed would likely result in a reduced comparability of the studies' results (see below).
Occurrence	Defined in simulation	Predefined by test circuit, same track segment	Yes, since it enhances the predictability of the take-over situation (locally).
Predictability (locally)	Low	High	Yes, since it potentially influences participants' (supervisory) behavior during the particular segment (e.g., more gazes towards the street and therefore faster reactions to the take-over situations.)
Predictability (time-wise)	Low	Low	/
Ego vehicle			
Speed	80 km/h	Max. 27 km/h	Yes, both the lower speeds driven and smaller distance to the lead vehicle during the test track study potentially influenced the experienced situation criticality (e.g., reduced criticality through reduced speed; however, partially counteracted through decreased distance between the vehicles). This might have reduced the comparability between the studies' results (e.g., longer gazes towards the NDRTs on the test track due to reduced speeds).
Distance to lead vehicle	70 m	Ego vehicle speed/ 2 + 7 m	
Longitudinal control	Automated	Automated	/
Lateral control	Automated	Wizard-of-Oz	No, the Wizard-of-Oz execution itself should not influence the results or the comparability of the studies' results, provided participants did not notice the researcher executing lateral control (which was not the case).
Lead Vehicle			
Speed	80 km/h	Max. 25 km/h,	No, the lead vehicle's speed generally is not a separate factor influencing the comparability of the studies' results. It is due to the reduced speeds driven on the test track. Hence, its influence is incorporated into the impact of the ego vehicle's speed and the distance to the lead vehicle.
Vehicle control	Simulator (preprogrammed)	Manually driven (with cruise control and motor deceleration/acceleration)	Yes, it reduces the standardization, comparability and reproducibility of the scenario and the take-over situations. Moreover, it influences the driving data (e.g., distance when the take-over situations were triggered). However, the participants' supervising behavior should not be affected.

On the corresponding test track segment (Figure 1, top row, left), the *lead vehicle deceleration* situation was employed as follows: During this segment, the lead vehicle was driven in activated cruise control mode and only slowed down when the motor decelerated after the cruise control was deactivated. The lead vehicle's brake lights did not activate. During this segment, the ego vehicle was not programmed to maintain distance to the lead vehicle and, therefore, moved closer until the participants intervened. During the *ego vehicle drifting* situation, a researcher sitting in the ego vehicle's passenger seat used a small steering wheel to execute the drifting (Figure 1, top row, right). To reduce human error likelihood and enhance reproducibility, the researcher always drifted the vehicle to the left.

As in the simulator studies, the participants needed to brake or steer in response to the take-over situation, although they could also stop the vehicle by merely touching the steering wheel. Unlike in the simulator studies, the two take-over situations did not result in a collision, even when the participants did not react. For this matter, several fallback solutions were included in case participants failed to intervene, such as programming the ego vehicle to stop automatically if a minimal safety distance is reached and a researcher who could stop the ego vehicle by employing the emergency brake.

2.2. Independent Variables

In Section 1.3, the requirement was formulated that the test protocol must be sensitive to the effects of different NDRTs (RQ1a) as well as to other relevant aspects to the PAD context, in this case different (in-vehicle) display locations (RQ1b). In the following two sections (Sections 2.2.1 and 2.2.2) and the corresponding subsections, the theoretical background as well as the specific implementation of the independent variables will be explained.

2.2.1. Non-Driving Related Tasks

Several studies have indicated that drivers tend to engage in NDRTs to reduce cognitive underload, boredom, and monotony resulting from the reduced driving demands during PAD e.g., [18–20]. Due to their potentially safety diminishing effects, the new test protocol must sensitively capture the different effects of these NDRTs on the drivers' supervisory and take-over capabilities to evaluate whether a certain NDRT is applicable during PAD.

Amongst other theories and models, the multiple resource model [37] is regularly used to differentiate between NDRTs based on their required modalities as well as between the different visual NDRT effects on drivers' performance and gaze behavior during manual driving e.g., [38,39]. Multiple resource theory, which builds the basis of the model, focuses on the idea that when executing multiple tasks simultaneously, it is necessary to share time and attention between these tasks [37]. Moreover, when these two tasks occupy the same modalities (e.g., both requiring visual attention), these tasks interfere with each other as resources and attention are divided [37,40]. This results in reduced (attentional) resources for both tasks compared to executing only one task at a time [40], thus decreasing performance for one or both tasks [37]. Based on the multiple resource model, it is assumed that visual NDRTs are especially distracting during driving e.g., [38] and cause decreased performance in the driving task, the NDRT, or both since the driving task itself is highly reliant on visual resources [41]. Therefore, visual NDRT execution seems especially problematic during manual driving and are thus given priority by NHTSA. In general, NHTSA focuses on visual tasks with a manual aspect, where the driver must manipulate a device to execute the task [14]. Since the driving task requires drivers to steer or shift gears, the manual NDRT component would likely interfere with these driving tasks as the resources would overlap. During PAD, the driver's main task is to vigilantly supervise the automated system and driving scene. As it is assumed that visual NDRTs would especially interfere with supervising, the current project focuses on visual tasks as well. In addition, drivers must regain vehicle control during take-over situations and resume steering for instance. Hence, visual NDRTs with a manual component are potentially problematic for PAD as well.

Moreover, the NHTSA guidelines prohibit certain visual tasks, known as *per se lock outs*, due to their distractive characteristics [14]. For instance, displaying photos or watching videos unrelated to the driving task, reading texts from books, the internet or social media as well as automatically scrolling texts or manually entering communication-based texts are prohibited during manual driving [14]. In addition, the guidelines propose that tasks should be interruptible at any time, completed within a maximum of 12 s total gaze time to the task and single gazes to the task should not last longer than 2 s [14]. Congruent with NHTSA, the current project incorporates visual-manual NDRTs. Even though NHTSA excludes the following from any examination, the current project focused on the effects of these per se lock out tasks on drivers' supervisory and take-over capabilities during PAD. The goal was to validate the new test protocol by using a broad range of guideline compliant to guideline non-compliant NDRTs. Regarding the latter group, a sensitive test protocol should yield strong effects concerning the drivers' supervisory and take-over capabilities.

Eventually, five NDRTs differing in guideline compliancy as well as similarity to everyday life/artificiality were chosen. Three of these tasks did not comply with the NHTSA guidelines, for instance due to presenting videos unrelated to the driving task. These three tasks included a browsing task, a video watching task and a text reading task, which were all similar to everyday life. The two tasks complying with the NHTSA guidelines included the artificial surrogate reference task (SuRT) [42] and the manual radio tuning task [34]. The latter task, a well-established reference task for manual driving, was designed to reduce the total gaze time of one trial to 20 s [34]. To match these specifications, the trials of the other tasks were designed to last no longer than 20 s as well.

During the browsing task, participants manually entered a departure point, a destination, two flight dates and the number of passengers. Participants received this information from the researchers. During the video task, participants viewed news video segments lasting 20 s and answered a question about the visual or general content of the video. The text reading task presented the participants with 70–100-character texts, which took approximately 20 s to read [43]. The participants had to scroll through the text to read its entirety. After finishing a text, the participants answered a question regarding its content. The SuRT task included finding a target (a bigger circle) amongst many distractors (smaller circles). During the manual radio tuning task, participants needed to set the radio to predefined frequencies.

2.2.2. Display Locations

In addition to executing NDRTs during manual driving and PAD, a related trend towards integrating increasing amounts of technology into vehicles has increased the potential of driver distraction and inattention during manual driving [44]. Another trend exists towards using increasing amounts of driving unrelated information [45] as well as smartphones during manual driving [46,47].

The main problem with different (in-vehicle) displays surrounds their proximity to the driving scene. Displays located further away from the windshield and driving scene are associated with enhanced reaction times [48]. For instance, head-up displays (HUD) were associated with significantly shorter reaction times as they are very close to the driving scene or may even overlay it. In contrast, display locations located further away from the driving scene were associated with shorter time to collisions [49]. Additionally, focusing on displays with less vertical proximity to the normal line of sight led to slower reactions than focusing on displays with equivalent horizontal proximity [49]. Moreover, several studies have found that the display location influenced drivers' gaze behavior during manual driving e.g., [45,48,50]. For instance, gazing away from the road towards a head-down display (HDD) was occurring significantly less often than towards a HUD [50]. In addition, gaze durations during HUD interactions increased compared to HDD (e.g., the instrument cluster or head unit) interaction [45]. However, when focusing on the HUD, driving performance was improved (e.g., fewer lane deviations) because the driving scene was visible peripherally [45].

Hence for manual driving, several studies have shown clear differences in gaze behavior and driving performance depending on the display's proximity to the driving scene. Currently there are no comparable studies examining the effects of different display locations on drivers during (partially) automated driving. Further, no studies exist examining how different NDRTs during PAD affect drivers across display locations. Therefore, the goal of the second study was to incorporate this aspect in the method developed in the first simulator study. The (in-vehicle) display locations were chosen to reflect well-established displays (i.e., instrument cluster, head unit) as well as newer, more innovative technologies (i.e., HUD) and to reflect displays close to the driving scene (i.e., HUD) vs. further away (i.e., instrument cluster, head unit). Moreover, since smartphone usage during manual driving has increased e.g., [46,47], the smartphone was included as a handheld and forbidden display location. In addition, the following three displays were chosen: a head unit, an instrument cluster, and a HUD.

2.2.3. Driving Simulator Implementation

In the first simulator study, all five tasks were executed on a touch display. Therefore, the adapted manual radio tuning task for touch-displays [51] was used. This display was situated in the center console, at the same position as the head unit.

In the second simulator study, the display location was included as an additional independent variable since the test protocols' ability to differentiate their effects and the opportunity to easily manipulate these displays could be safely validated. Given that the goal was to ensure an economic study design and given that participants performed comparably during the browsing and text reading task, the browsing task was excluded. The remaining tasks ranged from slightly visually distracting (i.e., SuRT and manual radio tuning task) to highly distracting (i.e., text reading task). The video watching task was considered in the middle of this range. With exception of the video watching task, the tasks were not adapted from the first simulator study. The results of the first study led to the assumption that participants were listening to more than looking at the video segments. Therefore, the questions following each video segment were adapted to focus solely on the video's visual content to highlight its importance and enhance the comparability to other, more compelling videos (e.g., blockbuster videos).

The four chosen display locations were implemented as follows. For the head unit, a well-established HDD, the same 9-inch pre-installed display in the driving simulator's fully equipped vehicle mockup, was used as in the first study. For the instrument cluster, also a well-established HDD, a 9-inch display was installed behind the fairing of the vehicle mockup's built-in displays. The installed display thereby covered the tachometer but not the speedometer. Due to the fairing, parts of the 9-inch display were covered, thus the presented information (i.e., NDRTs) had to be downsized. Regarding the head-up display, a glass plate with mirror foil used to retrofit HUDs in vehicles was installed on the dashboard since the vehicle mockup was without a windshield. A 9-inch display was positioned under the glass plate with its presented information reflected onto the mirror foil. For the smartphone, a Huawei P9 with Android was used. Participants needed to hold the smartphone close to the gearstick, simulating the realistic attempt to hide phone usage during manual driving. Therefore, the smartphone condition was considered a part of the HDD category as well. During take-over situations, participants had to put down the smartphone before regaining vehicle control. For an overview of the NDRTs and (in-vehicle) display locations assessed within the three studies, see Table 2.

Table 2. Overview of the independent and dependent variables addressed and assessed within the three studies.

		Study 1—Simulator	Study 2—Simulator	Study 3—Test Track
Independent variables	Non-driving related tasks	- Manual radio tuning task - Surrogate reference task - Video watching task - Text reading task - Browsing task	- Manual radio tuning task - Surrogate reference task - Video watching task - Text reading task	- Manual radio tuning task - Video watching task - Text reading task
	(In-vehicle) Display locations	- Head-unit	- Head-Up display - Head-unit - Instrument cluster - Smartphone	- Head-unit
Dependent variables	Supervisory capabilities	- Mean Gaze Duration - Total Gaze Duration - Maximum Gaze Duration - Number of Gazes - Number of Transitions	- Mean Gaze Duration - Total Gaze Duration - Number of Transitions	- Mean Gaze Duration - Total Gaze Duration - Number of Transitions
	Take-over capabilities	- Reaction time - Minimal distance to the lead vehicle at initial reaction - Maximal brake pressure - Maximal steering angle - Number of crashes	- Reaction time - Number of crashes	- Reaction time - Minimal distance to the lead vehicle at initial reaction

2.2.4. Test Track Implementation

Due to limited time resources, only a reduced NDRT selection was used and the display location aspect was excluded entirely. Additional reasons for excluding the latter included safety concerns for participants. The three NDRTs implemented in the study (i.e., manual radio tuning task, reading task, and video watching task) reflected a broad range of distractive potential (as determined during the two simulator studies) and a strong similarity to everyday life. The three tasks were executed on a tablet with touch control, attached at the head unit's position.

2.3. Dependent Variables

The new test protocol was required to be sensitive regarding the effects of diverse NDRTs and display locations on drivers' supervisory and take-over capabilities. For that matter, parameters are necessary that sensitively capture and evaluate the capabilities (RQ2). To meet the requirement and answer the research questions, extensive examinations of different parameters for the supervisory and take-over capabilities were completed.

For manual driving, NHTSA recommends analyzing gaze behavior in terms of mean and total gaze duration towards the NDRTs [14]. Congruent with NHTSA, these parameters were included when analyzing supervisory behavior during PAD. However, since PAD differs from manual driving and the supervisory tasks increased in importance, further parameters were examined to achieve a comprehensive view. The assumption behind the additional parameters, including for instance the number of gazes or transitions between certain areas of interest (AOIs), is that these were assumed to be useful parameters to judge the drivers' compensatory behavior. For example, if long gazes to the NDRTs occur but are accompanied by many transitions between the NDRT and driving scene, the length of these gazes is somewhat compensated. Thereby, the driver will likely know more about current driving events and might react better to system failures compared to a driver executing long gazes to the NDRTs with few transitions. Furthermore, parameters reflecting and examining the supervisory behavior during PAD based on the NHTSA guidelines and cut-off values (i.e., maximal 2 s per gaze towards the NDRT) were included, such as the maximum gaze duration towards the NDRTs. Hence, the following parameters for supervisory capabilities were of interest:

- the *mean gaze duration* towards the NDRT
- the *total gaze duration* towards the NDRT
- the *maximum gaze duration* towards the NDRT
- the *number of gazes* towards the NDRT
- the *number of transitions* between the driving scene and NDRT AOIs

Regarding the take-over capabilities, new parameters were proposed for the new test protocol given the automated system takes over the driving task during PAD. Firstly, reaction time indicated the criticality of the situation when the initial reaction occurred as well as the quality of the drivers' supervisory behavior. Longer reaction times would indicate reduced or insufficient supervision of the driving scene and system, probably due to the NDRT's greater distractive potential. Moreover, as reaction times increase, the criticality of the situation increases. For instance, the distance to the lead vehicle decreases each second, eventually making collision avoidance impossible. Four additional parameters were included to indicate situation criticality: the number of crashes, the minimal distance to the lead vehicle at initial reaction, the maximal brake pressure and maximal steering angle. For instance, more crashes or a small minimal distance to the lead vehicle would suggest a higher situation criticality. In addition, these variables were assumed to provide context to the reaction time and indications about potential compensatory behavior. For example, strong steering or braking responses might still prevent a collision even with a slow reaction time indicative of a critical situation. The parameters of interest were defined as follows:

- *Reaction time*—The time between the beginning of a take-over situation until the participants' initial reaction (braking or steering).
- *Number of crashes*—The number of collisions with guardrails (lateral) and the lead vehicle (longitudinal).
- *Minimal distance to the lead vehicle at initial reaction*—The distance between the two vehicles when participants initially reacted (braking). Applies only to the *lead vehicle deceleration* take-over situation.
- *Maximal brake pressure*—The highest administered brake pressure during the initial braking interval. Applies only to *lead vehicle deceleration*.
- *Maximal steering angle*—The greatest administered steering angle during the initial steering interval. Applies only to *ego-vehicle drifting*.

2.3.1. Driving Simulator and Test Track Implementation

The parameters were comparably assessed across the three studies. In the first study, all described parameters were assessed. However, based on the results of the first study some of the parameters were excluded from the following studies. The rationale behind this will be addressed in more detail within the result and discussion chapters. Table 2 gives a short overview of the parameters assessed within each of the three studies.

2.4. Equipment and Materials

In general, both study environments (driving simulator and actual vehicle) had to allow for scenario implementation (i.e., car-following) and independent variable examination. Hence, the following equipment was implemented within the three studies of the current project.

Firstly, at least two displays were required: One providing participants information about system states (i.e., instrument cluster) and one on which participants could execute the NDRTs (e.g., head unit). For the second study, the simulator had to allow to include further, controllable displays to examine the effects of display locations.

As discussed in Section 2.3, it was essential that the take-over capabilities and gaze behavior could be captured. Regarding the former, it was necessary to record participants' driving or take-over behavior. For this matter, the simulator software had to be programmed to record all relevant variables (see Section 2.3). The real vehicle had to contain data recording devices and the necessary sensors as well (e.g., LiDAR) to record data and compute relevant parameters.

Concerning the supervisory behavior, several methods to capture gaze behavior were employed of which the general (dis-)advantages will be discussed in the following section before describing the specific implementation within the three studies in the respective subchapters. Head-mounted eye trackers are a common tool to assess gaze behavior (e.g., Tobii Pro Glasses). Advantages of head-mounted eye trackers are, amongst others, the opportunity to analyze gaze data across different levels of detail (e.g., level of fixations or gazes). Moreover, AOIs are seen from the participants' perspective and their gazes are directly projected on to these AOIs. This allows for easy and reliable manual mapping of gazes towards AOIs, even for relatively small AOIs. In addition, the included eye tracking analysis software often provides the opportunity to automatically map raw gaze data on to relevant AOIs. However, it is still necessary to check the accuracy of the automatic mapping, and often manual remapping is required. An important disadvantage of head-mounted eye trackers is that most do not allow participants to wear glasses, thus these participants cannot take part in the study. This is especially problematic when examining older age groups as they are more likely to wear them. For instance, in 2014, 63.5% of all German citizens wore glasses and 92% of those older than 60 wore them compared to only 32% for those aged 20 to 29 and 38% for 30 to 44-year-old citizens [52]. Moreover, participants are highly aware of wearing these head-mounted eye-trackers and wearing them for prolonged times can be very uncomfortable.

Another method to assess supervisory behavior is using video annotations, whereby gaze behavior is annotated manually using multiple, synchronized videos facing the participants. This method is a non-invasive alternative to eye trackers since participants do not have to wear anything extra. This also allows participants wearing glasses to take part. Based on detailed annotation schemes, including descriptions of the AOIs that should be mapped and instructions on how to detect gazes to these AOIs, as well as the inclusion of training annotations with detailed feedback, it is possible to reliably annotate gaze behavior even across multiple researchers. However, in contrast to head-mounted eye tracking, where AOIs are seen from the participants' perspective, the videos are facing the participants. Therefore, only gaze *directions* towards a certain region representing AOIs (e.g., instrument cluster or street) can be annotated and differentiation between smaller AOIs closer together is difficult. Nevertheless, when relatively large AOIs (e.g., instrument cluster, head unit, mirrors or street) are of interest, this is less problematic. Another disadvantage is that video annotation does not allow the annotation of fixations. However, when focusing predominantly on gaze levels, as often done in this type of research e.g., [14], this disadvantage is less relevant.

In addition to the technical aspects, several formal aspects were necessary, such as information regarding the study, an informed consent and a data privacy statement. Furthermore, to standardize the information, participants received all instructions (e.g., concerning the partially automated system's activation and deactivation, the NDRT's system failures and execution) in written form.

To supplement the performance data, demographics such as participant age, gender and prior system experience were assessed. This allowed for an even distribution of gender and age as recommended by NHTSA [14], and this information could function as control variables during the analyses. Additionally, the participants' subjective experience regarding, for instance, the PAD or NDRT executions during PAD were assessed by questionnaires (e.g., Van-der-Laan-Acceptance-Scale, NASA TLX, and Trust in Automation). These subjective evaluations enriched the objective results or clarified aspects such as the participants' willingness to execute certain NDRTs during PAD before and after the study was completed.

To run the study as smoothly as possible, at least two researchers were deployed. One researcher focused on technical aspects (i.e., starting the simulator or driving the lead vehicle) and the other focused solely on supporting and supervising participants, including answering their questions or monitoring for simulator sickness.

2.4.1. Driving Simulator Implementation

To employ the developed test protocol in a simulated environment, the driving simulator used in the study included simulation software that presented the test track and scenario. The current project utilized a fixed-base driving simulator that consisted of a fully equipped mockup of the front of a vehicle (up to the B-pillar) with side- and rearview mirrors. Three connected screens presented a 180° horizontal field of view. In both studies, the driving simulator contained several cameras focused on the driving scene, the pedals and the driver (from two different angles). In the first study, drivers' gaze behavior was analyzed based on video annotations. The main reason for choosing this method was that no reliable eye tracker was available. Nevertheless, using non-invasive video annotations to examine gaze behavior enhanced participant comfort and allowed those with glasses to participate as well, thereby increasing the potential participant pool. This method could also reliably assess the relevant AOIs (e.g., the street, head unit or the instrument cluster). However, the second study incorporated the head-mounted Tobii Pro Glasses eye tracker [53] to record gaze behavior. Even though this method excluded glasses-wearing participants, it was very useful to assess more refined AOIs (e.g., handheld smartphone) and differentiate AOIs (e.g., differentiation within the instrument cluster between one part presenting system-related information and the other presenting the NDRT).

Moreover, during both studies, the instrument cluster presented the various states of the partially automated system (e.g., active, inactive, and deactivated) using a very minimal design. The second simulator study implemented a self-turning steering wheel inside the vehicle mockup. This reflected the actual PAD experience more closely since an actual vehicle's steering wheel moves during curved segments as well. The slight steering wheel movement during the *ego vehicle drifting* take-over situation could, however, lead to faster recognition of the situation and hence, faster reactions. Lastly, two researchers were present for both studies, one focusing on technical aspects and the other on participants. Participants received written instructions to enhance standardization and also received several questionnaires.

2.4.2. Test Track Implementation

To reproduce the test protocol in a real driving environment, it was necessary to include two vehicles: (a) A partially automated ego vehicle that was programmable to deliberately trigger system failures and enable the capturing of driving and take-over data, and (b) a lead vehicle with (if possible) advanced driving assistance systems (ADAS) such as cruise control (see Figure 1 (bottom row, right) in Section 2.1.2).

The lead vehicle's cruise control started at 25 km/h and was driven manually by a researcher. To ensure maximum comparability, the researchers received detailed instructions and absolved several training runs. With exception of the *lead vehicle deceleration* take-over situation, the researcher always drove the vehicle in second gear with the motor executing all necessary accelerations or decelerations to maintain a constant speed as much as possible (approximately 15 km/h).

A second, programmable vehicle served as the ego vehicle, equipped with various measurement technologies (e.g., Denso LiDAR, Novatel DGPS). The partial automation was achieved through combining genuine vehicle automation and Wizard-of-Oz techniques. The automation controlled longitudinal movement, vehicle speed (max. 27 km/h), and held constant the distance to the lead vehicle relative to speed except during the *lead vehicle deceleration* take-over situation. The distance was set to half of the ego vehicle's speed with an additional buffer of seven meters (ego vehicle speed/ 2 + 7 m). The Wizard-of-Oz techniques controlled lateral movement. The researcher in charge of programing the partial automation sat in the passenger's seat during the entire study to secure safety and execute lateral control (steering) using a small steering wheel unseen by participants (see Figure 1 (top row, right) in Section 2.1.2). Although participants did not notice the researcher steering the vehicle, this led to a major disadvantage in that the steering movements and lane keeping were not completely identical during each drive.

As in the simulator studies, the ego vehicle contained several cameras focused on the participants (from two angles), driving scenario, and vehicle interior. The recordings of the participants were used to analyze drivers' gaze behavior towards the AOIs (e.g., the street, head unit, and instrument cluster) based on video annotations.

Three researchers were present in this study: one focused on technical aspects and ego vehicle steering, one focused solely on driving the lead vehicle and one focused on supervising and supporting participants in between trials. Again, participants received written instructions and questionnaires with the discussed contents (See Section 2.4).

2.5. Experimental Design and Procedure

A within-subjects design was used to test the NDRTs, take-over situations, and other independent variables such as the different display locations. This approach allowed to reduce the number of participants necessary for high statistical power by directly comparing each participant to themselves and excluding any influences from interindividual confounding variables.

The participants experienced both take-over situations, all NDRTs and a reference trial without NDRT execution. During trials with NDRT execution, participants needed to continuously execute the task whenever the partial automated system was active and to only cease task execution during take-over situations. In the reference trial, participants drove partially automated and experienced both take-over situations. The trials with and without NDRTs were randomized and counterbalanced to reduce order effects.

Each trial included four take-over situations and started with a short familiarization segment (see Figure 2). The four take-over situations within each trial were to avoid predictability and, hence, a change in gaze behavior. With only two take-over situations, participants could easily predict the second take-over situation after experiencing the first one. It would also be problematic having three take-over situations if the first two were the same, the third would have been easily predictable. With four or more take-over situations within one trial, it was possible to make the order of the situations unpredictable. The two take-over situations were sequentially counterbalanced across the four occurrences to reduce predictability as well as order and learning effects. The first and third occurrences always included the two take-over situations: *lead vehicle deceleration* and *ego vehicle drifting*. The order of the two situations was alternated. For the other two occurrences, the two take-over situations were randomly assigned to reduce predictability. However, identical situations would not follow each other more than twice.

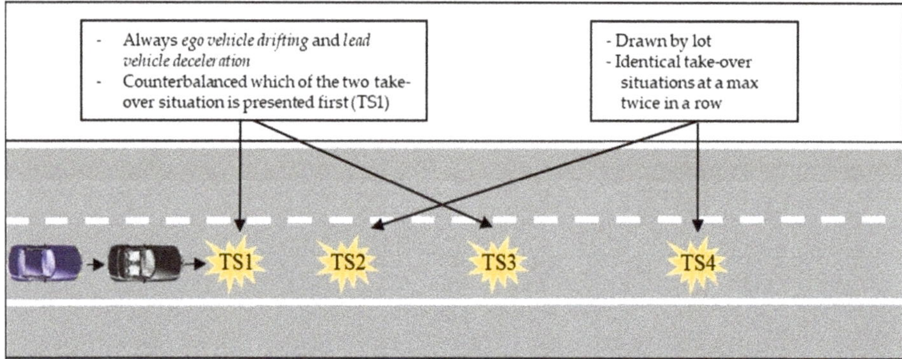

Figure 2. Overview of the four take-over situations presented to the participants per trial.

Participants began the studies receiving information regarding the goal of the study and an informed consent. They experienced manual and partially automated familiarization drives to get accustomed to the driving simulator or ego vehicle and partial automated system. Participants received written instructions for the partially automated system, the take-over situations and task priorities (i.e., giving the safe driving task the highest priority). Moreover, before each trial, the corresponding NDRT was introduced and explained to the participants and, if applicable, each display location as well. Congruent with NHTSA's guidelines, participants practiced the tasks during vehicle standstill to ensure a comparable level of understanding before the trial with data recording started.

2.5.1. Driving Simulator Implementation

In the first simulator study, the NDRTs were included as within-subjects factors, resulting in every participant executing all five NDRTs. In addition, all participants experienced the reference trial, leading to a total of six trials. The second study contained only four of the five NDRTs (see Table 2). The reference trial was excluded to ensure an economic study design. For the same reason, participants only experienced two out of four display locations. Three of the four NDRTs (i.e., the SuRT, text reading, and video watching tasks) were executed on these display locations. The manual radio tuning task was executed on an additional display representing the typical location for in-vehicle sound systems/radios. This resulted in seven trials per participant. The procedure is described in Section 2.5. Study participation took 2–2.5 h.

2.5.2. Test Track Implementation

In the test track study, three of the five NDRTs (see Table 2) were included as within-subjects factors. In addition, all participants experienced a reference trial, resulting in four trials total. The procedure was identical to the simulator studies (see Section 2.5) with one small exception: an additional manual familiarization drive used to accustom participants to the test track. This drive did not include take-over situations. The study lasted approximately two hours for each participant.

2.6. Data Preparation and Analysis

Regarding the supervisory behavior, either the data from the head-mounted eye tracker or the video annotations were used for further analysis. The data were prepared and analyzed using the specification of a gaze towards an AOI, following the ISO standard (EN ISO 15007-1) [54] definition of glance duration. This can be defined as the time from when a gaze initially moved towards an AOI to when it moved away towards another AOI, which would include all consecutive fixations towards this AOI during that time. This includes all saccades occurring within this time as well [54]. The three studies focused on gazes towards the following AOIs:

1. Driving scene—Gazes through the windshield, directed towards the driving scene
2. NDRT—Gazes inside the vehicle, towards where NDRTs were executed (i.e., towards the head unit in the first and third study or to different locations in the second study)
3. Instrument cluster + steering wheel—Gazes inside the vehicle, towards the instrument cluster and steering wheel
4. Vehicle interior—Gazes inside the vehicle that were not directed to the NDRT or other relevant locations (e.g., gazes to the researcher in the passengers' seat during the third study)

To analyze participants' supervisory capability during PAD, a predefined segment prior to a system failure and take-over situation occurring was examined (i.e., several seconds before the take-over situation occurred).

Data recorded by the simulator or vehicle were gathered for further analyses of the take-over capabilities.

Across the three studies, only the first and third take-over situations including both take-over situation types were further analyzed for each trial. This was done to reduce predictability of the upcoming situation and to ensure a consistent number of analyzed events for each participant and trial. As the first take-over situation was chosen randomly, it was completely unpredictable and thus participants' gaze behavior was assumed to be as natural as possible (i.e., checking for both possibilities, lane deviations or reduced distance to the lead vehicle). The second take-over situation was chosen randomly as well. Hence, the second situation could be the same take-over situation as the first one or it could be the other one. It was believed that participants would likely expect the other take-over situation they had not experienced to occur and hence adjust their gaze behavior accordingly (e.g., just checking for lane deviations). The third situation was always the take-over situation participants had not experienced in the beginning. It was thought that participants' will once again scan for both possible take-over situations (i.e., show natural gaze behavior) after realizing that there is no systematic presentation of the take-over situations (e.g., in an alternating manner). The fourth situation was again chosen randomly.

For take-over capabilities, the performance from when a take-over situation begins to when the participants' make their initial response was analyzed. If participants did not react, their performance from when the take-over situation commenced until a crash or an intervention of the vehicle or researcher occurred was analyzed. The take-over and supervisory capabilities during the NDRT execution trials were compared to each other, to the reference trial without NDRT and, additionally, to the manual radio tuning task.

2.6.1. Driving Simulator Implementation

In both simulator studies, a 1-km segment equal to a duration of approximately 45 s prior to the beginning of the first and third take-over situation for each trial was used to analyze the supervisory capabilities based on the relevant parameters (see Section 2.3). It was assumed that the gaze behavior prior to the take-over situations did not differ depending on the following situation due to the study's design implemented to reduce predictability (see Section 2.6). As the data supported this assumption, the supervisory behavior prior to the situations was averaged across both events. The take-over capability was analyzed from the start of the take-over situation until participants' initial reaction or a collision (see Section 2.6).

To examine the test protocol's ability of differentiating the effects across different NDRTs and display locations, as well as whether these differences are as expected (RQ1), repeated measures ANOVAs (rmANOVAs) were used that are highly robust, even with slight deviations from the assumption of normality [55].

2.6.2. Test Track Implementation

In the test track study, a 10-s segment prior to the beginning of a system failure and take-over situation was examined to determine participants' supervisory capabilities. The segment length was chosen to ensure that the previous take-over situation would not interfere with the analyzed segment. Hence, this required the previous take-over situation to be completed and the partially automated system to be active again so that the 10-s segment included only actual supervisory behavior during PAD. This was necessary because the two take-over situations could have occurred within one round with relatively little time in between.

The take-over capabilities were analyzed identically to those in the simulator studies with one exception: The analyzed interval would end with the researcher's intervention in the ego vehicle or with the ego vehicle itself in case participants did not react.

Due to the within-subjects design, robust repeated measures ANOVAs (rmANOVAs) were computed.

2.7. Participants

Following NHTSA's methodology, the goal of the three studies was to balance participants' gender and age across the four age groups described in the methodology [14]. This was to achieve a heterogeneous participant group, which allows for controlling and assessing any gender or age effects. Table 3 shows the distribution across age groups, gender, and the total number of participants in the studies and the total number participants actually analyzed.

Table 3. Number of participants, age, and gender distributions of participants per study.

Studies	N (Total)	N (Analyzed) [1]	Age Groups				Gender Distribution
			18–24	25–39	40–54	Older than 55	
Study 1—Simulator	57	47	12	15	12	8	27 male, 20 female
Study 2—Simulator	58	50	15	11	13	11	27 male, 23 female
Study 3—Test track	39	36	8	11	9	8	20 male, 16 female

[1] Exclusion of several participants due to, e.g., simulator sickness or technical problems, resulted in the reduced number of analyzed cases.

3. Results

This section shortly covers the results of the formulated research questions that examine the sensitivity of the test protocol and necessary parameters to sensitively evaluate participant take-over and supervisory capabilities. The current article focused on the examination of the proposed test protocol's suitability to sensitively evaluate the effects of NDRTs on the drivers during PAD and not on the NDRTs' particular effects on the drivers' take-over and supervisory capabilities. For a better understanding, an exemplary description of the supervisory capabilities (i.e., mean gaze duration) will be presented. The specific results of the supervisory and take-over capabilities across the NDRTs and display locations within the three studies will be described in more detail in separate papers e.g., [56].

Regarding the supervisory capabilities, several parameters were analyzed (see Section 2.3). Table 4 shows the effect sizes for the main effects of NDRTs and display locations across these parameters. Following the convention of Cohen [57], the effect sizes were categorized into weak ($\eta^2 < 0.06$), medium (η^2 between 0.06 and 0.14), and strong effects ($\eta^2 > 0.14$). All three studies revealed predominantly strong effects for the assessed parameters regarding the NDRTs. The results also corroborated the expectations. Figure 3 exemplary shows the effects of NDRT execution on the drivers' supervisory capabilities in terms of mean gaze duration towards the executed NDRT (in seconds) for the first and third study. In line with the expectations, the non-compliant browsing and text reading task resulted in considerably longer mean gaze durations towards the NDRTs than the manual radio tuning task. The guideline conform SuRT resulted in comparably long mean gaze durations towards the task as the reference task (i.e., manual radio tuning task). Less expectedly, the mean gaze duration towards the video watching task in the first study was only slightly longer and in the third study even shorter compared to the manual radio tuning task. The main effect of the display locations yielded strong effects as well. Again, the results were congruent with expectations. For instance, executing NDRTs with the smartphone was more captivating and resulted in less supervision of the driving scene than execution on the instrument cluster. Hence, with regard to the supervisory capabilities, the strong effects for both independent variables indicated the test protocols' ability to sensitively differentiate between NDRTs with different visually distractive potentials (RQ1a) as well as to sensitively detect differences between various (in-vehicle) display locations (RQ1b). Moreover, with exception of the video watching task, the results were in line with the expectations based on the literature (RQ1).

Table 4. Effects of the parameters used to examine supervisory capabilities across the three studies.

	Study 1—Simulator	Study 2—Simulator		Study 3—Test Track
	Non-Driving Related Tasks (NDRTs)	NDRTs	Display Locations	NDRTs
Mean Gaze Duration	Strong effect [1]	Medium effect	Strong effect	Strong effect
Total Gaze Duration	Strong effect	Strong effect	Strong effect	Strong effect
Maximum Gaze Duration	Strong effect	/	/	/
Number of Gazes	Strong effect	/	/	/
Number of Transitions	Strong effect	Strong effect	Strong effect	Strong effect

[1] Following Cohen [57], weak effects are defined as $\eta^2 = 0.00$–0.06, medium effects as $\eta^2 = 0.06$–0.14 and strong effects as $\eta^2 > 0.14$.

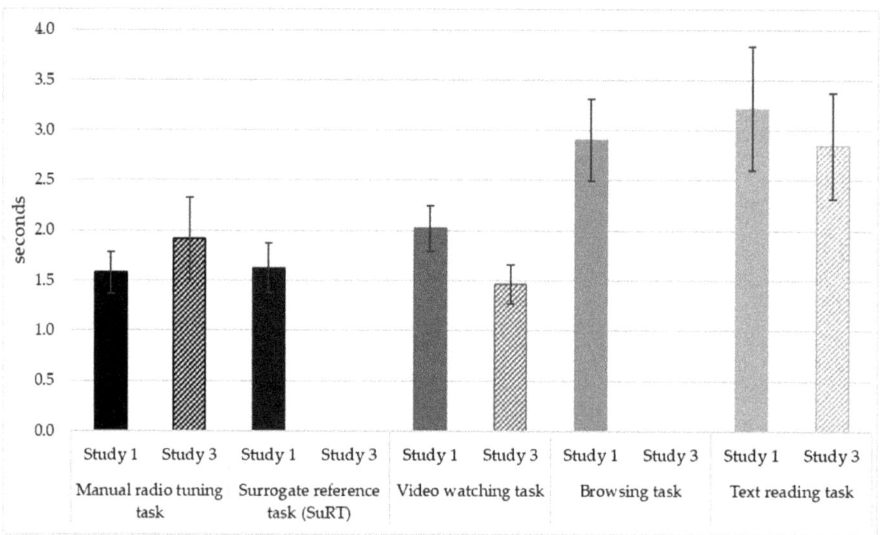

Figure 3. Mean gaze duration towards the NDRTs (in seconds) depending on the executed non-driving related task for study 1 in the simulator and study 3 on the test track.

Concerning the take-over capabilities, several parameters were analyzed. Table 5 shows effect sizes for the main effects of NDRTs and display locations. The first simulator study revealed predominantly strong effects regarding the main effects of NDRTs concerning reaction time. In the other two studies, NDRT effects regarding reaction time ranged from weak to strong depending on the particular display location on which the tasks were executed (or weak in the case of the third study). The results were also congruent with expectations. For instance, more distractive NDRTs (e.g., browsing task) resulted in impaired take-over capabilities including longer reaction times. Medium sized effects existed for the reaction time dependent on the display location. For example, the smartphone resulted in longer reaction times as was expected, indicating the test protocol's ability to sensitively detect these differences. The effect sizes regarding the other parameter of the first study were predominantly strong and weak regarding the third study. Hence, specific parameters of the test protocol (e.g., reaction time) were sensitive to NDRT effects with varying visually distractive potentials (RQ1a) and to some extent sensitive to display location effects (RQ1b). Further, the results were in line with the expectations based on the literature (RQ1).

Table 5. Effects of the parameters used to examine take-over capabilities across the three studies.

	Study 1—Simulator	Study 2—Simulator		Study 3—Test Track
	NDRTs	NDRTs	Display Locations	NDRTs
Reaction time	Strong effect [1]	Weak—strong effect	Medium effect	Weak effect
Minimal distance to the lead vehicle at initial reaction	Strong effect	/	/	Weak effect
Maximal brake pressure	Strong effect	/	/	/
Maximal steering angle	Weak effect	/	/	/
Number of crashes	Only descriptive analyses.			/

[1] Following Cohen [57], weak effects are defined as $\eta^2 < 0.06$, medium effects as $\eta^2 = 0.06$–0.14 and strong effects as $\eta^2 > 0.14$.

In sum, regarding the first research question (RQ1) and the sub questions (RQ1a and RQ1b), the results showed that the test protocol was sensitive to the effects of different NDRTs and (in-vehicle) display locations. Especially, the supervisory capabilities were proven very sensitive to these effects.

As mentioned earlier (see Section 2.3.1), all parameters described in Section 2.3 were assessed within the first study and all parameters concerning the supervisory capabilities yielded strong effects. However, the mean gaze duration and the maximum gaze duration were very similar in terms of their effect strengths (see Table 4) as well as in terms of the particular results of the effect of NDRT execution. More precisely, the maximum gaze duration presented very similar findings as the mean gaze duration as presented in Figure 3: visually more distracting tasks (i.e., the browsing and the text reading task) resulted in considerably higher mean and higher maximum gaze durations than visually less distracting tasks (i.e., manual radio tuning task and SuRT). Further, in contrast to the number of gazes towards one AOI, the number of transitions is more useful as it combines the information from the number of gazes towards two AOIs and is a good indicator of drivers' compensatory behavior. Hence, the maximum gaze duration and the number of gazes were not analyzed in Study 2 and 3.

Regarding the take-over capabilities, the effects found within the first study were predominantly strong as well. However, the minimal distance to the lead vehicle at initial reaction, the maximal brake pressure, and the maximal steering angle are logically connected with the reaction time. For instance, longer reaction times logically result in a reduced minimal distance towards the lead vehicle, hence, demanding stronger initial reactions (e.g., higher brake pressure). Therefore, these variables were not assessed in Study 2 and 3. The number of crashes, which was analyzed descriptively, was a useful addition to the reaction time.

Finally, a selection of the most useful parameters was chosen for the following studies based on the first study, including: mean gaze duration, total gaze duration, number of transitions between the driving scene and NDRT, reaction time and number of crashes (RQ2). However, the number of crashes, which was analyzed descriptively, was a useful addition to the reaction time for the second study, but could not be assessed within the third study to ensure participants' safety. In order to evaluate the test protocol in light of the changes made for the test track environment, the minimal distance to the lead vehicle at initial reaction was assessed within the third study again, but yielded only weak effects.

4. Discussion

This project's overarching goal was to take the initial steps towards developing a test protocol that systematically evaluates drivers' supervisory and take-over capabilities during PAD. The research questions addressed the test protocol's ability to sensitively detect differences (as expected based on the literature) in drivers' supervisory and take-over capabilities during PAD across different NDRTs (RQ1a) and display locations (RQ1b). Moreover, it was examined which parameters are sufficient to sensitively capture and evaluate drivers' take-over and supervisory capabilities (RQ2).

The three studies revealed mixed results concerning the test protocol's sensitivity to detect the effects of visual-manual NDRTs (RQ1a) and (in-vehicle) display locations (RQ1b) on drivers' supervisory and take-over capabilities during PAD. Regarding the supervisory capabilities, predominantly strong effects existed for most of the analyzed gaze parameters. This firmly indicates the test protocol's ability to sensitively detect differences in the drivers' supervisory behavior based on the executed visual-manual NDRT as well as display location (on which a NDRT is executed) (RQ1). As described in Section 3, the mean gaze duration, total gaze duration, and the number of transitions were deemed as the most useful parameters that sufficiently examine supervisory capabilities during PAD, since they yielded strong effects (RQ2). Additionally, these parameters still provide the required data to compute other parameters. For instance, the total gaze duration adds together all single gaze durations, from which the maximum gaze duration can be extracted. Moreover, with exception of the video watching task, the detected differences were congruent with the expectations as was shown exemplary for the mean gaze duration towards the NDRTs. For instance, more distractive tasks (e.g., browsing task) resulted in poorer supervision compared to the manual radio tuning task. In contrast, the video watching task appeared to be less visually distracting than expected, in terms of only slightly poorer supervision than during manual radio tuning task. However, the news video segments had low visual attraction and the content was predominantly presented aurally rather than visually. Other videos with greater visual attraction (e.g., blockbuster videos) might be more distractive, resulting in longer gazes that might influence drivers' supervisory and take-over capabilities more negatively. Nevertheless, the results of the supervisory capabilities, based on the examined parameters, can sufficiently answer the first research question and corresponding sub questions. However, the findings concerning take-over capabilities were less clear, especially in the third study. Even though reaction time yielded the strongest effect sizes for the differentiation between the NDRTs across the simulator studies and, therefore, seemed to be the best indicator of drivers' take-over capabilities and situation criticality (RQ2), this was not replicated in the closed test track study. However, the weaker effects were likely due to the changes and adaptations made to the test protocol for applicability to the test track scenario's limited space. Especially, having the take-over situations always being executed on the same track segment greatly increased the predictability of the take-over situations compared to the simulated environment. After the first trial, participants knew where the take-over situations would occur and were then likely more attentive during these track segments in the following trials. This likely resulted in weaker effects for NDRT differentiation.

Generally, the vigilant supervision of the driving scene and system enables the drivers to notice system failures in a timely fashion and prepares them to make any necessary and timely intervention if such a case arises [2,18]. Hence, despite the partially weaker effects regarding take-over capabilities, the supervisory capabilities are strongly related to the former. Therefore, the strong effects concerning the supervisory capabilities are promising and indicate that the test protocol is useful to differentiate between the effects of different visual-manual NDRTs on drivers during PAD (RQ1). Nevertheless, it is still necessary to examine the NDRTs' effects on parameters indicative of the situation's criticality and the drivers' take-over capability, such as reaction time and the number of crashes (RQ2). Both are relevant supplements to the supervisory parameters, when drawing conclusions about NDRTs' influence on drivers during PAD.

In general, the new test protocol should form the basis to assess how different NDRTs influence drivers' supervisory and take-over capabilities during PAD and, hence, to decide whether certain NDRTs are suitable for execution during PAD. Currently, conclusions can only be drawn based on the three studies conducted for this project. Nevertheless, based on comparing the tested NDRTs versus the manual radio tuning task across the three studies, some NDRTs seem less suitable than other tasks. For instance, the browsing and text reading tasks distracted drivers considerably more in terms of longer gazes towards the NDRTs and poorer take-over capability than seen with the manual radio tuning task. Following the NHTSA guidelines, stating that a task is not appropriate (for manual driving) when visual and driving performance are poorer than the manual radio tuning task, the browsing and text reading tasks would not be suitable for PAD. In contrast, the video watching task and SuRT showed similar results to the manual radio tuning task regarding drivers' supervisory behavior and take-over capabilities. Hence, the SuRT and video watching task might be rendered appropriate for PAD. However, final conclusions, especially regarding the suitability of the video watching task, should not yet be drawn. Moreover, conclusions regarding NDRT suitability during PAD should be handled cautiously since the test protocol is not yet broadly established.

4.1. Future Research

To draw conclusions concerning whether a NDRT is suitable for execution during PAD, some further steps are necessary. Firstly, further studies conducted in different environments using the developed test protocol are necessary to establish cut-off values for PAD comparable to those provided by the NHTSA for manual driving [14]. Secondly, the manual radio tuning task [34] needs to be evaluated regarding its suitability as a still reasonable reference task for PAD. Since drivers are relieved from parts of the driving task, other potentially more distractive tasks might possibly be executed during PAD without negative consequences compared to manual driving. If this is the case, the manual radio tuning task, which is perfectly congruent with the cut-off values for NDRT execution during manual driving (2 s per gaze, 12 s total gaze duration towards the NDRT), might be too conservative for PAD. Hence, if the new PAD cut-off values differ from those of manual driving in terms of longer gazes towards the NDRTs being allowed, the manual radio tuning task might render more NDRTs unsuitable due to being too conservative. Additionally, participants in these three studies were presented with the partially automated system and secondary tasks for only short periods. The effects of prolonged PAD periods should be examined to better understand the willingness and likelihood of NDRT execution during PAD as well as the development of supervisory behavior with increasing system experience.

For these further studies, the following sections include detailed recommendations regarding test protocol usage in both driving environments.

4.2. Recommendations Regarding Test Protocol Implementation

When using the developed test protocol for studies evaluating NDRT effects during PAD in a simulated or real driving environment, we, the authors, would like to provide the following recommendations. These are mainly based on the results and experiences we gathered during the three studies conducted for this project. In addition, further literature enriching these recommendations focused on standardized NDRT evaluation for manual driving e.g., [14] or higher automation levels for instance (i.e., SAE level 3) [31]. In the end, a table is provided giving an overview of the recommendations.

4.2.1. Driving Scenario and Take-Over Situations

The current project employed NHTSA's [14] well-established car-following scenario [14] and extended it to PAD. Given this scenario is implementable in a simulated or real driving environment (i.e., closed test track), we recommend its usage with the necessary extensions (i.e., take-over situations) for further PAD studies. Depending on the particular driving environment, certain adaptations might be necessary.

The recommended scenario extensions include take-over situations considered necessary to examine participants' take-over and supervisory capabilities during PAD. We suggest implementing at least two types of take-over situations addressing system limitations of lateral and longitudinal vehicle control, such as the two take-over situations (*lead vehicle deceleration* and *ego vehicle drifting*) used in the current project. Other take-over situations that realistically address limitations (e.g., losing lateral control due to a curve in the road, missing lane markings, or failing to detect a road obstacle) of the partially automated system can be implemented as well. Independent of the situation type, we advise excluding any warnings or take-over requests to realistically simulate PAD (SAE level 2) as well as any external cues (e.g., trees, houses or brake lights) to reduce predictability of the take-over situations.

The driving simulator scenario can be implemented nearly identically to the NHTSA [14] specifications (see Section 2.1). We highly recommend using predominantly straight road segments for identical implementation of the two take-over situations used in the current studies. If other take-over situations are used, the test track can include curved segments as well. However, these increase the chances of simulator sickness occurring and therefore should be implemented cautiously. Corroborating NHTSA's guidelines [14], we recommend incorporating multiple lanes (i.e., two lanes in each direction) as well, especially with take-over situations addressing lateral vehicle control. Additionally, a beginning segment without take-over situations is advisable to allow participants to start the scenario, activate the partially automated system and execute the NDRTs without time pressure.

As with NHTSA [14], we used a speed of 80 km/h and a distance of 70 m to the lead vehicle in the simulator. A seven second TTC was implemented for the two take-over situations. To enhance situation criticality and scenario validity, researchers can change the speed and distance specifications or use the lead vehicle's variable speed profile [14]. However, the latter can complicate detection of system failures. For greater situation compatibility, the adaptations should result in matching TTCs.

Even though NHTSA's guidelines [14] allow for sparse (oncoming) traffic, we excluded all traffic except for the lead vehicle to reduce potential distractions (especially during the reference trial) and to implement the take-over situations as described. For instance, when implementing the *ego-vehicle drifting* take-over situation, we recommend excluding other traffic during that interval to prohibit any traffic collisions. Other (oncoming) traffic can be included for a more realistic driving scenario or a higher situation criticality.

For real-world driving studies (e.g., closed test track) we recommend implementing the same driving scenario. Therefore, a test track allowing the application of scenario and take-over situations with similar speed or TTC specifications is highly recommended. For the current test protocol, we suggest using a straight track to implement both take-over situations as described. This also ensures that the *ego vehicle drifting* take-over situation is not mistaken for driving around a curve and that driving around a curve is not mistaken for a take-over situation itself. Provided other take-over situations are chosen, curved segments may be necessary.

The test track length depends on the number and timing of the take-over situations. Based on the simulator studies, when driving 80 km/h and implementing four take-over situations, we recommend using an 11-km test track. This allowed an analysis of a 45-s interval, equal to a 1-km route segment, prior to each take-over situation. However, combining four take-over situations on an 11-km test track results in a relatively high frequency of system failures, which might reduce external validity (see Subsection *Experimental Design* in Section 4.2.5). Hence, using an even longer test track is recommended to increase the time and distance between take-over situations to create a more realistic experience for the participants.

If such a test track is not available, adaptations become inevitable. If speed reductions are necessary, the TTC should be reduced in relation to the speed. When using a similar test track as in the current project, it is important to reduce predictability of the take-over situations in terms of time and location as much as possible, as this can strongly influence participants' supervisory and take-over behavior.

Several adaptations should always be made independent of the test track. Firstly, the take-over situations cannot result in a collision with, for instance, the lead vehicle or guardrails if the participants do not react. For this matter, fallback solutions, as described in Section 2.1.2 (i.e., programming of ego vehicle) are necessary to ensure participants' and involved researchers' safety at all times. For the same reason, additional traffic should be excluded as well or, at least, be controlled and reduced to a minimum.

4.2.2. Independent Variables

The test protocol was able to discover and distinguish expected differences between different visual-manual NDRTs. This allowed evaluating guideline compliant and non-compliant tasks as well as artificial tasks and those closer to everyday life. The number of tasks that can be examined is flexible; however, it is recommended to strive for an economic study design. Moreover, we recommend comparing the effects of a partially automated drive with and without NDRT execution or comparing a partially automated drive with NDRT execution to a drive while executing a reference task (e.g., manual radio tuning task [34]). Regarding the manual radio tuning task, we recommend using the version adapted for touch displays [51] to ensure comparable task execution. As in Schömig et al. [31] and NHTSA [14], we recommend predefining the start and finish of task execution when examining distractive effects on the drivers instead of spontaneous task execution. Moreover, participants should practice the tasks to achieve comparable task understanding before each trial see [14].

Furthermore, the current project showed the test protocol's ability to distinguish between the effects of NDRT execution on different display locations. Depending on the research question, different display locations of interest can be included. In the current studies, it was not always possible to use the built-in display locations to present the NDRTs to participants. Even though we attempted to present these NDRTs in similar positions as these built-in display locations occupy and use comparable control elements for execution (e.g., touch displays), using external displays might have reduced the realism of NDRT execution during PAD. It is recommended to use available, built-in displays as much as possible (which should be controlled in a similar manner) to strive for an economically designed study.

Moreover, it seems reasonable to validate the test protocol considering other independent variables that are meaningful for PAD (e.g., prior system experience or different HMI designs).

4.2.3. Analyzed Variables

As previously discussed, several different parameters can be analyzed to evaluate drivers' supervisory capabilities and all parameters that were evaluated, provided strong effects. However, to ensure an economic study design, we suggest using mean and total gaze duration towards the NDRT (and driving scene) and the number of transitions between the driving scene and NDRT as discussed in Section 3. These three parameters can sensitively examine and reflect the supervisory capabilities and compensatory behavior during NDRT execution.

The results showed that take-over capabilities yielded weaker effects than the supervisory capabilities. Nevertheless, take-over capabilities must still be assessed and therefore different parameters can be analyzed. We recommend using reaction time to measure situation criticality, which should be enriched by the number of crashes or lane deviations for example. Other parameters can be used as well (e.g., TTC), but these parameters should be chosen based on their ability to provide additional and valuable information.

4.2.4. Equipment and Materials

Driving Simulator and Test Vehicles

Depending on the study environment, either a driving simulator with a vehicle mock-up and corresponding simulation software or two vehicles (an ego and lead vehicle) are necessary to implement the driving scenario and take-over situations.

For both vehicle mock-up and actual ego vehicle, it is recommended that at least two (in-vehicle displays) are available, including the instrument cluster presenting (automated) system-related information and another display for NDRT execution (e.g., the head unit). The displays must be customizable for study relevant information and the participants must be able to smoothly interact with the display during NDRT execution. It is also suggested to equip the mock-up and actual vehicle with cameras facing participants, the driving scene, and the task to record study relevant behavior. Moreover, any driving input made by participants must be reflected by the simulator or ego vehicle and the corresponding partially automated system in a timely fashion to ensure a realistic system experience. This input includes braking, steering or system (de-)activation by pushing the corresponding buttons on the steering wheel for instance.

It might be useful to incorporate a self-turning steering wheel in the driving simulator to represent a more realistic PAD experience. However, this could cause participants to recognize the ego-vehicle's drifting faster than if there were no movement (especially when driving on a straight road). Moreover, in real driving environments, PAD includes hands-on warnings requiring drivers to leave one hand on the steering wheel at all times. In the current project, participants needed to remove their hands from the steering wheel to mimic an extreme situation. Both aspects must be considered based on the relevant research questions.

For the test track vehicles, we strongly recommend using high automation levels to ensure standardized and replicable driving scenario execution and take-over situations, as well as to reduce chances for human error. At a minimum, the ego vehicle should take over tasks controlled by the partially automated system and should be programmed to deliberately trigger the two take-over situations. If higher automation levels are not possible, Wizard-of-Oz approaches are reasonable alternatives; however, these reduce comparability. The ego vehicle must include sensors (e.g., LiDAR or Novatel DGPS) and devices to record driving data. The lead vehicle should at least include ADAS (i.e., cruise control), the drivers should be extensively trained on their tasks, and landmarks should exist for comparable execution of take-over situations. Additionally, it is recommended to synchronize the vehicles. This could include using walkie-talkies; however, programmed synchronization would be preferable for standardization and replication.

Human–Machine Interface

As mentioned by Schömig et al. [31] for SAE level 3 automation, the human–machine interface (HMI) should present participants with all relevant system states (e.g., active or inactive) and corresponding transitions between these states. The instrument cluster would be the most suitable since it presents drivers with further driving related information (e.g., speed). Additionally, the HMI must reflect participants' input (system activation and deactivation) in a timely fashion. When the goal is focusing on the effects of different visual-manual NDRTs, as with the current test protocol, we recommend using a minimal, intuitively understandable HMI that does not distract drivers from NDRT execution or cause mode confusion.

Eye Tracking

Driver's gaze behavior must be recorded to evaluate their supervisory capabilities. Depending on the detail level (e.g., AOIs, fixations) examined, study design (i.e., study length and environment), or test sample of interest (e.g., younger vs. older participants), the researcher must decide between using a head-mounted eye tracking system or video annotations (see Section 2.4 for a more detailed discussion of the (disadvantages of both methods).

Questionnaires

At the least, we highly suggest collecting participants' demographic information (e.g., age, gender, and prior system experiences). In addition, further questionnaires administered before and after trials with and without NDRT execution would supplement the objective data with subjective experiences, which would help shed light on possible explanations for their past or potentially future behavior such as willingness to execute NDRTs during PAD.

Instructions

As with NHTSA [14] and Schömig et al. [31], we recommend using written instructions regarding the following aspects to enhance standardization. Firstly, the NDRT execution should be clearly communicated, including the NDRT's goal, what constitutes successful execution, and when NDRT should be executed. When examining NDRT's distractive effects during PAD, we suggest instructing participants to continuously execute the NDRTs when the partially automated system is active and the situations allow it based on the participant's judgment, which corroborates Schömig et al. [31] and NHTSA's recommendations [14]. The instructions should also explain participants' task priorities, such as the safe execution of the driving task has the highest priority. Secondly, to ensure comparable system understanding, the partially automated system's usage and states should also be explained to participants. Only when researchers are interested in intuitive system interaction should these instructions be excluded e.g., [31]. In addition, similar to a partially automated vehicle manual, the system limits and corresponding take-over situations should be discussed with participants as well. Depending on the research questions, it might be useful to describe the most appropriate reaction to the situation, except when attempting to capture participants' spontaneous reactions.

In general, as with Schömig et al. [31] we recommend explaining system functionalities, limits and take-over situations in detail to reduce possible learning effects due to experiencing multiple take-over encounters that are recommended for the current test protocol. However, when focusing on initial contact with the system and take-over situation, reduced instructions are more suitable e.g., [31].

4.2.5. Experimental Design and Procedure

Experimental Design

For both study environments, the design depends on the research question. However, we recommend including a complete, within-subjects design limiting the number of independent variables to ensure an economic study design, reduce test sample size, enhance statistic power, enable direct comparisons of participant performance across the independent variables, and exclude interindividual confounding variables. Additionally, it is highly important to randomize and counterbalance trials to reduce learning and order effects.

Regarding the number of take-over situations, it is recommended to repeat the encounters and in order to reduce learning and first contact effects it is recommended to clearly instruct the participants regarding the system's functionality and limitations [31]. Regarding the number of take-over situations, aspects such as the length of the analyzed intervals as well as the influence of the take-over situations' number on the system evaluation [31], must be considered. For a duration of 8–12 min as in the current studies, we recommend a maximum of four encounters, which should be randomized and counterbalanced across timing and situation type to reduce predictability. However, this recommendation aims at maximizing the number of take-over situations to be analyzed. This high frequency of system failures potentially lowering external validity must be considered. Depending on the research question, the number of take-over situations should be reduced and the route length should be extended (e.g., to evaluate how supervisory and take-over capabilities evolve over time and with long periods without system failures).

Procedure

The actual procedure depends on, for instance, the study design, employed techniques, questionnaires, etc. Generally, we highly suggest including familiarization drives as mentioned in Schömig et al. [31]. In both study environments, participants should get accustomed driving manually in the simulator or actual vehicle if possible. For the former, this also allows checking for signs of simulator sickness. Depending on the research questions, participants should also be familiarized with partially automated driving and potentially with the take-over situations. We recommend familiarizing participants with partially automated driving but not with the take-over situations. This allows to achieve a comparable understanding of PAD across participants as well as to analyze the initial contact with these situations during NDRT execution. Nevertheless, the possibility of some take-over situations occurring during the trials and take-over situations itself should be described to the participants in the instructions.

Depending on the study's complexity, we suggest involving two researchers who can divide the technical tasks and participant supervision between each other to ensure a smoothly conducted study. In case of additional tasks (e.g., driving the lead vehicle), including another researcher is advisable. The researchers should receive detailed instructions and extensive trainings regarding their tasks, especially considering any driving tasks.

4.2.6. Data Preparation and Analyses

Regarding the supervisory behavior, the camera recordings or the eye tracking data must be annotated or mapped concerning the relevant AOIs: the NDRT, driving scene, instrument cluster, and vehicle interior. Other AOIs can be included if needed. The take-over capability data must be extracted from the simulator or ego vehicle and prepared for further analyses.

When examining supervisory capabilities during PAD, we recommend using an interval prior to the take-over situation. In that interval, the partially automated system must be active and should exclude any parts of earlier take-over situations. Therefore, the interval length depends on the time between the take-over situations. For instance, the current project included a 45-s segment in the simulator studies and a 10-s segment in the test track study, whereas Dogan et al. [25] chose a 15-s segment before a take-over situation occurred. In general, the interval length should be long enough to include at least one complete NDRT execution trial. In the current case, the NDRT trials were designed to take no longer than 20 s. Since participants are unlikely to complete a trial within 20 consecutive seconds, we recommend using a generous interval of 45 s for instance. Moreover, NHTSA [14] specifies that a NDRT trial should be completed within a total gaze duration of 12 s. With a 45 s interval, it should be possible to find these cumulative 12 s of total gaze duration as well. Moreover, if new PAD cut-off values are less conservative and result in longer total gaze durations towards the NDRTs, the 45-s intervals might also provide enough buffer for this. In contrast, to examine take-over capabilities, we suggest using an interval from the moment the take-over situation is triggered until participants' initial reaction. If participants do not react, the interval should last until the collision occurs or the researcher terminates the situation.

The current project analyzed the first and third situation (see Section 2.6). Depending on the research questions, other analyses can be done as well, such as comparing the first and last take-over situation or all take-over situations. However, the latter is only possible if the predictability of the take-over situations is low. Comparisons between the trials with and without NDRTs as well as between the trial with the reference task (e.g., manual radio tuning task) and the trials with other NDRTs are recommended to evaluate drivers' supervisory and take-over capabilities. The concrete analyses depend on the chosen research design.

4.2.7. Participants

Concerning the participants, the following aspects must be considered. Firstly, the *sample size*. NHTSA [14] recommends including 24 participants to examine the distractive effects of visual-manual NDRTs. For studies involving conditional automated driving (SAE level 3), a sample size of at least $n = 20$ is recommended when assessing the suitability of in-vehicle systems or at least $n = 12$ participants per experimental test condition [31]. In general, desired sample size depends on the research question and intended statistical power. As with Schömig et al. [31], it is recommended to include at least $n = 12$ participants per experimental test condition or $n = 20$ participants depending on the research design. Secondly, the age distribution must be considered. The current studies aimed to follow NHTSA's guidelines of distributing the participants evenly across four recommended age groups: 18–24 years, 25–39 years, 40–54 years, and older than 55 [14]. The age distribution had no effect in either of the three studies. Nevertheless, we recommend involving all relevant age groups in the sample to control for age effects and reflect on different levels of driving experience. As with Schömig et al. [31], it is advisable to use the four age groups NHTSA highlights to achieve a heterogeneous age group. However, it must be taken into account that evenly distributing participants across these four age groups does not realistically reflect the populations' age distribution. Thirdly, the gender distribution must be considered. NHTSA [14] recommends having an even gender distribution. Similar to the age distribution, gender did not affect the results in either of the current three studies. When examining subjective PAD or NDRT execution experiences, it might still be useful to obtain an even gender distribution as Schömig et al. [31] recommend. We also recommend including an even gender distribution to control for gender effects. In addition to these three aspects, it might be reasonable to examine other sample characteristics as well, such as prior system experience, depending on the research questions.

5. Conclusions

In conclusion, the current project's overarching goal was to fill the methodological gap and take initial steps towards developing a test protocol for the systematic evaluation of the effects of NDRT execution on the drivers' supervisory and take-over capabilities during PAD. We believe that the systematic evaluation of the NDRTs' effects during PAD using the new test protocol developed within this project enhances comparability between different studies and generalizability of the studies' results, as well as provides a basis for developing cut-off values for deciding whether certain NDRTs are applicable for PAD. For the matter of using the test protocol, we provide a summarizing overview of the most important recommendations in Table 6.

Table 6. Short summary of the recommendations for test protocol usage.

	Recommendations
Driving scenario	- Simulator: Driving scenario (car-following task) as described in Section 2.1, without other traffic and curved road segments
	- Test track: Same scenario as in simulator, requires appropriate test track (e.g., straight segment with minimum 11 km length)
Take-over situations	- Types: Responding to both lateral and longitudinal vehicle control (e.g., *deceleration of lead vehicle* and *ego vehicle drifting*) - Specifications: Exclusion of warnings and take-over requests, matching time to collisions (e.g., 7 s), multiple, counterbalanced, and randomized encounters
Non-driving related tasks (NDRTs)	- Types: Visual-manual NDRTs in comparison with reference task (e.g., manual radio tuning task [34]) and reference trial without NDRT execution - Specifications: Predefined start and finish, continuous execution while system is active
Analyzed variables	- Supervisory capabilities: mean gaze duration, total gaze duration, number of transitions - Take-over capabilities: reaction time and parameters for situation criticality (e.g., number of crashes)
Equipment	- Vehicle: Simulated mock-up or (automated) vehicles (see Section 2.4) - Human–machine interface: simple, intuitive design with relevant system states, timely reactions to input - Eye-tracking: head-mounted or video-based depending on level of detail required - Questionnaires: at a minimum demographic information - Instructions: written form, regarding partially automated system, take-over situations, NDRTs, task priorities
Design and procedure	- Study design: economic (e.g., limitation of number of NDRTs), for instance within-subjects design - Number of take-over situations: max. 4 per trial of a length of 8–12 min
Procedure	- Familiarization: Manual and partially automated drives (without take-over situations), NDRT execution
Data preparation and analysis	- Analyzed intervals: 45 s prior to take-over situations for supervisory capabilities, start of take-over situations until participant reaction, collision or fallback solutions - Analyzed take-over situations: in case of 4 take-over situations, first and third take-over situation
Participants	- Number: at least $n = 12$ per experimental group or at least $n = 20$ in total - Demographics: age and gender controlled (e.g., following NHTSA [14])

Author Contributions: Conceptualization, C.H., N.R., F.N., S.H., J.F.K., and A.K.; methodology, C.H., N.R., F.N., and S.H.; investigation, C.H.; formal analysis, C.H. and N.R.; resources, F.N., S.H., J.F.K., and A.K.; writing—original draft preparation, C.H.; writing—review and editing, N.R., F.N., S.H., J.F.K., and A.K.; supervision, J.F.K. and A.K.; project administration, C.H., N.R., F.N., and S.H.; funding acquisition, J.F.K., C.H., and N.R. All authors have read and agreed to the published version of the manuscript.

Funding: This study was funded and supported by the BMW Group, Germany.

Acknowledgments: All opinions expressed in this paper are those of the authors and not necessarily those of the BMW Group. We thank Sebastian Scholz for his technical support during the simulator studies and Benjamin Jähn for his technical support and his support during the execution of the test track study.

Conflicts of Interest: The authors declare no conflicts of interest. The funder (BMW Group) was left unmentioned during the acquisition of the participants and the study execution. Moreover, tasks such as the data analyses were done by the authors belonging to the Technical University of Chemnitz as an independent research institute with the authors belonging to the BMW Group providing advice and support from a solely scientific point of view.

References

1. Flemisch, F.; Kelsch, J.; Löper, C.; Schieben, A.; Schindler, J. Automation spectrum, inner/outer compatibility and other potentially useful human factors concepts for assistance and automation. In *Human Factors for Assistance and Automation*; de Waard, D., Oberheid, F.O., Flemisch, B., Lorenz, H., Brookhuis, K.A., Eds.; Shaker Publishing: Maastricht, The Netherlands, 2008; pp. 1–16.

2. Merat, N.; Jamson, A.H.; Lai, F.C.; Carsten, O. Highly automated driving, secondary task performance, and driver state. *Hum. Factors* **2012**, *54*, 762–771. [CrossRef] [PubMed]

3. SAE International. *Taxonomy and Definitions for Terms Related to Driving Automation Systems for On-Road Motor Vehicles (No. J3016)*; SAE International: Warrendale, PA, USA, 2016.

4. Stanton, N.A.; Young, M.S. Vehicle automation and driving performance. *Ergonomics* **1998**, *41*, 1014–1028. [CrossRef]

5. Van den Beukel, A.P.; van der Voort, M.C. Design considerations on user-interaction for semi-automated driving. In Proceedings of the 35th FISITA World Automotive Congress, Maastricht, The Netherlands, 2–6 June 2014; pp. 1–8.

6. Körber, M.; Weißgerber, T.; Kalb, L.; Blaschke, C.; Farid, M. Prediction of take-over time in highly automated driving by two psychometric tests. *DYNA* **2015**, *82*, 195–201. [CrossRef]

7. Saxby, D.J.; Matthews, G.; Warm, J.S.; Hitchcock, E.M.; Neubauer, C. Active and passive fatigue in simulated driving: Discriminating styles of workload regulation and their safety impacts. *J. Exp. Psychol. Appl.* **2013**, *19*, 287–300. [CrossRef] [PubMed]

8. Coughlin, J.F.; Reimer, B.; Mehler, B. Monitoring, managing, and motivating driver safety and well-being. *IEEE Pervasive Computing* **2011**, *10*, 14–21. [CrossRef]

9. Regan, M.A.; Hallett, C.; Gordon, C.P. Driver distraction and driver inattention: Definition, relationship and taxonomy. *Accid. Anal. Prev.* **2011**, *43*, 1771–1781. [CrossRef] [PubMed]

10. Spence, C.; Ho, C. Crossmodal information processing in driving. In *Human Factors of Visual and Cognitive Performance in Driving*; Castro, C., Ed.; CRC Press: Boca Raton, FL, USA, 2009; pp. 187–200.

11. Renge, K. The effects of driving experience on a driver's visual attention. An analysis of objects looked at: Using the 'verbal report' method. *Intern. Asso. Traffic Safety Sci. Res.* **1980**, *4*, 95–106.

12. Hughes, P.K.; Cole, B.L. What attracts attention when driving? *Ergonomics* **1986**, *29*, 377–391. [CrossRef]

13. Ahlstrom, C.; Kircher, K. Changes in glance behavior when using a visual eco-driving system – A field study. *Appl. Ergon.* **2017**, *58*, 414–423. [CrossRef]

14. National Highway Traffic Safety Administration. *Visual-Manual NHTSA Driver Distraction Guidelines for In-Vehicle Electronic Devices*; Department of Transportation: Washington, DC, USA, 2016.

15. Birrel, S.A.; Fowkes, M. Glance behaviours when using an in-vehicle smart driving aid: A real-world, on-road driving study. *Transp. Res. F Traffic Psychol. Behav.* **2014**, *22*, 113–125. [CrossRef]

16. Ayama, M.; Hasegawa, H.; Kawaguchi, M.; Ihata, N.; Ikegami, M.; Kasuga, M. A study to measure spare capacity of driver's attention payable to cognitive subtask. In Proceedings of the 5th IEEE International Conference on Intelligent Transportation Systems, Singapore, 6 September 2002; pp. 279–283. [CrossRef]

17. Uno, H.; Hiramatsu, K. Effects of auditory distractions on driving behavior during lane change course negotiation: Estimation of spare mental capacity as a index of attention distraction. *JSAE Rev.* **2000**, *21*, 219–224. [CrossRef]

18. Jamson, A.H.; Merat, N.; Carsten, O.M.; Lai, F.C.H. Behavioural changes in drivers experiencing highly-automated vehicle control in varying traffic conditions. *Transp. Res. C Emerg. Technol.* **2013**, *30*, 116–125. [CrossRef]

19. Llaneras, R.; Salinger, J.; Green, C. Human factors issues associated with limited ability autonomous driving systems: Drivers' allocation of visual attention to the forward roadway. In Proceedings of the 7th International Driving Symposium on Human Factors in Driver Assessment, Training and Vehicle Design, New York, NY, USA, 17–20 June 2013; pp. 92–98.

20. Carsten, O.; Lai, F.C.H.; Barnard, Y.; Jamson, A.H.; Merat, N. Control Task Substitution in Semiautomated Driving: Does It Matter What Aspects Are Automated? *Hum. Factors* **2012**, *54*, 747–761. [CrossRef] [PubMed]

21. Hergeth, S.; Lorenz, L.; Vilimek, R.; Krems, J.F. Keep Your Scanners Peeled: Gaze Behavior as a Measure of Automation Trust During Highly Automated Driving. *Hum. Factors* **2016**, *58*, 509–519. [CrossRef] [PubMed]

22. Cunningham, M.L.; Regan, M.A. Driver distraction and inattention in the realm of automated driving. *IET Intell. Transp. Syst.* **2017**, *12*, 1–7. [CrossRef]

23. Martens, M.; van den Beukel, A. The road to automated driving: Dual mode and human factors considerations. In Proceedings of the 16th International IEEE Annual Conference on Intelligent Transportation Systems, The Hague, The Netherlands, 6–9 October 2013; pp. 2262–2267.

24. Eriksson, A.; Stanton, N.A. Takeover time in highly automated vehicles: Noncritical transitions to and from manual control. *Hum. Factors* **2017**, *59*, 689–705. [CrossRef]

25. Dogan, E.; Rahal, M.C.; Deborne, R.; Delhomme, P.; Kemeny, A.; Perrin, J. Transition of control in a partially automated vehicle: Effects of anticipation and non-driving-related task involvement. *Transp. Res. F Traffic Psychol. Behav.* **2017**, *46*, 205–215. [CrossRef]

26. Lin, R.; Liu, N.; Ma, L.; Zhang, T.; Zhang, W. Exploring the self-regulation of secondary task engagement in the context of partially automated driving: A pilot study. *Transp. Res. F Traffic Psychol. Behav.* **2019**, *64*, 147–160. [CrossRef]

27. Large, D.R.; Banks, V.A.; Burnett, G.E.; Baverstock, S.; Skrypchuk, L. Exploring the behaviour of distracted drivers during different levels of automation in driving. In Proceedings of the 5th international conference on driver distraction and inattention (DDI2017), Paris, France, 20–22 March 2017.

28. Lee, J.; Hirano, T.; Hano, T.; Itoh, M. Conversation during partially automated driving: How attention arousal is effective on maintaining situation awareness. In Proceedings of the 2019 IEEE International Conference on Systems, Man and Cybernetics (SMC), Bari, Italy, 6–9 October 2019; pp. 3718–3723. [CrossRef]

29. Louw, T.; Kuo, J.; Romano, R.; Radhakrishnan, V.; Lenné, M.G.; Merat, N. Engaging in NDRTs affects drivers' responses and glance patterns after silent automation failures. *Transp. Res. F Traffic Psychol. Behav.* **2019**, *62*, 870–882. [CrossRef]

30. Naujoks, F.; Höfling, S.; Purucker, C.; Zeeb, K. From partial and high automation to manual driving: Relationship between non-driving related tasks, drowsiness and take-over performance. *Accid. Anal. Prev.* **2018**, *121*, 28–42. [CrossRef]

31. Schömig, N.; Befelein, D.; Wiedemann, K.; Neukum, A. *Methodische Aspekte und aktuelle inhaltliche Schwerpunkte bei der Konzeption experimenteller Studien zum hochautomatisierten Fahren*; [Methodological Aspects and Current Focus Areas Concerning the Conception of Studies for Highly Automated Driving]; FAT Forschungsvereinigung Automobiltechnik: Berlin, Germany, 2020.

32. Naujoks, F.; Wiedemann, K.; Schömig, N.; Hergeth, S.; Keinath, A. Towards guidelines and verification methods for automated vehicle HMIs. *Transp. Res. F Traffic Psychol. Behav.* **2019**, *60*, 121–136. [CrossRef]

33. Wiggerich, A. Development of a modular tool for safety assessments of human-machine-interaction for assisted driving functions (SAE level 2). In Proceedings of the 26th International Technical Conference on the Enhanced Safety of Vehicles (ESV), Eindhoven, The Netherlands, 10–13 June 2019.

34. Alliance of Automobile Manufacturers. *Statement of Principles, Criteria and Verification Procedures on Driver Interactions with Advanced In-Vehicle Information and Communication Systems Including*; Alliance of Automobile Manufacturers: Washington, DC, USA, 2006.

35. Lorenz, L.; Hergeth, S. Einfluss der Nebenaufgabe auf die Überwachungsleistung beim teilautomatisierten Fahren [Influence of secondary tasks on the supervisory performance during partially automated driving]. In *Der Fahrer im 21. Jahrhundert*; VDI, Ed.; VDI-Verlag: Düsseldorf, Germany, 2015; Volume 8, pp. 159–172.

36. Green, D.M.; Swets, J.A. *Signal Detection Theory and Psychophysics*; John Wiley and Sons: New York, NY, USA, 1966.

37. Wickens, C.D. Multiple resources and performance prediction. *Theor. Issues Ergon. Sci.* **2002**, *3*, 159–177. [CrossRef]

38. Horrey, W.J.; Wickens, C.D. Driving and side task performance: The effects of display clutter, separation, and modality. *Hum. Factors* **2004**, *46*, 611–624. [CrossRef] [PubMed]

39. Jamson, A.H.; Merat, N. Surrogate in-vehicle information systems and driver behaviour: Effects of visual and cognitive load in simulated rural driving. *Transp. Res. F Traffic Psychol. Behav.* **2005**, *8*, 79–96. [CrossRef]

40. Metz, B.; Schömig, N.; Krüger, H.P. Attention during visual secondary tasks in driving: Adaptation to the demands of the driving task. *Transp. Res. F Traffic Psychol. Behav.* **2011**, *14*, 369–380. [CrossRef]

41. Vollrath, M.; Krems, J. Fahren [Driving]. In *Verkehrspsychologie. Ein Lehrbuch für Psychologen, Ingenieure und Informatiker*; Vollrath, M., Krems, J., Eds.; Kohlhammer: Stuttgart, Germany, 2011; pp. 22–40.

42. Mattes, S. The lane change task as a tool for driver distraction evaluation. In *Quality of Work and Products in Enterprises of the Future*; Strasser, H., Kluth, K., Rausch, H., Bubb, H., Eds.; Ergonomia: Stuttgart, Germany, 2003; pp. 57–60.

43. Fry, E.B. *Teaching Faster Reading: A Manual*; Cambridge University Press: Cambridge, UK, 1963.

44. Victor, T.W.; Harbluk, J.L.; Engström, J.A. Sensitivity of eye-movement measures to in-vehicle task difficulty. *Transp. Res. F Traffic Psychol. Behav.* **2005**, *8*, 167–190. [CrossRef]

45. Ecker, R. Der verteilte Fahrerinteraktionsraum [The Distributed Driver Interaction Space]. Ph.D. Thesis, Ludwig-Maximilians-Universität (LMU) München, München, Germany, 14 June 2013.

46. Vollrath, M.; Huemer, A.K.; Teller, C.; Likhacheva, A.; Fricke, J. Do German drivers use their smartphones safely?—Not really! *Accid. Anal. Prev.* **2016**, *96*, 29–38. [CrossRef]

47. National Highway Traffic Safety Administration. *Traffic Safety Facts Research Note: Distracted Driving 2014*; Report No. DOT HS 812 260; National Highway Traffic Safety Administration: Washington, DC, USA, 2016. Available online: https://crashstats.nhtsa.dot.gov/Api/Public/ViewPublication/812260 (accessed on 5 May 2020).

48. Wittmann, M.; Kiss, M.; Gugg, P.; Steffen, A.; Fink, M.; Pöppel, E.; Kamiya, H. Effects of display position of a visual in-vehicle task on simulated driving. *Appl. Ergon.* **2006**, *37*, 187–199. [CrossRef]

49. Lamble, D.; Laakso, M.; Summala, H. Detection thresholds in car following situations and peripheral vision: Implications for positioning of visually demanding in-car displays. *Ergonomics* **1999**, *42*, 807–815. [CrossRef]

50. Weinberg, G.; Harsham, B.; Medenica, Z. Evaluating the usability of a head-up display for selection from choice lists in cars. In Proceedings of the 3rd International Conference on Automotive User Interfaces and Interactive Vehicular Applications, Salzburg, Austria, 30 November–2 December 2011; pp. 39–46. [CrossRef]

51. Krause, M.; Angerer, C.; Bengler, K. Evaluation of a Radio Tuning Task on Android while Driving. *Procedia Manuf.* **2015**, *3*, 2642–2649. [CrossRef]

52. Brillenstudie 2014 [Glasses Study 2014]. Available online: https://www.zva.de/brillenstudie (accessed on 20 March 2020).

53. *Tobii Pro Glasses 2 [Apparatus and software]*; Tobii AB: Stockholm, Sweden, 2016.

54. ISO 15007-1. *Road Vehicles—Measurement of Driver Visual Behaviour with Respect to Transport Information and Control Systems Part 1: Definitions and Parameters*; ISO: Geneva, Switzerland, 2014.

55. Sedlmeier, P.; Renkewitz, F. *Forschungsmethoden und Statistik: Ein Lehrbuch für Psychologen und Sozialwissenschaftler [Research Methods and Statistics: A Textbook for Psychologists and Social Scientists]*, 2nd ed.; Pearson: München, Germany, 2013.

56. Hensch, A.-C.; Rauh, N.; Schmidt, C.; Hergeth, S.; Naujoks, F.; Krems, J.F.; Keinath, A. Effects of secondary tasks and display position on glance behavior during partially automated driving. *Transp. Res. F Traffic Psychol. Behav.* **2020**, *68*, 23–32. [CrossRef]

57. Cohen, J. *Statistical Power Analysis for the Behavioral Sciences*, 2nd ed.; Laurence Erlbaum Associates: Hillsdale, NJ, USA, 1988.

 *information*

Article

Mode Awareness and Automated Driving—What Is It and How Can It Be Measured?

Christina Kurpiers *, Bianca Biebl *, Julia Mejia Hernandez and Florian Raisch

BMW Group, Knorrstrasse 147, 80788 München, Germany; Julia.Mejia-Hernandez@bmw.de (J.M.H.); Florian.Raisch@bmw.de (F.R.)
* Correspondence: christina.kurpiers@bmw.de (C.K.); Bianca.Biebl@bmw.de (B.B.)

Received: 27 April 2020; Accepted: 18 May 2020; Published: 21 May 2020

Abstract: In SAE (Society of Automotive Engineers) Level 2, the driver has to monitor the traffic situation and system performance at all times, whereas the system assumes responsibility within a certain operational design domain in SAE Level 3. The different responsibility allocation in these automation modes requires the driver to always be aware of the currently active system and its limits to ensure a safe drive. For that reason, current research focuses on identifying factors that might promote mode awareness. There is, however, no gold standard for measuring mode awareness and different approaches are used to assess this highly complex construct. This circumstance complicates the comparability and validity of study results. We thus propose a measurement method that combines the knowledge and the behavior pillar of mode awareness. The latter is represented by the relational attention ratio in manual, Level 2 and Level 3 driving as well as the controllability of a system limit in Level 2. The knowledge aspect of mode awareness is operationalized by a questionnaire on the mental model for the automation systems after an initial instruction as well as an extensive enquiry following the driving sequence. Further assessments of system trust, engagement in non-driving related tasks and subjective mode awareness are proposed.

Keywords: mode awareness; measurement method; automated driving; SAE Level 2; SAE Level 3

---

1. The Relevance of Automation

Within the next few years, technical advances will enable the development of vehicles that can transport users to their destination without human input. This exclusion of drivers from the control and guidance tasks eliminates human errors and as a result leads to increased road safety [1,2]. The technical complexity of such systems however does not allow a direct switch from manual to fully autonomous driving. Consequently, various car manufacturers are currently developing semi-autonomous systems that can manage some but not all driving functions. The level of automation in these systems is called partially automated driving (PAD; Level 2) according to the taxonomy by SAE International [3]. PAD systems can control longitudinal and lateral acceleration. Nevertheless, these systems require constant monitoring of their performance, traffic and the surrounding by the driver. In contrast to fully autonomous driving, PAD is still prone to human errors like inattention or distraction since the driver has the role of a supervisory controller who acts in collaboration with the system. This automation system cannot detect all its limits and errors, which is why the driver is responsible for intervening if necessary even without a preceding warning or take-over request [4]. This responsibility allocation will change with the introduction of conditionally automated driving (CAD; Level 3). These systems also control longitudinal and lateral acceleration and thus resemble PAD. Contrary to Level 2 however, CAD can detect all system limits itself and will request the user to take over within a certain time frame if necessary. As such, the driver is not required to be attentive to the system's status or traffic when CAD is active and he or she is then allowed to engage in non-driving related tasks (NDRT). Taken together, the main difference between Level 2 and Level 3 systems is the driver's responsibility for the driving task and the concomitant obligation to pay attention to the traffic situation in PAD but not CAD.

The safety of assisted driving functions is therefore reliant on the user's awareness of the currently active system and the knowledge about his or her responsibilities in this automated driving mode. This understanding is naturally aggravated if PAD and CAD are available within the same vehicle and if both systems are repeatedly activated within one drive [5,6]. It is especially safety critical if the user neglects the monitoring task during PAD because he or she might not notice the system reaching a limit in time. The danger for such an improper behavior is especially increased if the system works perfectly because the user might not expect any system limits [7,8]. In conclusion, it is of great importance to secure a good understanding and clear differentiation of the responsibilities in PAD and CAD. Various measures are currently being developed and tested to provide a so called mode awareness like the issuance of attention requests [9], hands On/Off options [10], the inclusion of one or multiple automation modes within a drive [5] and manual drives in-between periods of automated driving [11]. To assess the effect of such measures on mode awareness, it is however necessary to define mode awareness and to develop appropriate measurement methods first. This article aims to give an overview on the concept of mode awareness and to present a newly developed approach to measure mode awareness during alternating manual, PAD and CAD drives.

2. Constructs Concerning Monitoring Behavior

Before addressing the measurement of mode awareness, it is important to define this complex construct first and differentiate it from otherwise variables. The following chapter will provide an overview over all relevant constructs to mode awareness.

2.1. Situation Awareness

Mode awareness is similar but not identical to the concept of situation awareness. The latter is constituted of sufficient knowledge about the vehicle's surrounding, the current state of the automation, the system's task performance and the driver's own tasks and responsibilities [12]. If the driver lacks situation awareness, critical situations might be identified too late so that the driver cannot take compensating actions to resolve the situation [12]. According to Endsley and Kiris [13] the extent of situation awareness depends on three factors: automation information presentation; vigilance, monitoring and trust; engagement. Since systems with a higher reliability of autonomy go along with less attention on traffic and system performance, situation awareness is often reduced in higher automation levels [14]. That is why the driver should be given a sufficient amount of take-over time in order to get back in-the-loop before driving manually [15].

2.2. Mode Awareness

According to [16], there are two kinds of mode awareness: the awareness of the existence of different automation levels and the awareness of the currently active mode. While both aspects are necessary for mode compliant behavior, the latter is at particular risk when a vehicle incorporates two or more automation levels and is intended to be the focus of this paper [6]. Mode awareness is a subconstruct of situation awareness that merely excludes the knowledge about the current situation and surrounding [17]. It comprises the knowledge about the currently active automation system, its performance level and the driver's tasks and responsibilities [6]. Similar to situation awareness, mode awareness is established by the perception and correct interpretation of system information, the build-up of knowledge and finally the prediction of future system behavior [6,17,18]. A deficit can arise on any of these levels. Most common are however a misinterpretation of the systems' behavior and symbols (mode confusion) or a lack of knowledge about the systems (mental model).

2.3. Mode Confusion and Mode Errors

Mode confusion is one possible reason for deficient mode awareness [17,19]. It can be described as a kind of automation surprise, where the system does not behave according to the user's expectations. In the case of mode confusion, the user loses track of which system is currently active or what kind of behavior is appropriate for which mode. Mode confusion is safety critical [20] because it can lead to mode errors. This term describes behavior that fits the assumed but not the actual active automation level [6,21]. It results from an erroneous combination of information in the mental model [22]. Mode confusion can arise if a driver experiences two or more systems when changing between vehicles or when multiple systems are available within one vehicle, of which the latter represents a greater risk for mode confusion. The likelihood for mode confusion further increases if the systems appear similar for the user, e.g., in the case of PAD and CAD [6]. As a result, drivers might engage in NDRTs while driving in PAD and thus neglect their monitoring task. This can be highly dangerous if the system reaches its limit without the driver noticing, which can lead to collisions.

2.4. Mental Model

An awareness of the currently active automation mode itself is not sufficient for the creation of mode awareness. In addition, the user must have a correct mental model concerning the automation systems. Mental models are internal representations of a system that are formed by interacting with the system. These models do not need to contain correct technical details as long as users understand the functional characteristics of the system. Mental models can differ greatly in complexity depending on existing knowledge about the system, experience from interacting with the system and education [18,23].

2.5. Overtrust

As mentioned earlier, mode awareness is essential for a correct amount of monitoring and the controllability of system limits. Even if users have adequate declarative knowledge about the currently active system, its function and the users' own responsibility, they might however not behave according to the requirements of the automated system [24]. Next to fatigue, risk tolerance, boredom or extrinsic motivation, the greatest danger for such an improper behavior is an inappropriate level of trust in the system. In general, trust describes the attitude of users to let a system support them in situations characterized by uncertainty and potential danger [25]. This trust influences the usage of the automated system. In case of under-trust, users will tend to disuse the system because of the subjectively increased work load and risk [26]. This state should be avoided because a disuse of the automated system will decrease the customer value of the vehicle [27]. Van Loon and Martens [28] for example describe three factors that might be positively impacted by increased automated highway systems: a reduction of traffic congestions, a more economic driving style with a concomitant conservation of resources as well as increased traffic safety. According to the authors, the latter is especially improved after users get used to the new technologies or the system assumes increasing parts of the driving tasks.

In respect of safety in use, the more pertinent problem is over-trust. This blind trust in a seemingly perfect system can result in a misuse of the system beyond its functional limits and thus in a safety risk. In the case of PAD, users will presumably show a decreased monitoring of the driving scene and the system performance with an increased trust in automation because they assume that everything will be working properly [29,30]. This so called complacency is especially provoked if participants simultaneously have to perform multiple tasks which reduces the amount of cognitive resources available for monitoring [31]. Ironically, over-trust and its adjunctive misuse is elicited by a highly reliably functioning system because users will hardly ever experience the system limits they theoretically know about [8,32].

3. Measurement of Mode Awareness

The development of a measurement method for mode awareness is crucial for the serial implementation of automated systems since mode inappropriate behavior increases the risk for critical take-over scenarios and crashes. Principally, there are multiple methods to examine mode awareness. It is, however, very difficult to identify a technique which allows a measurement of all subjective and objective aspects of mode awareness. Various potential approaches will be presented and discussed in the following chapter.

3.1. Subjective Measurement Methods

Surely, the simplest way of getting insight into the user's mode awareness is by simply asking the driver via self-rating scales or interviews (e.g., [33,34]). Both methods give fast and explicit information about the user's state and can be used directly after a use case of interest or subsequent to the entire driving sequence. As with any subjective measurement method, it is however subject to a personal bias. Self-ratings on the user's assessment of his or her mode awareness are furthermore insufficient, because users might not understand the complexity of this construct in its entirety. Since misconceptions in the mental model can inherently not be detected by the users themselves, it is not advisable to use self-ratings as an indicator for mode awareness. An interview meets some of these flaws by allowing a standardized, and thus, partly objective assessment of mode awareness. In order to cover all aspects of mode unawareness however you need to identify all potential problems beforehand, which is not only time-consuming but also improbable. Additionally, interviews present multiple difficulties concerning study designs. If they are conducted while driving, the cognitive distraction might confound the driving performance. The conduction of multiple interviews (e.g., before and after an experimental manipulation) also poses the risk of influencing the mental model because the mere reproduction of information increases the knowledge level [35]. If the interview is conducted subsequent to the drive to avoid these confounding factors, the time interval between the driving scenario and the interview may lead to memory distortions, which in turn reduces the validity of information. One method to counteract the disadvantages of self-rating scales and interviews but maintain explicit information about the user's inner processes is a driver commentary (e.g., [36]). This method does not have the problem of memory loss or the need for predefined mode awareness deficits, because it aims to gather all thoughts of a user directly while using the system. The greatest benefit of this method is surely its flexibility towards individual and situational differences. The lack of standardization on the other hand complicates quantitative and comparative analyses. Furthermore, it might lower the ecological validity of a study because the simple instruction to formulate all thoughts might change these thoughts and interfere with the driving task.

3.2. Objective Measurement Methods

Another approach to investigate mode awareness is the use of objective measurements, which eliminates all subjective distortions and focuses on the actual user behavior. As illustrated previously, the main difference between PAD and CAD is the allocation of responsibility between the system and the driver and as such the required amount of monitoring [3,37]. Monitoring implies the placing of visual attention to the street or control instruments, which is most often accompanied by a corresponding eye movement [38]. Therefore, gaze behavior counts as a good indicator for mode awareness. To interpret gaze behavior in terms of mode awareness however you need a comparison value like a drive in another automation level. Another indicator of the user's knowledge about his or her responsibility is the interaction with NDRTs. The engagement in tasks like e.g., smartphone apps, in-vehicle information system, phoning or eating [39] will reduce the time of gaze spent on the traffic or system functionality. Consequently, it covers similar aspects concerning gaze behavior as mode awareness but is more restricted to the engagement in specific tasks. It must be noted that an engagement in NDRTs cannot always be implemented because of the study design or legal requirements.

Ultimately, the main interest in mode awareness does not lie in monitoring behavior, distractions or declarative knowledge itself, but in the consequential driving performance. This includes the take-over performance and the handling of critical situations, like e.g., the reaction time (time until gaze redirects from the NDRT to the road or control instruments; time until hands on the steering wheel and time until the first take-over reaction is performed), the time-to-collision (TTC), the maximum lateral and longitudinal acceleration and crash rate among others [40]. Possible take-over situations can range from uncritical switches between automation levels to undetected system failures in PAD (e.g., following a tar track instead of the line marking). The controllability of such situations is of special interest because of its safety implication. It must however be noted that driver behavior in such take-over situations is not specific to mode awareness and cannot indicate mode awareness problems on its own. It might, for example, be influenced by momentary inattention, fatigue, the familiarity with take-over situations or the individual participant's driving skills. It can thus only be interpreted against the backdrop of the attention ratio during the drive, the pre-existing knowledge and the post-enquiry.

That is why some studies (e.g., [41,42]) look for certain behavior patterns that are likely to be specific to mode unawareness. Mode confusion for example could become apparent when the system reaches a system limit. In addition, a user might grab the steering wheel during CAD, press random buttons repeatedly or show facial cues of confusion. These behavioral characteristics can however vary between participants and might not always occur during a drive, which is why their comparability is reduced. Furthermore, this behavior cannot be ascribed to mode awareness for sure without a follow-up interview. A user might for example put the hands on the steering wheel for comfort or by habit and not because of a misunderstanding of the currently active automation mode.

3.3. Combination of Measurement Methods

Mode awareness is a complex construct that circumferences sufficient knowledge about the system and its limits as well as behavioral aspects while driving. Currently, there is no gold standard for measuring all aspects of mode awareness, which is why most authors use a combination of multiple methods. Victor et al. [43], for instance, examined mode awareness during a PAD drive and added a take-over situation at the end of the drive because of an obstacle on the road. Mode awareness was operationalized by subjective as well as objective variables. The former consisted of a questionnaire on trust and open interview questions on impulse to intervene as well as the realization of the need to intervene. Objective data compromised response process variables and glance variables. Generally, a mixture of subjective and objective indicators is advisable for a valid interpretation of the declarative knowledge of users about the current system and its functionality as well as the behavior according to system requirements. In this case, the behavior aspect is assessed sufficiently by analyzing gaze behavior as well as take-over performance and the handling of critical situations. While the short interviews after the drive can give an impression on trust level and situation awareness, however they do not allow an evaluation of the user's mental model and mode confusion.

Another approach to measuring mode awareness can be found in a study by Wang and Söffker [44]. The authors investigated six driving scenarios. The implementation of both Level 2 and Three in these scenarios is reasonable for studying mode awareness in a worst case approach [5]. Mode awareness was operationalized by a situation awareness questionnaire but the authors also measured take-over time and quality in case of system failures and the engagement in NDRTs. While these measures do provide subjective and objective information, the lack of monitoring data does not allow a full objective interpretation of mode awareness. Furthermore, it has to be noted that the mid-questionnaire between the scenarios only included six questions on mode awareness, which is very little for such a complex construct.

Othersen [18] conducted a study on situation and mode awareness. Objective measures consisted of driving parameters, specifically reaction and take-over time, the quality of reactions and potential deactivating of the system. Furthermore, the author examined gaze behavior and video data as well as the performance in an audio-verbal NDRT. Subjective data circumferenced items on mode confusion, monitoring behavior, responsibilities during the drive and critical situations as well as the user's take-over performance. This approach covers many aspects of mode awareness like knowledge about the user's monitoring task (objective and subjective), awareness of the currently active mode (subjective) as well as the resulting take-over performance. The analysis of gaze behavior was however conducted absolutely without systematically comparing different drives to a baseline. In addition, the closed self-rating scale does not provide detailed information about the user's responsibilities. On the contrary, a comprehensive assessment of the user's mental model is crucial to define the cause of a potential lack of monitoring behavior.

4. A Subjective and Objective Measurement Method for Mode Awareness

In order to assess all aspects of mode awareness, we wanted to develop a new method that combines subjective and objective information in a worst case scenario. This approach allows the assessment of all major aspects of mode awareness (see Figure 1): the knowledge about which mode is currently active and the knowledge about the system's abilities and limits (knowledge pillar) as well as the resulting mode compliant behavior (behavior pillar).

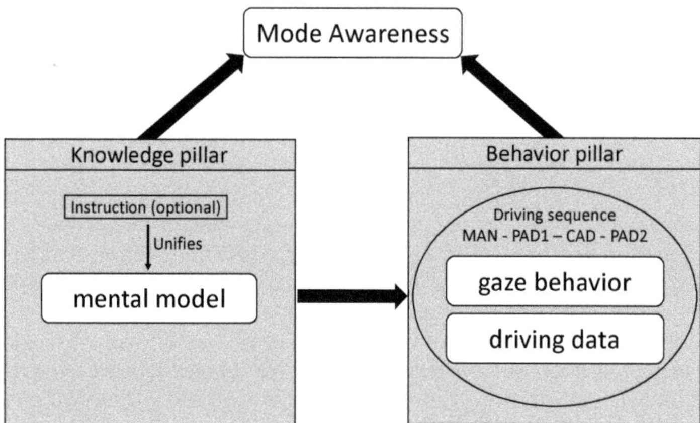

Figure 1. Mode awareness can be subdivided into a knowledge and a behavior pillar, which are measured separately in our proposed study design. All white frames represent dependent variables.

4.1. Knowledge Pillar

One aspect of the definition of mode awareness according to [6] is sufficient knowledge about the system. Therefore, an assessment of mode awareness should include the measurement of the participant's mental model, which should be conducted before the experimental drive. In studies questioning the effectiveness of certain methods to promote mode awareness, it is vital to first instruct the participants on the automation systems because a different amount of preknowledge can impact the effectiveness of such methods. The subsequent knowledge test can thus ensure a homogenous level of existing knowledge before the drive. Such an extensive instruction is, however, not advisable when studies aim to hedge mode awareness in order to get approval for an automated system. The initial knowledge test then serves as a first indicator of mode awareness during the drive.

It is, furthermore, important to include an extensive post-enquiry to test the amount of knowledge after the drive. That allows a conclusion on the knowledge on the systems' limits, human-machine interface (HMI) and the driver's responsibilities during the drive. This second conductance of the test is furthermore relevant in order to measure the change of the mental model due to a driving sequence or (if applicable) certain experimental manipulation.

The questionnaire for the mental model we developed consists of five parts. At the beginning, participants are asked to subjectively rate their knowledge about all assistance systems of interest on a 7-point Likert scale. This subjective rating is followed by an objective evaluation of the user's knowledge. They are first asked to formulate the two main aspects of each system. These statements are then evaluated by the examiner on the basis of a rating system, which categorizes information in mandatory and optional information. This is followed by detailed questions on various aspects of the assistance system, mainly the systems' limits and abilities as well as the responsibility of the driver. These statements have to be assigned to the respective assistance system or alternatively classified as true or false. Lastly, the participants are tested on their knowledge about the HMI and the handling of the systems' (de-)activation. That mainly functions as an indicator for mode confusion in the drive since insecurities about the corresponding icons for each mode as well as the correct button for activating and deactivating the systems can easily lead to confusion about the currently active system.

4.2. Behavior Pillar

4.2.1. Design

Declarative knowledge about the systems and their capabilities is necessary but not sufficient for mode awareness. Drivers might for instance technically be well aware of the currently active system and his or her responsibilities but still neglect the monitoring task because of over-trust [29]. A distracted or inattentive driver might then not notice a PAD system reaching its limit and thus crash. This use case is just one potential scenario and certainly represents a worst case setting. In order to ensure the safety of driver assistance systems in studies however a worst case approach is necessary [45].

We propose the following study design to validly measure mode compliant behavior in a worst case scenario (see Figure 2).

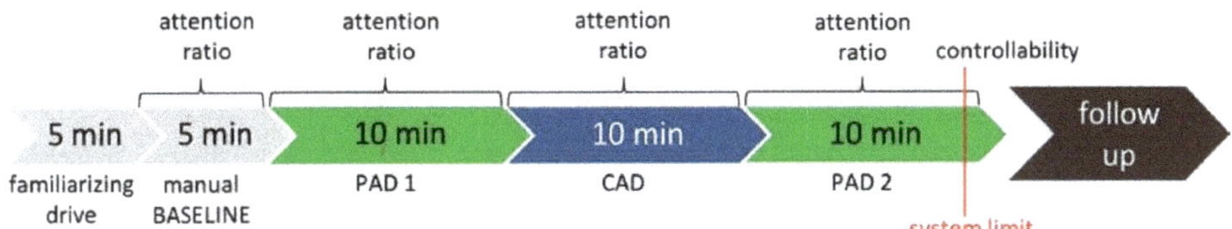

Figure 2. A schematic depiction of the driving sequences. A first familiarizing drive and a manual baseline are followed by a sequence of driving partially automated driving (PAD), conditionally automated driving (CAD) and then PAD again. Mode awareness is operationalized by the comparison of attention to driving related areas of interest during the drives and the controllability of a system limit at the end of the second PAD drive.

The drive starts subsequent to the theoretical instruction and the questionnaire on the mental model with a familiarizing drive. This drive is crucial to eliminate the influence of prior experiences with driver assistance systems, the make of the car or potential situational factors (e.g., being in a driving simulator). Depending on the research question, the familiarizing drive can contain short drives in all assistance modes including the switches between them or just a manual drive. It is advisable to then start a short period of driving manually as a baseline. The gaze behavior and driving data during this drive serve as comparison values for all subsequent automated drives. As mentioned earlier, a frequent switch between these automation modes is especially challenging for maintaining mode awareness, since the functions seem very similar for the user. In line with a worst case approach, we thus recommend including multiple switches between automation modes within the study. Between CAD and PAD, the latter has fewer situational requirements, which is why a first switch from manual driving to PAD is the most ecologically valid option. It also allows the assessment of the first contact of drivers with PAD as a baseline value. After a certain period of driving PAD, the system should then enable the switch to CAD. This drive should be terminated by a take-over request to initiate the last driving sequence in PAD. Until this point, both Level 2 and Level 3 would have worked perfectly without reaching any unexpected system limits. This is in line with a worst case approach since the high performance level makes it difficult to distinguish between both systems. By definition however, PAD systems might reach their limit without giving a warning or take-over request, e.g., because they accidentally follow a tar track instead of the actual lane. It is advisable to include such a scenario to assess the controllability of a potentially critical situation. That is crucial for driving safety and actually of higher importance than monitoring behavior. We recommend a silent system error at the end of the second PAD drive, by driving straight ahead instead of following the curved road. Without intervention of the driver the vehicle would then crash with the crash barrier or drive on the adjacent patch of grass.

The time frame of these drives can be chosen according to resources and research question. Generally, a longer time-frame will lead to more reliable data. A longer time-frame will however lead to increased driver fatigue [46]. Multiple internal studies showed that participants need 5 to 10 min to get used to the system. To avoid the influence of fatigue but ensure a sufficient amount of data we thus advise a duration of approximately 8 to 10 min per automated drive. Studies by Kurpiers et al. [9] and Feldhütter et al. [47] confirmed the assumption that this is an appropriate time frame to avoid insecurities when handling the system while simultaneously avoiding fatigue. Certain research questions and participant characteristics might however require the adaptation of these time slots. It must also be noted that the study design proposed in Figure 2 is only applicable in this form for studies in driving simulators. The uncontrollability of on-road studies may not allow the strict adherence to the proposed time frames because of interchanging road conditions and environments. This study design is however an appropriate basis for measuring mode awareness in a simulated environment. As such, it serves as a good tool to test changes in the automated function during development to ensure their security. Furthermore, it can be used to check the effectiveness of measures to increase mode awareness.

4.2.2. Attention Ratio

The aim of the manual-PAD-CAD-PAD sequence is to assess the participants' behavior in respect to the mode dependent responsibilities for the user. The requirement to keep the attention on the traffic at all times in PAD but not CAD is surely the greatest difference between these two automation systems and the monitoring behavior therefore a suitable operationalization for mode awareness. As most shifts in visual attention go along with a shift of gaze, glance behavior can be used for operationalization of mode awareness [48]. The most interesting metric of glance behavior is the attention ratio, which represents the percentage of time that a participant's gaze is directed in a certain area of interest (AOI) in relation to the total duration of each driving phase. One AOI of particular interest is the road center slightly below the horizon [49,50], since hazard and event detection requires focal vision [51]. Further driving relevant areas within the visual field are the instrument cluster to monitor the system's status as well as the lanes to the left and right, the side mirrors and the rearview mirror for traffic monitoring. If NDRTs are used (e.g., smartphone apps or games in the central information display), their location should be evaluated as a relevant non-driving related AOI to track the attention ratio to the NDRT.

4.2.3. Target and Actual Values

When PAD is active, the driver is assisted in the lateral and longitudinal guidance of the vehicle. Similar to manual driving, all other responsibility lies with the driver [3]. Since the driver must be able to detect all system limits and take-over at all times even without warning during PAD, the attention ratio to the traffic situation and the system's performance should not differ from that in manual driving. A mode aware driver should furthermore not show any significant differences in gaze behavior between the first and the second PAD drive despite the interposed CAD sequence. During CAD on the other side, it is expected to find a reduced amount of monitoring behavior compared to manual and PAD driving, since the driver is allowed and instructed to engage in NDRTs [3]. While the amount of monitoring in PAD is safety critical, a comparison between CAD and a manual or a PAD drive can mainly serves as an indicator for the quality of discrimination concerning the user's tasks. Lack of such a discrepancy in monitoring behavior is however not necessarily evidence for mode unawareness since users might also deliberately want to monitor the CAD function and their surroundings.

This proposed gaze behavior has been tested with similar designs in various studies [9,51]. In the static simulator study by Feldhütter et al. [47] for example, participants showed an attention ratio to road center of 89% during manual driving, which was significantly reduced to 51% and 18% in the first and second PAD drive respectively with a significant decrease from the first to the second PAD drive. This is a characteristic example for a mode awareness deficit, since the monitoring task was neglected during the PAD drives compared to manual driving, which was intensified by the CAD drive in between (schematic depiction in Figure 3).

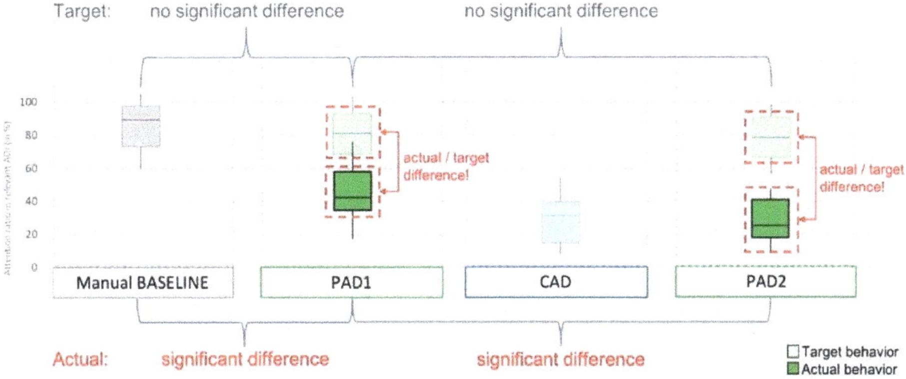

Figure 3. A schematic depiction of the target and actual gaze behavior during manual, PAD and CAD driving. Mode awareness can be assumed if there is no significant difference between the attention ratio in the manual and the first PAD drive or a significant decline in monitoring behavior from the first to the second PAD drive.

4.2.4. Controllability

Next to the monitoring behavior itself, the safety of automated vehicles in Level 2 and Level 3 is highly dependent on the user's ability to manage system limits. In the proposed worst case scenario of a silent system error in the second PAD drive, the car will keep driving straight ahead while the track makes a curve. The controllability of this situation can be assessed by the ability of the driver to keep the car on track. Potential parameters are the amount of the vehicle in surface area that has crossed the track before the driver intervenes and the crash rate. Feldhütter et al. [47] for example found that only 16% of participants intervened before the car had left the track and 29% did not take-over before the car had left the track completely. This clearly demonstrates the dangers of insufficient monitoring behavior during PAD that results from a deficit in mode awareness. It has to be noted however, that a bad performance in the take-over scenario cannot necessarily be ascribed to a lack of mode awareness. As a result, we advise a qualitative interrogation on the take-over scenario after the drive.

4.3. Additional Variables

The main problem when using gaze data is its lack of specificity since it can be influenced by factors like extrinsic motivation or boredom [24], risk tolerance [11], a faulty mental model and mode confusion [6] or over-trust [29]. Questionnaires before and after the drive are thus essential to ascribe a lack of monitoring and controllability to a concrete source. Next to the before mentioned knowledge test, one important assessment is the evaluation of trust in the automated system, (e.g., the automation trust scale (ATS) by Jian, Bisantz, and Drury [52]; the questionnaire on human-computer trust by Madsen and Gregor [53]), because deficits in mode awareness and over-trust cannot be distinguished without background information on the user's experience and mindset. In addition, a subjective test for mode awareness (like the one used by Othersen et al. [18]) might be useful in many cases. Furthermore, it is of great value to add various questions, e.g., concerning the perception of the events during the silent system error, the engagement in NDRTs during PAD, a lack of engagement in NDRTs during CAD, automation surprises and other subjective data. The specific choice of questions should be based on the individual characteristics of the driving behavior and the examiner's observations.

5. Limitations and Benefits

Despite this approach's theoretical soundness, the validity of the proposed study design has not been calculated yet. The data in [9,47] that result from study designs using the proposed method can give a first impression of its applicability but allows no testimony on the validity of the measurement approach. The main reason for this lies in the circumstance that there is no best practice for measuring mode awareness that could be compared to the results of the suggested approach. Furthermore, the interpretation of mode awareness in our study design is based on a number of different variables that need to be encountered as a whole. As a mixture of quantitative and qualitative measure, it is hardly possible to calculate one parameter for mode awareness that might be used in a validation process. In addition, this design for measuring mode awareness is not applicable in all study designs. First of all, the addition of a critical take-over scenario at the end of the second PAD drive is obviously impossible in on-road car studies. The only solution to evaluate controllability is to look for naturally arising system limits of PAD and assess the take-over quality of the participants. Any study in real traffic is, furthermore, liable to uncontrollable circumstances like weather, traffic and road works that might influence the availability of the assistance systems. Second, some research questions might call for a variation of drives compared to the proposed design, which might change the values of mode awareness. Furthermore, since the order of drives is essential to the assessment of mode awareness, it is not advisable to alter the sequence of the automated drives. That way, however, the attention ratio in the second PAD drive might already be reduced because of tiredness or exhaustion. That should be factored in by performing an objective sleepiness rating, e.g., the Karolinska sleepiness scale (KSS; [54]).

When testing the human-machine interaction of automated functions, the aim of most studies is to predict user behavior in the field. In order to secure the safety of the function, it is important to prove the robustness of the function even in worst cases. Wickens [45] actually argue that accidents in aviation are often caused by worst-case performers in worst-case situations. That is why extreme cases should not be treated as outliers in a normal distribution but considered for safety issues. Consequently, the proposed study design is an appropriate approach to testing mode awareness in PAD and CAD. In addition, this design is eligible for measuring mode awareness in different scenarios because the switch from PAD to CAD and back to PAD allows the relative comparison of attention ratio in Level 2. The use of absolute values on the other hand would lead to misinterpretations, since attention ratio itself will differ greatly between a simulator study without any actual danger and a real car study on public highways.

6. Conclusions

We propose a study design to assess mode awareness by focusing on its behavioral aspect, more precisely the attention ratio while driving in PAD, CAD and then PAD again in addition to the controllability of a critical take-over scenario at the end of the second PAD drive. Questionnaires and interviews on the mental mode, trust, the engagement in NDRTs and other observations during the drive will enable the examiner to extract the source of a potential negligence of the monitoring behavior during PAD. Taken together, we feel positive about the potential of this approach to cover all aspects of mode awareness while differentiating it from similar constructs. Further validation of the proposed design and assessment technique is required for a further evaluation.

Author Contributions: Conceptualization, C.K. and B.B.; Investigation, C.K.; Methodology, C.K., B.B. and J.M.H.; Project administration, C.K.; Writing—original draft, C.K. and B.B.; Writing—review & editing, C.K., B.B., J.M.H. and F.R. All authors have read and agreed to the published version of the manuscript.

Funding: This research received no external funding.

Acknowledgments: We thank the Chair of Ergonomics of the Technical University Munich for the collaboration. Especially, we would like to thank Klaus Bengler, Anna Feldhütter, Moritz Körber and Michael Rettenmaier for providing the static driving simulator of the Technical University Munich and for conducting three studies that investigated different measures for maintaining mode awareness with the developed measurement method.

Conflicts of Interest: The authors declare no conflict of interest.

References

1. Fagnant, D.J.; Kockelman, K. Preparing a nation for autonomous vehicles: Opportunities, barriers and policy recommendations. *Transp. Res. Part A Pol. Practice* **2015**, *77*, 167–181. [CrossRef]
2. Treat, J.; Tumbas, N.; McDonald, S.; Shinar, D.; Hume, R. *Tri-Level Study of the Causes of Traffic Accidents. Executive Summary*; U.S. Department of Transportation: Washington, DC, USA, 1979; pp. 1–328.
3. SAE International. *Taxonomy and Definitions for Terms Related to Driving Automation Systems for on-Road Motor Vehicles*; SAE International: Warrendale, PA, USA, 2018.
4. Endsley, M.R. From here to autonomy: Lessons learned from human-automation research. *Hum. Factors* **2017**, *59*, 5–27. [CrossRef] [PubMed]
5. Feldhütter, A.; Segler, C.; Bengler, K. Does shifting between conditionally and partially automated driving lead to a loss of mode awareness? In Proceedings of the International Conference on Applied Human Factors and Ergonomics, Los Angeles, CA, USA, 17–21 July 2017; Springer International Publishing: Basel, Switzerland, 2018. [CrossRef]
6. Sarter, N.B.; Woods, D.D. How in the world did we ever get into that mode? Mode error and awareness in supervisory control. *Hum. Factors* **1995**, *37*, 5–19. [CrossRef]
7. Strauch, B. Ironies of automation: Still unresolved after all these years. *IEEE T. Hum.-Mach. Syst.* **2017**, *48*, 419–433. [CrossRef]
8. Bainbridge, L. Ironies of automation. *IFAC Proc. Vol.* **1983**, *15*, 129–135. [CrossRef]
9. Kurpiers, C.; Lechner, D.; Raisch, F. The influence of a gaze direction based attention request to maintain mode awareness. In Proceedings of the 26th International Technical Conference on the Enhanced Safety of Vehicles, Eindhoven, The Netherlands, 10–13 June 2019.
10. Naujoks, F.; Purucker, C.; Neukum, A.; Wolter, S.; Steiger, R. Controllability of partially automated driving functions—Does it matter whether drivers are allowed to take their hands off the steering wheel? *Transp. Res. Part A Traff. Psych. Beh.* **2015**, *35*, 185–198. [CrossRef]
11. Boos, A. Habituation Effects in Automated Driving: The Influences of Exposure Duration and Exposure Frequency on Visual Monitoring Behaviour. Master's Thesis, Technical University Munich, Munich, Germany, 2018.
12. Endsley, M.R. Toward a theory of situation awareness in dynamic systems. *Hum. Factors* **1995**, *37*, 32–64. [CrossRef]
13. Endsley, M.R.; Kiris, E.O. The out-of-the-loop performance problem and level of control in automation. *Hum. Factors* **1995**, *37*, 381–394. [CrossRef]
14. Kaber, D.B.; Endsley, M.R. The effects of level of automation and adaptive automation on human performance, situation awareness and workload in a dynamic control task. *Theor. Issues Ergon.* **2004**, *5*, 113–153. [CrossRef]
15. Gold, C.; Damböck, D.; Lorenz, L.; Bengler, K. "Take over!" How long does it take to get the driver back into the loop? In Proceedings of the Human Factors and Ergonomics Society Annual Meeting, San Diego, CA, USA, 30 September–4 October 2013; Sage Publications Sage CA: Los Angeles, CA, USA, 2013. [CrossRef]
16. Monk, A. Mode errors: A user-centred analysis and some preventative measures using keying-contingent sound. *J. Man.-Mach. Stud.* **1986**, *24*, 313–327. [CrossRef]
17. Kolbig, M.; Müller, S. Mode Awareness im Fahrkontext: Eine theoretische Betrachtung. In Proceedings of the 10. Berlinger Werkstatt Mensch-Maschine Systeme, Berlin, Germany, 10–12 October 2013; Brandenburg, E., Doria, L., Gross, A., Günzler, T., Smieszek, H., Eds.; Universitätsverlag der TU Berlin: Berlin, Germany, 2013.
18. Othersen, I. *Vom Fahrer zum Denker und Teilzeitlenker*; Springer: Berlin/Heidelberg, Germany, 2016. [CrossRef]

19. Bredereke, J.; Lankenau, A. Safety-relevant mode confusions—Modelling and reducing them. *Reliab. Eng. Syst. Safe.* **2005**, *88*, 229–245. [CrossRef]
20. Bredereke, J.; Lankenau, A. A rigorous view of mode confusion. In Proceedings of the International Conference on Computer Safety, Reliability, and Security, Catania, Italy, 10–13 September 2002; Anderson, S., Felici, M., Bologna, S., Eds.; Springer: Berlin/Heidelberg, Germany, 2002. [CrossRef]
21. Sarter, N. Investigating mode errors on automated flight decks: Illustrating the problem-driven, cumulative, and interdisciplinary nature of human factors research. *Hum. Factors* **2008**, *50*, 506–510. [CrossRef] [PubMed]
22. Dekker, S. *The field Guide to Understanding Human Error*, 3rd ed.; Ashgate Publishing Ltd.: Aldershot, UK, 2006.
23. Lindberg, T. *Entwicklung Einer ABK-Metapher für Gruppierte Fahrerassistenzsysteme*; Dr. Hut: Berlin, Germany, 2012; pp. 1–329.
24. Rettenmaier, M. Mode Awareness im Fahrzeug: Absicherung von hochautomatisiertem Fahren. Master's Thesis, Technical University Munich, Munich, Germany, 2017.
25. Lee, J.D.; See, K.A. Trust in automation: Designing for appropriate reliance. *Hum. Factors* **2004**, *46*, 50–80. [CrossRef] [PubMed]
26. Parasuraman, R.; Riley, V. Humans and automation: Use, misuse, disuse, abuse. *Hum. Factors* **1997**, *39*, 230–253. [CrossRef]
27. Martinez von Bülow, R.; Raisch, F. Driver assistance systems: Highly automated driving—Acceptance and benefit for customers. In Proceedings of the VDA Technical Congress, Berlin, Germany, 27–28 February 2018.
28. Van Loon, R.J.; Martens, M.H. Automated driving and its effect on the safety ecosystem: How do compatibility issues affect the transition period? *Procedia Manuf.* **2015**, *3*, 3280–3285. [CrossRef]
29. Hergeth, S.; Lorenz, L.; Vilimek, R.; Krems, J.F. Keep your scanners peeled: Gaze behavior as a measure of automation trust during highly automated driving. *Hum. Factors* **2016**, *58*, 509–519. [CrossRef]
30. Parasuraman, R.; Manzey, D.H. Complacency and bias in human use of automation: An attentional integration. *Hum. Factors* **2010**, *52*, 381–410. [CrossRef]
31. Moray, N.; Inagaki, T. Attention and complacency. *Theor. Issues Ergon.* **2000**, *1*, 354–365. [CrossRef]
32. Manzey, D.; Reichenbach, J.; Onnasch, L. Human performance consequences of automated decision aids: The impact of degree of automation and system experience. *J. Cogn. Eng. Decis. Mak.* **2012**, *6*, 57–87. [CrossRef]
33. Forster, Y.; Naujoks, F.; Neukum, A. Your turn or my turn? Design of a human-machine interface for conditional automation. In Proceedings of the 8th International Conference on Automotive User Interfaces and Interactive Vehicular Applications, Ann Arbor, MI, USA, 24–26 October 2016. [CrossRef]
34. Naujoks, F.; Purucker, C.; Wiedemann, K.; Neukum, A.; Wolter, S.; Steiger, R. Driving performance at lateral system limits during partially automated driving. *Accid. Anal. Prev.* **2017**, *108*, 147–162. [CrossRef]
35. Kromann, C.B.; Jensen, M.L.; Ringsted, C. The effect of testing on skills learning. *Med. Educ.* **2009**, *43*, 21–27. [CrossRef] [PubMed]
36. Banks, V.A.; Stanton, N.A. Discovering driver-vehicle coordination problems in future automated control systems: Evidence from verbal commentaries. *Procedia Manuf.* **2015**, *3*, 2497–2504. [CrossRef]
37. Endsley, M.R. Autonomous driving systems: A preliminary naturalistic study of the Tesla Model S. *J. Cogn. Eng. Decis. Mak.* **2017**, *11*, 225–238. [CrossRef]
38. Klauer, S.G.; Dingus, T.A.; Neale, V.L.; Sudweeks, J.D.; Ramsey, D.J. The impact of driver inattention on near-crash/crash risk: An analysis using the 100-car naturalistic driving study data. *NHTSA* **2006**, *594*. [CrossRef]
39. Naujoks, F.; Befelein, D.; Wiedemann, K.; Neukum, A. A review of non-driving-related tasks used in studies on automated driving. In Proceedings of the International Conference on Applied Human Factors and Ergonomics, Prague, Czech Republic, 26–28 October 2016; Naujoks, F., Befelein, D., Wiedemann, K., Neukum, A., Eds.; Springer International Publishing: Basel, Switzerland, 2017. [CrossRef]
40. Feldhütter, A.; Gold, C.; Schneider, S.; Bengler, K. How the duration of automated driving influences take-over performance and gaze behavior. In *Advances in Ergonomic Design of Systems, Products and Processes*; Springer: Berlin/Heidelberg, Germany, 2017; pp. 309–318. [CrossRef]
41. De Winter, J.C.; Happee, R.; Martens, M.H.; Stanton, N.A. Effects of adaptive cruise control and highly automated driving on workload and situation awareness: A review of the empirical evidence. *Transp. Res. Part A Traff. Psych. Beh.* **2014**, *27*, 196–217. [CrossRef]

42. Toffetti, A.; Wilschut, E.S.; Martens, M.H.; Schieben, A.; Rambaldini, A.; Merat, N.; Flemisch, F. CityMobil: Human factor issues regarding highly automated vehicles on eLane. *Transp. Res. Rec.* **2009**, *2110*, 1–8. [CrossRef]

43. Victor, T.W.; Tivesten, E.; Gustavsson, P.; Johansson, J.; Sangberg, F.; Ljung Aust, M. Automation expectation mismatch: Incorrect prediction despite eyes on threat and hands on wheel. *Hum. Factors* **2018**, *60*, 1095–1116. [CrossRef]

44. Wang, J.; Söffker, D. Bridging gaps among human, assisted, and automated driving with DVIs: A conceptional experimental study. *IEEE T. Intell. Transp. Syst.* **2018**, *20*, 2096–2108. [CrossRef]

45. Wickens, C.D. Attention to safety and the psychology of surprise. In Proceedings of the 2001 Symposium on Aviation Psychology, Columbus, OH, USA, 5–8 March 2001.

46. Körber, M.; Cingel, A.; Zimmermann, M.; Bengler, K. Vigilance decrement and passive fatigue caused by monotony in automated driving. *Procedia Manuf.* **2015**, *3*, 2403–2409. [CrossRef]

47. Feldhütter, A.; Härtwig, N.; Kurpiers, C.; Hernandez, J.M.; Bengler, K. Effect on mode awareness when changing from conditionally to partially automated driving. In Proceedings of the Congress of the International Ergonomics Association, Florence, Italy, 26–30 August 2019; Springer: Berlin/Heidelberg, Germany, 2019. [CrossRef]

48. Crundall, D.E.; Underwood, G. Effects of experience and processing demands on visual information acquisition in drivers. *Ergonomics* **1998**, *41*, 448–458. [CrossRef]

49. Land, M.F.; Lee, D.N. Where we look when we steer. *Nature* **1994**, *369*, 742–744. [CrossRef] [PubMed]

50. Recarte, M.A.; Nunes, L.M. Effects of verbal and spatial-imagery tasks on eye fixations while driving. *J. Exp. Psychol. Appl.* **2000**, *6*, 31–43. [CrossRef] [PubMed]

51. Summala, H.; Nieminen, T.; Punto, M. Maintaining lane position with peripheral vision during in-vehicle tasks. *Hum. Factors* **1996**, *38*, 442–451. [CrossRef]

52. Jian, J.-Y.; Bisantz, A.M.; Drury, C.G. Foundations for an empirically determined scale of trust in automated systems. *Int. J. Cogn. Ergon.* **2000**, *4*, 53–71. [CrossRef]

53. Madsen, M.; Gregor, S. Measuring human-computer trust. In Proceedings of the 11th Australasian Conference on Information Systems, Brisbane, Australia, 6–8 December 2000.

54. Åkerstedt, T.; Gillberg, M. Subjective and objective sleepiness in the active individual. *Int. J. Neurosci.* **1990**, *52*, 29–37. [CrossRef]

 *information*

Article

Engagement in Non-Driving Related Tasks as a Non-Intrusive Measure for Mode Awareness: A Simulator Study

Yannick Forster [1],*, Viktoria Geisel [2], Sebastian Hergeth [1], Frederik Naujoks [1] and Andreas Keinath [1]

[1] BMW Group, Knorrstr. 147, 80937 Munich, Germany; sebastian.hergeth@bmw.de (S.H.); frederik.naujoks@bmw.de (F.N.); andreas.keinath@bmw.de (A.K.)

[2] Department of Mechanical Engineering, Technical University Munich, Boltzmannstr. 15, 85748 Garching, Germany; viktoria.geisel@gmail.com

* Correspondence: Yannick.forster@bmw.de

Received: 8 April 2020; Accepted: 23 April 2020; Published: 28 April 2020

Abstract: Research on the role of non-driving related tasks (NDRT) in the area of automated driving is indispensable. At the same time, the construct mode awareness has received considerable interest in regard to human–machine interface (HMI) evaluation. Based on the expectation that HMI design and practice with different levels of driving automation influence NDRT engagement, a driving simulator study was conducted. In a 2×5 (automation level x block) design, $N = 49$ participants completed several transitions of control. They were told that they could engage in an NDRT if they felt safe and comfortable to do so. The NDRT was the Surrogate Reference Task (SuRT) as a representative of a wide range of visual–manual NDRTs. Engagement (i.e., number of inputs on the NDRT interface) was assessed at the onset of a respective episode of automated driving (i.e., after transition) and during ongoing automation (i.e., before subsequent transition). Results revealed that over time, NDRT engagement increased during both L2 and L3 automation until stable engagement at the third block. This trend was observed for both onset and ongoing NDRT engagement. The overall engagement level and the increase in engagement are significantly stronger for L3 automation compared to L2 automation. These results outline the potential of NDRT engagement as an online non-intrusive measure for mode awareness. Moreover, repeated interaction is necessary until users are familiar with the automated system and its HMI to engage in NDRTs. These results provide researchers and practitioners with indications about users' minimum degree of familiarity with driving automation and HMIs for mode awareness testing.

Keywords: automated driving; human-machine interface; mode awareness

1. Introduction

The market introduction of vehicles equipped with SAE Level 3 (L3) automated driving systems (ADS) is only a matter of time. Automated driving promises numerous benefits: among others, it is expected to foster efficiency in terms of time usage. The driver may divert his/her attention to non-driving related activities while the ADS is executing vehicle guidance. SAE Level 2 (L2) driving automation—which is already commercially available—is also capable of controlling vehicle guidance while the driver still has to constantly monitor the system functioning [1]. L3 automated driving systems differ from L2 automation in such a manner that the driver has to be readily available as a fallback performer in case the system requests a transition to manual control. Thus, with the transition from L2 to L3 automation, the human driver's role shifts from that of an active system supervisor to a fallback-ready user who may engage in non-driving related tasks (NDRT). The availability of different driving modes (i.e., L1, L2, and L3) in one vehicle poses additional challenges to the driver to understand his/her role accordingly and not to confuse different automation modes and levels. Mode awareness as a critical issue in driving automation requires further research efforts for ensuring safe operation of different automated driving functions. Knowledge on the assessment of mode awareness, however, is scarce. Addressing this issue, the present study examines engagement in a representative visual–manual NDRT during different levels of automated driving as a non-intrusive measure for mode awareness. In the following, we first outline theoretical backgrounds on mode awareness and methodology to assess this construct. Subsequently, the research question and hypotheses are derived based on the preceding considerations.

2. Background

In the automotive context, the evaluation of HMIs has a long history. The distraction potential of in-vehicle information systems (IVIS) is the main focus for manual driving (SAE L0). Here, test procedures to assess visual workload associated with the IVIS have already been established [2,3]. However, the change of the driver's role from manual driver to supervisor in L2 and fallback performer in L3 automation renders the application of these methods unfeasible. For example, NHTSA distraction guidelines only permit 2 s per glance and 12 s total glance duration on IVIS. It might be questionable whether these numbers as they were proposed for manual driving are also suitable for L2 automation. In addition, with the driving automation executing longitudinal and lateral vehicle control, distance and lane keeping are not applicable measures for indicating the suitability of an HMI in this particular context. In contrast, a variety of constructs related to the safe driver–automation interaction such as trust [4–7] controllability [8–10], understanding in form of mental models [11–13], or usability [14] could be used as criteria. Research has shown that these pose challenges to the design and evaluation of automated vehicle HMIs. For an outline of evaluation methods for automated vehicle HMIs see [15]. One further step towards an ADS method validation concerns the investigation of mode awareness. This term was proposed by Sarter and Woods [16]. The authors report that even pilots who can be considered highly skilled and trained operators of flight automation can face situations where they are not certain of roles and responsibilities for the aircraft operation task. Such situations can lead to dangerous outcomes and consequently a safety-related assessment is indispensable.

Mode awareness is a central aspect for appropriate and safe human–automation interaction in general and in the context of driving automation in particular. For example, Gopinath and Johansen [17] outline that mode awareness of operators is of crucial importance for safety when interacting with production robots. By appropriate design of the automation and according HMIs, safety risks can be mitigated (e.g., [18]). In the driving automation context, Feldhuetter, Segler and Bengler [19] provide evidence that drivers' mode awareness is reduced when the vehicle is equipped with additional driving automation functions (see also [20]). Similar to the proposal by Gopinath and Johansen [17], they investigated whether an adaptive HMI design could support mode awareness, but could not find an effect. Other research supports their hypothesis that HMI design can affects drivers' visual behavior. For example, Kraft, Naujoks, Woerle and Neukum [21] report the impact of the HMI design on glance distributions during active L2 automation. In this study, a reduced and simple display produced positive effects in terms of distraction on both a self-reported and behavioral level. In addition, familiarity-dependent practice effects occurred for glance patterns. In general, behavioral adaptation to automated driving can be expected as outlined in [22]. An appropriate design of L3 automated vehicle HMIs can support self-reported usability and trust in automation (Hergeth, 2016). Since trust is expected to determine reliance behavior [6,23], we assume that such HMI variations can also affect behavioral parameters concerning NDRT engagement. This influence of HMI design on user behavior is of high importance since it must convey information about the driver's role during active L2 and L3 functioning. Investigating mode awareness between driving episodes, Feldhuetter and colleagues [24] tested whether manual driving episodes as intermittent features between transitions of L2 and L3 automation can help to promote mode awareness. In this experiment, they operationalized mode awareness via the visual attention towards driving-relevant areas and engagement in NDRTs. The study shows that there is a difference of visual attention allocation and NDRT engagement. However, it remains unknown whether this observation is stable or prone to changes over time. As there is research indicating behavioral changes in interaction with driving automation when interacting repeatedly [14,21], NDRT-related behavior might also change. Especially findings of more accurate mental models over time [11–13] lead to the question whether mode awareness is also dependent on the familiarity with the driving automation.

As indicated above, reliance behavior is suggested to be closely tied to NDRT engagement during automated driving [7]. The difference between L2 and L3 is that the driver is responsible for supervising the automation in L2 whereas he/she has to be readily available to perform driving task fallback in L3. For the HMI design, this indicates that L2 automation systems require a feature ensuring that drivers are attentive to the supervising task either by steering wheel input or gaze tracking to the forward roadway (see e.g., [25]). By issuing a so called "hands-on request" or "attention request", the system draws the driver's attention back towards the supervising task. In comparison, such interface features are not part of a L3 system as it allows NDRT engagement. L3 systems only request driver input at operational design domain (ODD) limits or system malfunctions [26]. Thus, NDRT-related behavior should differ depending on the understanding of the current level of automation (i.e., mode awareness) given an interface is designed in accordance with the prior considerations. The design of automated vehicle HMIs is therefore a crucial aspect for the facilitation of visual attention towards relevant events inside or outside the vehicle [27,28]. A study by Llaneras and colleagues [29] found that drivers tend to engage in NDRTs during reliable L2 automation that does not monitor or restrict behavior. This leads to risky driving and diverts attention away from the roadway and supervision of the system. Therefore, investigation and comparison of NDRT engagement during L2 and L3 automation is of high importance. It is expected that HMI features such as hands-on or attention requests during L2 automation should consequently lead to improved mode awareness with better understanding of his/her roles and responsibilities (i.e., supervising during L2). This understanding eventually translates in observable behavior of less NDRT engagement during L2 as compared to L3 automation.

The study outlined above shows that there is a growing body of research on mode awareness in the driving automation domain. Additionally, HMI considerations outlined above suggest that NDRT engagement can serve as an indicator of mode awareness. However, commonly agreed methodological approaches are still missing. In relation to the theoretical and conceptual developments, the present study's aim was to investigate how mode awareness can be assessed in a non-intrusive way. It seeks to extend the findings on understanding as reported in [13]. Results of this publication showed that the general understanding of roles and responsibilities (i.e., mode awareness) was high for both L2 and L3 automation. However, the question remained whether this understanding also translates in observable behavior. Non-intrusive measurements of mode awareness bear both advantages for researchers and practitioners as well as for the real-world application of driver-monitoring systems. On the one hand, during the development and evaluation of automated vehicle HMIs, mode awareness represents a critical issue that needs to be assessed. With the availability of a non-intrusive measure, research methodology benefits from the present research. On the other hand, real-world application could use driver monitoring technology to detect potential losses of mode awareness based on the driver's current behavior. Thus, an ADS might undertake necessary precautions such as displaying warning messages which are already in effect today for fatigue detection.

Research Question and Hypotheses

From theoretical considerations outlined above, the following research question is derived: How does NDRT engagement calibrate for different levels of automation (i.e., for different graphical HMI designs) and with rising system experience? The following two hypotheses are formulated for this research question:

Hypothesis 1 (H1). *Drivers change their engagement in NDRTs over time;*

Hypothesis 2 (H2). *There is more NDRT engagement during an active L3 ADS compared to an active L2 driving automation.*

3. Method

3.1. Sample

A total of N = 59 participants took part in the driving simulation experiment. N = 10 drop-outs occurred because four participants did not complete the experimental procedure and six incomplete datasets were collected. This left N = 49 (13 female, 36 male) participants for data analysis. Mean age of the final sample was 30.96 years (SD = 9.08, MAX = 62, MIN = 21). All participants were BMW Group employees, held a German driver's license, and had normal or were corrected to normal vision.

3.2. Driving Simulation and Non-Driving Related Task

The study was conducted in a moving-base driving simulator (see Figure 1, left). The integrated vehicle's console contained all necessary instrumentation and was identical to a BMW 5 series with automatic transmission. Seven 1080p projectors provided a 240° horizontal × 45° vertical frontal field of view. One LCD screen positioned behind the back inside the vehicle mockup seats and two outside projections with the same specifications served as rear view. The motion system consisted of a hydraulic hexapod with six degrees of freedom, capable of up to 7 m/s^2 transitional acceleration and 4.9 m/s^2 continuous acceleration. The Surrogate Reference Task [30] was displayed on a 12.3″ tablet mounted on the center stack console and was active during the entire experimental drive (see Figure 1, right). NDRT engagement is measured using a task that is representative for many NDRTs in terms of demands and distraction potential to obtain high external validity. The Surrogate Reference Task (SuRT, [31]) is such a representative task since it is used as a generic visual–manual secondary task in distraction studies. In addition to these, it has also been used for an NDRT in automated driving studies [7,9,32]. The SuRT requires participants to identify a target stimulus (i.e., large circle) within an array of distractors (i.e., small circles). By varying the amount of distractors and size difference between target and distractors, the NDRT demand and resulting workload can be adjusted specifically. An advantage of the SuRT is its potential to support high experimental control while on the downside, it is not a naturalistic NDRT and thus motivation to extensively engage in the SuRT could be limited.

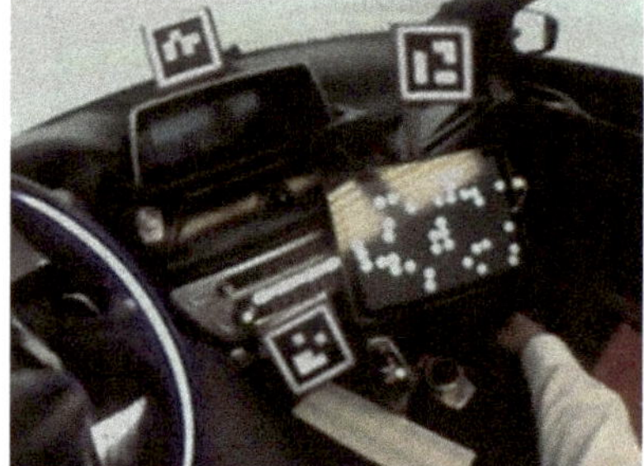

Figure 1. Dynamic driving simulator from the outside (**left**) and mockup interior with the Surrogate Reference Task (SuRT) tablet used in the current study (**right**).

The interface on which the SuRT was presented did not display a score to the drivers to make NDRT engagement completely voluntary and free of a potential competitive character. The circles could be selected by touching the surface with a finger. When the participant selected the correct circle, it turned green before the subsequent pattern emerged. In case the wrong target was selected, it turned red and the pattern stayed until it was solved correctly.

3.3. Study Design and Procedure

The study employed a 2×5 mixed within–between subjects design. The within-subject factor "block" had five levels from the first to the fifth block of use cases. The between-subjects factor "feedback" had two levels where participants either received feedback on their interaction success after each use case or not. Because the between-subjects factor was out of scope for the present research question, this research reports results of the within-subject factor "block".

Upon arrival, participants were welcomed and gave informed consent. After a brief explanation of the study purpose, the experimenter led them to the vehicle mockup. To accustom themselves with the simulator setup, participants had to complete at least two correct trials with the SuRT at standstill. Subsequently, they completed a five-minute manual familiarization drive without NDRT engagement. Prior to the experimental drive, the experimenter outlined the procedure and explained that participants would encounter two automated systems that are a L2 driving automation and a L3 ADS. They also received information stating that they would not have to constantly monitor the correct functioning of the L3 ADS. Concerning NDRT engagement, participants were instructed before each block that they could freely decide whether to engage in the NDRT when the automation was active. In doing so, the experimenter did not specify the level of automation or explicitly named any of the two functions. Furthermore, there was no additional incentive for executing the NDRT. The subsequent experimental drive included five blocks, each consisting of six driver initiated control transitions. After the successful completion of each interaction, there was a 20-s time window where users' NDRT-related behavior was observed. Table 2 additionally provides an overview of the windows of observation for NDRT-related behavior. Subsequently, there was a brief inquiry during the drive that occurred six times for each block [33]. Having finished use case specific questions, there was another time window of at least 20 s up to one minute where users could freely engage in the NDRT before the upcoming instruction of the next use case. After each block, participants were told to pull over to the right shoulder, stop there, and complete the block inquiry. Participants completed the drive on a three-lane highway with low to medium traffic density. Surrounding vehicles drove with an average of 150 km/h on the center lane and an average of 180 km/h on the left lane. Vehicles on the right lane drove with an average of 130 km/h. The conditions were good with clear visibility at daytime and a dry road. The highway itself was in good condition without potholes or construction areas. The experimental drive lasted approximately 60 min. Figure 2 schematically depicts the procedure.

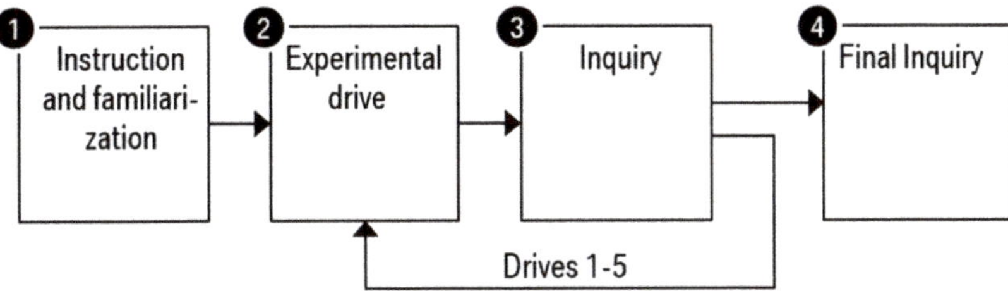

Figure 2. Schematic outline of experimental procedure.

3.4. Use Cases

The present experiment included driver initiated transitions between manual, L2, and L3 automated driving [34] as use cases (UCs). Considering both upward and downward transitions, one experimental block consisted of six use cases. For the present analysis, only transitions to an automated driving mode are of interest. Consequently, transitions to manual are not considered here. The use cases with transition type, automation level at use case initiation, target automation level, and use case numbering are shown in Table 1. To counteract sequential effects, participants were randomly assigned to one of six possible block sequences that were created using a Latin square. Each block consisted of six trials. In total, each participant completed 30 use cases. To standardize instructions, we recorded samples for each use case that were triggered by the experimenter.

Table 1. Overview of use cases for one experimental block.

Transition Type	Scenario	Automation Level at UC Initiation	Automation Target Level	Use Case Number
	Activation L3	L0	L3	1
Upward transition	Activation L3	L2	L3	3
	Activation L2	L0	L2	2
Downward transition	Deactivation L3	L3	L2	4

3.5. Automated Driving System

As soon as the driver activated the respective function, it carried out longitudinal and lateral vehicle guidance. The longitudinal and lateral vehicle guidance of the L2 and L3 automation was identical. The L3 ADS was capable of executing independent lane change maneuvers (e.g., overtaking slower vehicles ahead, pulling back to the right lane). The L2 driving automation set speed was the current velocity and could be adjusted without restrictions. The L3 ADS set speed was 130 km/h and could be adjusted to slower speeds. If adjusted to a faster speed than 130 km/h, it deactivated the L3 ADS and activated the L2 driving automation. Vehicle following distance (time headway) to a lead vehicle was 2 s.

3.6. Human–Machine Interface

The visual HMI was shown on the instrument cluster. It showed the vehicle and its surroundings in both L2 and L3 automated driving. The HMI for automated driving resembled a combination of adaptive cruise control and additional steering assistance [35]. The present HMI constitutes a representative solution for an automated system due to the conceptual similarity to solutions in prior research [4,36]. The L2 vehicle surroundings and L3 vehicle surroundings differed in (1) their informational content (i.e., higher level of detail in L3: visibility of adjacent lanes and vehicles) and (2) their perspective (i.e., larger field of view in L3). Thus, specifically the distance between the eye point and the vehicle, the angle between the direct line of sight and the road, and the opening angle of the field of view were manipulated. Figure 3 schematically depicts the configurations for L2 and L3 automation of the vehicle surround views from a profile perspective. An activated L2 automation was colored in green while an activated L3 ADS was colored in blue. In addition, during activated L3 ADS, the steering wheel was illuminated in blue color. The L2 driving automation displayed a hands-on request (HOR) after 15 s of hands-free driving. The HOR was displayed as hands grabbing a steering wheel [37,38] and yellow pulses on the illuminated steering wheel. The system functions could be activated with a button on the left side of the steering wheel for both levels of automation. For a more comprehensive description of the operating elements, see [14].

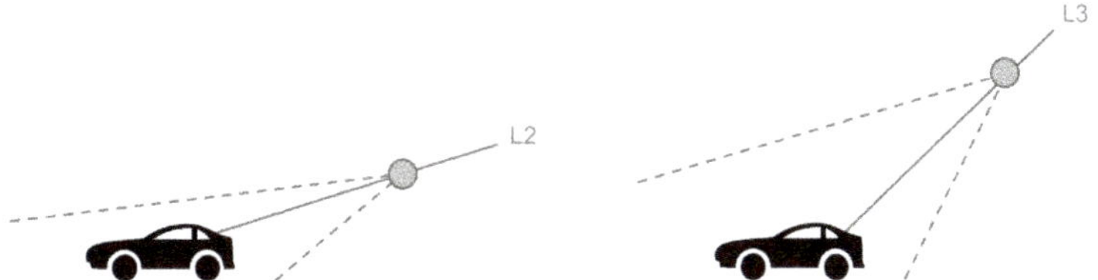

Figure 3. Schematic depiction of vehicle surroundings point of view for L2 (**left**) and L3 automation (**right**). The gray dot represents the eye point.

3.7. Dependent Variables

The present study operationalized NDRT engagement as input with the finger on the NDRT surface. Table 2 visualizes the windows of observation for the dependent variables. To find out about the onset of engagement, we counted the total number of inputs on the surface for a time interval of 20 s after successful completion of each use case (NDRT observation window 1). Since it can be assumed that it takes some time for the NDRT engagement to set in and then to stabilize, we also investigated NDRT-related behavior at the end of an automated driving episode where the onset had most likely occurred and NDRT engagement was on a stable level. For that purpose, there was another window of observation covering the 20 s just before the onset of the subsequent use case (NDRT observation window 2).

Table 2. Schematic outline of experimental procedure for each use case. The two observation windows are colored in blue.

Step	Standardized Experimenter Instruction	Task Completion Time	NDRT Observation Window 1	UC Specific Inquiry	NDRT Engagement	NDRT Observation Window 2
Duration	5 s	0–60 s	20 s	10–30 s	0–20 s	20 s

3.8. Statistical Procedure and Data Analysis

NDRT data were pre-processed and visualized using Matlab Version 2015 (Mathworks Inc., Natis, MA, USA). Statistical tests were calculated using IBM SPSS Statistics Version 23 (IBM, Armonk, NY, USA). For observation window 1, means and standard deviations (SD) were computed for onset NDRT input frequency by use case and block. In contrast, when observation window 2 started, the transition of control already dated back too far so that a comparison of NDRT-related behavior on use case level (i.e., considering the respective previous level of automation) would not be useful for that period of time. Therefore, we compared NDRT engagement during observation window 2 only in regard to the level of automation that was active at that time. For that purpose, the sum of NDRT inputs during active L2 automation (after UC2 and UC4) and active L3 ADS (after UC1 and UC3), respectively, was calculated for each participant and block. Means and standard deviations (SD) were computed for these ongoing input sums. A significance level of $\alpha = 0.05$ was applied for inferential testing unless stated otherwise. To control for alpha inflation due to multiple testing, correction after [39] was applied if necessary.

4. Results

4.1. Onset Input Frequency

Table 3 shows descriptive statistics (i.e., *M*, *SD*) of NDRT input frequency within the 20 s after UC completion by use case and block. Means and standard errors of onset input frequency by use case and block are depicted in Figure 4. Descriptive values revealed that the overall number of NDRT inputs during the 20 s after task completion was on a low level with mean input frequency not exceeding a number of two. Furthermore, there was a tendency towards more NDRT engagement with increasing system experience in all four use cases. However, the observed increase was stronger for transitions to L3 automation (UC1 and UC3) than for transitions to L2 automation (UC2 and UC4). Independent from the block, descriptive data showed considerably more NDRT engagement after transitions to L3 than after transitions to L2.

Table 3. Descriptive statistics (i.e., M, SD) of onset input frequency for the four use cases (UCs) by block.

UC	Block 1	Block 2	Block 3	Block 4	Block 5
UC1	0.39 (0.73)	1.16 (1.18)	1.53 (1.21)	1.41 (1.19)	1.57 (1.26)
UC2	0.06 (0.32)	0.31 (0.68)	0.35 (0.81)	0.55 (0.94)	0.47 (0.96)
UC3	0.67 (1.01)	0.98 (1.09)	1.35 (1.13)	1.51 (1.10)	1.27 (0.93)
UC4	0.06 (0.32)	0.31 (0.77)	0.33 (0.77)	0.53 (0.98)	0.51 (0.89)

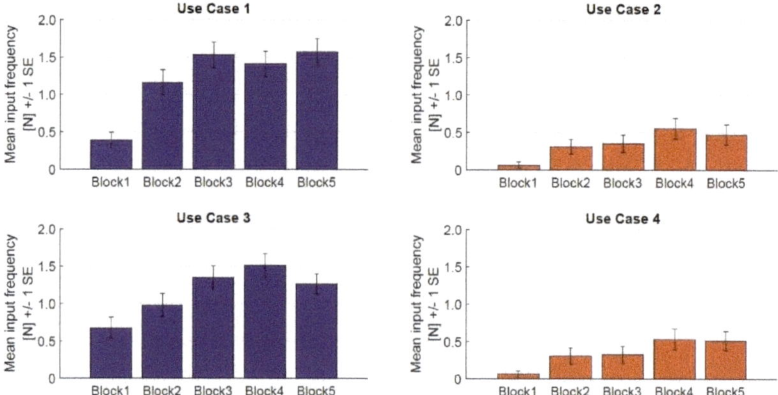

Figure 4. Means and SE of onset input frequency by UC and block (blue: transitions to L3 automation, red: transitions to L2 automation).

A 4×5 (UC $\times$ block) repeated measures analysis of variance (ANOVA) was conducted for onset input frequency. Results revealed significant main effects for both use case and block as well as a significant interaction effect (see Table 4). These inferential results indicate that mean input frequency differed significantly over time and for the different use cases, but the effect of the block depended on the respective use case. The effect sizes showed large effects ([40]; see Table 4). To examine these effects in detail, planned contrast analyses were performed to compare onset input frequency for the two different levels of automation (L2: after UC2 and UC4; L3: after UC1 and UC3) and for consecutive blocks. Results are displayed in Table 5. Regarding the two levels of automation, results revealed that there was significantly more NDRT engagement during active L3 than during active L2 automation; the effect size (see Table 5) indicated a strong effect [40]. Comparisons between consecutive blocks showed a mixed picture: Mean NDRT input frequency was significantly higher in block 2 than in block 1. There were also significantly more NDRT inputs in block 3 as compared to block 2; medium to large effect sizes were obtained [40] (Cohen, 1988). The remaining contrasts between successive blocks did not reach significance (see Table 5). The results of the planned contrast analyses indicate that NDRT engagement increased within the first three system encounters and stabilized in subsequent system encounters.

Table 4. Inferential statistics (i.e., F, df1, df2, p, $\eta_p{}^2$-value) of main and interaction effects for onset input frequency. Statistically significant effects are colored in gray.

Effect	F	df1	df2	p	$\eta_p{}^2$
Use Case	37.378	3	46	<0.001	0.709
Block	12.885	4	45	<0.001	0.534
Use Case * Block	2.609	12	37	<0.05	0.458

Table 5. Inferential statistics (i.e., F, df1, df2, p, $\eta_p{}^2$-value, and 95% CI limits) of planned contrast analyses for L2 (after UC2 and UC4) vs. L3 automation (after UC1 and UC3) and successive blocks for onset input frequency. Statistically significant effects are colored in gray.

Contrast	F	df1	df2	p	$\eta_p{}^2$	95% CI
L2 vs. L3	112.989	1	48	<0.001	0.702	[6.785; 9.950]
Block 1 vs. Block 2	19.755	1	48	<0.001	0.292	[0.861; 2.282]
Block 2 vs. Block 3	5.399	1	48	<0.05	0.101	[0.107; 1.485]
Block 3 vs. Block 4	1.039	1	48	0.313	0.021	[−0.436; 1.334]
Block 4 vs. Block 5	0.297	1	48	0.588	0.006	[−0.862; 0.494]

4.2. Ongoing Input Frequency

Descriptive statistics (i.e., *M, SD*) of ongoing NDRT input sums within the 20 s before the onset of the upcoming use case by level of automation (L2: after UC2 and UC4; L3: after UC1 and UC3) and block can be found in Table 6. Figure 5 depicts means and standard errors of ongoing NDRT inputs by level of automation and block. The descriptive values showed similar tendencies as for onset NDRT engagement: The overall number of inputs during the 20 s before onset of the upcoming use case summed for active L2 and L3 automation, respectively, was relatively small with means not exceeding a number of four. Furthermore, a trend towards more NDRT engagement with rising system experience could be observed for both levels of automation with a seemingly weaker upward trend for L2 automation. However, descriptive NDRT engagement tended to stabilize after the first three system encounters. Descriptive data also indicated notably more ongoing NDRT engagement during active L3 automation than during active L2 automation in all five blocks.

Table 6. Descriptive statistics (i.e., M, SD) of ongoing input frequency summed for L2 (after UC2 and UC4) and L3 automation (after UC1 and UC3) by block.

	Block 1	Block 2	Block 3	Block 4	Block 5
L2	0.12 (0.49)	0.59 (1.22)	0.78 (1.87)	1.29 (2.03)	1.25 (2.21)
L3	0.84 (1.07)	1.80 (1.95)	2.74 (2.74)	3.31 (2.36)	3.25 (2.43)

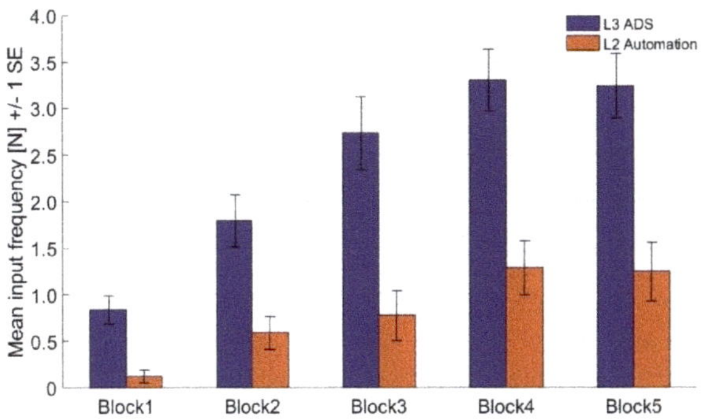

Figure 5. Means and SE of ongoing input frequency summed for L2 and L3 automation by block.

A 2 × 5 (level of automation × block) repeated measures ANOVA was performed for ongoing NDRT engagement to examine main and interaction effects of the level of automation. Results are displayed in Table 7. There was a significant main effect of level of automation as well as of block. This means that ongoing NDRT engagement was significantly higher during L3 automation than during L2 automation and differed over time. Furthermore, there was a significant interaction effect indicating that the effect of block on NDRT engagement depended on the level of automation that was active. The effect sizes (see Table 7) showed large effects [40].

Table 7. Inferential statistics (i.e., F, df1, df2, p, η_p^2-value) of main and interaction effects for ongoing input frequency summed for L2 and L3 automation. Statistically significant effects are colored in gray.

Effect	F	df1	df2	p	η_p^2
Level of Automation	54.652	1	48	<0.001	0.532
Block	15.105	4	45	<0.001	0.573
Level of Automation * Block	5.085	4	45	<0.05	0.311

5. Discussion and Conclusions

This research investigated the analysis of NDRT engagement at different levels of automated driving. The results of $N = 49$ participants showed that the levels of driving automation and accordingly designed HMIs lead to differences in NDRT engagement. An increase of NDRT engagement over time was observed for both automation levels whereas this increase was stronger in L3 as compared to L2 automation. These results indicate that users' behavioral adaptation occurs during initial system encounters. It also shows that the HMI design that follows considerations for L2 and L3 driving automation leads to specific behavioral patterns. The following section discusses the obtained results and relates them to prior considerations about NDRT engagement and mode awareness.

Overall, there were differences in NDRT engagement between the L3 and the L2 automation with significantly more engagement in L3 as compared to L2 automation as indicated by statistically significant main effects in Tables 4 and 7. Thus, these differences can be traced back to two sources. First, the L3 HMI permitted hands-free driving while the L2 HMI included hands-on requests. Second, the HMI designs differed in adaptations of informational content and perspective. Eventually, there is no final statement possible which HMI variation led to the differences in the observed behavior between the automation levels. Referring back to initial considerations of the HMI design for automated vehicles, it is important to include a form of feedback for L2 automation that prompts the drivers to supervise the driving automation. If these are not present (as in the present L3 case), there is high NDRT engagement. This observation supports the results by Llaneras and colleagues [29] The difference between NDRT engagement during L2 and L3 automation was observed for both the onset (see Figure 4) and ongoing (see Figure 5) NDRT engagement. These observations are in accordance with the findings reported in [19]. The results reported herein extend their findings by repeatedly observing the engagement in an NDRT. Here, similar results were obtained for L2 and L3 automation. Namely, engagement in NDRTs at initial contacts with driving automation—independent of the level of automation—is on a low level. The engagement rises in both instances as indicated by significant main effects for the block factor in both Tables 4 and 7. However, the rise in NDRT engagement was much stronger for L3 automation as compared to L2 automation as indicated by the significant interaction effects in the same tables. These results show that mode awareness might not only be captured by users' NDRT engagement in one block but also over the time course (e.g., five repetitions). The behavioral adaptation of NDRT engagement corresponds to related research that investigated human–automation interaction across repeated interactions [13,14,21]. A closer investigation of differences between the blocks by means of planned contrast analysis (see Table 5) showed that a change over time is present from the first up to the third encounter. From then on, stable engagement in NDRTs can be assumed. This has implications for study designs concerning automated driving and engagements in NDRTs. When setting up a study, researchers should be aware that behavioral adaptation requires a certain number of repeated trials until reliable user behavior is present. One example is the study by Hergeth and colleagues [7], where the authors investigated whether NDRT engagement and according glance behavior could be an indicator of reliance behavior and marker for trust in automation. Indeed, they considered familiarization with NDRT and automated driving system including $N = 8$ repeated NDRT engagements.

NDRT engagement was also present at L2 driving automation. By definition, users of L2 driving automation are responsible for supervising the driving task at all times and may not leave the control loop [1]. Even though NDRT engagement during L2 automation was on a descriptively low level, there were participants that diverted their attention away from supervising the driving automation. This observation has implications for the design of L2 automation. It has to be noted, that secondary task activities occur even in manual driving [41]. Such distraction during manual driving (i.e., engaging in NDRTs) is considered a safety risk and should be minimized [1]. In contrast, there is first evidence that this tendency can be used in a beneficial way during automated driving as it might be turned into controlled engagement. For example, Paetzold and colleagues [42] did not find differences in reaction time to automation errors between participants that were either engaged or not engaged in an NDRT. In the same vein, Hensch and colleagues [43] found effects of display position and secondary task on the driver's glance behavior in both automated and manual driving. They especially report longer eyes-on display time for NDRTs in head-up display configurations. However, due to its proximity to the driving environment it might enable a faster identification of and reaction to critical situations such as system failures. Thus, there are still challenges for conceptual developments of a HMI design for L2 automated vehicle HMIs.

Eventually, this study supports that NDRT-related behavior can be used to distinguish between levels of automation and their HMI conceptualization. Indeed, drivers' differences in behavior in NDRTs support the conclusion that mode awareness for the HMIs in L2 and L3 automation was on a high level. This difference is not only apparent overall, but also by differences in changes over time. Moreover, the study showed a methodological aspect on how to evaluate NDRT behavior during an episode (i.e., onset vs. ongoing) which led to similar results. Especially the fact that NDRT engagement changes over time implies that research needs to focus on prolonged periods and that drivers need to adapt to this technology first before it can be used appropriately.

Limitations and Future Research

This study comes with a number of limitations. First, there were no incentives for engaging in the NDRT. In real-road driving, drivers might disengage only if the NDRT has a rewarding character. It remains therefore unknown whether the NDRT engagement in especially L2 automation would remain at such a low level if rewards would have been applied in this study. Second, the NDRT consisted of the SuRT alone, which is a standardized method for visual–manual distraction. This NDRT does, on the one hand, only cover two modalities of distraction (i.e., visual and manual) and, on the other hand, it might not be a very motivating NDRT. For example, Purucker and colleagues [44] have used a more naturalistic set of NDRTs for their study that increases external validity of the findings. Third, the NDRT was mounted in a fixed way in the center console. It might be that engagement is increased if the NDRT is located closer to the line of sight [43]. Thus, future research has to determine how the NDRT-related behavior in a different level of automation evolves for differing activities, modalities, and locations in the vehicle interior. Moreover, the present research only supports insights on the group level that support the predictive character of the SuRT as a measure for mode awareness. However, this does not permit inferences on the individual level. There is still room for future research to determine whether and how predictive the engagement in the SuRT is for mode awareness on an individual level.

Author Contributions: Conceptualization, Y.F., S.H., F.N.; methodology, Y.F., S.H., and F.N.; formal analysis, Y.F., V.G.; data curation, Y.F., V.G.; writing—original draft preparation, Y.F., V.G.; writing—review and editing, Y.F.; visualization, Y.F., V.G.; supervision, S.H., A.K. All authors have read and agreed to the published version of the manuscript.

Funding: This research received no external funding.

Conflicts of Interest: The authors declare no conflict of interests.

References

1. SAE. *Taxonomy and Definitions for Terms Related to On-Road Motor Vehicle Automated Driving Systems (No. J3016R)*; SAE: Warrendale, PA, USA, 2018.
2. AAM. *Statement of Principles, Criteria and Verification Procedures on Driver Interactions with Advanced In-Vehicle Information and Communication Systems*; Alliance of Automobile Manufactures: Washington, DC, USA, 2006.
3. NHTSA. *Visual-Manual NHTSA Driver Distraction Guidelines for In-Vehicle Electronic Devices*; National Highway Traffic Safety Administration (NHTSA), Department of Transportation (DOT): Washington, DC, USA, 2012.
4. Forster, Y.; Naujoks, F.; Neukum, A. Your Turn or My Turn? Design of a Human-Machine Interface for Conditional Automation. In Proceedings of the 8th International Conference on Automotive User Interfaces and Interactive Vehicular Applications, Ann Arbor, MI, USA, 24–28 October 2016.
5. Forster, Y.; Naujoks, F.; Neukum, A. Increasing anthropomorphism and trust in automated driving functions by adding speech output. In Proceedings of the 2017 IEEE Intelligent Vehicles Symposium (IV), Redondo Beach, CA, USA, 11–14 June 2017.
6. Hergeth, S. Automation Trust in Conditional Automated Driving Systems: Approaches to Operationalization and Design. Ph.D. Thesis, Technische Universität Chemnitz, Chemnitz, Germany, 2016.
7. Hergeth, S.; Lorenz, L.; Vilimek, R.; Krems, J.F. Keep your scanners peeled: Gaze behavior as a measure of automation trust during highly automated driving. *Hum. Factors* **2016**, *58*, 509–519. [CrossRef] [PubMed]
8. Gold, C.; Damböck, D.; Lorenz, L.; Bengler, K. "Take over!" How long does it take to get the driver back into the loop? *Proc. Hum. Factors Ergon. Soc. Annu. Meet.* **2013**, *57*, 1938–1942. [CrossRef]
9. Happee, R.; Gold, C.; Radlmayr, J.; Hergeth, S.; Bengler, K. Take-over performance in evasive manoeuvres. *Accid. Anal. Prev.* **2017**, *106*, 211–222. [CrossRef] [PubMed]
10. Naujoks, F.; Mai, C.; Neukum, A. The effect of urgency take-over requests during highly automated driving under distraction conditions. *Adv. Hum. Asp. Trans.* **2014**, *7 Pt I*, 431.
11. Beggiato, M.; Krems, J.F. The evolution of mental model, trust and acceptance of adaptive cruise control in relation to initial information. *Trans. Res. Part F Traffic Psychol. Behav.* **2013**, *18*, 47–57. [CrossRef]
12. Beggiato, M.; Pereira, M.; Petzoldt, T.; Krems, J.F. Learning and development of trust, acceptance and the mental model of ACC. A longitudinal on-road study. *Trans. Res. Part F Traffic Psychol. Behav.* **2015**, *35*, 75–84. [CrossRef]
13. Forster, Y.; Hergeth, S.; Naujoks, F.; Beggiato, M.; Krems, J.F.; Keinath, A. Learning and Development of Mental Models in Interaction with Driving Automation: A Simulator Study. In Proceedings of the Driving Assessment Conference, Santa Fe, NM, USA, 24–27 June 2019.
14. Forster, Y.; Hergeth, S.; Naujoks, F.; Beggiato, M.; Krems, J.F.; Keinath, A. Learning to Use Automation: Behavioral Changes in Interaction with Automated Driving Systems. *Trans. Res. Part F Traffic Psychol. Behav.* **2019**, *62*, 599–614. [CrossRef]
15. Naujoks, F.; Hergeth, S.; Wiedemann, K.; Schömig, N.; Forster, Y.; Keinath, A. Test procedure for evaluating the human–machine interface of vehicles with automated driving systems. *Traffic Inj. Prev.* **2019**, *20* (Suppl. 1), S146–S151. [CrossRef]
16. Sarter, N.B.; Woods, D.D. How in the world did we ever get into that mode? Mode error and awareness in supervisory control. *Hum. Factors* **1995**, *37*, 5–19. [CrossRef]
17. Gopinath, V.; Johansen, K. Understanding situational and mode awareness for safe human-robot collaboration: Case studies on assembly applications. *Prod. Eng.* **2019**, *13*, 1–9. [CrossRef]
18. Reilhac, P.; Hottelart, K.; Diederichs, F.; Nowakowski, C. User experience with increasing levels of vehicle automation: Overview of the challenges and opportunities as vehicles progress from partial to high automation. In *Automotive user interfaces*; Springer: Berlin/Heidelberg, Germany, 2017; pp. 457–482.

19. Feldhütter, A.; Segler, C.; Bengler, K. Does Shifting Between Conditionally and Partially Automated Driving Lead to a Loss of Mode Awareness? In Proceedings of the International Conference on Applied Human Factors and Ergonomics, Los Angeles, CA, USA, 17–21 July 2017.

20. Seppelt, B.; Reimer, B.; Russo, L.; Mehler, B.; Fisher, J.; Friedman, D. Consumer confusion with levels of vehicle automation. In Proceedings of the 10th International Driving Symposium on Human Factors in Driver Assessment, Training, and Vehicle Design, Santa Fe, NM, USA, 27 June 2019.

21. Kraft, A.-K.; Naujoks, F.; Wörle, J.; Neukum, A. The impact of an in-vehicle display on glance distribution in partially automated driving in an on-road experiment. *Trans. Res. Part F Traffic Psychol. Behav.* **2018**, *52*, 40–50. [CrossRef]

22. Martens, M.H.; Jenssen, G.D. Behavioural adaptation and acceptance. In *Handbook Intelligent Vehicles*; Springer: London, UK, 2012.

23. Lee, J.D.; See, K.A. Trust in automation: Designing for appropriate reliance. *Hum. Factors* **2004**, *46*, 50–80. [CrossRef] [PubMed]

24. Feldhütter, A.; Härtwig, N.; Kurpiers, C.; Hernandez, J.M.; Bengler, K. Effect on Mode Awareness When Changing from Conditionally to Partially Automated Driving. In *Advances in Intelligent Systems and Computing: Vol. 823. Proceedings of the 20th Congress of the International Ergonomics Association (IEA 2018)*; Bagnara, S., Tartaglia, R., Albolino, S., Alexander, T., Fujita, Y., Eds.; Springer: Cham, Switzerland, 2019; Volume 823, pp. 314–324. [CrossRef]

25. Schömig, N.; Wiedemann, K.; Hergeth, S.; Forster, Y.; Muttart, J.; Eriksson, A.; Naujoks, F. Checklist for expert evaluation of automated vehicles HMIs–discussions on its value and adaptions of the method within an expert workshop. *Information* **2020**, *11*, 233. [CrossRef]

26. DeGuzman, C.; Hopkins, S.; Donmez, B. Driver Takeover Performance and Monitoring Behaviour with Driving Automation at System-Limit versus System-Malfunction Failures. *Trans. Res. Rec. J. Trans. Res. Board* **2020**. [CrossRef]

27. Louw, T.; Madigan, R.; Carsten, O.; Merat, N. Were they in the loop during automated driving? Links between visual attention and crash potential. *Inj. Prev.* **2016**, *23*, 281–286. [CrossRef]

28. Morando, A.; Victor, T.W.; Dozza, M. Reference model for driver attention in automation: Glance behavior changes during lateral and longitudinal assistance. *IEEE Trans. Intell. Trans. Syst.* **2019**, *20*, 2999–3009. [CrossRef]

29. Llaneras, R.E.; Salinger, J.; Green, C.A. Human factors issues associated with limited ability autonomous driving systems: Drivers' allocation of visual attention to the forward roadway. In Proceedings of the 7th International Driving Symposium on Human Factors in Driver Assessment, Training and Vehicle Design, Bolton Landing, New York, NY, USA, 17–20 June 2013.

30. ISO. *Road Vehicles—Ergonomic Aspects of Transport Information and Control Systems—Calibration Tasks for Methods which Assess Driver Demand due to the Use of In-Vehicle Systems*; (ISO, 14198); ISO: Geneva, Switzerland, 2012.

31. ISO. *Road Vehicles—Ergonomic Aspects of Transport Information and Control Systems—Specifications and Test Procedures for In-Vehicle Visual Presentation*; (15008); International Organization for Standardization: Geneva, Switzerland, 2017.

32. Radlmayr, J.; Gold, C.; Lorenz, L.; Farid, M.; Bengler, K. How Traffic Situations and Non-Driving Related Tasks Affect the Take-Over Quality in Highly Automated Driving. *Proc. Hum. Factors Ergon. Soc. Annu. Meet.* **2014**, *58*, 2063–2067. [CrossRef]

33. Forster, Y.; Hergeth, S.; Naujoks, F.; Krems, J.F.; Keinath, A. Tell them how they did: Feedback on operator performance helps calibrate perceived ease of use in automated driving. *Multimodal Technol. Interact.* **2019**, *3*, 29. [CrossRef]

34. Naujoks, F.; Hergeth, S.; Keinath, A.; Wiedemann, K.; Schömig, N. Use Cases for Assessing, Testing, and Validating the Human Machine Interface of Automated Driving Systems. In Proceedings of the Human Factors and Ergonomics Society Annual Meeting, Philadelphia, PA, USA, 1–5 October 2018.

35. Naujoks, F.; Purucker, C.; Neukum, A.; Wolter, S.; Steiger, R. Controllability of Partially Automated Driving functions–Does it matter whether drivers are allowed to take their hands off the steering wheel? *Trans. Res. Part F Traffic Psychol. Behav.* **2015**, *35*, 185–198. [CrossRef]

36. Manca, L.; de Winter, J.C.F.; Happee, R. Visual Displays for Automated Driving: A Survey. In Proceedings of the 7th International Conference on Automotive User Interfaces and Interactive Vehicular Applications, Nottingham, UK, 1–3 September 2015.

37. Forster, Y.; Hergeth, S.; Naujoks, F.; Krems, J.F.; Keinath, A. Empirical Validation of a Checklist for Heuristic Evaluation of Automated Vehicle HMIs. In Proceedings of the 10th International Conference on Applied Human Factors and Ergonomics, Washington, DC, USA, 24–28 July 2019.

38. Jarosch, O.; Kuhnt, M.; Paradies, S.; Bengler, K. It's Out of Our Hands Now! Effects of Non-Driving Related Tasks During Highly Automated Driving on Drivers' Fatigue. In Proceedings of the Driving Assessment Conference, Manchester Village, VT, USA, 26–29 June 2017.

39. Benjamini, Y.; Hochberg, Y. Controlling the False Discovery Rate: A Practical and Powerful Approach to Multiple Testing. *J. R. Stat. Soc. Ser. B (Methodol.)* **1995**, *57*, 189–300. [CrossRef]

40. Cohen, J. *Statistical Power Analysis for the Behavioral Sciences*, 2nd ed.; Routledge: Hillsdale, NJ, USA, 1988.

41. Dingus, T.A.; Klauer, S.G.; Neale, V.L.; Petersen, A.; Lee, S.E.; Sudweeks, J.D.; Gupta, S. *The 100-car Naturalistic Driving Study, Phase II-Results of the 100-Car Field Experiment*; U.S. Department of Transportation: Washington, DC, USA, 2006.

42. Pätzold, A.; Schmidt, C.; Rauh, N.; Cocron, P.; Hergeth, S.; Keinath, A.; Krems, J.F. From distraction to controlled engagement: How secondary tasks affect drivers' supervisory and fall-back performance of the driving task while using SAE level 2 driving automation. In Proceedings of the Europe Chapter Human Factors and Ergonomics Society Annual Meeting 2017, Rome, Italy, 28–30 September 2017.

43. Hensch, A.-C.; Rauh, N.; Schmidt, C.; Hergeth, S.; Naujoks, F.; Krems, J.F.; Keinath, A. Effects of secondary tasks and display position on glance behavior during partially automated driving. *Trans. Res. Part F Traffic Psychol. Behav.* **2020**, *68*, 23–32. [CrossRef]

44. Purucker, C.; Naujoks, F.; Wiedemann, K.; Neukum, A.; Marberger, C. Effects of Secondary Tasks on Conditional Automation State Transitions While Driving on Freeways: Judgements and Observations of Driver Workload. In Proceedings of the 6th International Conference on Driver Distraction and Inattention, Gothenburg, Sweden, 15–17 October 2018.

 *information*

Article

Methodological Considerations Concerning Motion Sickness Investigations during Automated Driving

Dominik Mühlbacher [1,*], **Markus Tomzig** [1], **Katharina Reinmüller** [2] and **Lena Rittger** [2]

[1]   WIVW GmbH, D-97209 Veitshöchheim, Germany; tomzig@wivw.de
[2]   AUDI AG, D-85045 Ingolstadt, Germany; katharina.reinmueller@audi.de (K.R.); lena.rittger@audi.de (L.R.)
*   Correspondence: muehlbacher@wivw.de

Received: 22 April 2020; Accepted: 11 May 2020; Published: 13 May 2020

Abstract: Automated driving vehicles will allow all occupants to spend their time with various non-driving related tasks like relaxing, working, or reading during the journey. However, a significant percentage of people is susceptible to motion sickness, which limits the comfort of engaging in those tasks during automated driving. Therefore, it is necessary to investigate the phenomenon of motion sickness during automated driving and to develop countermeasures. As most existing studies concerning motion sickness are fundamental research studies, a methodology for driving studies is yet missing. This paper discusses methodological aspects for investigating motion sickness in the context of driving including measurement tools, test environments, sample, and ethical restrictions. Additionally, methodological considerations guided by different underlying research questions and hypotheses are provided. Selected results from own studies concerning motion sickness during automated driving which were conducted in a motion-based driving simulation and a real vehicle are used to support the discussion.

Keywords: motion sickness; automated driving; methodology; driving comfort

---

1. Introduction and Overview

Motion sickness is well known amongst users of any kind of transportation. Sea sickness, airplane sickness, even space sickness have been investigated over the past 100 years [1]. Today, depending on the considered reference, up to 60% of Americans suffer from car sickness [2,3]. At the same time, original equipment manufacturers (OEMs) are developing towards automated driving, allowing drivers to hand over full control to the vehicle and by that engaging in non-driving related tasks while driving. Moreover, automated vehicles may include new cabin designs that enable different human postures and thus support the execution of non-driving related activities. With that, the possibility of suffering from motion sickness expands from passengers to drivers. To use the "value of time" generated by automated vehicles, users expect to engage in a large variety of tasks during driving, ranging from reading, working, playing video games, watching movies and many more [4,5]. However, while enabling such tasks in the vehicle is promising in terms of user satisfaction, it is exactly those activities that increase the probability of motion sickness [2,6]. Hence, within the context of automated driving, higher incidence numbers and more severe symptoms of motion sickness can be expected [7,8], which will impair the user experience in the vehicle. Besides negative subjective experiences, it is, until now, unclear how motion sickness influences take over and driving performance in case the automation system reaches its limitations and drivers have to take over the driving task. Regarding professionals at sea, a study found up to 60% impaired performance due to sea sickness [7]. Hence, motion sickness could not only lead to a decreased acceptance of automated vehicles but also to a decrease in driving safety. Consequently, there has been an increasing demand of investigation in motion sickness in the context of automated driving. In particular, two research questions are of interest:

(1)   The first goal is to get valid estimations of the prevalence, symptoms, and symptom evolution, to understand the influencing factors and development over time.

(2)   The second main issue is the development of countermeasures for motion sickness that are applicable in the vehicle.

To answer these research questions, controlled empirical studies are necessary. Yet, there has only been a small amount of research conducted in realistic vehicle settings.

This paper discusses basic methodological considerations for designing and conducting empirical studies on motion sickness in the vehicle context. It shall support applied researchers to decide on the right methods, measures, samples, and ethical considerations. The paper furthermore includes unpublished data confirming the methodological approaches.

The data presented in the subsequent chapters is based on a study conducted in the high-level driving simulator of the Wuerzburg Institute for Traffic Sciences (WIVW GmbH) and an AUDI A8 serial vehicle. A total of N = 24 participants took part in the study. The study had a within-subjects design, i.e., every participant took part in four separate sessions, of which two were drives in the real vehicle and two were drives in the driving simulator. In the real driving part, an Audi A8L equipped with SAE Level 2 functionality was used as the test vehicle. A trained experimenter drove the car on an Autobahn track while the participant was sitting in the front passenger seat. The experimenter used the Level 2 functions whenever possible. Passing maneuvers were performed similar to an autonomous vehicle. The simulator runs with the driving simulation software SILAB®. The motion system uses a hexapod with six degrees of freedom and can briefly display a linear acceleration up to 5 m/s^2 or 100°/s^2 on a rotary scale. It consists of 6 electro-pneumatic actuators (stroke ± 60 cm; inclination ± 10°). The mockup is created with a BMW 520i with automatic transmission. As the visual system of the WIVW simulator is defined for the driver only, the participants were sitting in the driver's seat during the run in the driving simulator. The driving behavior of the simulated vehicle was defined as comparable as possible to the driving behavior of the real vehicle. Additionally, the road geometry of the real Autobahn track was implemented precisely in the driving simulation. The participant's task was to watch a video during the rides of approx. 40 min. The four runs occurred in a counterbalanced order with a minimum of two days between each day of participation. In the study, motion sickness was measured via the misery-scale (MISC) [9] every two minutes during the run. After the run, a symptom questionnaire was used. It included a list with symptoms of the motion sickness questionnaire (MSQ) [10] and from the simulator sickness questionnaire [11], which were rated on a scale with four categories ranging from "none" to "severe". After the last run, the participants had to compare both test settings (real vehicle vs. driving simulator) in form of several questions. Physiological data (participants' temperature, electrodermal activity, electrogastrogram) were recorded with a Varioport Polygraph (Becker Meditec).

2. The Phenomenon of Motion Sickness

2.1. Symptoms, Prevalence, and Time Course of Motion Sickness

The main symptom of motion sickness is nausea leading up to vomiting [12]. However, nausea is typically preceded and accompanied by symptoms like burping, (cold) sweat, pallor, fatigue, headache, or dizziness [9,13,14]. Appearance and chronology of the symptoms varies a lot between different persons and between the different types of motion sickness (e.g., carsickness, seasickness, simulator sickness) [7]. For example, oculomotor symptoms (such as eye strain or difficulties in focusing) are more often to be found in situations in which sickness is induced by visual stimuli (e.g., simulators) than in situations in which sickness is primarily elicited by movements (e.g., sea travel) [12].

It is difficult to make statements about the prevalence of motion sickness because its occurrence depends on multiple factors such as the mean of transport (car, bus, ship, train, etc.) but also on duration and intensity of provocation (driving through curves vs. driving on a straight highway). However, it has repeatedly been demonstrated in laboratory studies that motion sickness occurrence and intensity highly depend on the frequency of accelerations which act upon the passenger. Frequencies of about 0.16 to 0.2 Hz are particularly provocative to elicit motion sickness [15–24]. Furthermore, there is an effect of age: young children up to two years are immune to motion sickness [25]. Afterwards, mean motion sickness susceptibility raises to a peak at the age of 16 to 20 years [26]. Subsequently, the susceptibility decreases with increasing age. Concerning gender, women are generally more prone to motion sickness than men [25,27–31]. The reasons for the gender difference are not clear. However, hormonal factors or a lower threshold to admit motion sickness symptoms in women are discussed as possible explanations [25,27–31].

Similarly to the prevalence, the time course of motion sickness also depends on various factors like the individual susceptibility as well as the type and intensity of sickness provocation. The first symptoms may be perceived immediately after onset in highly provoking conditions. In less severe conditions, motion sickness may occur after 10 to 20 min in susceptible participants. Depending on the study design, symptoms often intensify linearly with progressing provocation duration and decrease rapidly after offset [32–34].

2.2. Motion Sickness Theories

There are various models and theories trying to explain the mechanisms leading to motion sickness, e.g., the toxin-hypothesis [35], postural instability theory [36], negative reinforcement model [37], or the rule of thumb [38]. More popular than these models is the theory of sensory conflict and rearrangement [14] or its revision, the neural mismatch theory [39]. It states that motion sickness occurs if there is a discord between different sensory inputs, i.e., the visual and the vestibular system. For example, if a passenger is reading a book during a drive, the eyes register a static environment and give feedback that the person is not moving. However, the vestibular organs register the longitudinal and lateral accelerations of the vehicle and give feedback that the person is moving. In this situation, the probability for motion sickness is higher than in a passenger who looks ahead and thus has no contradictory impressions [34]. In addition, motion sickness depends on the type of task [40,41].

Reason and Brand extended the theory by the component of expected sensory impressions [14]. The sensory rearrangement theory states that motion sickness increases when the actual visual and vestibular impressions differ from the expected ones, i.e., when future movements cannot be anticipated. Within this context, effects of habituation may also be relevant (neural mismatch theory) [39]. In general, the better the passenger's view ahead, the lower the risk of motion sickness [34]. For these reasons, motion sickness is more likely to occur when the passenger is sitting in the back seat compared to sitting in the front seat.

3. Methods for Investigating Motion Sickness in Autonomous Vehicles

3.1. The Study Setting

In general, two study methods are applicable for motion sickness studies concerning autonomous driving: field experiments with real vehicles and studies in a driving simulator.

In field experiments, the participant is passenger of a real vehicle in a realistic road environment or on a test track. Naturally, the experimenter has no full control of the dynamic events happening and the experiences the participants make in a field study, resulting in reductions of internal validity. However, there are different ways to control for this. First of all, the driving style of the used vehicle should be standardized. In future, this can be realized by using automated functions that perform driving manoeuvers in the same way with high reliability. Until these automated vehicles are commonly available for this kind of study, human drivers need to drive the test vehicles. High levels of realism for future automated vehicles can be achieved by Wizard-Of-Oz settings, in which the automation is simulated by a human driver, e.g., [42–44]. It is necessary that human drivers are instructed or trained towards a specified and thus reproducible driving style [45]. In the presented setting (cf. Chapter 1), we used assistance systems like adaptive cruise control (ACC) and lane keeping for standardization.

The trained drivers learned how to perform lane change maneuvers with the necessary step of actions (for example setting indicator before moving steering wheel, changing lanes in six to seven seconds). Along with that, there should be a low number of experimental drivers in order to avoid inter-individual differences in driving style between experimenters. Finally, after completing data collection, it is recommended to analyze the dynamic driving data to identify any conspicuousness within the actual realized driving behavior. If possible, the data can be systematically compared to the vehicle dynamics measured (1) within the same study to check for internal validity and (2) in other settings or situations in order to check for external validity.

However, even if the vehicle dynamics are kept as standardized as possible, external factors like traffic or weather conditions cannot be kept constant or manipulated consciously. However, these aspects can affect motion sickness: A high traffic density can lead to an increased number of braking and overtaking maneuvers due to slower vehicles. This driving behavior can lead to stronger symptoms of motion sickness. In contrast, a low traffic density enables homogenous driving with less accelerations and decelerations, which reduces the probability for motion sickness.

In contrast to field studies, driving simulators enable conducting studies in a highly controlled environment. They are used since the 1960s to investigate driving performance and behavior and are classified into three categories [46]:

- High-level simulators incorporate a motion system and full vehicle cabs;
- Mid-level simulators are static simulators with a full vehicle cab;
- Low-level simulators are built around simple components such as game controllers and computer monitors.

As most researchers attribute motion sickness in vehicles to contradictory impressions between the vestibular system (which perceives motion) and the visual system (which perceives no motion, e.g., while reading a book), the use of a high-level simulator with a motion system is recommended. In mid-level and low-level simulators, in contrast, only visual induced motion sickness can be investigated. Basically, research questions concerning countermeasures or physiological correlates are conceivable in these simulators. However, it remains unclear how the results of these studies in simulators without motion system would be applicable for automated driving.

The most common motion platform of high-level simulators is a hexapod which provides motion in six degrees of freedom (x, y, z, roll, pitch, yaw). Compared to travelling in a real vehicle, longitudinal and lateral accelerations are different. The feeling for realistic accelerations is generated by hacks like tilting the presented scenery. More elaborated simulators mount the hexapod on an x-y table on which the simulation cabin is moved to produce more realistic accelerations. According to Carsten and Jamson [47], however, even a large motion system is not capable to provide realistic accelerations in special driving situations like negotiating a long curve.

Probably the most important benefit of driving simulation is the ability to create repeatable scenarios which are tailored to a certain research question. Depending on the research question, motion-sickness provoking scenarios with many strong lateral and longitudinal accelerations are possible as well as more homogenous driving scenarios with few accelerations only (e.g., highway scenarios). Additionally, the researcher is free in the selection of the driving behavior of the autonomous vehicle: each imaginable driving style is feasible even if this driving behavior is not possible in a real autonomous vehicle yet. Another benefit of driving simulation is the availability of data: the simulator provides all data that would be provided by a real test vehicle (e.g., velocity, acceleration) as well as data of the traffic environment (e.g., surrounding traffic, road geometry). Besides, the participant's behavior (e.g., head movement, glance behavior) and physiological data can be monitored and recorded in a simple way: the laboratory conditions make video recordings easier due to constant light conditions and physiological data recording more precise due to less disturbing artifacts of the environment (e.g., temperature, humidity).

On the other hand, there are disadvantages of driving simulation. Some participants of simulator studies suffer simulator sickness, which is a subtype of motion sickness in simulated environments. The phenomenon occurs in all types of simulators—it also appears in fixed-base simulators without motion system due to visual stimuli only. Similar to motion sickness, it is caused by a mismatch between the visual perception and the vestibular sensation of acceleration and deceleration [14,48]. For a motion sickness study, this means that the results for motion sickness can be confounded with simulator sickness. For studies regarding prevalence or development of motion sickness it is recommended to exclude participants who have shown symptoms of simulator sickness in previous studies in order to diminish this artifact. However, as simulator sickness and motion sickness are related and show similar symptoms due to similar reasons, it is possible that some countermeasures are effective against both symptoms. Therefore, it has to be discussed if participants with simulator sickness problems are allowed in a study concerning motion sickness countermeasures. However, this issue has to be decided for each countermeasure or research question separately.

An important issue of driving simulation is the validity. A distinction that has been made on simulator validity is between absolute and relative validity [49]. Relative validity exists when effects in the simulator and under the same road conditions are in the same order and direction. In contrast, absolute validity is present when the numerical values are about equal in both systems. A lot of validation studies were carried out in various simulators. They compared various parameters of the driver's behavior (e.g., velocity, lateral displacement, braking behavior, gaze direction) between driving in a simulator and driving in a real vehicle. In most cases, the studies showed that relative validity exists while absolute validity was only rarely verified [50]. However, these results do not provide evidence that validity is given for motion sickness studies. In a motion sickness study, behavior of a driver is not relevant—moreover, the occupants' visual and vestibular perceptions are important.

To the authors' knowledge, there have not yet been studies comparing an occupant's motion sickness in a driving simulator to his/her motion sickness in a real vehicle. Therefore, we conducted the study design as described above.

The results showed that the progress of motion sickness was comparable in both conditions. After a general rise at the beginning of the run (approx. first 12 min), the sickness ratings increased more slowly in the second and last third. Compared to real driving, self-reported motion sickness was slightly higher in the simulation compared to the real vehicle (Figure 1). However, the maximum sickness values during the runs do not differ (Wilcoxon signed-rank test: $Z = 1.40$, $p = 0.162$). The sessions of $n = 3$ drivers had to be aborted due to high sickness ratings in the simulator. In the field study, the run of $n = 1$ driver was terminated before the end of the test course. According to the symptom questionnaire, most symptoms occurred in a similar frequency and intensity in both runs (Figure 2 left). However, three symptoms differed significantly concerning their intensity: in the driving simulator, participants had higher general discomfort, more difficulties concerning focusing, and increased appetite (Figure 2 right). In a final interview after both runs, the participants stated that the motion sickness symptoms were more distinct in the driving simulator compared to the real vehicle ($t(23) = 5.65$, $p < 0.001$).

These results indicate that relative validity is given for the high-level simulator of the WIVW GmbH concerning motion sickness as the progression during the runs was comparable and the occurrence of frequent symptoms was similar. In contrast, absolute validity cannot be verified, as some of the self-reported symptoms were more distinct in the simulator.

The recommendation for the most appropriate study setting depends on the research questions: A field experiment offers the highest validity and should be used for studies which investigate the prevalence and the development of motion sickness. In this case, a conduction on public roads should be selected. The realistic test track could represent a highway, rural road or inner-city track. Previous studies used driving on highways and inner city roads to identify if and how strong motion sickness occurs. In these studies, the participants performed different tasks in the vehicle [51].

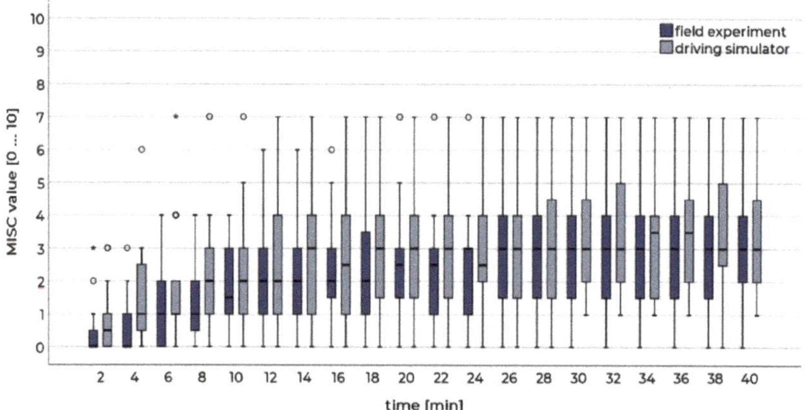

Figure 1. Misery scale (MISC) rating over time during the runs in a real vehicle and in the high-level driving simulator. Sessions were terminated when a value of 7 = moderate nausea was given. These participants were assigned continuing values of 7 for the purpose of this illustration. Boxplots are shown.

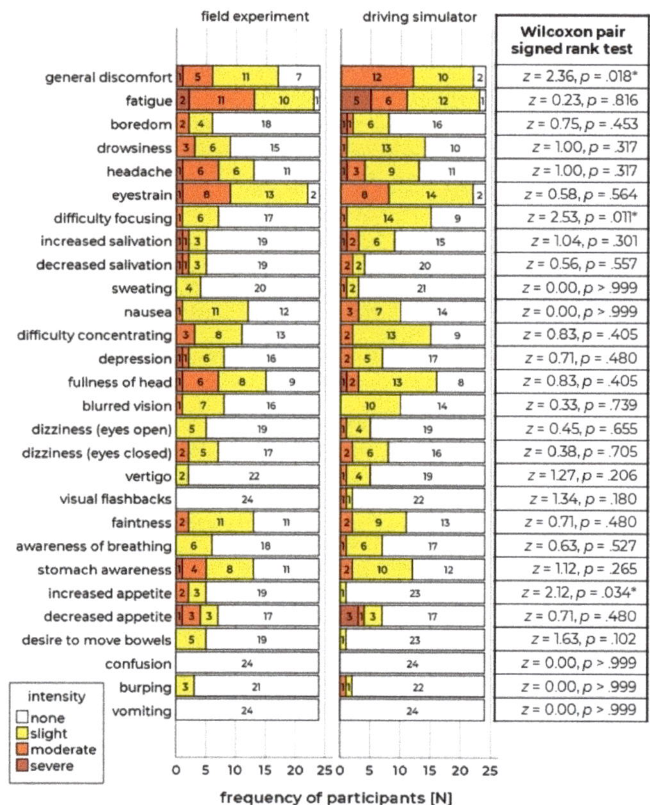

Figure 2. Frequencies of symptom judgments in field experiment and driving simulator. The color indicates the symptoms' intensity (**left**). Results of a Wilcoxon pair signed rank test, significant results ($p < 0.05$) are indicated by * (**right**).

In contrast, in case the research question covers the investigation of countermeasures avoiding or reducing the symptoms of motion sickness, it is crucial to choose a test setting that causes motion sickness in the participants quickly and with a high probability. In the vehicle context, this setting was mainly realized on test tracks, on which high provoking maneuvers were driven by the experimenters (e.g., driving in the shape of an eight, or constant stop and go). Other researchers made use of placing the participant rearwards in a vehicle driving on urban roads [51,52]. Within this setting, a comparison between a baseline trial and a repetition of the same condition with potential countermeasures allows to investigate the effectiveness in avoiding symptoms. In particular, considering the efforts put into these kinds of participant studies, an efficient and reliable creation of provoking situations needs to be considered in the study design. Besides, a simulator study using a motion sickness provoking scenario can also be conducted when investigating countermeasures. A requirement for this option is the validity of the driving simulator. The presented study shows that a high level driving simulator without x-y table can also offer relative validity—however, as driving simulators are very different this has to be tested for each simulator individually.

3.2. The Participant's Task

In general, automated driving will enable the driver to engage in various non-driving related activities. In motion sickness research, one relevant research question refers to specifically examining the different non-driving related tasks (NDRTs) for their potential to cause motion sickness. In respective investigations, subjects could either be free to engage in realistic everyday NDRTs of their choice or be presented with a specific NDRT. While many standardized tasks exist in the context of manual driving, such standardization is widely missing in the context of automated driving. Therefore, it would be desirable to also evaluate secondary tasks that cover certain groups of conceivable NDRTs in the future. For our setting, we chose a naturalistic NDRT. Based on previous research [5], it can be expected that the use case watching a video in an automated vehicle has some external validity.

Concerning other research questions such as the evaluation of countermeasures, it may also be relevant to induce motion sickness in a targeted manner or to investigate an extreme scenario. In this case, NDRTs that are characterized by highly limited peripheral and external vision of motion are required as hints about the vehicle's future motions can counteract motion sickness [53–55]. Therefore, a mainly visual NDRT should be presented in a way that assures gazing away from the road scene. To ensure standardization of the amount of peripheral vision across participants, visual material should be presented at a fixed location, e.g., by means of displays instead of providing handheld devices such as tablets. Naturally, fixed display positions also lead to more standardized participant movements. Since peripheral vision can be manipulated by both display position and size [7], to prevent the participant from using peripheral vision, a visual NDRT could be presented at a downward angle or on a large display. Further, to promote continuous task engagement, it is recommended to choose an NDRT that is difficult to interrupt or provides instructions and incentives for subjects to focus on the task and refrain from road glances (e.g., concentrating on visual tasks like reading or watching a movie during the drive increases the risk of motion sickness). Artificial, standardized NDRTs can therefore be suitable for this. Please note that engaging in a visual NDRT may cause visual problems such as strained eyes or blurred vision, which cannot be differentiated from symptoms of motion sickness. To better control for this, visual task characteristics and the duration of the task engagement may be considered. Similarly, fatigue may occur due to the experimental session's duration or as a motion sickness symptom. Further research should examine both the relationship between motion sickness and fatigue as well as methods to control for confounding effects.

For the selection of an adequate task for empirical studies on motion sickness, classifications of NDRTs, provide relevant dimensions such as the primary modality, the locality, the possibility of road glances, the need for sustained attention, and incentives to continue the task [56]. In addition, the presented material should be controlled for emotionality of content when motion sickness is measured using physiological correlates. Therefore, in our study, subjects watched a movie on a display positioned below the central information display. We further instructed subjects to refrain from road glances. The videos contained documentaries, which were interesting but not emotionally arousing. Other examples for such tasks may be reading a text or answering a quiz that is presented visually. Finally, participant posture should be considered in motion sickness studies given that the risk of motion sickness is also higher when the passenger is sitting on a rearward facing seat compared to a forward facing seat [52]. Moreover, for postures facing in the driving direction, a regular driving posture may increase the risk of motion sickness compared to a reclined posture [57].

3.3. Sample and Recruitment

In order to investigate motion sickness in autonomous vehicles a participant study is recommended. The requirements for the recruitment depend on the study's research question.

For a large variety of research questions, it is necessary that a significant part of the sample suffers from motion sickness during the study. For example, the effect of a countermeasure for motion sickness during travelling can only be demonstrated when a control condition in a between- or within-subjects design exists in which motion sickness occurs. In contrast, people who are not susceptible to motion sickness do not need countermeasures and are not relevant for the study question. It is only possible to identify physiological correlates of motion sickness when the participants have phases with and without motion sickness. Therefore, the selection of participants is crucial for the study's success as not all people are susceptible to motion sickness. This consideration leads to the next question regarding participants' recruitment: how to identify participants who are susceptible to motion sickness?

A common instrument to predict motion sickness susceptibility is the MSSQ (Motion Sickness Susceptibility Questionnaire) [14,58]. This tool queries how often several means of transport (e.g., cars, busses, airplanes) and amusement rides (e.g., carousels, rollercoasters) were used in the past and how often sickness occurred. The answers result in a motion sickness susceptibility score. However, the results of our own study indicate that the MSSQ total score is not appropriate to identify subjects who are susceptible to motion sickness while travelling in a car. There was no significant correlation (Spearman $r(24) = 0.266$; $p = 0.210$) between the MSSQ total score and the suffered motion sickness (measured via a misery scale according to [9]) in a real driving study on the Autobahn in which the $N = 24$ participants were passengers and had to watch a video during the drive (see Figure 3 left). The MSSQ probably covers too many means of transport—respondents with no motion sickness problems in cars can also achieve high MSSQ scores when having symptoms, for instance, in trains and airplanes. In contrast, respondents who compensate for their motion sickness in real driving situations might reach lower MSSQ scores than would be intended: people who know that they are susceptible to motion sickness might not engage in NDRTs in provoking situations and therefore did not experience any severe motion sickness in the past years.

However, the more specific MSSQ item "Over the last 10 years, how often you felt sick or nauseated in cars?" also showed no significant correlation (Spearman $r(24) = 0.212$; $p = 0.319$) to the suffered motion sickness in the study (see Figure 3 right). The question is very inaccurate as it does not differ between driving in an urban or rural area or on a highway. In addition, it summarizes travelling in a car while reading or texting on the back seat as well as being a co-driver who is attentive to the traffic situation. As the prevalence depends on the individual threshold to motion stimulation and varies under different situations [59], a curvy rural road can lead to symptoms for some people while other people suffer from motion sickness in urban scenarios only. Therefore, it is recommended to use a highly specific question with the exact test scenario as a screening question for the participants' recruitment (e.g., "Do you get symptoms of motion sickness as a co-driver while reading on the Autobahn?").

Concerning other research questions, a more common sample is required. A representative sample is necessary to investigate the prevalence of motion sickness. The sample should be representative concerning all aspects which can affect the prevalence of motion sickness, e.g., age [60,61] and gender [27,29].

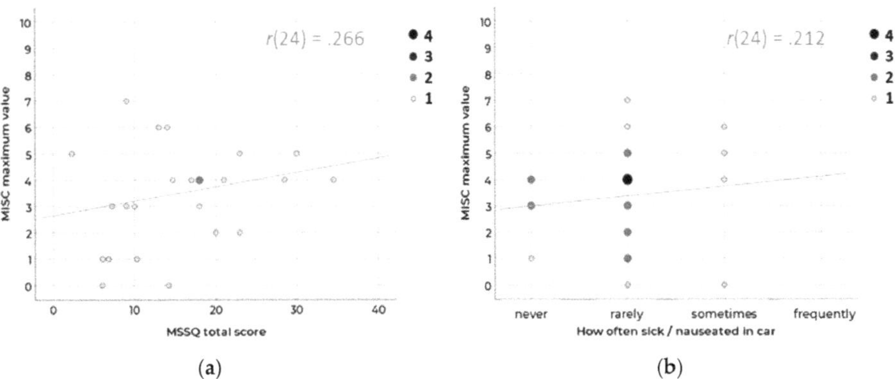

Figure 3. Spearman correlation between the maximum value on the misery scale during the session and (**a**) the MSSQ total score and (**b**) the single item concerning sickness in cars. Size and color of the dot indicate the number of respondents.

3.4. Measurement of Motion Sickness

3.4.1. Subjectively Perceived Motion Sickness

Subjective participant ratings via questionnaires are the most common method to measure motion sickness and to validate other measurement tools like physiological or behavioral measures. Within the subjective measurement approaches, there are two basic principles: either the participants are asked to evaluate their overall motion sickness in a single rating or the participants are questioned in detail about multiple or even all potential motion sickness symptoms and their intensity. Short questionnaires allow for a continuous online assessment of motion sickness during the test drive, which enables describing the time course of motion sickness development. In contrast, detailed questionnaires are suitable for pre-post evaluations to determine if and to what extent a certain condition has led to motion sickness.

One example for a short overall rating is the fast motion sickness scale (FMS) [62]. The FMS is a verbal rating scale ranging from 0 (no motion sickness at all) to 20 (severe sickness). Participants are asked to evaluate the current motion sickness and to focus on nausea, general discomfort and stomach problems. However, the scale of the FMS is unanchored. Hence, it is not possible to verbally describe what the distinct values on the scale stand for. Further, it is uncertain if the values on the scale actually represent the same degree of subjectively perceived motion sickness for each participant. It thus remains concealed if e.g., a value of 15 is associated with nausea and if this is valid for every participant of the sample. Therefore, unanchored scales do not deliver information about the characteristics of motion sickness. Due to its unspecific character, the rating may further be biased by other comfort restrictive factors, like boredom or fatigue.

Another tool to quickly measure subjective motion sickness is the misery-scale (MISC) [9]. It is an 11-point scale trying to capture the quantitative and qualitative degree of motion sickness within one combined rating. For this purpose, the scale's numeric values are assigned to more or less specific motion sickness symptoms and their intensity. The scale comprises the following gradation: 0 (no problems), 1 (uneasiness without specific symptoms), 2–5 (slightly to severely perceived specific symptoms like dizziness, headache, stomach awareness, etc.), 6–9 (nausea from slight to severe/retching), and 10 (vomiting). Thus, in contrast to scales like the FMS, the MISC values can be interpreted descriptively and it is assumed that every single value is interpreted similarly by all participants. Like the FMS, the MISC is able to assess motion sickness quickly, in short intervals and during motion sickness induction. The MISC suggests that nausea is perceived as more inconvenient than all other motion sickness symptoms. This, however, neglects that other symptoms like severe headache may also be perceived as very unpleasant. Without experiencing nausea, the MISC does not allow the participant to reach high motion sickness scores, even if the driving comfort has largely decreased. Therefore, strictly speaking MISC data cannot be considered as interval scaled. This impedes the analysis and interpretation of the results.

For these reasons, it may be useful to let the participants evaluate different specific symptoms on separate Likert scales. In addition to nausea, it would be plausible to include headache, general discomfort, dizziness, and—depending on the study design—also fatigue (especially during long or uneventful drives). In our study, these symptoms have been observed frequently after a 40-min Autobahn drive (71% of participants stated general discomfort, 96% fatigue) or are assumed to be perceived as particularly inconvenient (nausea, headache, dizziness). However, it should be ensured that the interrogation remains short.

In contrast to these quick and efficient methods, the motion sickness questionnaire (MSQ) [10] represents an approach to capture multiple or even all potential motion sickness symptoms and their intensity. There are different versions of the MSQ with different numbers of items [11]. The questionnaire consists of a checklist with items that are evaluated either concerning their presence (symptom present vs. not present) or concerning their intensity (none, slight, moderate, severe). Thus, the MSQ provides an extensive impression of the participant's current motion sickness. However, completing the questionnaire is relatively time-consuming and is thus not suitable for frequent motion sickness interrogations. It is, therefore, recommended to use it at the end of the driving study or during breaks (directly after provocation offset). Hence, the scale is rather suitable for pre-post evaluations and may be combined with a short online-questionnaire like the FMS, MISC or symptom-specific Likert scales. A comparative overview of the four discussed tools is given in Table 1.

Table 1. Comparison of the four discussed motion sickness assessment tools (++ very good; + good; o okay; − weak).

	FMS	MISC	Symptom-Specific Likert-Scales	MSQ
Duration of application	++	++	+	-
Informational content	-	o	+	++
Scale	interval	ordinal	interval	ordinal

It is important to add that subjective ratings may be prone to several biases, such as demand characteristics or social desirability as discussed in Chapter 4. Further, the participant's mental model of the own susceptibility may affect the ratings (i.e., self-fulfilling prophecy). For example, participants believing to be highly susceptible may indicate higher motion sickness ratings, not only because they feel motion sick, but also because they expect to do so and in that sense to confirm their own beliefs. In addition, directly asking participants about their motion sickness symptoms may lead to a very conscious introspection of perceived motion sickness symptoms. Thus, participants may "discover" symptoms which would not have been perceived consciously otherwise. Further research is needed to determine if and to what extend these potential biases affect subjective motion sickness ratings. Nonetheless we consider it important to directly ask participants about their sickness symptoms because motion sickness and discomfort highly depends on the subjective evaluation.

3.4.2. Physiological Correlates

Because subjective ratings may be prone to biases, research has tried to measure motion sickness objectively. Over the last decades, there have been many attempts to describe motion sickness with physiological correlates. Among others, heart rate, blood pressure, respiration rate, gastrointestinal reactions, and skin conductance parameters have been investigated, e.g., [63–67]. However, until now there has been no reliable success in correlating physiological measures with subjectively perceived motion sickness. Reasons are the high variability of motion sickness provoking stimuli as well as the high individual specificity of reactions. For example, there are rather individual correlations between subjectively reported motion sickness and heart rate or blood pressure [68].

Three measures in which a correlation with motion sickness has been shown across multiple laboratory studies are body temperature [69], skin conductance [69,70], and electrogastrogram [71,72]. Hereinafter it shall be discussed to what extent these three measures are applicable to capture motion sickness in a driving study under naturalistic conditions.

Temperature

In previous studies, it was shown that motion sickness affects the human thermoregulation [69]. Nobel and colleagues demonstrated that in cold water body temperature decreases faster in motion sickness induced participants than in control participants [73]. Similarly, in a thermo-neutral environment body temperature was lower in motion sick participants than in control participants [74]. In the latter study, for example, the mean difference was about 0.4 °C between control participants and such who stated to be "very nauseous/almost vomiting". In the cited studies, body temperature was measured by a rectal thermistor. Not surprisingly, this procedure is perceived as an unreasonable imposition by many participants and may be doubtful for ethical reasons. One of multiple alternatives to make temperature measurement more convenient for the participants is to place the thermistor under the armpit. The participants should not move their arm during the measurement. It should be considered that mean axillary temperature is some tenth °C lower than rectal body temperature [75]. Within this procedure, body and skin temperature cannot be clearly distinguished, although they should not be equated. In some previous studies, differences in body temperature were not necessarily accompanied by significant differences in skin temperature [67,73]. Further, skin temperature can be biased, e.g., by perspiration, environmental temperature, or participants' clothing (warm/light). However, most biases can be controlled easily by the experimenter. Temperature and ventilation in the test vehicle can be held constant by air condition and participants can be instructed to wear comparable types of warm/light clothes. Further, the measured signal can be controlled easily by the experimenter since the range of value is relatively constant across participants (approx. between 36 and 38 degrees Celsius), which makes it easy to detect technical signal disturbances. Moreover, the signal is relatively stable and hardly susceptible to artifacts (e.g., movements, speaking; see Figure 4). As body temperature seems to react relatively slowly to influences, it is to expect that it does so with regard to motion sickness. Consequently, to detect potential effects, heavy provocation and/or a long measurement period might be necessary.

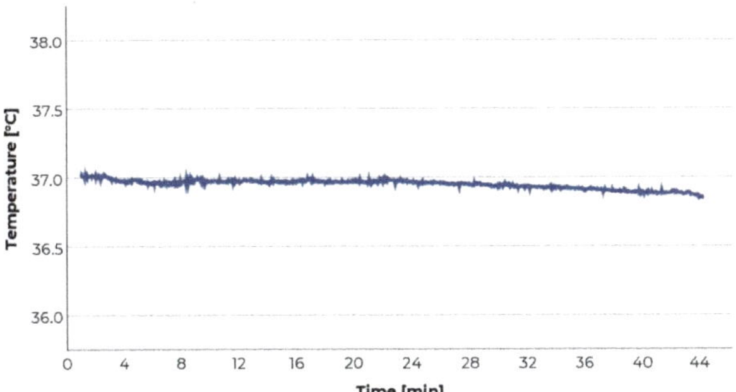

Figure 4. Exemplary raw temperature data of a participant during a 44-min drive as passenger on a German Autobahn. In contrast to other physiological data, temperature is hardly affected by artifacts (see Figures 5 and 6).

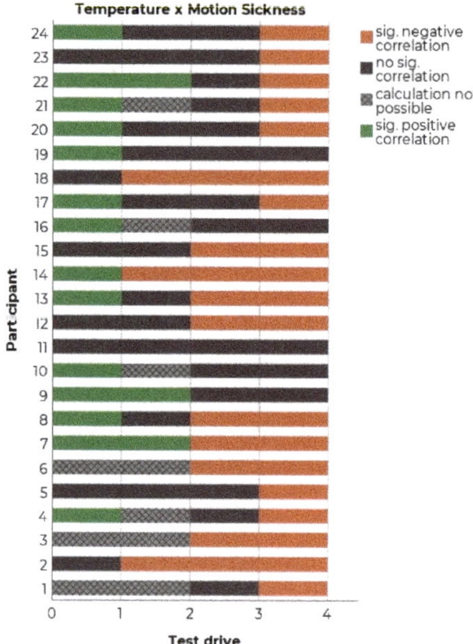

Figure 5. Count of significant Spearman correlations between temperature and motion sickness for each test drive and each participant.

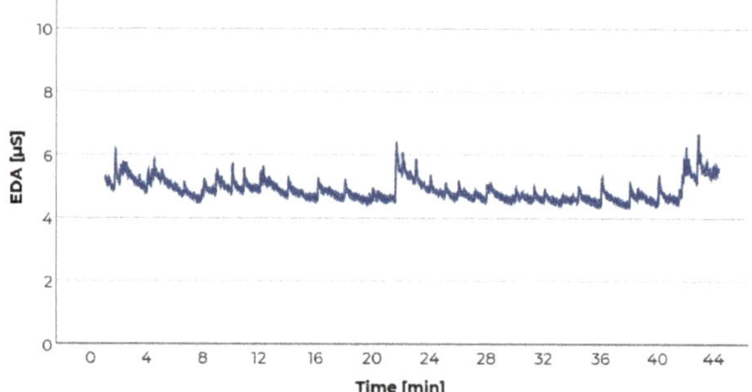

Figure 6. Exemplary raw electrodermal activity (EDA) data of a participant during a 44-min drive as a passenger on a German Autobahn. The numerous peaks in the chart indicate external events like braking, participant's movements, and motion sickness rating procedures. Since these events are not necessarily related to motion sickness in a naturalistic test setting, these peaks should be considered as artifacts.

In our study, the temperature's median was calculated for each interval of two minutes and served as the dependent measure for the subsequent analyses. The temperature was recorded under the armpit and correlated with the likewise every two minutes recorded MISC-ratings. Because a high inter-individual variability was expected [68], the number of significant positive or negative correlations between temperature and subjective measurement of motion sickness every two minutes was counted for each participant and each run (two-tailed testing). In 57.6% of all cases, a significant positive (i.e., temperature increases with motion sickness rating) or negative correlation (i.e., temperature decreases with an increasing motion sickness rating) between temperature and subjective motion sickness was observed (Figure 5).

In order to estimate if the found correlations are stable within each participant, the possibility to replicate the found correlations was checked. However, only $n = 3$ participants showed significant negative correlations between temperature and motion sickness in more than two test drives (i.e., temperature decreased with increasing motion sickness ratings). The results indicate not only a high inter-individual, but also a high intra-individual variability of the found correlations. The variability may also derive from confounding factors like driving time or time of day.

Electrodermal Activity

Another measure which has frequently been investigated with regard to motion sickness is skin conductance. Derived from the observations of "cold sweating" [76], a positive correlation between motion sickness and electrodermal activity (EDA) seems quite plausible and has been shown in several studies [69,70]. Like temperature measurement, EDA recording is technically simple. The procedure is hardly unpleasant for the participants because the electrodes are fixed on the hands (frequently index and middle finger). The electrodes can be attached by the experimenter; hence, it is ensured that the electrodes are pinned correctly and identically across all participants. The measurement can be monitored by the experimenter because whether the measurement is working properly is apparent from the raw signal.

However, EDA is very susceptible to external influences and artifacts. This is a major obstacle in recording EDA under natural driving conditions. Unexpected stimuli strongly affect the EDA. These include, for example, motion perceptions resulting from longitudinal and lateral accelerations, which emerge naturally during driving. Additionally, EDA is affected by speaking and movements of the participants (see Figure 6). Therefore, participants should not move or speak during the drive—this should particularly be considered when asking participants about their current motion sickness. Instead of orally answering questions, it is possible to capture the participants' responses via e.g., a numeric keypad. Alternatively, intervals of motion sickness provocation and intervals of interrogation can be separated, and the latter be excluded from the statistical analysis. However, a temporal separated recording of subjective and physiological data impairs correlation analyses. Beside artifacts, effects deriving from the driving time can bias EDA.

EDA measurement and analysis is characterized into two types: first, the (tonic) skin conductance level (SCL) which describes the slowly changing conductance of the skin and can be analyzed by computing and comparing means or medians per time interval. The tonic level is overlaid by the second type—the (phasic) skin conductance reactions (SCR)—which are referred to discrete stimuli (e.g., sound, motion perception) and can be seen as sudden peaks in the raw signal. In a naturalistic setting, these phasic reactions frequently represent artifacts which are not directly associated with motion sickness but rather surprise or arousal [77] and are therefore not a suitable measure to detect motion sickness in driving. Therefore the more robust SCL should be analyzed if EDA is recorded.

To assess if the EDA is associated with motion sickness, our study also investigated the effects of motion sickness on skin conductance. EDA was recorded on the participants' index and middle fingers (left hand in simulator, right hand in real vehicle). The EDA's median was calculated for each interval of two minutes and served as the dependent measure for the correlations with the MISC ratings. A rise of the EDA was observed at the beginning of the test drive. Therefore, the first eight minutes of the 40- to 45-min test drive were excluded. Additionally, intervals with tight curves were also excluded from the analysis to minimize artifacts deriving from the traffic scenario. Like in the temperature analysis, for each participant and each condition it was counted whether there is a significant positive (i.e., EDA increases with motion sickness rating) or negative correlation (i.e., EDA decreases with increasing motion sickness rating) with subjectively measured motion sickness. In 38.6% of all cases, a significant positive or negative correlation between EDA and motion sickness was observed. Again, the possibility to replicate the found correlations was checked in order to estimate if the found correlations are stable within each participant. However, as shown in Figure 7, no participant showed replicable positive or negative correlations between EDA and motion sickness in more than two test drives. Again, the results indicate not only a high inter-individual but also a high intra-individual variability of the found correlations. As described above, we observed that SCL rose at the beginning of the test drive (probably due to excitement) and then fell over time, independently of perceived motion sickness (probably due to habituation to the study setting). Thus, contrary to the temperature findings, it is highly probable that the found variability derives from confounding factors like driving time or the appearance of external events (e.g., sudden brakes), which emerge naturally during a realistic test drive. These biases may conceal potential effects from motion sickness on SCL. Altogether, there are several confounding effects which affect EDA in a natural driving setting. These should be considered and carefully controlled within the study.

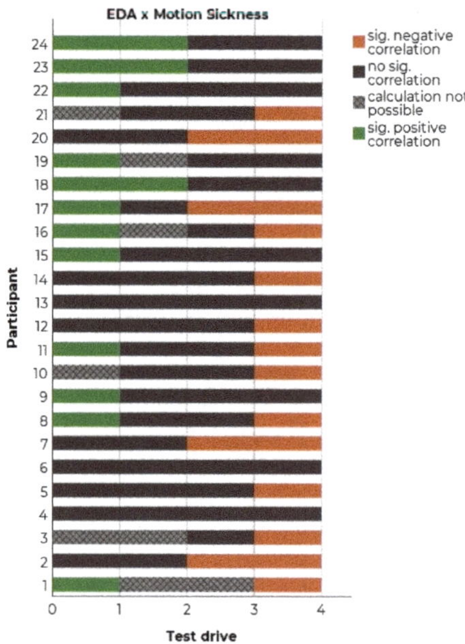

Figure 7. Count of significant Spearman correlations between EDA and motion sickness for each test drive and each participant.

Electrogastrography

Electrogastrography (EGG) is another method which has been investigated to measure motion sickness. The EGG measures pacemaker potentials in the stomach which coordinate the gastric contractions [78]. Thus, the EGG does not capture the actual motility of the stomach but rather the efforts to actuate. Corresponding to typical motion sickness symptoms like nausea or awareness of the stomach, Stern and colleagues found changes in this pacemaker potential, namely a decrease in amplitude and an increase in frequency from 3 to 5–7 cycles per minute in motion sick participants [71,72]. Even if this correlation seems to be rather individual [79], it could nonetheless be shown across different studies, as for example [71,72,79–81]. Therefore, the EGG seems to be a promising signal for a physiological measurement of motion sickness. The EGG is a very weak signal which is easily overlaid by movements (e.g., of the abdominal muscles; see Figure 8) [78,82]. Therefore, it is very important that participants do not move or speak during EGG recording. [71,72] used an optokinetic drum to induce motion sickness by vection. With this method it is possible to induce motion sickness without participants moving or being moved. In the context of driving, however, the application of EGG is naturally more challenging. In a naturalistic drive, participants are moved by the vehicle. The resulting acceleration forces may elicit unconscious movements of the participants like e.g., muscle tensions to compensate centrifugal forces in a curve. Similar to SCL, the circumstance that participants should not speak or move makes it difficult to ask them about their current motion sickness. However, motion artifacts have a different impact on EGG-analysis in comparison to the impact they have on SCL-analysis. SCL is analyzed by computing and comparing means or medians. Therefore motion artifacts reduce the interpretability of the results. In contrast, EGG is analyzed by spectral analysis which can be entirely ruled out by frequent or unnoticed motion artifacts [78,82]. In addition, the EGG raw signal is overlaid by other signals (e.g., from respiration, activity of intestine, etc.) [78] which are filtered later on. Hence, the experimenter cannot monitor any interpretable raw-signal during the test drive.

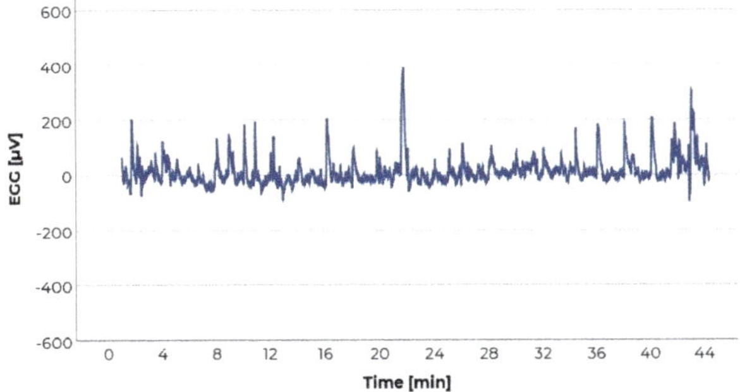

Figure 8. Exemplary raw electrogastrography (EGG) data of a participant during a 44-min drive as passenger on a German Autobahn. Like with EDA (see Figure 6), the peaks in the chart indicate artifacts. In EGG data, these derive mainly from participant's movements.

In our study, EGG was recorded on the participant's abdominal surface. The electrodes were positioned according to the recommendations of Yin and Chen [82] and were attached by the participants themselves. Despite the instruction not to move, we found a high number and frequency of motion artifacts in most participants (an example is given in Figure 8). Therefore, a meaningful analysis was not possible and we refrain from reporting results.

Beside these methodological issues, some ethical aspects should be considered when EGG is recorded. The restriction not to move or speak might withhold participants from reporting when they feel very ill or when they wish to quit the study. In addition, for some participants it can be uncomfortable to have electrodes placed on the abdominal surface by an experimenter. To avoid this, it is possible to let the participants attach the electrodes themselves. Then, however, the experimenter has no control over whether the electrodes are placed correctly. Preparing the skin for attaching the electrodes [82] can also result in unpleasant feelings for participants. Additionally, amount and time of the last meal have to be controlled because this affects the stomach's activity and the development of motion sickness [83]. To avoid this, it is possible to ask participants to be fasted when EGG is recorded or to provide a standardized meal at some time before the start of the test drive.

Altogether, the EGG is hardly suitable to be applied in motion sickness studies under naturalistic driving conditions from the standpoint of current measurement techniques.

3.5. Data Analyses in General

Due to ethical reasons (see Chapter 4), participants must be able to terminate participation at any stage of the study. Furthermore, the experimenter has to terminate the session in cases of conspicuous suffering of the participant. Therefore, a researcher has to expect dropouts during the conduction of a motion sickness study. In driving studies concerning other topics (e.g., acceptance of a new driver assistance system) these dropout participants are often replaced by other participants so that each condition consists of a sufficient and equal number of data, which facilitates the statistical analysis.

In a motion sickness study, however, the occurrence of a dropout is very important as it indicates that motion sickness was too distinct.

Concerning post-study questionnaires (e.g., MSQ), dropouts are not a problem for data analysis as all participants—regardless of cancelling or completing the session—can fill it out. However, all data collected during the runs are sensitive to dropouts during the session. On the one hand, this influences the statistical data analysis and might necessitate the usage of tests which can handle dropouts and missing data. On the other hand, however, researchers can use dropout rates as dependent variables, investigating which conditions caused how many people to abort the trials due to sickness. Furthermore, dropouts enable time-based parameters describing the progress of motion sickness: How long does it take until the dropouts occur? Does this time differ between the test conditions? Therefore, researchers should not see dropouts as a problem (like in other research issues), but rather as an increase of information.

In general, time-based parameters describing the progress of motion sickness are important for motion sickness studies: if a continuous online assessment of motion sickness is conducted (e.g., via FMS, MISC, or symptom-specific Likert scales), it is possible to use parameters which define the time until a participant reaches a specific symptom (e.g., "time to nausea" or "time to sweating"). These data are helpful for the description of motion sickness and the effect of countermeasures.

4. Ethics in Motion Sickness Studies

The American Psychological Association has released a code of conduct that is relevant to research in psychology and other sciences [84]. It includes five fundamental principles which define how to treat participants in scientific investigations. The first principle "beneficence and nonmaleficence" states that researchers should take care of their participants and their wellbeing.

This principle is violated by studies concerning motion sickness as unpleasant symptoms like headache, nausea, or sweating are provoked in these studies. Regarding this aspect, motion sickness research has similarities to pain research: research on a specific topic requires undesirable physical effects and uncomfortable situations for the study participants. Concerning pain studies, the Committee on Ethical Issues of the International Association for the Study of Pain (IASP) has published ethical guidelines for pain research [85]. According to the authors, "health, safety and dignity of human subjects have the highest priority in pain research"—of course, this is also applicable for motion sickness research. Researchers of motion sickness can orientate and adjust their procedure to these guidelines, in particular concerning the following principles:

"Potential participants should be informed fully of the goals, procedures, and risks of the study before giving their consent". In a motion sickness study, participants must fill out an informed consent prior to the study. In particular, research on motion sickness has to be mentioned as the study's aim (i.e., no cover story) and the participant has to be informed that undesirable physical effects of the study (e.g., headache, sickness, sweating) are likely.

"Participants must be able to decline, or to terminate, participation at any stage without risk or penalty. Stimuli should never exceed a subject's tolerance limit and subjects should be able to escape or terminate a painful stimulus at will". In a motion sickness study, the participant is allowed to leave the study anytime. The experimenter has to stop the run immediately or as soon as possible. Of course, it is not allowed to exert pressure on the participants to continue the test session.

"The minimal intensity of noxious stimulus necessary to achieve goals of the study should be established and not exceeded." In a motion sickness study, the researcher must consider criteria when to break off a session: is it really necessary that the participants get strong motion sickness until vomiting? For most research questions it should be sufficient that participants feel first or moderate symptoms of nausea (e.g., for the evaluation of an intervention's effect) as several studies have shown that the motion sickness process is linear with further provocation [32,33]. Besides, even weaker symptoms are experienced as uncomfortable and are not desired during autonomous driving. The break-off criterion could be a predefined participant judgement on a scale measuring well-being, which is given regularly during the session. Additionally, a continuous monitoring through the experimenter can also help to evaluate the participants' well-being: In cases of conspicuous suffering (e.g., moaning, convulsing) the experimenter has to terminate the experiment.

After deciding to stop the experiment due to the participant's wish or a participant's rating over a predefined threshold or conspicuous suffering, the experimenter must stop the session immediately. After the participant has left the sickness provoking situation, the experimenter has to offer various options to the participant in order to relieve her/his motion sickness: e.g., breathe fresh air, visit a restroom, have a cold or warm drink; for emergency cases like a circulatory collapse the participants should have the option to lie down.

At the end of the study the participants' well-being should be evaluated again. If the participants still suffer from motion sickness symptoms, they should be strongly encouraged not to drive a car for safety reasons. In this case, the researcher should provide a shuttle back home or organize a taxi transfer and take on its costs.

The experimenter must be trained in all these mentioned aspects to ensure a good treatment of the participant. A high degree of empathy and training in the detection of motion sickness signals is especially important in order to avoid artefacts of the study situation. Some participants might play down the symptom severity because (1) they form an interpretation of the experiment's purpose and adjust their judgments to fit that interpretation (demand characteristics) or (2) they see high severity judgments as an indicator for weakness (social desirability). The experimenter must break off the experiment in both cases to impede further suffering of the participant.

5. Conclusions

Automated vehicles have the potential to provide significant benefits for the occupants as they can spend their time with various non-driving related activities during the journey. However, this scenario increases the risk of motion sickness and requires an investigation of the phenomenon of motion sickness in the context of automated driving. The present paper discusses methodological aspects for studies investigating the two main research questions: (1) what is the prevalence of motion sickness in a specific scenario (e.g., autonomous driving on a highway) and how do the symptoms develop? (2) Which countermeasures are effective in the prevention and reduction of motion sickness?

If researchers are interested in the prevalence and development of motion sickness in a specific scenario, we suggest conducting a field study in a setting which is as natural as possible. The test vehicle should be driving autonomously or operated by a trained experimenter (Wizard-Of-Oz setting) on public roads in order to achieve external validity. The participants should deal with an NDRT which is likely to be used in an autonomous vehicle in a future setting (e.g., reading or texting). This task should be self-paced so that the participants can interrupt the task when they want to and are able to glance up at the road. As the prevalence of motion sickness in this scenario is of interest, the researchers should select a representative sample concerning all aspects which can affect the prevalence of motion sickness, e.g., age and gender.

In contrast, the setting of a study investigating countermeasures for motion sickness is more standardized. This is necessary as the comparison between runs with the countermeasure (treatment run) and runs without the countermeasure (baseline run) has to be conducted under controlled conditions in order to achieve a high degree of internal validity. The influences of extraneous variables to the measurement should be minimized or removed. Therefore, the study must be conducted in a standardized setting, either on a test track or in a driving simulator. The scenario should provoke motion sickness in the baseline run as a positive effect in the treatment run can only be detected under these conditions. On a test track, standardized maneuvers like driving in a figure eight or constant stop-and-go are recommended. The maneuvers should be driven by a trained experimenter. In the driving simulator, a more naturalistic test course like a winding rural road is possible. The participants should deal with a standardized NDRT which controls glances on the road or totally impedes them. The researchers should select participants who are susceptible to motion sickness in the investigated setting. For this purpose, specific screening questions are more useful than general tools like the MSSQ. Table 2 gives an overview of the recommendations for studies concerning the two main research questions.

Table 2. Overview of the recommendations for studies concerning the two main research questions.

	Prevalence and Development of Motion Sickness	Countermeasures for Motion Sickness
Setting	• Field study on public roads • Naturalistic environment • External validity	• Standardized setting (test track or driving simulator) • Motion sickness provoking scenarios / test course • Internal validity
Participants' task	• Naturalistic NDRT • Self-paced NDRT	• Artificial NDRT possible • Standardized NDRT
Sample	• Representative sample	• Susceptible sample

Of course, the two research questions concerning prevalence/development and countermeasures are not distinct opposites which require an "either-or decision" in the study design. Mixed research questions are imaginable, e.g., when investigating which of two countermeasures is the most effective one in a naturalistic setting. These studies require a mix of methods from both directions.

Independent of the research question, subjective measurement tools like questionnaires and inquiries are necessary to determine motion sickness. Quick and efficient tools like the MISC scale or symptom-specific Likert scales are recommended to assess the intensity of the symptoms during driving. In contrast, comprehensive questionnaires like the MSQ are appropriate to capture a lot of motion sickness symptoms and their intensity after a run. The usage of physiological measurements to detect motion sickness is difficult under non-laboratory conditions. Existing literature reports a high degree of inter-individual variance in physiological reactions—additionally, we found a high intra-individual variance during the study with four test sessions. Furthermore, most data are affected by external events like breaking or a change of posture. It will be challenging to detect physiological correlates of motion sickness which can be assessed reliably and practicably during autonomous driving in realistic settings.

When planning a study concerning motion sickness during autonomous driving, it is imperative that the researchers consider ethical principles. Especially a comprehensive informed consent, predefined break-off criteria, and a protective treatment by trained experimenters is necessary to conduct the study in an appropriate manner.

In sum, more research is necessary for the investigation of motion sickness and possible countermeasures. This paper contributes to solving methodological questions during this research.

Author Contributions: Conceptualization, D.M., K.R., L.R., M.T.; methodology, D.M., K.R., LR., M.T.; software, D.M. and M.T.; validation, D.M., K.R., LR., MT.; formal analysis, D.M. and M.T.; investigation, D.M. and M.T.; resources, K.R. and L.R.; data curation, D.M. and M.T.; writing—original draft preparation, D.M. and M.T.; writing—review and editing, K.R. and L.R.; visualization, D.M. and M.T.; supervision, not relevant; project administration, K.R. and L.R.; funding acquisition, not relevant. All authors have read and agreed to the published version of the manuscript.

Funding: This research was funded and supported by the AUDI AG.

Acknowledgments: We want to thank Alex Neukum, Michael Herter and Jörg Oberglock for administrative support in our experiment. We would like to thank Anna Posset for her support in data collection and for correcting English in the final version of this manuscript.

Conflicts of Interest: This study was supported by the AUDI AG. Katharina Reinmueller and Lena Rittger are employees of this company. They contributed in the design of the study, the review and edition process of the manuscript and the decision to publish the results.

References

1. Tyler, D.B.; Bard, P. Motion sickness. *Physiol. Rev.* **1949**, *29*, 311–369. [CrossRef] [PubMed]
2. Sivak, M.; Schoettle, B. *Motion Sickness in Self Driving Vehicles*; UMTRI-2015-12; University of Michigan: Ann Arbor, MI, USA, 2015.
3. Diels, C. Carsickness: Preventive measures. In *TRL Published Project Report*; TRA: London, UK, 2009.
4. Dungs, J.; Herrmann, F.; Duwe, D.; Schmidt, A.; Stegmüller, S.; Gaydoul, R.; Peters, P.; Sohl, M. *The Value of Time: Nutzerbezogene Service-Potenziale Durch Autonomes Fahren*; Fraunhofer-Institut für Arbeitswirtschaft und Organisation IAO, Horváth & Partners: Stuttgart, Germany, 2016.
5. Yang, Y.; Klinkner, J.N.; Bengler, K. How will the driver sit in an automated vehicle?–The qualitative and quantitative descriptions of non-driving postures (NDPs) when non-driving-related-tasks (NDRTs) are conducted. In Proceedings of the Congress of the International Ergonomics Association, Florence, Italy, 26–30 August 2018; pp. 409–420.
6. Kato, K.; Kitazaki, S. *Improvement of Ease of Viewing Images on an In-vehicle Display and Reduction of Carsickness*; 2008-01-0565; SAE: Warrendale, PA, USA, 2008.
7. Diels, C.; Bos, J.E. Self-driving carsickness. *Appl. Ergon.* **2016**, *53 Pt B*, 374–382. [CrossRef] [PubMed]
8. Wada, T. Motion sickness in automated vehicles. In Proceedings of the 13th International Symposium on Advanced Vehicle Control (AVEC'16), Munich, Germany, 13–16 September 2015; pp. 169–174.
9. Bos, J.E.; MacKinnon, S.N.; Patterson, A. Motion sickness symptoms in a ship motion simulator: Effects of inside, outside, and no view. *Aviat. Space Environ. Med.* **2005**, *76*, 1111–1118. [PubMed]
10. Kennedy, R.S.; Graybiel, A. *The Dial Test: A Standardized Procedure for the Experimental Production of Canal Sickness Symptomatology in a Rotating Environment (NASA-CR-69664)*; 113 NSAM-930; Naval School of Aviation Medicine: Pensacola, FL, USA, 1965.
11. Kennedy, R.S.; Lane, N.E.; Berbaum, K.S.; Lilienthal, M.G. Simulator Sickness Questionnaire: An enhanced method for quantifying simulator sickness. *Int. J. Aviat. Psychol.* **1993**, *3*, 203–220. [CrossRef]
12. Golding, J.F. Motion sickness. *Handb. Clin. Neurol.* **2016**, *137*, 371–390. [CrossRef]
13. Graybiel, A.; Wood, C.D.; Miller, E.F.; Cramer, D.B. Diagnostic criteria for grading the severity of acute motion sickness. *Aerosp. Med.* **1968**, *39*, 453–455.
14. Reason, J.T.; Brand, J.J. *Motion Sickness*; Academic Press: London, UK, 1975.
15. O'Hanlon, J.F.; McCauley, M.E. Motion sickness incidence as a function of the frequency and acceleration of vertical sinusoidal motion. *Aerosp. Med.* **1974**, *45*, 366–369.
16. Dai, M.; Sofroniou, S.; Kunin, M.; Raphan, T.; Cohen, B. Motion sickness induced by off-vertical axis rotation (OVAR). *Exp. Brain Res.* **2010**, *204*, 207–222. [CrossRef]
17. Donohew, B.E.; Griffin, M.J. Motion sickness: Effect of the frequency of lateral oscillation. *Aviat. Space Environ. Med.* **2004**, *75*, 649–656.
18. Golding, J.F.; Mueller, A.G.; Gresty, M.A. A motion sickness maximum around the 0.2 Hz frequency range of horizontal translational oscillation. *Aviat. Space Environ. Med.* **2001**, *72*, 188–192.
19. Griffin, M.J.; Mills, K.L. Effect of frequency and direction of horizontal oscillation on motion sickness. *Aviat. Space Environ. Med.* **2002**, *73*, 537–543. [PubMed]
20. Howarth, H.V.; Griffin, M.J. Effect of roll oscillation frequency on motion sickness. *Aviat. Space Environ. Med.* **2003**, *74*, 326–331. [PubMed]
21. Lawther, A.; Griffin, M.J. Prediction of the incidence of motion sickness from the magnitude, frequency, and duration of vertical oscillation. *J. Acoust. Soc. Am.* **1987**, *82*, 957–966. [CrossRef] [PubMed]
22. McCauley, M.E.; Royal, J.W.; Wylie, C.D.; O'Hanlon, J.F.; Mackie, R.R. *Motion Sickness Incidence: Exploratory Studies of Habituation, Pitch, Roll, and the Refinement of a Mathematical Model*; Human Factors Research: Santa Barbara, CA, USA, 1976.
23. Golding, J.F.; Arun, S.; Wortley, E.; Wotton-Hamrioui, K.; Cousins, S.; Gresty, M.A. Off-vertical axis rotation of the visual field and nauseogenicity. *Aviat. Space Environ. Med.* **2009**, *80*, 516–521. [CrossRef]
24. Donohew, B.E.; Griffin, M.J. Motion sickness with fully roll-compensated lateral oscillation: Effect of oscillation frequency. *Aviat. Space Environ. Med.* **2009**, *80*, 94–101. [CrossRef]
25. Reason, J.T. Motion sickness: Some theoretical and practical considerations. *Appl. Ergon.* **1978**, *9*, 163–167. [CrossRef]

26. Bos, J.E. Motion Perception and Sickness, Eye Movevements and Human Performance. Available online: http://www.jeltebos.info/perception_sickness.htm (accessed on 28 August 2018).
27. Cheung, B.; Hofer, K. Lack of gender difference in motion sickness induced by vestibular coriolis cross-coupling. *J. Vestib. Res.* **2002**, *12*, 191–200.
28. Dobie, T.G.; McBride, D.; Dobie, T., Jr.; May, J. The effects of age and sex on susceptibility to motion sickness. *Aviat. Space Environ. Med.* **2001**, *72*, 13–20.
29. Flanagan, M.B.; May, J.G.; Dobie, T.G. Sex differences in tolerance to visually-induced motion sickness. *Aviat. Space Environ. Med.* **2005**, *76*, 642–646.
30. Lentz, J.M.; Collins, W.E. Motion sickness susceptibility and related behavioral characteristics in men and women. *Aviat. Space Environ. Med.* **1977**, *48*, 316–322.
31. Park, A.H.; Hu, S. Gender differences in motion sickness history and susceptibility to optokinetic rotation-induced motion sickness. *Aviat. Space Environ. Med.* **1999**, *70*, 1077–1080. [PubMed]
32. Förstberg, J. *Ride Comfort and Motion Sickness in Tilting Trains—Human Responses to Motion Evironments in Train and Simulator Experiments*; Royal Institute of Technology: Stockholm, Sweden, 2000.
33. Golding, J.F.; Kerguelen, M. A comparison of the nauseogenic potential of low-frequency vertical versus horizontal linear oscillation. *Aviat. Space Environ. Med.* **1992**, *63*, 491–497. [PubMed]
34. Griffin, M.J.; Newman, M.M. Visual field effects on motion sickness in cars. *Aviat. Space Environ. Med.* **2004**, *75*, 739–748.
35. Treisman, M. Motion sickness: An evolutionary hypothesis. *Science* **1977**, *197*, 493–495. [CrossRef] [PubMed]
36. Riccio, G.E.; Stoffregen, T.A. An ecological theory of motion sickness und postural instability. *Ecol. Psychol.* **1991**, *3*, 195–240. [CrossRef]
37. Bowins, B. Motion sickness: A negative reinforcement model. *Brain Res. Bull.* **2010**, *81*, 7–11. [CrossRef]
38. Stott, J.R.R. Mechanisms and Treatment of Motion Illness. In *Nausea and Vomiting: Mechanisms and Treatment*; Davis, C.J., Lake-Bakaar, G.V., Grahame-Smith, D.G., Eds.; Springer: Berlin, Germany, 1986; pp. 110–129.
39. Reason, J.T. Motion sickness adaptation: A neural mismatch model. *J. R. Soc. Med.* **1978**, *71*, 819–829. [CrossRef]
40. Ramulu, S.H.S. Relationship of Motion Sickness with Mental Demand and Cold Sweating. Master's Thesis, University of Technology, Eindhoven, The Netherlands, 2019.
41. Bohrmann, D.; Lehnert, K.; Scholly, U.; Bengler, K. Kinetosis as a Challenge of Future Mobility Concepts and Highly Automated Vehicles. In Proceedings of the 27th Aachen Colloquium Automobile and Engine Technology, Aachen, Germany, 26 November 2018.
42. Baltodano, S.; Sibi, S.; Martelaro, N.; Gowda, N.; Ju, W. The RRADS platform: A real road autonomous driving simulator. In Proceedings of the 7th International Conference on Automotive User Interfaces and Interactive Vehicular Applications, Nottingham, UK, 1–3 September 2015; pp. 281–288.
43. Naujoks, F.; Purucker, C.; Wiedemann, K.; Marberger, C. Noncritical State Transitions during Conditionally Automated Driving on German Freeways: Effects of Non–Driving Related Tasks on Takeover Time and Takeover Quality. *Hum. Factors* **2019**, *61*, 596–613. [CrossRef]
44. Kiss, M.; Schmidt, G.; Babbel, E. Das Wizard of Oz Fahrzeug: Rapid Prototyping und Usability Testing von zukünftigen Fahrerassistenzsystemen. In Proceedings of the 3. Tagung Aktive Sicherheit durch Fahrerassistenz, Garching, Germany, 7–8 April 2008.
45. Bengler, K.; Omozik, K.; Müller, A.I. The Renaissance of Wizard of Oz (WoOz)—Using the WoOz methodology to prototype automated vehicles. In *Proceedings of the Human Factors and Ergonomics Society Europe Chapter 2019 Annual Conference*; de Waard, D., Toffetti, A., Pietrantoni, L., Franke, T., Petiot, J.-F., Dumas, C., Botzer, A., Onnasch, L., Milleville, L., Mars, F., Eds.; HFES: Nantes, France, 2020.
46. Kaptein, N.A.; Theeuwes, J.; Van Der Horst, R. Driving simulator validity: Some considerations. *Transp. Res. Record* **1996**, *1550*, 30–36. [CrossRef]
47. Carsten, O.; Jamson, A.H. Driving simulators as research tools in traffic psychology. In *Handbook of Traffic Psychology*; Porter, B.E., Ed.; Elsevier Academic Press: San Diego, CA, USA, 2011; pp. 87–96. [CrossRef]
48. Stoner, H.A.; Fisher, D.L.; Mollenhauer, M. Simulator and scenario factors influencing simulator sickness. In *Handbook of Driving Simulation for Engineering, Medicine, and Psychology*; Fisher, D.L., Rizzo, M., Caird, J., Lee, J.D., Eds.; Taylor & Francis: Boca Raton, FL, USA, 2011; pp. 13.11–13.18.
49. Blaauw, G.J. Driving experience and task demands in simulator and instrumented car: A validation study. *Hum. Factors* **1982**, *24*, 473–486. [CrossRef]

50. Mullen, N.; Charlton, J.; Devlin, A.; Bedard, M. Simulator validity: Behaviors observed on the simulator and on the road. In *Handbook of Driving Simulation for Engineering, Medicine and Psychology*; Fisher, D.L., Rizzo, M., Caird, J., Lee, J.D., Eds.; Taylor & Francis: Boca Raton, FL, USA, 2011; pp. 13.11–13.18.

51. Schmidt, E.; Emmermann, B.; Venrooij, J.; Reinprecht, K. Occurrence of motion sickness during highway and inner-city drives. In *Proceedings of the Human Factors and Ergonomics Society Europe Chapter 2018 Annual Conference*; de Waard, D., Brookhuis, K., Coelho, D., Fairclough, S., Manzey, D., Naumann, A., Onnasch, L., Röttger, S., Toffetti, A., Wiczorek, R., Eds.; HFES: Berlin, Germany, 2019.

52. Salter, S.; Diels, C.; Herriotts, P.; Kanarachos, S.; Thake, D. Motion sickness in automated vehicles with forward and rearward facing seating orientations. *Appl. Ergon.* **2019**, *78*, 54–61. [CrossRef] [PubMed]

53. Feenstra, P.J.; Bos, J.E.; van Gent, R.N.H.W. A visual display enhancing comfort by counteracting airsickness. *Displays* **2011**, *32*, 194–200. [CrossRef]

54. Karjanto, J.; Yusof, N.M.; Wang, C.; Terken, J.; Delbressine, F.; Rauterberg, M. The effect of peripheral visual feedforward system in enhancing situation awareness and mitigating motion sickness in fully automated driving. *Transp. Res. Part F* **2018**, *58*, 678–692. [CrossRef]

55. Kuiper, O.X.; Bos, J.E.; Schmidt, E.A.; Diels, C.; Wolter, S. Knowing What's Coming: Unpredictable Motion Causes More Motion Sickness. *Hum. Factors* **2019**. [CrossRef] [PubMed]

56. Naujoks, F.; Befelein, D.; Wiedemann, K.; Neukum, A. A review of non-driving-related tasks used in studies on automated driving. In Proceedings of the International Conference on Applied Human Factors and Ergonomics, Los Angeles, CA, USA, 17–21 July 2017; pp. 525–537.

57. Bohrmann, D. Probandenstudie-Vom Fahrer zum Passagier. *ATZextra* **2019**, *24*, 36–39. [CrossRef]

58. Reason, J.T. Relations between motion sickness susceptibility, the spiral after-effect and loudness estimation. *Br. J. Psychol.* **1968**, *59*, 385–393. [CrossRef]

59. Zhang, L.L.; Wang, J.Q.; Qi, R.R.; Pan, L.L.; Li, M.; Cai, Y.L. Motion Sickness: Current Knowledge and Recent Advance. *CNS Neurosci. Ther.* **2015**, *22*, 15–24. [CrossRef]

60. Mirabile, C.S., Jr.; Glueck, B.C.; Stroebel, C.F.; Pitblado, C. Susceptibility to motion sickness and ego closeness, ego distance as measured by the autokinetic response tendency. *Neuropsychobiology* **1977**, *3*, 193–198. [CrossRef]

61. Paillard, A.C.; Quarck, G.; Paolino, F.; Denise, P.; Paolino, M.; Golding, J.F.; Ghulyan-Bedikian, V. Motion sickness susceptibility in healthy subjects and vestibular patients: Effects of gender, age and trait-anxiety. *J. Vestib. Res. Equilib. Orientat.* **2013**, *23*, 203–209. [CrossRef]

62. Keshavarz, B.; Hecht, H. Validating an efficient method to quantify motion sickness. *Hum. Factors* **2011**, *53*, 415–426. [CrossRef]

63. Cowings, P.S.; Suter, S.; Toscano, W.B.; Kamiya, J.; Naifeh, K. General autonomic components of motion sickness. *Psychophysiology* **1986**, *23*, 542–551. [CrossRef] [PubMed]

64. Hu, S.; Stern, R.M.; Vasey, M.W.; Koch, K.L. Motion sickness and gastric myoelectric activity as a function of speed of rotation of a circular vection drum. *Aviat. Space Environ. Med.* **1989**, *60*, 411–414. [PubMed]

65. Kim, Y.Y.; Kim, H.J.; Kim, E.N.; Ko, H.D.; Kim, H.T. Characteristic changes in the physiological components of cybersickness. *Psychophysiology* **2005**, *42*, 616–625. [CrossRef] [PubMed]

66. LaCount, L.T.; Barbieri, R.; Park, K.; Kim, J.; Brown, E.N.; Kuo, B.; Napadow, V. Static and dynamic autonomic response with increasing nausea perception. *Aviat. Space Environ. Med.* **2011**, *82*, 424–433.

67. Mekjavic, I.B.; Tipton, M.J.; Gennser, M.; Eiken, O. Motion sickness potentiates core cooling during immersion in humans. *J. Physiol.* **2001**, *535*, 619–623. [CrossRef]

68. Benson, A.J. Motion Sickness. In *Medical Aspects of Harsh Environments*; Pandolf, K., Burr, R., Eds.; Walter Reed Army Medical Center: Washington, DC, USA, 2002; Volume 2, pp. 611–1204.

69. Nalivaiko, E.; Rudd, J.A.; So, R.H. Motion sickness, nausea and thermoregulation: The "toxic" hypothesis. *Temperature* **2014**, *1*, 164–171. [CrossRef]

70. Meusel, C.R. Exploring Mental Effort and Nausea via Electrodermal Activity within Scenario-Based Tasks. Master's Thesis, Iowa State University, Ames, IA, USA, 2014.

71. Stern, R.M.; Koch, K.L.; Leibowitz, H.W.; Lindblad, I.M.; Shupert, C.L.; Stewart, W.R. Tachygastria and motion sickness. *Aviat. Space Environ. Med.* **1985**, *56*, 1074–1077.

72. Stern, R.M.; Koch, K.L.; Stewart, W.R.; Lindblad, I.M. Spectral analysis of tachygastria recorded during motion sickness. *Gastroenterology* **1987**, *92*, 92–97. [CrossRef]

73. Nobel, G.; Eiken, O.; Tribukait, A.; Kölegård, R.; Mekjavic, I.B. Motion sickness increases the risk of accidental hypothermia. *Eur. J. Appl. Physiol.* **2006**, *98*, 48–55. [CrossRef]
74. Nobel, G.; Tribukait, A.; Mekjavic, I.B.; Eiken, O. Effects of motion sickness on thermoregulatory responses in a thermoneutral air environment. *Eur. J. Appl. Physiol.* **2012**, *112*, 1717–1723. [CrossRef]
75. Sund-Levander, M.; Forsberg, C.; Wahren, L.K. Normal oral, rectal, tympanic and axillary body temperature in adult men and women: A systematic literature review. *Scand. J. Caring Sci.* **2002**, *16*, 122–128. [CrossRef] [PubMed]
76. Nobel, G. *Effects of Motion Sickness on Human Thermoregulatory Mechanisms*; KTH: Stockholm, Sweden, 2010.
77. Boucsein, W. *Electrodermal Activity*; Springer Science & Business Media: New York, NY, USA, 2012.
78. Verhagen, M.A.; Van Schelven, L.J.; Samsom, M.; Smout, A.J. Pitfalls in the analysis of electrogastrographic recordings. *Gastroenterology* **1999**, *117*, 453–460. [CrossRef] [PubMed]
79. Cheung, B.; Vaitkus, P. Perspectives of electrogastrography and motion sickness. *Brain Res. Bull.* **1998**, *47*, 421–431. [CrossRef]
80. Farmer, A.D.; Ban, V.F.; Coen, S.J.; Sanger, G.J.; Barker, G.J.; Gresty, M.A.; Giampietro, V.P.; Williams, S.C.; Webb, D.L.; Hellström, P.M. Visually induced nausea causes characteristic changes in cerebral, autonomic and endocrine function in humans. *J. Physiol.* **2015**, *593*, 1183–1196. [CrossRef] [PubMed]
81. Imai, K.; Kitakoji, H.; Sakita, M. Gastric arrhythmia and nausea of motion sickness induced in healthy Japanese subjects viewing an optokinetic rotating drum. *J. Physiol. Sci.* **2006**, *56*, 341–345. [CrossRef] [PubMed]
82. Yin, J.; Chen, J.D. Electrogastrography: methodology, validation and applications. *J. Neurogastroenterol. Motil.* **2013**, *19*, 5. [CrossRef]
83. Uijtdehaage, S.H.; Stern, R.M.; Koch, K.L. Effects of Scopolamine on Autonomic Profiles Underlying Motion Sickness Susceptibility. *Aviat. Space Environ. Med.* **1993**, *64*, 1–8.
84. American Psychological Association. Ethical Principles of Psychologists and Code of Conduct. Available online: https://www.apa.org/ethics/code/index (accessed on 4 December 2019).
85. Charlton, E. Ethical guidelines for pain research in humans. *Pain* **1995**, *63*, 277–278. [CrossRef]

 *information*

Article

Supporting Drivers of Partially Automated Cars through an Adaptive Digital In-Car Tutor

Anika Boelhouwer [1],*, Arie Paul van den Beukel [2], Mascha C. van der Voort [2], Willem B. Verwey [3] and Marieke H. Martens [4,5]

[1] Transport Engineering and Management, University of Twente, Drienerlolaan 5, 7522 NB Enschede, The Netherlands
[2] Department of Design, Production and Management, University of Twente, Drienerlolaan 5, 7522 NB Enschede, The Netherlands
[3] Department Cognitive Psychology and Ergonomics, University of Twente, Drienerlolaan 5, 7522 NB Enschede, The Netherlands
[4] TNO Traffic & Transport, Anna van Buerenplein 1, 2496 RZ The Hague, The Netherlands
[5] Department of Industrial Design, Eindhoven University of Technology, Groene Loper 3, 5612 AE Eindhoven, The Netherlands
* Correspondence: a.boelhouwer@utwente.nl

Received: 28 February 2020; Accepted: 27 March 2020; Published: 30 March 2020

Abstract: Drivers struggle to understand how, and when, to safely use their cars' complex automated functions. Training is necessary but costly and time consuming. A Digital In-Car Tutor (DIT) is proposed to support drivers in learning about, and trying out, their car automation during regular drives. During this driving simulator study, we investigated the effects of a DIT prototype on appropriate automation use and take-over quality. The study had three sessions, each containing multiple driving scenarios. Participants needed to use the automation when they thought that it was safe, and turn it off if it was not. The control group read an information brochure before driving, while the experiment group received the DIT during the first driving session. DIT users showed more correct automation use and a better take-over quality during the first driving session. The DIT especially reduced inappropriate reliance behaviour throughout all sessions. Users of the DIT did show some under-trust during the last driving session. Overall, the concept of a DIT shows potential as a low-cost and time-saving solution for safe guided learning in partially automated cars.

Keywords: Adaptive HMI; automated driving; automotive user interfaces; driver behaviour

---

1. Introduction

Although commercial cars are increasingly equipped with combinations of automated functions such as Adaptive Cruise Control (ACC) and Lane Keeping Systems (LK), drivers appear to have a hard time getting used to them. Many drivers do not know which Advanced Driver Assistance Systems (ADAS) their car has, what they do, and how to safely use them [1,2]. Several aspects appear to contribute to the confusion about car automation among drivers. First, different car brands are introducing automated systems with similar names but with different functions, or different system names for similar functions [3,4]. Second, research showed that at least a quarter of all drivers do not receive any information about ADAS from their salesman when they buy a car equipped with such a system [5,6]. Furthermore, only a small proportion of drivers gets to actually drive with the automated functions at their sales point. This is worrisome as drivers need multiple interactions with an automated system to properly understand it [7,8]. Third, current driver-car interfaces often fail to follow widely accepted human factors and human machine interaction guidelines [4], leading to misinterpretations of the system's capabilities. Co-driving (alternatively referred to as cooperative-

or shared control) (see, for example, [9–11]) has been suggested to reduce the need for frequent and complete control switches. Although this may take many forms, co-driving entails the shared control of the vehicle. Some responsibilities are allocated to the driver, while others are allocated to the car. Still, even in co-driving, a driver still needs to know how this shared control works, what the car's capabilities and limitations are, and when they are responsible for what particular driving task. All in all, a lack of understanding about ADAS may reduce traffic safety [12–15] and limit any prospected benefits of automated driving [16–20]. Drivers need to be supported in learning when it is (not) safe to use the automation in their car [21].

Several solutions have been proposed to support drivers in understanding, and safely using, the automation in their car. The first one is to stimulate the use of owners' manuals. However, not only are these usually long and complicated, studies suggest that practise is required to fully support safe automation use [22–24]. Driving simulators in particular allow drivers to practise with rare but critical driving situations [25–27]. The main downside to all these options is that additional training at a driving school or at a facility with a simulator requires high investments, both financially and time-wise.

1.1. Digital In-Car Tutor (DIT)

In the present study, we explore the potential of a Digital In-car Tutor (DIT) to support drivers in using in-vehicle automation. A DIT guides drivers through the different automated systems in their own cars, during regular drives. While a DIT may take various forms, we particularly studied a DIT prototype using audio and an Augmented Reality (AR) overlay on the windscreen (see Section 2.2.3). The DIT is designed to be used in real cars during regular drives. The following three steps illustrate the core functionalities of our DIT prototype. First, the DIT introduces one of the automated car systems while the driver is driving manually. New systems are only introduced when the driver is in a low complex situation [28], like an empty straight road on a clear day. Such an introduction concerns the system's functionalities, handling, capabilities and limitations, and equipment. Second, the driver can try out the functionality while the DIT provides immediate feedback. Third, the DIT reminds drivers about specific systems capabilities and limitations when a related situation is encountered. Furthermore, rare situations are addressed when driving in similar, but more frequent, situations to keep the driver's mental model up to date [7]. A new system is introduced as the driver has safely driven with it for a certain amount of kilometres (for example 500 km), and the cycle repeats itself. A DIT could have many benefits over regular driving lessons, simulator training, and the use of owners' manuals. First, it is less time consuming and costly, as it is active in the driver's own car during regular drives. Second, a DIT allows for continuous and situated support over a longer period of time. Last, a DIT can be brand- and model-specific, and can be adjusted when automated functions are changed by software updates.

1.2. Adaptive Communication

To facilitate learning and avoid an excessive cognitive demand, a DIT should be adaptive in various ways. First, instructions by the DIT should concern the current driving situation so that the driver is able to immediately process and apply them. Furthermore, the modality, timing, and duration of the communication needs to be adjusted to the demand of the driving situation to avoid overload. Studies on the cognitive demands of feedback suggest that tutoring in highly complex driving situations should be condensed and action-based. Elaborate theory and reflection can be presented during low complexity situations [29–31]. Last, the feedback needs to adapt to the driver's performance, to update his or her mental model. This includes both direct but short feedback, and elaborate reflection after the situation. For example, drivers may need to be informed if they turn on the automation outside of its Operational Design Domain (ODD) [32]. These tutor strategies were implemented in our DIT prototype.

Earlier, Simon [33] studied an auditory digital tutoring system for Adaptive Cruise Control (ACC). The tutor content was adapted to the traffic situation in general and to the driver's preferred maximum deceleration. However, the timing and duration did not adapt, nor was the information adjusted to the complexity of the traffic situation. These characteristics may, however, be required in a tutor system, as they may help to prevent driver overload. Simon [33] did find benefits to the tutor in terms of driving safety and a more efficient use of the ACC. However, with the introduction of a variety of automated systems, such research needs to be extended towards cars with multiple systems as these drastically increase the learning difficulty for drivers.

1.3. Present Study

In the current driving simulator study, we compared the effects of a DIT prototype (DIT group) with those of an information brochure (IB group) on the use of complex car automation during three driving sessions. In all driving scenarios, participants were required to decide whether they could rely on the automation or not. In the specific scenarios that required drivers to turn off the automation, the take-over quality was analysed. During the first driving session, the DIT group was supported by the DIT prototype in learning about the various automated car systems. In contrast, the IB group familiarized itself with the automation by reading an information brochure (IB group) before driving in the simulator. Two more driving sessions followed, one directly after the first and one after two weeks. During these sessions, the DIT was no longer active for the DIT group. The additional sessions were introduced to investigate how any effects of the DIT lasted over time. Last, multiple acceptance elements (e.g., ease of use) of DIT were assessed through a questionnaire.

Overall, we expected the DIT to provide drivers with a better understanding, and safer use, of the automation. Our first hypothesis was that using the DIT would result in more correct automation use. That is, drivers would only rely on the automation if it could deal with the situation safely, and take back control if it could not. A second hypothesis was that drivers were expected to show a better take-over performance in critical situations. A better take-over performance was defined as: taking-over earlier, braking less intensely, and showing a more stable vehicle control.

In conclusion, we examined whether a DIT was more beneficial for supporting drivers in safely using car automation, compared to drivers that received an information brochure. DITs may provide a more time- and cost-efficient solution to driver training of partially automated cars compared to training in driving simulators or on the road with driving instructors. Furthermore, it allows for situated and repeated learning. Lastly, any over-the-air updates of the automation can be directly integrated in the DIT, allowing for tailored instructions about the latest version of the automation. The results of this study allow us to gain insight in whether or not a DIT is an appropriate method to increase appropriate car automation use.

2. Materials and Methods

2.1. Participants

38 participants (23 female, 15 male) took part in the driving simulator study. 19 participants were part of the control condition (IB group) and 19 were part of the experimental condition (DIT group). All participants were students or employees of the University of Twente. All subjects gave their informed consent for inclusion before they participated in the study. The study was conducted in accordance with the Declaration of Helsinki, and the protocol was approved by the University of Twente BMS Ethics Committee (nr. 191220). Their average age was 27.5 years ($SD = 13.1$ years, range = 18–65 years). On average, participants possessed their driver's license for 9.2 years ($SD = 10.81$, range: 1–47). Eight participants drove almost every day, and 15 drove multiple times a week. Eight participants drove once a week, and seven drove less than once per week. Most had experience with Cruise Control ($N = 29$). Seven participants had experience with Adaptive Cruise Control, and two with Lane Assist. The Affinity for Technology Interaction (ATI) scale [34,35] was used to determine

the level of general affinity with technology of the participants. On this scale of 1 (low affinity with technology) to 6 (high affinity with technology), the participants scored an average of 3.9 (*SD* = 0.77). The groups did not significantly differ on any of these characteristics. Participants had to speak and understand English fluently to be able to participate as the experiment was conducted in English.

2.2. Research Design

2.2.1. Driving Simulator & Simulated Automated Car

The experiment took place in the driving simulator of the University of Twente (Figure 1). This simulator includes a car mock-up with a steering wheel and pedals. Three beamers project the simulation on a 7.8 m by 1.95 m screen with a view angle of approximately 180 degrees. Rear- and side mirrors were projected on the screen. A tablet displayed the speedometer, tachometer, and an icon that showed whether the automation was on. The simulated car was equipped with level 2 automation which included (1) Adaptive Cruise Control (ACC), (2) Lane Keeping (LK), (3) Obstacle Detection (OD), (4) Traffic Light and Priority Sign Detection (TS), and (5) Priority Road Markings Detection (RM). These systems were designed specifically for this experiment and did not resemble a particular car model to prevent transfer from existing cars. Participants were informed about this. The steering wheel included a blue button to turn all automation on and off. Participants could not turn the automation off by braking or steering.

Figure 1. The fixed-base driving simulator of the University of Twente.

2.2.2. Experimental Condition: Information Brochure Training (IB Group)

At the start of the first driving session, participants in the IB group received a paper brochure on the five automated systems. They read this information for 10 min before driving. This brochure included the functions, handling, equipment, capabilities, and limitations of each system. It contained the same system information that the DIT group received from the DIT. However, as the information was given prior to the practise scenarios, it did not include any situation- and driver-adaptive feedback.

2.2.3. Experimental Condition: Digital In-Car Tutor (DIT Group)

The DIT prototype introduced the five automated systems to the participants though auditory and visual information (ACC, LK, OD, TS, and RM). All visual information was projected as an overlay on the windscreen (Figure 2). This reduced the need for drivers to look away from the road and allowed the information to be directly related to the driving situation. All visual information was accompanied by verbal explanations. The digital standard Google Assistant voice was used for the verbal communication, and had been pre-recorded. This voice was female with a British accent.

Figure 2. Examples of the Digital In-car Tutor visuals. (**A**) Visuals while the digital tutor verbally explained Lane Keeping. (**B**) Visuals when the digital tutor verbally explained that the automation cannot deal with overly complex lane markings. (**C**) Visuals when the digital tutor reminded the driver that the automation had trouble driving in bad weather conditions such as heavy rain and fog, and that the weather would be changing.

Procedure. The DIT followed the following steps during Session 1 in the experiment. The DIT first introduced a specific automated system (e.g., Adaptive Cruise Control) at the start of the scenarios. This was always on a straight road without traffic. The DIT would verbally explain the functions, handling, equipment, capabilities, and limitations of this system (Figure 2A,B). The verbal explanations were supported by illustrations which were projected on the windscreen. The DIT then told participants to use the automation if they thought that it was safe. As participants approached the situation where they needed to either turn off the automation or leave it on, the DIT would remind the participant of the system capabilities and limitations that applied to the specific situation (Figure 2C).

Adaptivity. The information from the DIT was expected to put some cognitive demand on drivers [36,37]. To avoid driver overload, the length and type of DIT messages were adapted to the complexity of the driving situation. This could be considered a 'safety filter' for our DIT as described by Van Gent et al. [29]. The communication was longer and more detailed in low complex situations, while it was condensed during highly complex situations. Furthermore, discussing theory and reflecting upon situations only occurred during low complex situations. This included the system introductions on the simple straight road at the start of each scenario [28], and reflection after each critical situation. As an example, the ACC introduction was: "ACC keeps the car at a set speed, and automatically speeds up, and slows down the car, to keep a set distance to the car ahead. The car has several cameras which are used to detect a car ahead of you." If the driver correctly left the automation on in this scenario (ACC1), the reflection was "Great job. The ACC detected the cars in front of you and slowed down to keep the set speed". These strategies were based upon studies that investigated tutoring strategies by driving instructors [38,39]. In a similar way that studies have used human processing and decision-making strategies as a base for robotics or intelligent vehicles with artificial processing and decision-making skills [40], we implemented the observed feedback strategies of human tutors in a digital tutor.

The DIT also adapted to the driving situation by reminding drivers of the system's capabilities and limitations specific to the current situation. In combination with the overlay visuals, this meant that the driver could directly perceive and process the information in their specific context. Drivers did not have to interpret information in an artificial context (e.g., a screen with a simplified visualisation of the situation) and then apply it to the current driving situation. For example, when the weather changed for the worst in a scenario, the DIT reminded the driver that the car cannot function reliably in heavy fog and rain (Figure 2C). It is important to note that the DIT never explicitly told the driver that it was safe to leave the automation on, or that the automation needed to be turned off. This was decided as it would be unrealistic in a real-world driving scenario (driving a level 2 vehicle) both for safety and reliability issues. Similarly, the DIT is not intended to be used as a warning system. Rather, the DIT identifies some situations to provide situated tutoring and learning.

Last, the DIT adapted its feedback on the current performance of the driver. If the automation was used outside its ODD, the DIT reflected afterwards on why this was not safe. If the automation was unnecessarily turned off, the DIT would also reflect on this. The DIT would add that the driver's judgement was the most important, and that the automation should only be used if the driver thought that it could safely cope. The feedback was manually activated by the researcher.

2.2.4. Set-Up and Procedure

The experiment was a between-subjects design with an experimental condition (DIT group) and a control group (IB group). Both groups drove in three sessions (Table 1), which each containing multiple scenarios. All participants were given the following task for each scenario: "You can start the scenario by driving manually. Turn on the automation whenever you think that the car can safely cope, and turn (or leave) it off if it cannot. The car can't cope with a situation if: traffic regulations have to be violated or the car will damage something or harm someone".

Participants were informed at the start of each session that they remained responsible for their safety and that of their fellow road users while using the automation. They also needed to adhere to the general traffic rules and speed limits. If the participant hit something or someone, a crash sound was played and the scenario ended. After each scenario, participants were asked by the researcher whether they thought that the car could safely cope with the previous situation and why.

At the start of Session 1, all participants received a written overview of the experiment procedure and filled out an informed consent form and a demographics questionnaire. Participants could get used to the simulator in a 10-min demo scenario. Overall, Session 1 consisted of 10 scenarios and lasted 1 h. The DIT provided information and feedback during all scenarios in session 1 (see Section 2.2.3), while the IB group read a brochure about the automation for 10 min before driving. Participants were reminded of their task before each scenario (mentioned above). Session 2 started after a 10-min break. This session contained 8 scenarios and lasted 30 min. Again, participants were reminded of their task before each scenario. The DIT was disengaged for all participants in this session. All participants were asked to participate in Session 3, which took place after two weeks. However, as not all participants were able to come back due to work or school commitments, each group contained 11 participants during Session 3. The set-up for Session 3 was identical to that of Session 2. This last session was included to investigate how any potential effects of the DIT evolved after repeated interaction with the automation.

The order of the scenarios was randomized in Sessions 2 and 3. The scenarios in Session 1 were not randomized and followed the order as depicted in Table 1. This way, the DIT could introduce the different automated systems in a realistic and logical order to the DIT group. The same order of scenarios was adhered to for the IB group to avoid that different orders between groups might influence the results.

Table 1. Overview of the experiment set-up for the Digital-in Car Tutor (DIT) group and the Information Brochure (IB) group. Descriptions of all abbreviated driving scenarios are available in Tables 2 and 3.

		Session 1 (60 min, $N = 38$)	Session 2 (30 min, $N = 38$)	Session 3 (30 min, $N = 22$)
IB group (Control)	Information Brochure	Driving scenarios	Driving scenarios	Driving scenarios
		ACC1 ACC2 LK1 LK2 OD1 OD2 TS1 TS2 RM1 RM2	T1 T2 T3 T4 T5 T6 T7 T8	T1 T2 T3 T4 T5 T6 T7 T8
DIT group		Driving scenarios + Tutor Guidance	Driving scenarios	Driving scenarios
		ACC1 ACC2 LK1 LK2 OD1 OD2 TS1 TS2 RM1 RM2	T1 T2 T3 T4 T5 T6 T7 T8	T1 T2 T3 T4 T5 T6 T7 T8

2.2.5. Scenarios

All scenarios started with a straight road without traffic so drivers could calmly start driving manually and turn on the automation if they thought that it was safe to do so. Furthermore, during Session 1, the DIT introduced a new system to the DIT group on this road as they were still driving manually. After the straight road, the specific driving scenario started. All scenarios contained an event area during which the automation should be on or off.

Session 1 contained 10 driving scenarios (Table 2) of 3 to 4 min each. Each of the five automated systems described in Section 2.2.1 had two dedicated scenarios that addressed a particular capability or limitation of that system. Each system contained one scenario in which the automation could cope, and one in which the automation could not. During the first system-specific scenario, the DIT would explain the basic functionalities, capabilities, and limitations of the particular system. During the second scenario, the DIT would further elaborate on the limitations of the system. Sessions 2 and 3 both contained eight scenarios of 2 to 3 min each (Table 3). In each session, four scenarios required a take-over, and four did not. The scenarios in Session 3 were the same as those in Session 2 but with considerable changes to the environment. It made them look different to the participants, but still allowed for a comparison with Session 2. If a participant did not take back control in situations that the automation could not cope with, the car would crash and the scenario would end.

Table 2. An overview of all scenarios during Session 1. Each scenario addresses a particular automated system (e.g., Adaptive Cruise Control).

Driving scenarios in Session 1			
ID	Scenario	Need to turn off the automation?	Description
ACC1	Straight highway	No	Straight highway without any traffic.
ACC2	Fog	Yes	Straight highway with fog coming up. Driver needs to switch off automation before the fog and brake for slow cars within the fog section. The car's cameras do not function well in fog. Car crashes if the automation remains on.
LK1	Curved Rural	No	Curved rural road without any traffic.
LK2	Roadworks	Yes	Highway with roadworks. Driver needs to switch off the automation before the roadworks and follow the yellow road markings. The automation cannot deal with overly complex road markings. Car crashes if the automation remains on.
OD1	Jaywalker	No	City road with a pedestrian crossing the road.
OD2	Pedestrian obstructed view	Yes	City road with a pedestrian crossing the road from behind a bus. Driver needs to switch off the automation when driving past the bus. Car cannot detect the pedestrian behind the bus. Car crashes into the pedestrian if the automation remains on.
TS1	Priority signs	No	Rural road and simple signalised intersection.
TS2	Unsignalised intersection	Yes	City road and intersection without traffic signs or lights. The car's view is blocked by houses and it cannot detect oncoming traffic from the right. Driver needs to switch off the automation before the intersection. Car crashes if the automation remains on.
RM1	Pedestrian crossing	No	City road with pedestrian crossing on a zebra path.
RM2	Road markings missing	Yes	Highway with curved section without road markings. Driver needs to switch off automation before the section without road markings. Lane keeping cannot function without visible road markings. Car crashes if the automation remains on.

Table 3. An overview of all scenarios during Sessions 2 and 3.

Driving scenarios in Sessions 2 and 3			
ID	Scenario	Need to turn off the automation?	Description
T1	Curved rural	No	Rural road with gentle curves.
T2	Stationary car	Yes	Rural road with broken-down car in the middle of the road. Driver has to switch off automation when approaching and drive around the car. The speed difference is too large, the car cannot detect the stationary car and brake in time. Car crashes if the automation remains on.
T3	Emergency vehicle	Yes	Signalised intersections with emergency vehicles running the red light. The driver has to switch off the automation before the intersection. The automation cannot adapt its priority rules to emergency vehicles and other road users that break the general traffic rules. Car crashes if the automation remains on.
T4	Jaywalker	No	City road with a pedestrian crossing the road.
T5	Obstructed view	Yes	City road with a pedestrian crossing the road from behind a large construction vehicle. Driver needs to switch off the automation before driving past the construction vehicle. The car's view is obstructed by the construction vehicle and can therefore not detect the pedestrian. Car crashes if the automation remains on.
T6	Priority signs	No	Intersection with priority traffic signs and crossing traffic.
T7	Fog	Yes	Straight highway with fog coming up. Driver needs to switch off automation before the fog and brake for slow cars within the fog section. The car's cameras do not function well in fog. Car crashes if the automation remains on.
T8	Highway traffic	No	Highway with gentle curves and several cars.

2.2.6. Variables

This study contained two independent variables: Training Method (DIT versus information brochure), and Session (Sessions 1, 2, and 3). Three dependent variables were measured during the experiment: acceptance, appropriate automation use, and take-over quality.

Acceptance. Participants indicated their acceptance of their training method in a questionnaire at the end of the first session. This questionnaire was a slight adaptation of the Technology Acceptance Questionnaire [41] and addressed six core aspects of technology acceptance: perceived ease of use, perceived usefulness, attitude, intention to use, self-efficacy, and social norm [42–46] (Appendix A).

Appropriate automation use. Each scenario contained an 'event area' during which the automation should be on or off. For events that required the automation to be off, the event area started at the latest moment the participant could turn off the automation and brake to avoid a crash. For example, when the participant was driving 100 km/h, the event area started 76 m before the point where the car would crash into something or someone (members.home.nl/johngrimbergen/remwegformule.htm). For scenarios in which the automation could be (left) on, the event area started directly after the straight road at the start of the specific scenario. Whether a scenario required the automation to be off was determined before the experiment, based on the system information used in the driver training. Four subcategories were used to specify the type of automation use during the event areas: (1) Correct take-over, the automation is off when necessary, (2) Correct reliance, the automation is on while it is safe, (3) Incorrect take-over, the automation is off while this is not necessary, (4) Incorrect reliance, the automation is on when this is not safe. It was decided not to include a knowledge test to determine the participants' explicit knowledge about the automated systems. In our previous studies [22], we found that a good score on the initial knowledge test did not predict actual use of the automation in the driving simulator study.

Take-over quality. In scenarios that required the automation to be (turned) off, three following take-over quality variables were measured from the moment the driver turned off the automation until the location of a possible collision: Time To Collision (TTC) (s), deceleration rate (m/s^2), and lateral acceleration (m/s^2) [47,48].

Appropriate automation use and take-over quality were already used as performance measures during Session 1. As the DIT is intended to be used by drivers in real cars during regular trips, Session 1 represented drivers' first on-road experience with the automation. For the DIT condition this would be when the DIT provides situated training to the driver while he or she is driving with the automation for the first time. For the IB group, this would be when the driver is driving with the automation for the first time after reading the information brochure. Careful assessment of the automation use was therefore already necessary during the first session as drivers need to be able to safely use the automation as soon as they start driving.

2.2.7. Analysis

The frequency data on 'appropriate automation use' was first analysed using a Chi-Square test. Next, we investigated how the 'appropriate automation use' evolved over time for each of the training methods. This was achieved through a mixed model approach, specifically Generalized Estimating Equation model (GEE). A Generalized Estimating Equation model was created as: our study was a 2 × 2 repeated measures design, the independent variable was binary, and we wanted to control for variations between scenarios [49,50]. In order to closer evaluate the specific types of (correct) automation use, a multinomial logistic regression model was created [51,52] to allow categorical response variables with more than two options. The response variable was 'automation use type' (correct take-over, correct reliance, incorrect take-over, and incorrect reliance).

The average lateral acceleration and deceleration rates were determined for the scenarios that required a take-over, starting directly after the participant turned off the automation until the end of the scenario. Then, any group differences on 'vehicle control' were analysed with unpaired independent t-tests. All research data is freely available in the Supplementary Materials and in the following data repository https://osf.io/xebrw/?view_only=eb59ffbbddc04bdf8f18d811f74d65ab.

3. Results

3.1. Appropriate Automation Use

3.1.1. Collisions

The total number of collisions appeared higher for the IB group in Session 1 (N_{IB} = 24, N_{DIT} = 20), Session 2 (N_{IB} = 10, N_{DIT} = 5), and Session 3 (N_{IB} = 5, N_{DIT} = 1). However, the Chi-Square tests did not indicate significant differences in the individual sessions (all p > 0.05). Two specific scenarios showed a significantly higher number of collisions for the IB group on a 0.1 level. These were OD2 (N_{IB} = 5, N_{DIT} = 1, χ^2 (1, N = 38) = 3.167, p = 0.075) and TS2 (N_{IB} = 3, N_{DIT} = 0, χ^2 (1, N = 38) = 3.257, p = 0.071).

3.1.2. Correct Take-Over and Reliance Behaviour

During the first session, the IB group used the automation incorrectly (either incorrect reliance or incorrect take-over) more often than the DIT group (N_{IB} = 65, N_{DIT} = 46) (Table 4). This difference was significant overall (χ^2 (1, N = 379) = 4.285, p = 0.025), and also for the specific scenarios OD2 (χ^2 (1, N = 38) = 8.992, p = 0.003) and RM2 (χ^2 (1, N = 38) = 7.795, p = 0.006). In the scenario OD2, a pedestrian crossed the street from behind a large bus that is blocking the view of the car's cameras.

In RM2, the lane markings are missing just before a sharp curve. No significant differences were found in Session 2 ($N_{IB} = 32$, $N_{DIT} = 26$) (χ^2 (1, $N = 301$) = 0.720, $p = 0.240$) and Session 3 ($N_{IB} = 13$, $N_{DIT} = 17$) (χ^2 (1, $N = 176$) = 0.643, $p = 0.274$). The observed power was sufficient for the Chi-Square tests per session (1-β >.8, d = 0.3, α = 0.05), but insufficient for between group comparisons in specific scenarios (1- β < 0.6, d = 0.3, α = 0.05). Consequently, if we control for the number of scenarios through a rather conservative Bonferroni correction ($\alpha_{adjusted}$ = 0.05/26 = 0.002), the differences found in individual scenarios are no longer significant (all $p > 0.002$).

Table 4. Overview of incorrect automation use (N) per scenario.

Session 1 [2] ($N_{IBgroup} = 19$, $N_{DITgroup} = 19$).	ACC1	ACC2	LK1	LK2	OD1	OD2 [2]	RM1	RM2 [2]	TS1	TS2	Total
IB group	16	2	3	2	7	12	5 [1]	10	6	1	64
DIT group	18	6	7	2	3	3	2	2	3	0	46
Total	34	8	10	4	10	15	8	12	9	1	110
Required take-over	N	Y	N	Y	N	Y	N	Y	N	Y	

Session 2	T1	T2	T3	T4	T5	T6	T7	T8	Total
IB group	1	2 [1]	6	2	3	10	2	6	32
DIT group	0	0	7	3	0	10 [1]	1	5	26
Total	1	1	11	5	13	20	3	11	58
Required take-over	N	Y	Y	N	Y	N	Y	N	

Session 3	T1	T2	T3	T4	T5	T6	T7	T8	Total
IB group	0	1	3	2	2	4	0	1	13
DIT group	1	0	2	3	0	6	0	5	17
Total	1	1	5	5	2	10	0	6	30
Required take-over	N	Y	Y	N	Y	N	Y	N	

[1] = 1 missing participant. [2] = Significant difference between groups on a 0.05 significance level.

Some specific scenarios appeared to show particularly more incorrect automation uses compared to the other scenarios: ACC1 and T6. ACC1 ($N = 34$) was the very first scenario that any of the participants encountered during this study. T6 contained a signalized intersection with intersecting traffic ($N_{session2} = 20$, $N_{session3} = 10$). The car would stop for the crossing traffic through traffic signs and continue after all traffic had passed. Multiple participants indicated that they thought the buildings were too close to the intersection and might block the view of the cameras.

Next, a Generalized Estimating Equation procedure followed (Section 2.2.7). The dependent variable was correct automation use. The random effects were the participants and scenarios. The fixed effects were the groups and sessions (Table 5). The chosen working correlation matrix type was 'exchangeable', as this resulted in the lowest Quasi Likelihood under the Independence Model Criterion ($QIC = 917.230$) [50]. The binary logit model showed a significant effect of sessions (χ^2 (1, $N = 856$) = 17.158, $p < 0.001$), but no overall effect of groups (χ^2 (1, $N = 856$) = 0.249, $p = 0.618$), nor an overall interaction effect (χ^2 (2, $N = 856$) = 4.186, $p = 0.123$). However, there were near significant effects on a 0.05 significance level of group in Session 1 (χ^2 (1, $N = 379$) = 3.835, $p = 0.050$) and Session 2 (χ^2 (1, $N = 301$) = 3.688, $p = 0.055$).

Table 5. The Generalized Estimating Equations model that was developed. The working correlation matrix was exchangeable. The random effects were the participants and scenarios, while the fixed effects were the groups and sessions.

Parameter	β	95% CI	SE	p
Intercept	1.375	1.016–1.735	0.183	0.000
IB group	0.369	−0.447–1.185	0.416	0.375
Session 1	−0.234	−0.727–0.259	0.252	0.352
Session 2	0.190	−0.173–0.553	0.185	0.306
IB group * Session 1	−0.840	−1.681–0.001	0.429	0.050
IB group * Session 2	−0.621	−1.255–0.013	0.324	0.055

Note. The DIT group and Session 3 statistics are not included as these were the baseline.

Looking at the specific types of incorrect automation use (incorrect take-over or incorrect reliance), it appeared that the IB group had more incorrect reliance decisions in Session 1 (N_{IB} = 27, N_{DIT} = 13), Session 2 (N_{IB} = 16, N_{DIT} = 12), and Session 3 (N_{IB} = 6, N_{DIT} = 2) (Figure 3). A Chi-Square analysis confirmed a difference between groups in incorrect reliance decisions but only for Sessions 1 (χ^2 (1, N = 190) = 6.20, p = 0.020). The DIT group had more incorrect take-overs in Session 3 (N_{IB} = 7, N_{DIT} = 15) (χ^2 (1, N = 88) = 3.879, p = 0.049). That is, they did not rely on the car when it was safe to do so more often than the IB group. The observed power for these Chi-Square tests was sufficient at > 0.8 (d = 0.3, α = 0.05). A multinomial logistic regression model was created next (Table 6). Similar to the GEE analysis, the fixed effects of the multinomial logistic regression were group and session, and the random effects were participant and scenario. The analysis confirmed an effect of both session and group on the specific types of automation use. Participants in the IB group were more likely to show an incorrect reliance behaviour (p = 0.030). Furthermore, participants were more likely to show incorrect reliance (p = 0.014) and incorrect take-overs (p = 0.044) during Session 1. No interaction effects of groups and sessions were found (all p > 0.05).

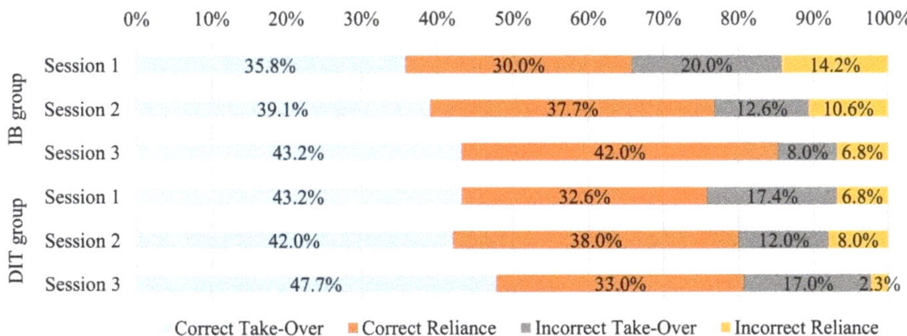

Figure 3. Overview of the different types of (in)correct automation use. Incorrect take-over means that the driver unnecessarily turned off the automation. Incorrect reliance indicates that the automation was on when it was not safe.

Table 6. Multinomial logistic regression model in which the response variable was 'automation use type', the fixed effects were 'group' and 'session', and the random effects were 'participant' and 'scenario'.

Parameter		β	95% CI	SE	p
Correct Take-over	Intercept	0.264		0.184	0.151
	IB group	−0.145	0.635–1.178	0.158	0.357
	Session 1	0.038	0.692–1.557	0.207	0.855
	Session 2	−0.126	0.583–1.335	0.211	0.552
Incorrect Take-Over	Intercept	−1.075		0.268	0.000 *
	IB group	−0.048	0.630–1.442	0.211	0.822
	Session 1	0.582	1.017–3.148	0.311	0.044 *
	Session 2	−0.027	0.530–1.789	0.311	0.931
Incorrect Reliance	Intercept	−2.449		0.411	0.000 *
	IB group	0.581	1.059–3.017	0.267	0.030 *
	Session 1	1.026	1.231–6.320	0.417	0.014 *
	Session 2	0.710	0.875–4.730	0.431	0.099

Note. The automation use type 'correct reliance, the DIT group, and Session 3 were not included as these were the baseline. * = significant effect on a 0.05 level. The interaction effects were all non-significant (all $p > 0.05$) and were excluded from this table for readability purposes.

Summary. Overall, the DIT group appeared to have a more correct automation use than the IB group during Sessions 1 and 2. However, a significant difference was only confirmed for Session 1. Considering the specific types of automation use, the DIT group consistently showed less incorrect reliance behaviour than the IB group throughout all sessions. This difference was confirmed through a multinomial regression. Surprisingly, however, the DIT group unnecessarily took back control (incorrect take-over) more often than the IB group in Session 3.

3.2. Take-Over Quality and Vehicle Control

During the first driving session, the DIT group showed larger Times To Collision (TTC) at take-over in three (ACC2, OD2, and RM2) out of five scenarios that required a take-over (Figure 4). For the scenario ACC2, the DIT group took back control significantly earlier ($M_{DIT} = 11.30$, $SD_{DIT} = 7.54$) than the IB group ($M_{IB} = 3.48$, $SD_{IB} = 3.57$) ($t(20.59) = 3.80$, $p = 0.001$). The DIT group also took back control significantly earlier in the scenario OD2 ($t(27) = 2.45$, $p = 0.025$), with a mean TTC of 6.19 s for the DIT group ($SD = 2.55$) and 3.67 s for the IB group ($SD = 2.92$). Similarly, the DIT group took back control significantly earlier in scenario RM2 ($t(21.63) = 2.27$, $p = 0.034$). In this scenario, the mean take-over distance was even negative for the IB group, indicating that take-over after the collision location had already passed ($M_{IB} = -0.03$, $SD_{IB} = 2.12$) ($M_{DIT} = 1.24$, $SD_{DIT} = 0.93$). In Sessions 2 and 3, it still appeared that the IB group took back control later in most scenarios that require a take-over; however, these results were not significant.

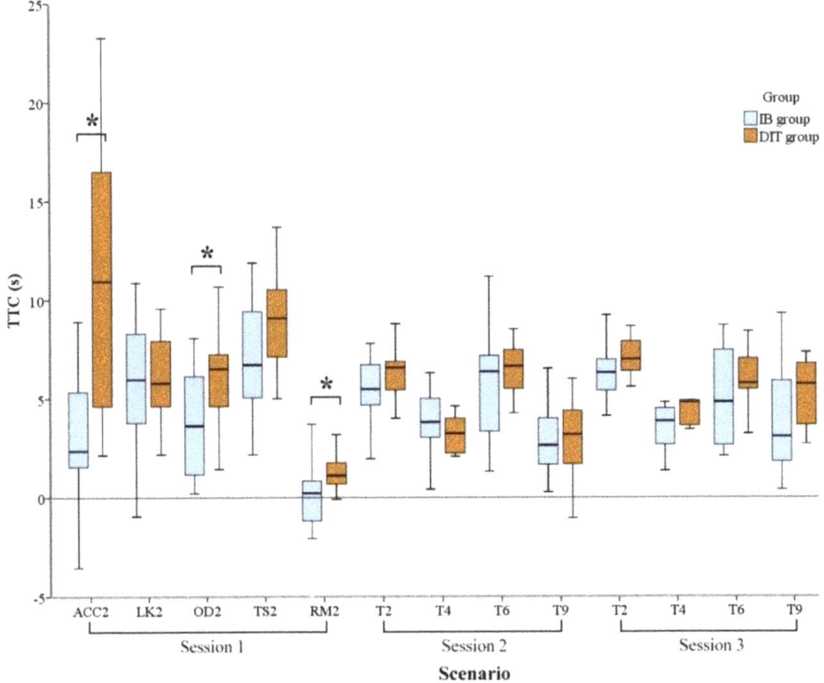

Figure 4. TTC when participants took back control.

During Session 1, the deceleration rate (m/s^2) was higher for the IB group in the same three scenarios in which the IB group showed later take-overs (ACC2, OD2, and RM2) (Figure 5). In scenarios ACC2 (M_{IB} = 1.79, SD_{IB} = 0.79, M_{DIT} = 0.88, SD_{DIT} = 0.44) ($t(34)$ = 4.12, $p < 0.001$) and RM2 (M_{IB} = 0.83, SD_{IB} = 0.38, M_{DIT} = 0.61, SD_{DIT} = 0.23) ($t(34)$ = 2.10, p = 0.043), the IB group decelerated significantly faster. This was also the case in scenario OD2, but only on a 0.1 significance level (M_{IB} = 2.25, SD_{IB} = 2.89, M_{DIT} = 0.92, SD_{DIT} = 0.61) ($t(28)$ = 1.85, p = 0.075). During the second session, only scenario Test 6 showed a difference between groups on the deceleration rate on a 0.1 significance level (M_{DIT} = 0.68, SD_{DIT} = 1.97, M_{IB} = 0.89, SD_{IB} = 2.51) ($t(36)$ = 1.72 p = 0.093). None of the scenarios in Session 3 showed significant differences on the deceleration rate between groups.

In Sessions 1 and 2, none of the scenarios showed a significant difference between groups on the average lateral acceleration after take-over. In Session 3, only one scenario (Test 9) showed a significant difference between groups on the average lateral acceleration after take-over ($t(19)$ = −2.38, p = 0.028). In this particular scenario, the DIT group showed a higher average lateral acceleration (M_{DIT} = 0.57, SD_{DIT} = 0.18, M_{IB} = 0.36, SD_{IB} = 0.22).

Summary. Overall, the DIT group showed significantly larger TTCs and smaller deceleration rates during the first session. This indicates earlier and consequently more gentle take-overs by the DIT group. While this still appeared to be the case in Sessions 2 and 3, the differences were no longer significant. Only one scenario across all sessions showed a difference between groups in the lateral acceleration. In this case, the DIT group showed a larger lateral acceleration. The possibility of Type II errors needs to be taken into account for the take-over quality and vehicle control variables, as the power was < 0.8 for these tests (d = 0.5, α = 0.05) [53].

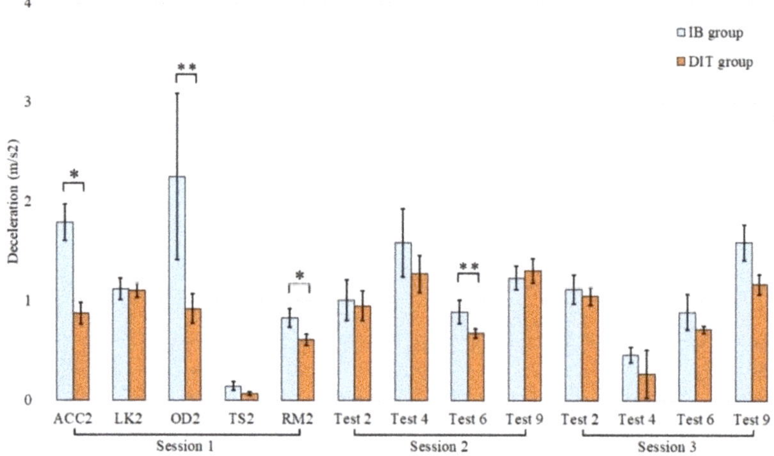

Figure 5. Deceleration rate after the participants took back control. * = significant on a 0.05 level. ** = significant on a 0.1 level. The error bars represent the Standard Error.

3.3. Acceptance

At the end of the first session, participants rated their agreement to several statements about their training on a scale of 1 (Strongly disagree) to 7 (Strongly agree) (Figure 6). Overall, the participants of the DIT group agreed that the DIT was easy to use ($M = 5.79$, $SD = 0.93$, 95% $CI = 5.34$–6.24) and useful ($M = 5.72$, $SD = 1.18$, 95% $CI = 5.15$–6.29). Participants were positive towards the DIT ($M = 5.74$, $SD = 1.11$, 95% $CI = 5.20$–6.27), and disagreed that it was annoying or frustrating ($M = 2.63$, $SD = 1.28$, 95% $CI = 2.02$–3.25). Furthermore, participants showed the intent to use the DIT if it was in their partially automated car ($M = 5.05$, $SD = 1.65$, 95% $CI = 4.26$–5.85), and felt that they were capable of using it ($M = 5.87$, $SD = 0.47$, 95% $CI = 5.64$–6.09). Participants disagreed that people who are important to them think that they should use the DIT ($M = 3.79$, $SD = 2.12$, 95% $CI = 2.77$–4.81). This seems logical as their friends and family most likely do not know about the system. The acceptance ratings could not be compared as each group only experienced one training method.

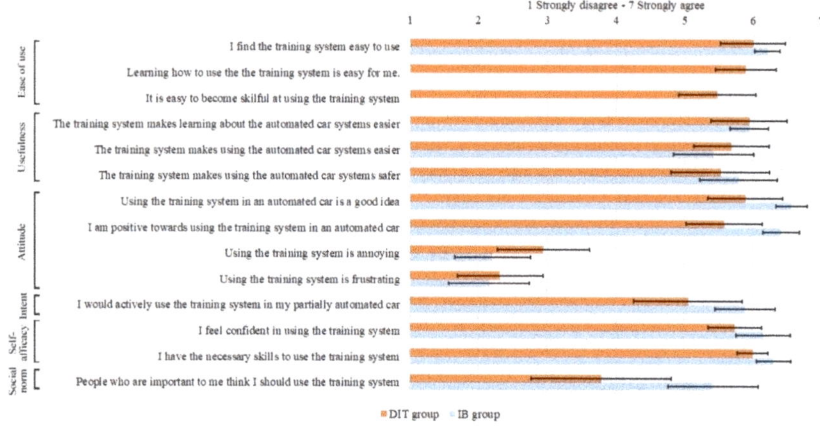

Figure 6. Overview of the acceptance ratings. For the IB group, the words 'training system' were replaced by 'training'. Two 'ease of use' questions did not apply to the IB group. The error bars indicate the 95% Confidence Intervals.

4. Discussion

A Digital In-car Tutor (DIT) is proposed as a situated, low-cost, and time efficient method for drivers to learn about their partially automated car during regular driving trips. In this study, we evaluated a DIT prototype for a complex (simulated) partially automated car. It was hypothesized that the DIT prototype would support drivers in deciding when it is safe to use the automation, and consequently lead to better vehicle control when taking back control. To study this, we compared appropriate automation use and take-over quality, in two groups over three driving sessions. The control group received information about the car automation through a brochure (IB group), while the experimental group received the information from the DIT prototype during the first driving session (DIT group). The DIT provided situated information about the systems' capabilities and limitations. Drivers were instructed to turn on the automation whenever they thought that the car could safely cope with the situation, and turn (or leave) it off if they thought that it could not. Each scenario contained an event in which it was either safe or unsafe to use the automation. This way, the automation use could be classified as follows: (1) Correct take-over, the automation is off when necessary, (2) Correct reliance, the automation is on while it is safe, (3) Incorrect take-over, the automation is off while this is not necessary, and (4) Incorrect reliance, the automation is on when this is not safe. It is important to note that the DIT is *not* a warning system that prompts all upcoming events. Rather, it identifies certain scenarios to support situated learning. Furthermore, the DIT never stated that it was safe to leave the automation on, or that it was necessary to take back control. For technical, safety, and liability reasons, this would be unrealistic to expect if the DIT were to be implemented in commercial cars.

Correct automation use. During the first driving session, the DIT group showed overall a more correct automation use (combined correct take-overs and correct reliance) compared to the IB group. During the second session, in which the DIT was no longer active, this still appeared to be the case, but the difference was no longer significant. During the third session, the two groups showed a similar level of correct automation use. Although a significant difference could only be confirmed for the first session, this still has implications for traffic safety. As the DIT should be used in real cars during normal trips, drivers need to be able to use the automation appropriately and safely from the start without any possible confusion. In simulator training, one could require drivers to go through multiple driving sessions to get to a desired performance level (although we did still see more inappropriate reliance behaviour in the control group after three driving sessions, which we will discuss soon). But as drivers are using the DIT during regular driving in their own car, initial appropriate automation use is critical for traffic safety. Still, although most learning is believed to occur during the initial interaction [7,8,54], it may still be necessary to increase the duration of the DIT to obtain a higher final performance level, especially since multiple studies, like those by Beggiato [7,54] and Forster [8], have shown that the learning curve stabilizes after approximately five interactions (or 3.5 h) [7,8]. Extended DIT support may also be necessary as situations that have not been experienced for a long time can fade from the driver's mental model [7]. Longer (but not necessarily continuous) DIT support provides the option to highlight rare situations in similar frequently occurring situations. This needs further investigation in a more longitudinal study.

Incorrect reliance. The DIT group already showed less incorrect reliance during the first session, compared to the IB group. By the third session, the amount of incorrect reliances of the DIT group had further decreased to around two and a half percent of all interactions. While the IB group also showed a decrease in incorrect reliances over time, both the initial and final amount of incorrect reliances appeared to be higher compared to the DIT group. During the third session, the brochure group still showed around seven percent of incorrect reliances out of all interactions. Further analysis confirmed that the IB group was more likely to show incorrect reliance behaviour. These results follow our expectations based on both established and more recent models that describe the interaction between automation feedback and automation use. These include, amongst others, Lee and See [55], Seppelt [56,57], and Revell [58]. All these interaction models suggest that (external) information about the automation, as well as repeated interactions and automation feedback all affect automation use (and reliance). The results suggest that by combining all these elements in the DIT, it was effective in specifically decreasing inappropriate reliance behaviour. This is an important implication of the prototype as inappropriate reliance can lead to severe safety issues.

Incorrect take-over. Both groups had a similar number of unnecessary (incorrect) take-overs during the first driving session. While the number of unnecessary take-overs decreased over time for the IB group, this was not the case for the DIT group. It seems that the DIT group was more careful to rely on the automation throughout the driving sessions. These results are unexpected as they are not in line with the statement that repeated interactions, feedback, and background information lead to improved mental models and consequently appropriate automation use. Similarly, they are not in line with the research on a digital tutor for ACC by Simon [33], which showed fewer unnecessary take-overs from users of the digital tutor. However, interestingly, that study also showed a slight increase of unnecessary take-overs during the third driving session in specific scenarios. One would expect that the feedback of the DIT would in this case lead to fewer unnecessary take-overs, just as the lack of feedback for the IB group should lead to an over- or under-reliance depending on the experience of safe driving situations or crashes.

The amount of unnecessary take-overs for the DIT group might be explained by the Signal Detection Theory [59–61]. In our study, correct take-over and correct reliance correspond respectively to 'hit' and 'correct rejection', while incorrect take-over and incorrect reliance correspond to 'false alarm' and 'miss'. The information and explicit feedback by the DIT repeatedly stressed the limitations of the automation. This may have made drivers change their criterion and take a more conservative attitude when judging situations as being inside the ODD of the automation, consequently increasing the number of incorrect take-overs (false alarms) and reducing the amount of incorrect reliance (misses). Another explanation is that drivers were still in the phase of forming their core mental models about the automation by the third session [33]. It is important to realize that unnecessary take-overs are not necessarily dangerous and are arguably preferred in ambiguous situations. Still, unnecessary take-overs need to be limited so that the automation can be used to its full potential. If drivers are constantly disengaging the automation when it is unnecessary, potential benefits of the automation such as increased traffic safety and driver comfort may not be achieved.

Challenging scenarios. Two particular driving situations were very difficult for both groups: ACC1 and T6 (see Section 2.2.5). It was safe to leave the automation on in both situations. ACC1 was the very first scenario that all drivers encountered during the study. As discussed earlier, drivers need repeated experience and feedback to develop a calibrated level of trust [7,8,62]. While reassurance feedback may support a higher initial level of trust, a DIT should never suggest that the automation can perfectly handle a situation. Scenario T6 was a signalized intersection with crossing traffic. The automated car would detect the priority signs and stop to let the crossing cars pass. Drivers did not rely on the car as they thought that the houses were too close to the street and might block the view of the car's cameras. This suggests that the drivers were well aware of the limitations (blocked cameras) and capabilities (detecting priority signs) of the automation. However, as no specific camera ranges were provided during the training, this particular situation became ambiguous for the drivers. Taking back control was then arguably the safest decision.

Vehicle control. We expected to see better vehicle control for the DIT group after disengaging the automation in situations that required to take back control [63,64]. For example, Simon [33] found less intense braking behaviour for users of the digital ACC tutor. In our study, the DIT group took back control significantly earlier, and braked less hard, than the IB group during the first session. However, no significant differences were found between the groups in the second and third sessions. Still, the minimum Time To Collision at take-over was consistently larger, and the maximum deceleration was smaller, for the DIT group. While overall no differences between groups were found for the lateral acceleration after take-over, one scenario surprisingly showed a larger lateral acceleration for the DIT group. The possibility for Type II errors needs to be taken into consideration for the vehicle control variables as these tests had limited power.

Acceptance. Our results show that participants found the DIT easy to use. Participants also indicated that the DIT made learning about, and using, the automation easier. They felt positively about the DIT and confident in using it. Participants indicated an intent to use the DIT, but did not think that their peers and family felt that they should use it.

4.1. Limitations

Certain limitations concerning this study have to be taken into account. First, participants in the control group were asked to read the brochure carefully before entering the driving simulator. However, in real life, a large share of drivers does not read the owner's manual, nor looks up any other information about the automation in their car [1,5]. Therefore, the group will not be representative of all drivers. A brochure was chosen for the control group as this is often used by car sellers as the main (and only) method of providing customers with information about the automation in their new car [5]. An additional study with a control group that does not receive any information about the automation before driving may be required for an improved representation of current drivers.

Second, it may be that the visual cues have contributed to the differences between groups during Session 1 due to a priming effect. Although the visuals were a core part of the DIT prototype as they allowed to address the systems' limitations in the current driving situation, further research is necessary to determine how the way that the information is presented influences learning. For example, it is unclear if a DIT that is strictly auditory will have similar effects.

Third, participants could only turn off the automation by pressing a button on the steering wheel. It is possible that the inability to disengage the automation through the brake has caused confusion among drivers in time-critical situations. However, participants were reminded that they had to disengage the automation through the button, and not the pedals, multiple times throughout the driving sessions.

Last, the current between-subject set-up did not allow us to compare the acceptance between the DIT and an information brochure. Additional studies with a within-subject design are required to examine the acceptance of the DIT more extensively.

4.2. Future Research

The results of this study provide multiple opportunities for further research. First, it is necessary to further investigate the specific information that needs to be included during the introduction of a new system. For example, it is unclear if it is necessary to include the technical equipment specifications.

Second, the effects of a DIT on driver distraction need to be assessed. By projecting the transparent images on the windscreen, the driver does not have to continuously shift his attention from the road to a secondary screen. However, the images are still expected to introduce glances away from the centre of the road and take up cognitive resources. They therefore need to be further refined so that they facilitate optimal learning while limiting distraction from the road. For example, the images may need to be located closer to the centre of the driver's field of view, without causing visual clutter [65,66], to adhere to the NHTSA guidelines on the number and duration of glances away from the centre of the road [67,68].

Last, while the concept prototype used the entire windscreen to project the images on, more practical implementations need to be explored. For example, the DIT may be implemented in an off-the-shelf head-up display device.

5. Conclusions

During the first driving session, in which the DIT was active for the experimental group, users of the DIT showed a more correct automation use (correct reliance and correct take-overs) and higher-quality take-overs. This first driving session represented the initial on-road contact with both the automation and DIT. However, the differences in correct automation use were reduced over time and disappeared by the last driving session, which took place two weeks after the first session. The IB group appeared to catch up with the DIT group and came to a similar level of correct automation use. Still, as the DIT is used in drivers' cars during regular drives, safe automation use is extremely important directly from the start. The DIT specifically led to less incorrect reliance behaviour throughout the driving sessions, something that would otherwise lead to immediate safety issues. While the IB and DIT groups both showed a decrease in incorrect reliance over the course of the driving sessions, the overall incorrect reliance was significantly lower in the DIT group throughout the sessions. That means that drivers relied less on the automation in situations that were outside of its Operational Design Domain. Still, further research is necessary on the precise required content of a DIT, and how the way of presenting the DIT information exactly influences learning. The results further indicated a possible under-trust of the automation among users of the DIT. While under-trust may be less dangerous, it may hinder the adoption (and proposed benefits) of automated driving. It is therefore necessary to investigate how to address the under-trust without the risk of creating overreliance. Finally, drivers found the DIT easy to use, useful, and felt confident in using it. Overall, this study provides an initial insight into the effects of a Digital In-Car Tutor on the appropriate use of complex car automation. The concept of a DIT shows some potential as a low-cost, time-efficient, situated, and long-term method for learning about partially automated cars, with additional benefits for instructing drivers after overnight software updates. Therefore, additional research is advised to further explore DIT content and form.

Supplementary Materials: The data collected during the study are freely available at www.mdpi.com/xxx/s1.

Author Contributions: Conceptualization, A.B. and A.P.v.d.B.; Data curation, A.B.; Formal analysis, A.B.; Investigation, A.B.; Methodology, A.B., A.P.v.d.B., M.C.v.d.V., W.B.V. and M.H.M.; Project administration, A.B.; Supervision, A.P.v.d.B., M.C.v.d.V., W.B.V. and M.H.M.; Visualization, A.B.; Writing – original draft, A.B. and A.P.v.d.B.; Writing – review & editing, A.B., A.P.v.d.B., M.C.v.d.V., W.B.V. and M.H.M. All authors have read and agreed to the published version of the manuscript.

Funding: This research is funded by the Dutch Domain Applied and Engineering Sciences, which is part of the Netherlands Organisation for Scientific Research (NWO), and which is partly funded by the Ministry of Economic Affairs (grant number 14896).

Conflicts of Interest: The authors declared no potential conflict of interest with respect to the research, authorship, and/or publication of this article.

Appendix A —Acceptance Questionnaire

The following acceptance questionnaire was completed by participants of the DIT group after the first session.

The following questions are specifically about the training system you experienced!

Perceived ease of use.

Please indicate for each statement to what extent you (dis)agree. (1- Strongly agree, 7- Strongly disagree)

1. I find the training system easy to use
2. Learning how to use the training system is easy for me
3. It is easy to become skillful at using the training system

Perceived usefulness.

Please indicate for each statement to what extent you (dis)agree. (1- Strongly agree, 7- Strongly disagree)

4. The training system makes learning about the automated car systems easier
5. The training system makes using the automated car systems easier
6. The training system makes using the automated car systems safer

Attitude.

Please indicate for each statement to what extent you (dis)agree. (1- Strongly agree, 7- Strongly disagree)

7.   Using the training system in an automated car is a good idea
8.   I am positive towards using the training system in an automated car
9.   Using the training system is annoying
10. Using the training system is frustrating

Intention to use.
Imagine that you own the partially automated car that you experienced today.
Please indicate for each statement to what extent you (dis)agree.   (1- Strongly agree, 7- Strongly disagree)

11. I would actively use the training system in my partially automated car

Self-efficacy.
Please indicate for each statement to what extent you (dis)agree.   (1- Strongly agree, 7- Strongly disagree)

12. I feel confident in using the training system
13. I have the necessary skills to use the training system

Social norm.
Imagine that you own the partially automated car that you experienced today.
Please indicate for each statement to what extent you (dis)agree.   (1- Strongly agree, 7- Strongly disagree)

14. People who are important to me think I should use the training system

References

1.   Harms, I.; Dekker, G.M. *ADAS: From Owner to User*. 2017. Available online: http://www.verkeerskunde.nl/Uploads/2017/11/ADAS-from-owner-to-user-lowres.pdf (accessed on 28 March 2020).
2.   McDonald, A.; Carney, C.; McGehee, D.V. *Vehicle Owners' Experiences with and Reactions to Advanced Driver Assistance Systems*. 2018. Available online: https://aaafoundation.org/vehicle-owners-experiences-reactions-advanced-driver-assistance-systems/ (accessed on 28 March 2020).
3.   Abraham, H.; Seppelt, B.; Mehler, B.; Reimer, B. What's in a Name: Vehicle Technology Branding & Consumer Expectations for Automation. In Proceedings of the ACM 9th International Conference on Automotive User Interfaces and Interactive Vehicular Applications, Oldenburg, Germany, 24–27 September 2017; pp. 226–234. [CrossRef]
4.   Carsten, O.; Martens, M.H. How Can Humans Understand Their Automated Cars? HMI Principles, Problems and Solutions. *Cogn. Technol. Work* **2018**, *21*, 1–18. [CrossRef]
5.   Boelhouwer, A.; Van Der Voort, M.C.; Hottentot, C.; De Wit, R.Q.; Martens, M.H. How are Car Buyers and Car Sellers Currently Informed about ADAS? An Investigation among Drivers and Car Sellers in The Netherlands. *Transp. Res. Interdiscip. Perspect.* **2020**, in press. [CrossRef]
6.   Abraham, H.; Reimer, B.; Mehler, B. Learning to Use In-Vehicle Technologies: Consumer Preferences and Effects on Understanding. In Proceedings of the Human Factors and Ergonomics Society 2018 Annual Meeting, Philadelphia, PA, USA, 1–5 October 2018; pp. 1589–1593. [CrossRef]
7.   Beggiato, M.; Pereira, M.; Petzoldt, T.; Krems, J. Learning and Development of Trust, Acceptance and the Mental Model of ACC. A Longitudinal On-road Study. *Transp. Res. Part F Psychol. Behav.* **2015**, *35*, 75–84. [CrossRef]
8.   Forster, Y.; Hergeth, S.; Naujoks, F.; Beggiato, M.; Krems, J.F.; Keinath, A. Learning and Development of Mental Models During Interactions with Driving Automation: A Simulator Study. In Proceedings of the Tenth International Driving Symposium on Human Factors in Driver Assessment, Training and Vehicle Design, Santa Fe, NM, USA, 24–27 June 2019; pp. 398–404. [CrossRef]
9.   Gao, H.; Yu, H.; Xie, G.; Ma, H.; Xu, Y.; Li, D. Hardware and Software Architecture of Intelligent Vehicles and Road Verification in Typical Traffic Scenarios. *IET Intell. Transp. Syst.* **2019**, *13*, 960–966. [CrossRef]

10. Flemisch, F.; Heesen, M.; Hesse, T.; Kelsch, J.; Schieben, A.; Beller, J. Towards a Dynamic Balance between Humans and Automation: Authority, Ability, Responsibility and Control in Shared and Cooperative Control Situations. *Cogn. Technol. Work* **2012**, *14*, 3–18. [CrossRef]

11. Abbink, D.A.; Mulder, M.; Boer, E.R. Haptic Shared Control: Smoothly Shifting Control Authority? *Cogn. Technol. Work* **2012**, *14*, 19–28. [CrossRef]

12. Martens, M.H.; van den Beukel, A.P. The Road to Automated Driving: Dual Mode and Human Factors Considerations. In Proceedings of the IEEE Conference on Intelligent Transportation Systems, The Hague, The Netherlands, 6–9 October 2013; pp. 2262–2267. [CrossRef]

13. Parasuraman, R.; Riley, V. Humans and Automation: Use, Misuse, Disuse, Abuse. *Hum. Factors* **1997**, *39*, 230–253. [CrossRef]

14. Lee, J.D.; Seppelt, B.D. Human Factors in Automation Design. In *Handbook of Automation*; Nof, S., Ed.; Springer: Berlin, Germany, 2009; pp. 417–436. [CrossRef]

15. Dickie, D.A.; Boyle, L.N. Drivers' Understanding of Adaptive Cruise Control Limitations. In Proceedings of the Human Factors and Ergonomics Society 53rd Annual Meeting, San Antonio, TX, USA, 19–23 October 2009; pp. 1806–1810. [CrossRef]

16. Fagnant, D.J.; Kockelman, K. Preparing a Nation for Autonomous Vehicles: Opportunities, Barriers and Policy Recommendations. *Transp. Res. Part A Policy Pract.* **2015**, *77*, 167–181. [CrossRef]

17. Van Wee, B.; Annema, J.A.; Banister, D. *The Transport System and Transport Policy, an Introduction*; Edward Elgar Publishing Limited: Cheltenham, UK, 2013.

18. Anderson, J.M.; Kalra, N.; Stanley, K.D.; Sorensen, P.; Samaras, C.; Oluwatola, O.A. *Autonomous Vehicle Technology A Guide for Policymakers*; RAND Corporation: Santa Monica, CA, USA, 2016. [CrossRef]

19. Davilla, A. *SARTRE Report on Fuel Consumption (Report No. D.4.3)*; SARTRE: Barcelona, Spain, 2013.

20. Luo, L.; Liu, H.; Li, P.; Wang, H. Model Predictive Control for Adaptive Cruise Control with Multi-objectives: Comfort, Fuel-economy, Safety and Car-following. *J. Zhejiang Univ. Sci. A* **2010**, *11*, 191–201. [CrossRef]

21. National Highway Traffic Safety Administration Preliminary Statement of Policy Concerning Automated Vehicles America. 2013. Available online: https://www.nhtsa.gov/staticfiles/r (accessed on 28 March 2020).

22. Boelhouwer, A.; van den Beukel, A.P.; Van Der Voort, M.C.; Martens, M.H. Should I Take Over? Does System Knowledge Help Drivers in Making Take-over Decisions while Driving a Partially Automated Car? *Transp. Res. Part F Traffic Psychol. Behav.* **2019**, *60*, 669–684. [CrossRef]

23. Forster, Y.; Hergeth, S.; Naujoks, F.; Krems, J.; Keinath, A. User Education in Automated driving: Owner's Manual and Interactive Tutorial Support Mental Model Formation and Human-automation Interaction. *Information* **2019**, *10*, 143. [CrossRef]

24. McDonald, A.B.; Reyes, M.L.; Roe, C.A.; Friberg, J.E.; Faust, K.S.; McGehee, D.V. *University of Iowa Technology Demonstration Study*. 2016. Available online: http://www.nads-sc.uiowa.edu/publicationStorage/20161480695480.N2016-021_Technology%20Demonstra.pdf (accessed on 28 March 2020).

25. Panou, M.; Bekiaris, E.D.; Touliou, A.A. ADAS module in driving simulation for training young drivers. In Proceedings of the Annual Conference on Intelligent Transportation Systems, Madeira Island, Portugal, 19–22 September 2010; pp. 1582–1587. [CrossRef]

26. Payre, W.; Cestac, J.; Dang, N.T.; Vienne, F.; Delhomme, P. Impact of Training and In-vehicle Task Performance on Manual Control Recovery in an Automated Car. *Transp. Res. Part F Traffic Psychol. Behav.* **2017**, *46*, 216–227. [CrossRef]

27. Ropelato, S.; Zünd, F.; Sumner, R.W. Adaptive Tutoring on a Virtual Reality Driving Simulator. In Proceedings of the 10th International Workshop on Semantic Ambient Media Experiences, Bangkok, Thailand, 27 November 2017; pp. 12–17. [CrossRef]

28. Boelhouwer, A.; van den Beukel, A.P.; van der Voort, M.C.; Martens, M.H. Determining Environment Factors That Increase the Complexity of Driving Situations. In Proceedings of the 8th International Conference on Human Factors in Transportation, San Diego, CA, USA, 16–20 July 2019. (In Press).

29. van Gent, P.; Farah, H.; van Nes, N.; van Arem, B. A Conceptual Model for Persuasive In-vehicle Technology to Influence Tactical Level Driver Behaviour. *Transp. Res. Part F Traffic Psychol. Behav.* **2019**, *60*, 202–216. [CrossRef]

30. Wilkison, B.D.; Fisk, A.D.; Rogers, W.A. Effects of Mental Model Quality on Collaborative System Performance. In Proceedings of the Human Factors and Ergonomics Society 51st Annual Meeting, Baltimore, MD, USA, 1–5 October 2007; pp. 1506–1510. [CrossRef]

31. Boelhouwer, A.; van den Beukel, A.P.; Casner, S.M.; Van Der Voort, M.C.; Martens, M.H. Adaptive Feedback Patterns in Driving Instructors: Towards an Adaptive Digital In-Car Tutor for Drivers of Complex Partially Automated Cars. (Submitted for Publication).

32. Forster, Y.; Hergeth, S.; Naujoks, F.; Krems, J. Tell Them How They Did: Feedback on Operator Performance Helps Calibrate Perceived Ease of Use in Automated Driving. *Multimodal Technol. Interact* **2019**, *3*, 29. [CrossRef]

33. Simon, J.H. Learning to Drive with Advanced Driver Assistance Systems, Technical University Chemnitz. 2005. Available online: https://d-nb.info/980929709/34 (accessed on 28 March 2020).

34. Franke, T.; Attig, C.; Wessel, D. Affinity for Technology Interaction (ATI) Scale. *Int. J. Human–Computer Interact.* **2018**, *2018*. [CrossRef]

35. Franke, T.; Attig, C.; Wessel, D. A Personal Resource for Technology Interaction: Development and Validation of the Affinity for Technology Interaction (ATI) Scale. *Int. J. Human–Computer Interact.* **2019**, *35*, 456–467. [CrossRef]

36. Rasmussen, J. Skills, Rules, and Knowledge; Signals, Signs and Symbols, and Other Distinctions in Human Performance Models. *IEEE Trans. Syst. Man. Cybern.* **1983**, *13*, 257–266. [CrossRef]

37. Birrel, S.; Young, M.; Stanton, N.A.; Jennings, P. Using Adaptive Interfaces to Encourage Smart Driving and Their Effect on Driver Workload. In Proceedings of the AHFE 2016 International Conference on Human Factors in Transportation, Walt Disney World, Bay Lake, FL, USA, 27–31 July 2016; pp. 31–43. [CrossRef]

38. Watson-Brown, N.; Scott-Parker, B.; Senserrick, T. Development of a Higher-order Instruction Coding Taxonomy for Observational data: Initial Application to Professional Driving Instruction. *Appl. Ergon.* **2018**, *70*, 88–97. [CrossRef]

39. Boelhouwer, A.; van den Beukel, A.P.; van der Voort, M.C.; Martens, M.H. Designing a Naturalistic In-Car Tutor System for the Initial Use of Partially Automated Cars: Taking Inspiration from Driving Instructors. In Proceedings of the 11th International Conference on Automotive User Interfaces and Interactive Vehicular Applications: Adjunct Proceedings, Utrecht, The Netherlands, 22–25 September 2019; pp. 410–414. [CrossRef]

40. Li, D.; Gao, H.A. Hardware Platform Framework for an Intelligent Vehicle Based on a Driving Brain. *Engineering* **2018**, *4*, 464–470. [CrossRef]

41. Park, S.Y. An Analysis of the Technology Acceptance Model in Understanding University Students' Behavioral Intention to Use e-Learning Research hypotheses. *Educ. Technol. Soc.* **2009**, *12*, 150–162.

42. Venkatesh, V.; Davis, F.D. A Model of the Antecedents of Perceived Ease of Use: Development and Test. *Decis. Sci.* **1996**, *27*, 451–481. [CrossRef]

43. Davis, F.D. Perceived Usefulness, Perceived Ease of Use, and User Acceptance of Information Technology. *Manag. Inf. Syst.* **1989**, *13*, 319–340. [CrossRef]

44. Venkatesh, V.; Davis, F.D. A Theoretical Extension of the Technology Acceptance Model: Four Longitudinal Field Studies. *Manage. Sci.* **2000**, *46*, 186–204. [CrossRef]

45. Ajzen, I. The Theory of Planned Behavior. *Organ. Behav. Hum. Decis. Process.* **1991**, *50*, 179–211. [CrossRef]

46. Featherman, M. Extending the Technology Acceptance Model by Inclusion of Perceived Risk. In Proceedings of the 2001 Americas Conference on Information Systems, Boston, MA, USA, 3–5 August 2001; pp. 758–760.

47. Gold, C.; Körber, M.; Lechner, D.; Bengler, K.J. Taking Over Control From Highly Automated Vehicles in Complex Traffic Situations. *Hum. Factors J. Hum. Factors Ergon. Soc.* **2016**, *58*, 642–652. [CrossRef]

48. Merat, N.; Jamson, A.H.; Lai, F.C.H.; Daly, M.; Carsten, O. Transition to Manual: Driver Behaviour when Resuming Control from a Highly Automated Vehicle. *Transp. Res. Part F Traffic Psychol. Behav.* **2014**, *27*, 274–282. [CrossRef]

49. Sainani, K. GEE and Mixed Models for Longitudinal Data. Available online: https://goo.gl/Ndcv7b (accessed on 28 March 2020).

50. Ziegler, A.; Vens, M. Generalized Estimating Equations: Notes on the Choice of the working Correlation Matrix. *Methods Inf. Med.* **2010**, *49*, 421–425. [CrossRef]

51. Osborne, J.W. Multinomial and Ordinal Logistic Regression. In *Best Practices in Logistic Regression*; SAGE Publications: Thousand Oaks, CA, USA, 2017; pp. 388–433. [CrossRef]

52. Twisk, J.W.R. *Applied Longitudinal Data Analysis for Epidemiology*; Cambridge University Press: Cambridge, UK, 2004. [CrossRef]

53. Cohen, J. Quantitative Methods in Psychology. *Psychol. Bull.* **1992**, *112*, 155–159. [CrossRef]

54. Beggiato, M. Changes in Motivational and Higher Level Cognitive Processes When Interacting with In-Vehicle Automation, Technischen Universität Chemnitz. 2014. Available online: https://monarch.qucosa.de/api/qucosa%3A20246/attachment/ATT-0/ (accessed on 28 March 2020).

55. Lee, J.D.; See, K. Trust in Automation: Designing for Appropriate Reliance. *Hum. Factors* **2004**, *46*, 50–80. [CrossRef]

56. Seppelt, B.D. Supporting Operator Reliance on Automation Through Continuous Feedback. Ph.D. Thesis, University of Iowa, Iowa City, IA, USA, 2009. [CrossRef]

57. Seppelt, B.D.; Lee, J.D. Keeping the driver in the loop: Dynamic Feedback to Support Appropriate Use of Imperfect Vehicle Control Automation. *Int. J. Hum. Comput. Stud.* **2019**, *125*, 66–80. [CrossRef]

58. Revell, K.M.A.; Bradley, M. Breaking the Cycle of Frustration: Applying Neisser's Perceptual Cycle Model to Drivers of Semi-Autonomous Vehicles. *Appl. Ergon.* **2020**. [CrossRef] [PubMed]

59. Wickens, T.D. *Elementary Signal Detection Theory*; Oxford University Press: New York, NY, USA, 2002.

60. Sheridan, T.B. Extending Three Existing Models to Analysis of Trust in Automation: Signal Detection, Statistical Parameter Estimation, and Model-Based Control. *Hum. Factors* **2019**, *61*, 1162–1170. [CrossRef] [PubMed]

61. Green, D.M.; Swets, J.A. *Signal Detection Theory and Psychophysics.*; Wiley: New York, NY, USA, 1966.

62. Walker, F.; Boelhouwer, A.; Alkim, T.; Verwey, W.; Martens, M.H. Changes in Trust after Driving Level 2 Automated Cars. *J. Adv. Transp.* **2018**, *2018*. [CrossRef]

63. Gold, C.; Damböck, D.; Lorenz, L.; Bengler, K.J. "Take over!" How Long Does it Take to get the Driver Back into the Loop? In Proceedings of the Human Factors and Ergonomics Society 57st Annual Meeting, San Diego, CA, USA, 30 September–4 October 2013; pp. 1938–1942. [CrossRef]

64. Rudin-Brown, C.M.; Parker, H.A. Behavioural Adaptation to Adaptive Cruise Control (ACC): Implications for Preventive Strategies. *Transp. Res. Part F Traffic Psychol. Behav.* **2004**, *7*, 59–76. [CrossRef]

65. Kroon, E.C.M.; Martens, M.H.; Brookhuis, K.; Hagenzieker, M.; Alferdinck, J.W.A.M.; Harms, I.; Hof, T. *Human Factor Guidelines for the Design of Safe In-Car Traffic Information Services*. 2016. Available online: https://repository.tudelft.nl/islandora/object/uuid:782025b2-1250-4581-8215-f2497455ee01/datastream/OBJ/download (accessed on 28 March 2020).

66. Stevens, A.; Quimby, A.; Board, A.; Kersloot, T.; Burns, P. *Design Guidelines for Safety of In-Vehicle Information Systems, (PA3721/01)*; Transport Research Laboratory TRL: Crowthorne, UK, 2002; pp. 1–55.

67. National Highway Traffic Safety Administration. 2016 Update to "Preliminary Statement of Policy Concerning Automated Vehicles". 2016. Available online: http://www.aamva.org/NHTSADOTAutVehPolicyUpdate_Jan2016 (accessed on 28 March 2020).

68. National Highway Traffic Safety Administration. Visual-Manual NHTSA Driver Distraction Guidelines for In-Vehicle Electronic Devices. 2013. Available online: https://www.federalregister.gov/documents/2014/09/16/2014-21991/visual-manual-nhtsa-driver-distraction-guidelines-for-in-vehicle-electronic-devices (accessed on 28 March 2020).

 *information*

Article

The Impact of Situational Complexity and Familiarity on Takeover Quality in Uncritical Highly Automated Driving Scenarios

Marlene Susanne Lisa Scharfe [1,*,†], **Kathrin Zeeb** [1] and **Nele Russwinkel** [2]

[1]   Robert Bosch GmbH, 74232 Abstatt, Germany; kathrin.zeeb@de.bosch.com
[2]   Department of Psychology and Ergonomics, Technical University Berlin, 10587 Berlin, Germany; nele.russwinkel@tu-berlin.de
*   Correspondence: marlene-susanne-lisa.scharfe@de.bosch.com
†   Current address: Robert-Bosch-Allee 1, 74232 Abstatt, Germany.

Received: 30 January 2020; Accepted: 17 February 2020; Published: 20 February 2020

Abstract: In the development of highly automated driving systems (L3 and 4), much research has been done on the subject of driver takeover. Strong focus has been placed on the takeover quality. Previous research has shown that one of the main influencing factors is the complexity of a traffic situation that has not been sufficiently addressed so far, as different approaches towards complexity exist. This paper differentiates between the objective complexity and the subjectively perceived complexity. In addition, the familiarity with a takeover situation is examined. Gold et al. show that repetition of takeover scenarios strongly influences the take-over performance. Yet, both complexity and familiarity have not been considered at the same time. Therefore, the aim of the present study is to examine the impact of objective complexity and familiarity on the subjectively perceived complexity and the resulting takeover quality. In a driving simulator study, participants are requested to take over vehicle control in an uncritical situation. Familiarity and objective complexity are varied by the number of surrounding vehicles and scenario repetitions. Subjective complexity is measured using the NASA-TLX; the takeover quality is gathered using the take-over controllability rating (TOC-Rating). The statistical evaluation results show that the parameters significantly influence the takeover quality. This is an important finding for the design of cognitive assistance systems for future highly automated and intelligent vehicles.

Keywords: highly automated driving; HAD; takeover; conditional automation; intelligent vehicles; objective complexity; subjective complexity; familiarity; cognitive assistance; takeover quality

---

1. Introduction

Within recent years, human factors have become an important research topic in automating driving [1]. Approaching the Level 3 of automation [2], the driver may shift attention to a non driving related task (NDRT) during the automated drive. Still, the driver remains as fallback if the automation requests a takeover (TOR; [2]). Most takeover requests in Level 3 highly automated driving [2] will be non-critical [3], giving the driver sufficient comfortable transition time [4]. The focus in this study lies on non-critical takeover situations in different scenarios and the resulting takeover quality. In contrast to critical takeover situations, where drivers abbreviate the takeover process, the driver has enough time to properly perceive the driving environment before performing a maneuver. During the automated mode, the driver can engage into a non-driving related task. The takeover is a complex task. As soon as a TOR is triggered, the driver has to shift the attention back to the driving environment, perceive the surrounding traffic environment and take over the driving task. Hands and feet have to be relocated, situation awareness regained and the driving task has to be executed [5,6].

In addition the in-vehicle environment has to be perceived and filtered for relevant information. All these processes happen after the driver has been out-of the loop. In a small amount of time and a dynamic environment, these are several cognitive and motoric processes that have to happen in a very small amount of time. It is thus important to investigate aspects that affect a safe and comfortable takeover. In this paper, four relevant factors that influence the takeover quality are examined. In the following, the four factors are described separately and distinguished. Still, they are not independent from each other. First, the takeover process is influenced by the complexity of the surrounding traffic environment that can be defined as objective complexity. The objective complexity mainly varies in its amount of relevant objects in the surrounding environment. However, other factors, such as weather conditions, road structure and relative speed can also add up to objective complexity. Especially when taking over the driving task, the objective complexity can impact the quality of the takeover. Different studies [7,8] found that high traffic density leads to a reduced takeover quality when a lane change is required. A reason for this is that the choice of lane change is more complex than just braking as vehicles on the other lanes have to be perceived and time gaps and relative speeds estimated. Second, besides the objective complexity, individual differences have to be taken into account [8]. Not only the traffic situation but also the current state of the individual driver (e.g., stress level, vigilance, workload of non-driving related task) may differ in every takeover situation. This is called subjective complexity. The subjective complexity is task- and resource-dependent and describes an individuals' subjective perception of complexity in a certain traffic situation [9]. Depending on the current attentional state of the driver, the perception of complex situations can vary. While one driver might be familiar and thus very comfortable with high traffic density and rate complexity of the situation as low, another driver might perceive the situation as more complex. Third, such an individual perception of complexity is influenced by the familiarity. Due to common driving routes of individual drivers, the familiarity with roads and therefore traffic situations (traffic jam, urban roads, villages etc.) varies. Reference [10] show that the overall response time is significantly lower for drivers who are familiar with the system. In unfamiliar situations, drivers thus have higher response times. This is highly important when dealing with safety aspects for takeover situations, as the takeover quality can be enhanced when lower reaction times are needed in familiar situations. Fourth, stable driver variables, such as driving style, driving frequency, driving routes and driving duration have an impact on the takeover quality. To improve the takeover quality, cognitive assistance systems can support the driver during a takeover. By integrating information about the surrounding traffic environment (objective complexity), the current state of the driver (subjective complexity), the customary traffic situations of individual drivers (familiarity) and stable driver variables, such as the driving style, the HMI as well as vehicle dynamics can be adapted. In a situation with high objective complexity and an unfamiliar driver, who perceives the situation as very complex, only relevant and supportive information would be presented to the driver (e.g., projection of best maneuver trajectory) and the automation would hand over the driving task gradually (e.g., handing over the steering but keeping adaptive cruise control activated). In highly familiar situations with low complexity, additional information, such as a radio channel, playing the favorite song or the time schedule of the next appointment could be presented to the driver to keep vigilance low. As it is already shown that the drivers' familiarity with a situation and the objective complexity of the current traffic situation influence the subjective complexity [9,11,12], this study investigates the impact of the situational variables familiarity, objective complexity and subjective complexity on the takeover quality. Furthermore, stable individual variables are integrated. All variables are related to each other in different ways. Figure 1 represents the relationships that are investigated in the present study. Based on this, cognitive assistance systems can be developed to support individual drivers accordingly. The following hypotheses are examined in this study:

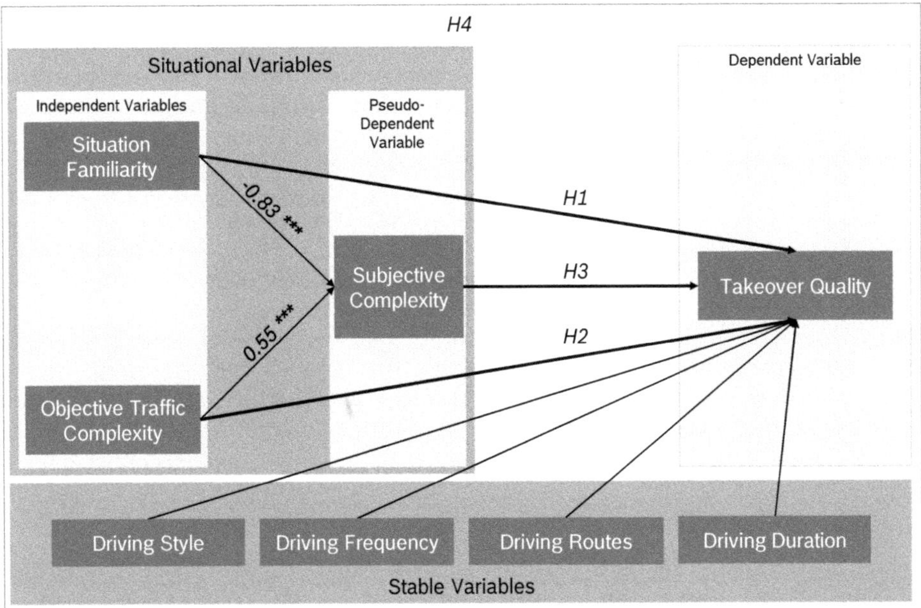

Figure 1. Hypothesised relationships between situational variables, stable variables and the takeover quality. The impact on subjective complexity as shown in [9].

Hypothesis 1. *Higher familiarity with the situation is related to increased quality of a takeover.*

Hypothesis 2. *Higher objective complexity is related to a decreased quality of the takeover.*

Hypothesis 3. *Higher subjective complexity is related to a decreased quality of the takeover.*

Hypothesis 4. *Situational and stable driver variables (driving style, driving frequency, driving routes and driving duration) together can best explain variance in takeover quality.*

2. Methods

To rate the takeover quality, this study evaluates videos of a driving simulator study. The driving simulator consists of six monitors that create a 360° surround view and a moveable driving unit to create a more realistic driving simulation. Six different traffic scenarios are built using the driving simulation SILAB [13]. Participants are tested in a controlled environment to enable measurements under exactly the same traffic conditions. A ten minutes learning session prior to the study is included for participants to get acquainted with simulator dynamics, notifications and the takeover itself. The implementation of the study is approved by the ethics committee of the TU Berlin in April 2019 and Robert Bosch GmbH.

2.1. Study Design

The study includes six scenarios with a different amount of relevant vehicles in the surrounding traffic environment (Section 2.3.1). In three blocks, each scenario is repeated once per block in randomized order. Overall, participants took over the driving task 18 times after an automated drive. Depending on each participant the global study duration lies between 90 and 120 min. After the mandatory documents, participants are theoretically instructed into the study (20–30 min). This is followed by a test drive in which participants get used to the simulator (5 min). Their main task is to drive onto the highway (starting from a parking lot) and onto the center lane, where they turn on the automation as soon as it is available. During the automated drive, they are instructed to play a quiz on a mounted tablet next to the center console until a takeover request is triggered. Each automated drive lasts around 2 min. As soon as the takeover is triggered, participants are instructed to immediately stop the quiz and take over. The takeover request is always triggered when the ego vehicle is driving on the center lane with a speed of 120 km/h. Participants are instructed to take over the driving task using the levers, and keep the speed at ca. 120 km/h. Each scenario triggers a certain maneuver that is the best solution in the given situation. Depending on the traffic situation (speed and position of relevant vehicles), participants should stick to the obligation to drive on the right and try not to break or accelerate enormously. Due to this, always one maneuver is most useful (right when the right lane is free; follow when the right is occupied and the leading vehicle faster or at the same speed; left when the right is occupied and the leading vehicle certainly slower than the ego vehicle). As soon as participants take an action decision the corresponding decision has to be indicated aloud. After each takeover, participants drive onto a parking lot to answer a rating sheet for subjective complexity (NASA-TLX; Section 2.3.3). From the parking lot, the next scenario starts as soon as participants finish the rating sheet. Depending on the time participants took to answer the rating sheet, each scenario lasts three to five minutes.

2.2. Participants

The simulator study took place in May and April 2019 after a successful pre-testing. Statistical evaluations base on $N = 20$ (13 male, 7 female) participants with a mean age of $M = 26.2$ years ($SD = 2.69$) who took part in the study. Most participants drive on average 30 min on a daily basis. They drive mostly on highways and indicate a moderate driving style (Figure 2).

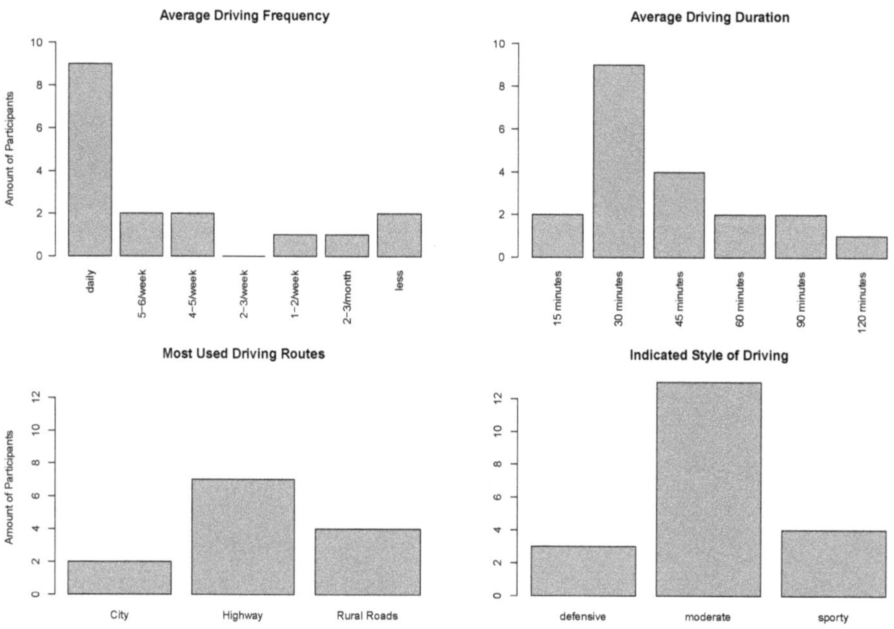

Figure 2. Distributions of driving statistics of the participants ($N = 20$).

2.3. Variables and Measurements

The study is designed to measure four main variables that are important for the takeover in highly automated driving. The connection between those variables is depicted in Figure 1. Variables and measurement methods are described in detail below.

2.3.1. Objective Complexity

The objective complexity is an independent variable (Figure 1) and based on the amount of relevant vehicles in the traffic environment. A vehicle is defined as relevant when it has a direct impact on the ego vehicle. Such a direct impact is either the necessity to react, the reason for a maneuver or a safety critical vehicle that has to be regarded during a maneuver (e.g., overtaking vehicles during a lane change to the left). Three different maneuver options are set up in the traffic simulation. The takeover is always triggered when the ego vehicle is in the highly automated mode on the center lane. Maneuver options are thus a lane change to the left, a lane change to the right or car following. Based on the obligation to drive on the right, the traffic environment is set up to trigger all three maneuvers. For every maneuver a complex and an easy traffic scenario exists. This results in overall six different scenarios that vary in their complexity based on the amount of vehicles relevant for the maneuver (0, 1, 2, 3, 6; Figure 3). Two scenarios have two relevant vehicles in the surrounding traffic environment that are similarly integrated into statistical analysis.

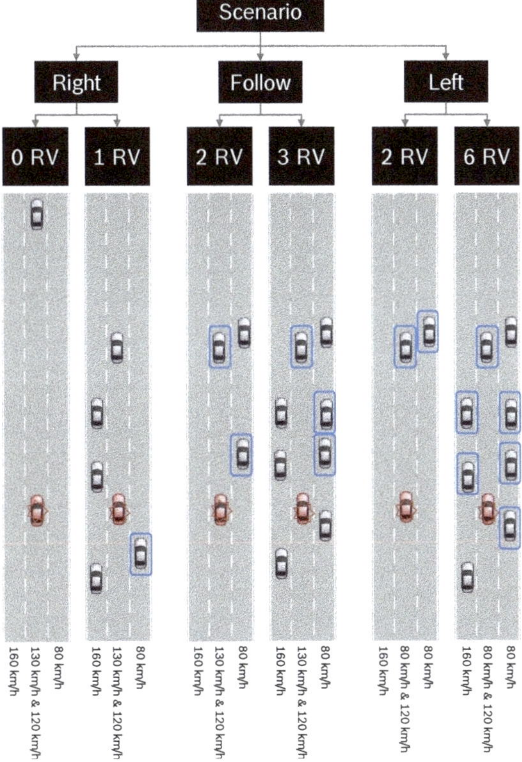

Figure 3. Traffic scenarios during the takeover request. Blue squares mark relevant vehicles in the given scenario situation, the red star marks the ego vehicle.

2.3.2. Familiarity

The second independent variable is the familiarity with a certain traffic situation (Figure 1). It is implemented by a repetition of the scenarios. Each scenario is represented three times for each participant in a randomized order. Therefore, the habituation to general traffic situations rises with repeated exposure.

2.3.3. Subjective Complexity

Subjective complexity is not a direct independent variable as it is not manipulated throughout the experiment. It indicates how complex participants perceive the scenario (individual perception of complexity in terms of "has this been a complex environment **for you**"). It is influenced by the objective complexity and the familiarity (Figure 1; [9]). To assess the subjective complexity, the multidimensional rating sheet NASA-Task Load IndeX /NASA-TLX; [14]) is used after each takeover. Originally the NASA-TLX is a rating scale in which information about magnitude and sources of six workload-related factors are combined to derive an estimate of workload. Due to its six sub-scales, the questionnaire is the most suitable to measure subjective complexity in takeover situations. On a 20-point likert scale, six different sub-scales are rated. The six sub-scales measure mental demand, physical demand, temporal demand, performance, effort and frustration. A weighting of the items as in [14] has been criticized in the past [15]. Reference [16] states that without the weighting of the scales a better differentiation and higher reliability can be achieved. Furthermore, it is stated that the weighting of the scales provides little informative value [17]. Another shortcoming of the weighting is the aspect of time that is additionally needed for the weighting. Based on this, the weighting is not used in this study. Participants are instructed to rate the complexity of the situation using the NASA-TLX after every trial, resulting in overall 18 ratings (six scenarios, three times each).

2.3.4. Takeover Quality

The takeover quality is the dependent variable (Figure 1). Both complexities and the familiarity are assumed to influence the takeover quality. The quality of the takeover is rated using the take-over controllability rating (TOC; [18]). The TOC is a procedure for an assessment of control transitions from automated to manual driving. It provides a standardized rating scheme on a scale from one to ten. Furthermore, it allows the integration of different aspects of driving performance during control transitions into a global measure when evaluating video material of a driving situation. The sub-scales of the TOC include braking response, longitudinal vehicle control, lateral vehicle control, lane change/lane choices, securing/communication, vehicle/system operation and the facial expression of the driver. The last sub-scale (facial expression of the driver) is not rated in this study as the video material does not include the face of the driver [18]. The sub-scales are rated on a 10-point scale. A perfect quality is rated with one. Values of two or three indicate imprecision. Those include jerky steering movement or imprecise lane keeping on the sub-scale of lateral vehicle control, unnecessary/wrong use of indicator on the sub-scale securing/communication, imprecision for vehicle/system operation and visible emotions on the sub-scale facial expression of the driver. Driving errors are rated between four and six, depending on the strength of the error. The following items indicate errors: too strong, too weak, too late, missing (braking response), safety distance too low, inadequate speed (longitudinal vehicle control), safety-distance too low, strong oscillation, crossing lane markings (lateral vehicle control), hesitant/interrupted, too late, missing, wrong lane (lane change/lane choices), missing/too late use of indicator, missing/too late control glance (securing/communication) and problems (vehicle/system operation). Endangerment is rated between seven and nine, including endangerment of others and self-endangerment over all the sub-scales. In cases of non-controllable events, the takeover is rated with a ten, including collision, lane departure/leaving road or loss of vehicle control over all the sub-scales [18]. Low values indicate a faultless takeover (= 1) and high quality. Higher values on the other hand indicate a bad quality of the takeover (10 = uncontrolled).

3. Results

Regression analysis is used to examine the influence of the independent variables on the dependent variable takeover quality. Residual vs. fitted, normal Q-Q, scale-location and residual vs. leverage plots are used to test on the model, normal distribution, homoscedasticity and outliers. To test on multicollinearity, the variance inflation factor is used. Mediation and moderation effects are tested as well, but no significant effects are found.

3.1. The Impact of Familiarity on Takeover Quality (H1)

To evaluate the impact that the familiarity with a traffic scenario has on the takeover quality, regression analysis is used. Results show that with a rise in familiarity, the quality of the takeover significantly improves ($\beta = -0.24, R^2 = 0.01, t(311) = -2, p < 0.05$; Figure 4, right). The slope of the regression is with -0.24 not very high and only one percent of variance in the takeover quality can be explained by familiarity. This shows that familiarity has a significant impact on the takeover quality, but only a small one. It has to be stated though that all participants are regular highway drivers. Hence, the familiarity may have been high already.

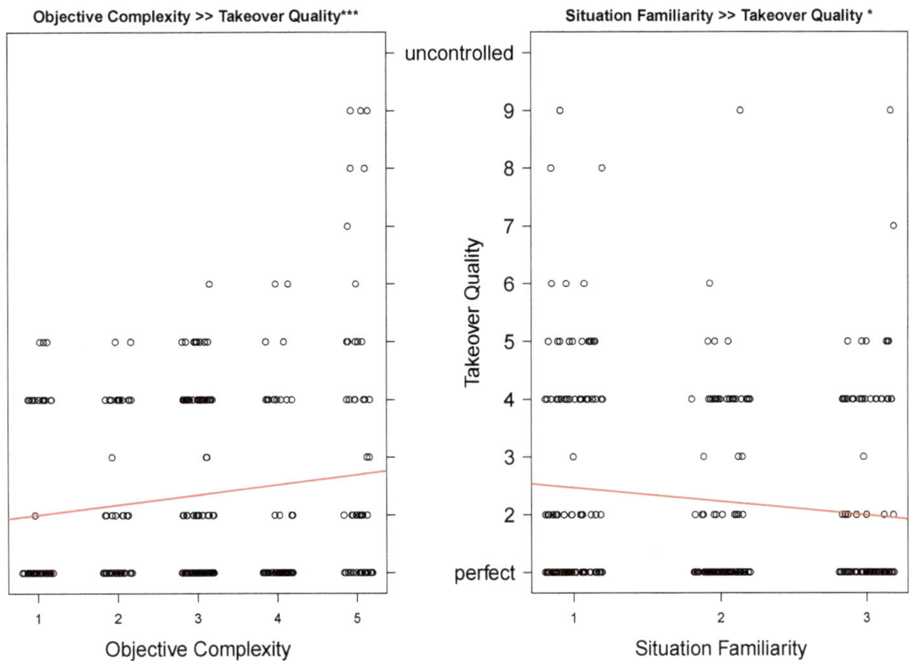

Figure 4. The relation between takeover quality and the objective complexity as relevant vehicles in the surroundingtraffic environment (**left**) and between takeover quality and situation familiarity (**right**). Red lines indicate the regression line (significance codes: 0 '***' 0.001 '**' 0.01 '*' 0.05).

3.2. The Impact of Objective Complexity on Takeover Quality (H2)

Additionally, the takeover quality is significantly influenced by the objective traffic complexity (Figure 4, left). With more relevant vehicles in the surrounding traffic environment, the takeover quality becomes worse. In scenarios that have a bad TOC rating, drivers do not hold enough safety distance, changed the lane very hesitant and interrupted, did not use the indicator or did not do the control glance. In cases where the low safety distance could have lead to a collision in real traffic, endangerment of self and others is rated. Wrong decisions did not influence the takeover quality when the maneuver was executed perfectly. Results show that the slope of the regression is 0.17. Three percent of variance can be explained by the amount of relevant vehicles in the surrounding traffic environment ($\beta = 0.17, R^2 = 0.03, t(311) = 3.44, p < 0.001$). The small amount of variance that can be explained can again be due to the participants driving history. As all drivers are used to highway situations where the objective complexity is usually high, the impact might be reduced due to the increased familiarity. Furthermore, other aspects that add up to objective complexity (e.g., traffic signs) may also play an important role.

3.3. The Impact of Subjective Complexity on Takeover Quality (H3)

The subjective complexity measures how complex each individual perceives the situation. It is significantly influenced by the objective complexity of the environment ($\beta = 0.55, p < 0.001$) and the familiarity with the situation ($\beta = -0.83, p < 0.001$; Figure 1; [9]). In addition, the aggregated subjective complexity has a significant impact on the takeover quality ($\beta = 0.07, p < 0.05$). A driver who perceives a situation as highly complex has a worse quality of the takeover (Figure 5). Although the impact is significant, only one percent of variance can be explained by the aggregated subjective complexity ($R^2 = 0.01, t(311) = 2.33, p < 0.05$). Subjective complexity consists of the six different sub-scales mental demand, physical demand, temporal demand, performance, effort and frustration. Mental demand and physical demand do not influence the takeover quality significantly. However, with a rise in temporal demand, the takeover quality decreases significantly ($\beta = 0.1, t(306) = 2.26, p < 0.05$). In addition, the takeover quality decreases with a rise in frustration ($\beta = 0.11, t(306) = 2.86, p < 0.01$). Surprisingly, with a rise in the perceived performance, the actual takeover quality also decreases ($\beta = 0.08, t(306) = 2.54, p < 0.05$). Furthermore, the effort has a positive effect on the takeover quality ($\beta = -0.15, t(306) = -4.36, p < 0.001$). Multiple linear regression analysis of the sub-scales can explain ten percent of variance in takeover quality ($R^2 = 0.1$; Figure 5). Figure 5 shows that many scores lie on the fourth marker. In the TOC rating, driving errors are rated between four and six. After taking over in this study, a lot of drivers make driving errors. These errors are mostly not enough distance, too strong braking, a missing use of indicators or a missing control glance. As these errors are not severe (e.g., low distance but no cutting in on other vehicles) in these cases, the lowest driving error rating is chosen.

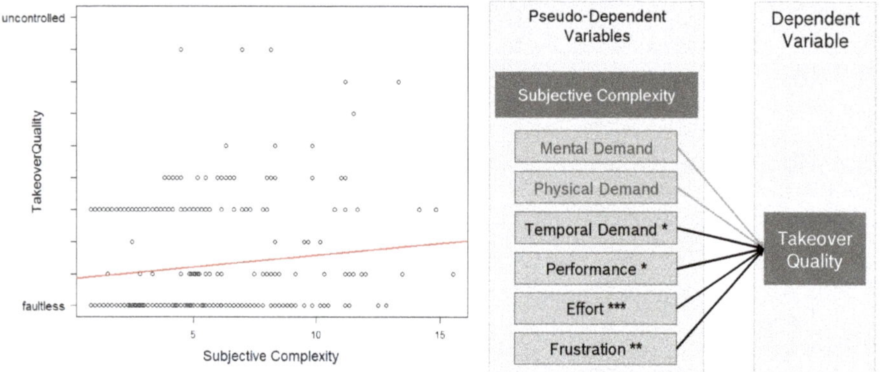

Figure 5. The relation between subjective complexity (left), its sub-scales (right) and the takeover quality. Red lines indicate the regression line (significance codes: 0 '***' 0.001 '**' 0.01 '*' 0.05).

3.4. Multiple Regression Analysis on Takeover Quality Including Stable Driver Variables (H4)

Separately, the variables show significant relationships, but the amount of variance in takeover quality that can be explained is not high. To estimate the impact of the combination of the variables, multiple regression analysis is used (Figure 6). Results show that a combination of stable (e.g., driving style) and situational variables (e.g., objective complexity) increases the amount of variance in takeover quality that can be explained to 58 percent. The stable variables that significantly influence takeover quality are indicated driving style, average driving frequency, most used driving routes and average driving duration. The takeover quality decreases with a more defensive driving style ($\beta = 0.77, t(183) = 5.85, p < 0.001$), less driving frequency ($\beta = -0.41, t(183) = -8.03, p < 0.001$) and longer average driving duration ($\beta = 0.44, t(183) = 4.77, p < 0.01$). More frequent highway usage is related to a better takeover quality ($\beta = 1.17, t(183) = 8.4, p < 0.001$). Situation familiarity is not significant in the multiple linear regression anymore. Similarly, the objective complexity is only significant on a .1 level ($\beta = 0.09, t(183) = 1.95, p < 0.1$). The sub-scales temporal demand, effort and frustration from subjective complexity add to the multiple linear regression. The higher temporal demand ($\beta = 0.09, t(183) = 3.16, p < 0.01$) and frustration ($\beta = 0.1, t(183) = 3.19, p < 0.01$), the lower is the resulting takeover quality. The more effort is spent during a takeover on the other hand, the better is the resulting quality ($\beta = -0.14, t(183) = -4.67, p < 0.001$). In contrast to the simple linear regressions, multiple linear regression shows that the combination of the above mentioned variables give a better understanding on how the variables influence takeover quality (Figure 6). Regression results can be used to compute predictions of takeover quality, depending on the input data that is available.

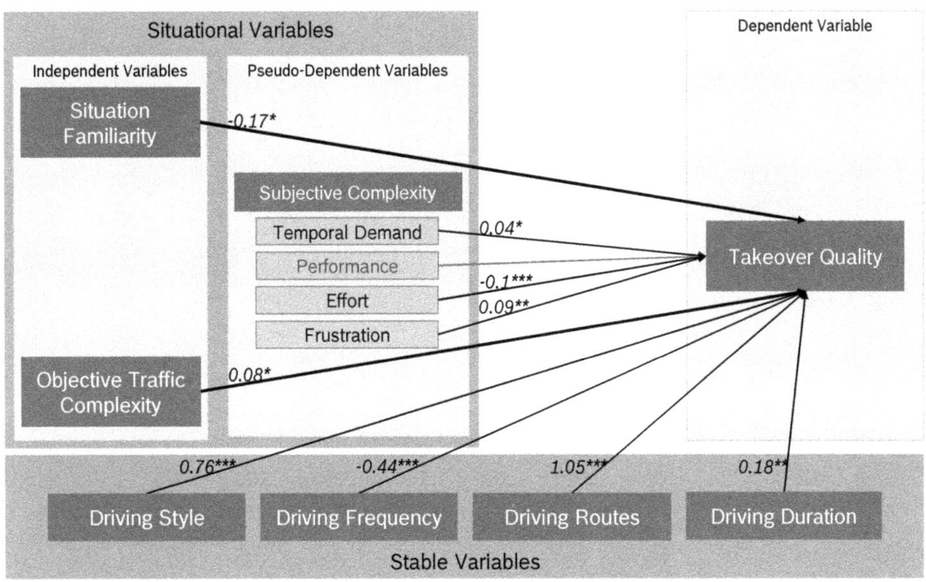

Figure 6. Multiple linear regression results for stable and situational variables on takeover quality. β coefficients indicate the slope of the relationship in the multiple regression (significance codes: 0 '***' 0.001 '**' 0.01 '*' 0.05).

4. Discussion

Results show that a combination of stable and situational variables can be used to explain 58 percent of variance in takeover quality. This new finding is important for the development of highly automated driving. Depending on the variables that can be assessed, a prediction of the takeover quality can now be made and cognitive assistance systems for highly automated driving adapted accordingly. In a user profile for example, the stable driver characteristics can be stored and used for predictions. Based on previous rides, the profile can adapt and store information about the drivers familiarity with certain situations. In combination with that, sensors of highly automated vehicles are able to provide information about the objective complexity of the current traffic environment. In contrast to these variables, measuring subjective complexity is more challenging. To integrate subjective complexity measurements into such a system, a faster and more easily manageable measurement method than the NASA-TLX rating sheet is needed. A way to measure subjective complexity is via eye-tracking (e.g., saccade distance [19], fixation times [20]) or physiological data (e.g., heart rate [21], skin conductance [22]). However, eye-tracking has to be supported in the corresponding vehicle or the driver wears a smart-watch featuring health tracking. Considering the current trend, these two measurement techniques are very likely. By integrating eye-tracking or physiological data, information about the current subjective complexity can be collected. In combination with measurements of the other situational and stable variables, good predictions about the current situation and the drivers state can be made. Based on this, cognitive assistance can support the driver during a takeover situation. Vehicle dynamics and the HMI can be adapted to increase the takeover quality. The results of the study already provide a very good basis for variables that are relevant for the takeover quality. For future research, it is important to consider further variables that might be important. Other distracting objects in the environment that are not vehicles (e.g., traffic signs, roadside environment), current stress level, vigilance and other variables are important to consider in future research. Furthermore, investigation in eye-tracking and physiological measurement methods to capture subjective complexity is important. If these methods are able to measure subjective complexity validly, a next step towards cognitive assistance systems that can be adapted based on the needs of the individual on hand is made.

5. Conclusions

In sum, it can be shown that already 58 percent of variance in takeover quality can be explained by the observed variables of this study. Those are the stable variables driving style, driving frequency, driving routes and driving duration as well as aspects of the situational variable subjective complexity. Objective complexity and familiarity did not become significant in the multiple regression analysis, but show a significant impact when taken separately. In future research it is thus still important to consider these variables. Stable variables can easily be stored in a user profile. Situational variables on the other hand have to be updated and integrated permanently. Different measurement methods have to be used and their output combined to validly display situational variables. Such a combination could be for example the integration of high traffic density (high objective complexity), a high heart rate or skin conductance level and a low saccade distance (high subjective complexity), low familiarity and a defensive driving style. Based on this combination, cognitive assistance would support with relevant information (e.g., projection of optimal driving trajectory), but suppress irrelevant information (e.g., radio or weather information). In addition, the automation would adapt vehicle parameters, such as decelerating while handing over or handing over step by step (e.g., first lateral dynamics—steering, second longitudinal dynamics—acceleration and deceleration). This process has to be very fast as takeover times are short and cognitive assistance has to be given as soon as possible. This paper gives an important selection of relevant variables that influence takeover quality. Based on this it is important to consider valid and fast measurement methods for situational variables and find further variables that influence the takeover quality. Then, cognitive assistance can be developed, individualized and adapted instantaneously.

Author Contributions: The individual contributions have been distributed as listed in the following (M.S.L.S., K.Z., N.R.): conceptualization, K.Z, N.R. and M.S.L.S.; data curation, M.S.L.S.; formal analysis, M.S.L.S.; investigation, M.S.L.S.; methodology, K.Z., N.R. and M.S.L.S.; project administration, M.S.L.S.; software, M.S.L.S.; supervision, K.Z. and N.R.; visualization, M.S.L.S.; writing—original draft preparation, M.S.L.S.; writing—review and editing, N.R. and M.S.L.S. All authors have read and agreed to the published version of the manuscript.

Funding: This research received no external funding.

Acknowledgments: I wish to acknowledge the help provided by my supervisor Michael Schulz at Robert Bosch GmbH for the support with the simulator, my department at TU Berlin and my second supervisor Klaus Bengler from the TU Munich. This work is part of the public promoted project PAKoS in which the Robert Bosch GmbH participated.

Conflicts of Interest: The authors declare no conflict of interest.

Abbreviations

The following abbreviations are used in this manuscript:

HMI	Human Machine Interface
NASA-TLX	NASA Task Load Index
NDRT	Non driving related task
TOC	Take-over controllability rating
TOR	Takeover request

References

1. Gold, C.; Happee, R.; Bengler, K. Modeling take-over performance in level 3 conditionally automated vehicles. *Accid. Anal. Prev.* **2018**, *116*, 3–13. [CrossRef] [PubMed]
2. SAE-International. Taxonomy and Definitions for Terms Related to On-Road Motor Vehicle Automated Driving Systems. Available online: https://www.sae.org/standards/content/j3016_201806/ (accessed on 18 February 2020).
3. Eriksson, A.; Stanton, N.A. Takeover time in highly automated vehicles: Noncritical transitions to and from manual control. *Hum. Factors* **2017**, *59*, 689–705. [CrossRef] [PubMed]
4. National Highway Traffic Safety Administration. *Preliminary Statement of Policy Concerning Automated Vehicles*; National Highway Traffic Safety Administration: Washington, DC, USA, 2013; pp. 1–14.
5. Kerschbaum, P.; Lorenz, L.; Bengler, K. A transforming steering wheel for highly automated cars. In Proceedings of the 2015 IEEE Intelligent Vehicles Symposium (IV), Seoul, Korea, 28 June–1 July 2015; pp. 1287–1292.
6. Zeeb, K.; Buchner, A.; Schrauf, M. What determines the take-over time? An integrated model approach of driver take-over after automated driving. *Accid. Anal. Prev.* **2015**, *78*, 212–221. [CrossRef] [PubMed]
7. Gold, C.; Körber, M.; Lechner, D.; Bengler, K. Taking over control from highly automated vehicles in complex traffic situations: The role of traffic density. *Hum. Factors* **2016**, *58*, 642–652. [CrossRef] [PubMed]
8. Radlmayr, J.; Gold, C.; Lorenz, L.; Farid, M.; Bengler, K. How Traffic Situations and Non-Driving Related Tasks Affect the Take-Over Quality in Highly Automated Driving. *Sage J.* **2014**, *58*, 2063–2067. [CrossRef]
9. Scharfe, M.; Russwinkel, N. The Individual in the Loop—The Influence of Familiarity and Complexity during a Takeover in Highly Automated Driving. **2020**, in press.
10. Larsson, A.F.; Kircher, K.; Hultgren, J.A. Learning from experience: Familiarity with ACC and responding to a cut-in situation in automated driving. *Transp. Res. Part F Traffic Psychol. Behav.* **2014**, *27*, 229–237. [CrossRef]
11. Paxion, J.; Galy, E.; Berthelon, C. Overload depending on driving experience and situation complexity: Which strategies faced with a pedestrian crossing? *Appl. Ergon.* **2015**, *51*, 343–349. [CrossRef] [PubMed]
12. Haerem, T.; Rau, D. The influence of degree of expertise and objective task complexity on perceived task complexity and performance. *J. Appl. Psychol.* **2007**, *92*, 1320. [CrossRef] [PubMed]
13. WIVW-GmbH. Driving Simulation and SILAB. 2014. Available onlien: https://wivw.de/en/silab (accessed on 7 September 2019).
14. Hart, S.G. *NASA Task load Index (TLX)*; NASA's—Ames Research Center: Mountain View, CA, USA, 1986; Volume 1.
15. Gross, U. Bestimmung Von Schwierigkeitsgraden in Einer Zu Entwickelnden Versuchsumgebung. Master's Thesis, Humboldt Universität zu Berlin, Berlin, Germany, November, 2004.
16. Pfendler, C. *Vergleichende Bewertung der NASA-TLX-Skala und der Zeis-Skala bei der Erfassung von Lernprozessen*; Technische Informationsbibliothek (TIB): Hannover, Germany, 1991.
17. Nygren, T. Psychometric properties of subjective workload measurement techniques: Implications for their use in the assessment of perceived mental workload. *Hum. Factors* **1991**, *33*, 17–33 [CrossRef]
18. Naujoks, F.; Wiedemann, K.; Schömig, N.; Jarosch, O.; Gold, C. Expert-based controllability assessment of control transitions from automated to manual driving. *MethodsX* **2018**, *5*, 579–592. [CrossRef] [PubMed]

19. Hayhoe, M. Advances in relating eye movements and cognition. *Infancy* **2004**, *6*, 267–274. [CrossRef]

20. Recarte, M.A. Nunes, L.M. Effects of verbal and spatial-imagery tasks on eye fixations while driving. *J. Exp. Psychol. Appl.* **2000**, *6*, 31. [CrossRef]

21. Miyaji, M.; Kawanaka, H.; Oguri, K. Driver's cognitive distraction detection using physiological features by the adaboost. In Proceedings of the 12th International IEEE Conference on Intelligent Transportation Systems (ITSC'09), St. Louis, MO, USA, 4–7 October 2009; pp. 1–6.

22. De Winter, J.C.; Happee, R.; Martens, M.H.; Stanton, N.A. Effects of adaptive cruise control and highly automated driving on workload and situation awareness: A review of the empirical evidence. *Transp. Res. Part F Traffic Psychol. Behav.* **2014**, *27*, 196–217. [CrossRef]

 *information*

Article

Repeated Usage of an L3 Motorway Chauffeur: Change of Evaluation and Usage

Barbara Metz *, Johanna Wörle, Michael Hanig, Marcus Schmitt and Aaron Lutz

WIVW GmbH, 97202 Veitshöchheim, Germany; woerle@wivw.de (J.W.); hanig@wivw.de (M.H.); schmitt@wivw.de (M.S.); aaronlutz@gmx.de (A.L.)
* Correspondence: metz@wivw.de

Received: 8 January 2020; Accepted: 14 February 2020; Published: 18 February 2020

Abstract: Most studies on users' perception of highly automated driving functions are focused on first contact/single usage. Nevertheless, it is expected that with repeated usage, acceptance and usage of automated driving functions might change this perception (behavioural adaptation). Changes can occur in drivers' evaluation, in function usage and in drivers' reactions to take-over situations. In a driving simulator study, N = 30 drivers used a level 3 (L3) automated driving function for motorways during six experimental sessions. They were free to activate/deactivate that system as they liked and to spend driving time on self-chosen side tasks. Results already show an increase of experienced trust and safety, together with an increase of time spent on side tasks between the first and fourth sessions. Furthermore, attention directed to the road decreases with growing experience with the system. The results are discussed with regard to the theory of behavioural adaptation. Results indicate that the adaptation of acceptance and usage of the highly automated driving function occurs rather quickly. At the same time, no behavioural adaptation for the reaction to take-over situations could be found.

Keywords: behavioural adaptation; SAE L3 motorway chauffeur; system usage; acceptance; attention; secondary task

1. Introduction

As discussed in the media, vehicle manufacturers plan to introduce (partly-) self-driving cars in the (near) future. According to the classification of the Society of Automobile Engineers (SAE) [1], drivers will be allowed to use the time while the system is active for non-driving related activities (NDRAs) starting from level 3 automation onwards (L3, conditional automation). By definition, in L3, all aspects of the driving task are executed by the automated driving system (ADS). Consequently, with L3 ADS, the role of the driver fundamentally changes when compared to manual driving. Even though the driver remains the fall-back option in the event of system limits, there is no need to monitor the driving environment or the system's performance while the ADS is driving. The driver is allowed to engage in NDRAs, such as browsing the internet or watching movies. However, in the event of a take-over request (TOR) by the system, the driver has to be able to retake control of the vehicle within a certain time frame. Therefore, from the driver's perspective, L3 is the first level of automation where vehicle automation can be experienced as completely self-driving within system boundaries with all the expected benefits.

The H2020 EU-funded project L3Pilot deals with L3/L4 vehicle automation (https://www.l3pilot. eu/). The overall objective of L3Pilot is to test and study the viability of automated driving as a safe and efficient means of transportation and to explore and promote new service concepts to provide inclusive mobility. Besides testing and evaluating current prototype versions of L3/L4 functions in on-road rests, one part of the project deals with the change of drivers' acceptance and usage of L3/L4-systems with repeated usage. Due to the real-world systems still being at a prototype stage, this cannot be done in on-road tests, e.g., due to safety reasons. Instead, a study in a driving simulator is conducted in which drivers have the possibility to use an L3-motorway ADS during several drives. The main goals of this study are to gain insight into the change of user attitudes and trust with repeated experience of an L3-function, to identify user and situational factors that affect driver behaviour and acceptance and to investigate changes in driver behaviour in terms of engagement in non-driving related activities, take-over performance and mode awareness.

1.1. Behavioural Adaptation

With increasing vehicle automation, the role of the driver changes fundamentally. On a technological level vehicle automation is progressing rapidly. Still, there are challenges regarding the interaction between human drivers and automated systems. This includes the impact of automated systems on the driver's mental workload, situation awareness and acceptance of automated driving, as well as trust and reliance issues [2]. Another important aspect that has to be considered is that drivers may change their behaviour due to automation. This phenomenon is referred to as "behavioural adaptation", which is defined as "behaviours which may occur following the introduction of changes to the road–vehicle–user system and which were not intended by the initiators of the change" [3]. It is known that in the past, many road safety measures did not have the expected safety benefit in terms of a decrease in accidents. It is argued that drivers react to a lower safety risk, for instance, by increasing their traffic intensity and increasing their travel speed [4]. The first theory explaining changes in driver behaviour following changes in the vehicle or road infrastructure was the theory of risk homeostasis [5]. The theory assumes that individuals have a target level of accepted risk in driving situations and that they adjust their behaviour such that the perceived level of risk matches their target level of risk. After the introduction of safety-promoting vehicle technology, drivers might increase risky driving behaviour to adjust their perceived level of risk to their original target level. More recent theories shift away from the concept of 'risk perception' as an underlying mechanism. Two more recent models of behavioural adaptation might explain drivers' behaviour when using advanced driver assistance systems (ADAS) or ADSs: The driver-in-control model [6] considers the driver and the vehicle as a unit, the joint driver–vehicle system. The model describes a cycle of intentions, actions and outcomes. The probably most applicable model in the context of automated driving is the qualitative model of behavioural adaptation [7] which is specially designed to explain behavioural adaptation to in-vehicle driver assistance systems. One main factor in this model is trust in the system. Trust is affected by the driver's personality (especially the variables locus of control and sensation-seeking). The model seems highly applicable since there is evidence suggesting that trust is one of the main factors in driver behaviour when using ADSs [8,9].

The manifestation of behavioural adaptation highly depends on system functionality. For instance, it was found that when using adaptive cruise control (ACC), drivers reacted more slowly to a hazardous situation and had a greater deviation from lane position than when driving without ACC [7] and that drivers increased their maximum speed when provided a congestion tail. In addition, drivers' engagement in a secondary task increased when driving with a congestion tail warning system [10].

Compared to the investigation of behavioural adaptation for ADAS, a slightly different approach needs to be chosen for studying behaviour adaptation for ADSs. During automated driving, driving parameters such as speed or steering behaviour do not depend on the driver but rather on the system implementation. Hence, simply comparing these parameters when using the system compared to manual driving is not applicable. Therefore, other indicators need to be defined to measure behavioural changes to highly ADS, e.g., in terms of lane choice and secondary task engagement [11]. It is also reported that drivers visually focus less on the centre of the road when driving in the automated mode.

For ADAS, six more high-level categories of changes in drivers' behaviour are proposed to study behavioural adaptation [12]. Those categories seem relevant when investigating behavioural adaptation to ADSs, too. Behavioural changes are defined in terms of:

- Perceptive changes (seeing, hearing, feeling).
- Cognitive changes (comprehending, interpreting, prioritizing, selecting, deciding).
- Performance changes (driving, system handling, error).
- Driver state changes (attentiveness/awareness, workload, stress, drowsiness).
- Attitudinal changes (acceptance, rejection, overreliance, mistrust).
- Changes in the adaptation to environmental conditions (weather, visibility, etc.).

It can be hypothesized that behavioural changes on different levels are interconnected. An increase in trust in the ADS (attitudinal change—acceptance), for instance, may lead to a higher willingness to engage in secondary activities (cognitive change—prioritizing), which could then lead to a decreased perception of the environment (driver state changes—attentiveness) or a decreasing performance in case of a TOR (performance changes—driving). Such links must be considered in the assessment of behavioural adaptation to ADSs.

1.2. Usage and Evaluation of ADS

One of the major preconditions of acceptance and usage of ADSs is the drivers' trust in the system. If drivers do not trust the automation, they will not use it (*disuse*). On the other hand, if drivers over-rely on the automated system, this might lead to decision errors, for example, in terms of not responding appropriately to takeover requests (TOR) [13]. Increasing acceptance of ADSs can already be found after the first drive. Drivers who have experienced crashes or safety-critical situations report lower trust levels [9]. Trust is therefore closely tied to the perceived reliability of an automated system. If the perceived reliability increases, trust is likely to increase as well.

The acceptance of ADS is also highly related to its perceived usefulness. The perceived usefulness of an ADS for the user might increase along with the increasing automation level. When drivers are not required to monitor the system's performance and are allowed to engage in other activities, they will perceive the system as more useful. Several surveys have been conducted on the NDRAs which drivers want to engage in while driving in the automated mode. The perceived usefulness of the ADS depends on the extent to which drivers are able to perform these activities [14]. NDRAs that drivers would like to engage in include eating, interacting with passengers, phoning, observing the scenery, emailing, etc. [15].

Another relevant aspect arises from the rather passive role of the driver while driving with the ADS: fatigue or sleepiness. Due to the monotony of the situation, while being driven by the car, drivers experience fatigue much earlier than in manual driving and at much higher levels [11,16]. The generation of fatigue during highly automated driving might in extreme cases even cause the driver to fall asleep while driving in automated driving (AD) mode. In a simulator study on the assessment of trust in automation, two participants fell asleep while driving in AD mode [9].

1.3. Change with Repeated Usage

Studies investigating user experiences of ADAS and ADSs mostly assess the drivers' behaviour and attitudes when they first encounter the new technology. In most studies, for practical reasons, only the first 1–2 h of using a new technology is investigated. However, it is likely that after a certain time of using and experiencing the behaviour of the system in various use cases, drivers will adapt their behaviour accordingly. However, changes with repeated usage are assessed very rarely since this is rather complex and expensive.

Theories on behavioural adaptation distinguish different phases: The learning process is crucial for drivers to gain an appropriate understanding of the system's functionality as well as system limits and helps to build an appropriate level of trust. The learning process takes some time and requires an experience of the system in different situations and different environments. Two phases in the learning process are suggested: in the "learning phase", the driver learns how to operate the system, identifies system limits and internalizes the system functionality. The learning phase heavily depends on the way the system is introduced to the driver. In the second stage, the "integration phase", the driver integrates the system into the management of the overall driving task by increasing experience in different situations [17].

When testing ADAS in the AIDE project [17], the focus was on directly observable behavioural changes due to the ADAS, mainly in terms of changes in driving parameters. However, when assessing L3 ADS, the approach must be adapted. Since the vehicle is controlled by the ADS most of the time, changes in human driving behaviour can only be assessed to a limited extent. However, attitudes towards automation can change dramatically over time, for instance when experiencing the system in different traffic situations.

The term 'behavioural adaptation' is said to have an inherent association with time because it suggests that changes in behaviour are a result of being exposed to e.g., a certain ADAS/ADS and experiencing it in different situations [18]. From a methodological point of view, it is therefore crucial not only to consider a single usage of a system but sufficiently long exposure. The question is: How long is long enough to capture behavioural adaptation? For the investigation of ADAS (like ACC or lane departure warnings), a few hours to a few weeks are considered to be short-term usage whereas long-term usage is meant to last at least 6 months [18].

In another approach, five phases of behavioural adaptation to ADAS are distinguished with defined durations [12]:

- First encounter: First day (1–6 h).
- Learning: 3–4 weeks.
- Trust: 1–6 months.
- Adjustment: 6–12 months.
- Readjustment: 1–2 years.

The *First encounter phase* depends greatly on how intuitive and self-explaining the human-machine interface (HMI) is. The *Learning phase* still depends highly on the HMI, especially in terms of required system input. The timely dimension of the learning phases is empirically supported by studies on e.g., electronic speed control [7]. The *Trust phase* is mainly characterized by a shift in the locus of control [19] from the driver to the vehicle. Related problems might be overreliance, passivity and drowsiness. In the phases *Adjustment* and *Readjustment*, drivers adjust their adapted behaviour depending on their experience of (critical) situations and system limitations. It can be expected that trust plays an important role in the behavioural adaptation to ADS and indeed, for the overall acceptance of the system. According to Muir [20], trust depends on the degree of experience with automation and thus can be expected to change over time.

The durations given in the literature for the different phases of behavioural adaptation relate to the time period during which an equipped vehicle is available to the driver and the system can be used. The required period of actual system usage can be expected to be much less. From the literature, it is not known how many hours of driving with an active system or how many occurrences of a certain system intervention/warning is needed to study behavioural adaptation. For our research, it also has to be considered that the phases defined in Martens and Jenssen [12] refer to behavioural adaptation to ADAS, not to ADSs. The learning phase for a system that only intervenes very occasionally can be expected to be much longer than for a system whose behaviour can be experienced continuously by the driver. For ADSs, it can be expected that the learning process is much faster. It seems likely that the phases of behavioural adaptation defined by Martens and Jenssen [12] only give a rough estimate and do not apply for behavioural adaptation to high automation.

One study is known that investigated secondary task engagement during highly automated driving with a repeated usage perspective. Six drivers were invited to undertake five 30-min journeys with a highly automated system in a driving simulator. They were encouraged to use the system just as they would in a real automated vehicle. Participants were asked to bring with them any objects or devices that they would be willing to engage with during the drives. The most common activities during the drives were reading articles or magazines, using mobile devices for social networking activities, web browsing and watching programmes or films on a laptop. Although the study was set up with a focus on repeated usage, no findings on changes in behaviour over time were reported [21].

1.4. Objective

In summary, there is literature that discusses the concept of behavioural adaptation especially with the focus of usage of ADAS. But even for ADAS, studies investigating behavioural adaptation are rare, probably due to the fact that such research is time-consuming and expensive. For L3/L4-ADSs it seems reasonable to assume that behavioural adaptation will have a relevant impact e.g., on function usage as soon as there are functions on the market. Nevertheless, experimental results are still lacking on that topic. The aim of our research is to study behavioural adaptation to an L3 motorway ADS with repeated usage. The focus is on

- *Attitudinal changes*, that is the change of e.g., acceptance and trust measured via questionnaires.
- *Cognitive changes*, that is the change of prioritizing and selecting side tasks, measured via indicators derived from driver behaviour and via questionnaires.
- *Driver state changes*, that is the change of attentiveness/awareness, stress, drowsiness, measured via objective indicators and via questionnaires.
- *Performance changes*, that is the change of system handling in take-over situations measured via objective indicators and via questionnaires.

Due to practical limitations, it is not possible to study the full process of behavioural adaptation where changes are still expected to occur even after several months. Instead, the focus is on the beginning of this process including the first encounter, the learning phase and maybe the beginning of the trust phase. It is expected that during the learning phase there is still some change of behaviour. For the trust phase, a more constant level is expected.

2. Materials and Methods

The study was conducted in the high-fidelity moving base driving simulator of the WIVW GmbH (see Figure 1). The mock-up consisted of a production type BMW 520i. The motion system used six degrees of freedom and could display a linear acceleration up to 5 m/s^2. All vehicle dynamics and noises were displayed realistically. The simulation software was SILAB® Version 6.0 (WIVW GmbH, Veitshöchheim).

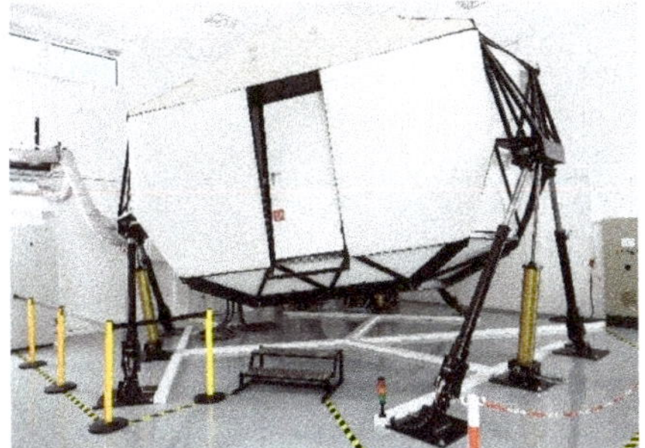

Figure 1. The high-fidelity driving simulator at WIVW.

Drivers were invited to participate in a study on the long-term effects of an L3-motorway ADS (L3ADS). The study consisted of six drives on a motorway during which the L3ADS could be used. The drives took place on six different days. In all drives, the drivers were free to use the L3ADS as they liked, meaning they could activate and deactivate it and attend to NDRAs as they wished. Drivers were instructed that while in the automated mode, they were not required to pay attention to the driving task and they were allowed to engage in other activities. However, when the system issued a TOR, they had to retake the vehicle guidance and were responsible for the driving task. For the description of the system and the responsibility of the driver, the actual wording of §1b of the German Road Transport Law [22] defining the responsibility of the driver when driving with an ADS, was used.

2.1. System Implementation

The study focused on acceptance, evaluation and usage of an L3ADS by ordinary, non-professional drivers. Therefore, participants tested a simulated L3ADS that worked realistically in the motorway scenarios included in the six test drives. The system was implemented based on the descriptions of L3 motorway systems to be tested in the on-road tests of L3Pilot [23]. It was designed to work in the driving scenarios tested in the study by using controllers already available in SILAB®.

The L3ADS had an operational design domain (ODD) that is similar to the ODD of highway ADSs tested in the on-road experiments in L3Pilot [23]:

- The implemented L3ADS had a speed range of 0 to 130 km/h. The system adopted the driven speed to the surrounding traffic as well as to speed limits along the road. This means that the system set the maximum speed situationally adapted based on the current speed limit and in sections with no speed limit, the system kept a maximum speed of 130 km/h. In case a lead vehicle was present, the system obtained a safe distance to the lead vehicle and adjusted speed accordingly. The regulation of speed and distance was based on a standard ACC-controller implemented in SILAB®.
- The system was able to execute lane changes automatically. Lane changes to the left were initiated when a slower vehicle was detected on the own lane and the adjacent lane on the left was free. The vehicle changed lanes back as soon as slower vehicles were passed and the lane to the right was free again. For decisions on lane changes, a simple controller was used that was tuned in a way that in the implemented experimental drives the behaviour of the ADS seemed reasonable and felt smooth. The trajectory during a lane change was defined such that the lane change itself felt smooth; the trajectory was not situationally adapted.
- The following situations were outside the ODD and therefore led to a TOR: highway exits and entries, construction sites, adverse weather conditions (i.e., heavy rain) and missing lane markings.
- All TORs were issued with a time budget of 15 s (based on take-over times of functions tested in on-road tests in L3Pilot [23]). Although not mandatory for L3-functions, a safe stop manoeuvre was performed in case the driver did not take control back during the take-over time.

2.2. Test Scenarios

Four of the six experimental drives had a duration of 30–35 min (drive 1 to drive 4 in Table 1). In those drives, it was taken care that the driving environment was not too boring and that traffic density and driving situations changed within and between the drives. As can be seen in Table 1, all four drives contained sections with low traffic density and changing speed limits (in three of them also unlimited), in three of the four drives, sections lasting between five and ten minutes with traffic jams occurred. The number of TORs varied between two and five per drive. Reasons for TORs were missing lane markings, approaches to construction sites, highway intersections and at the end of every drive the approach to the exit. Table 2 gives more details on the takeover scenarios. All scenarios were defined in a way that included common non-critical driving situations. Very critical, unusual or rare scenarios were avoided because the focus of the study was on simulating potential everyday usage of the L3ADS.

Table 1. Content of the six experimental drives. Results from the drives in bold are included in this paper. The two other sessions are excluded to avoid confusion of the effects of repeated usage and driver state.

Drive	Driving in ODD	Driving Outside ODD	Reasons for Take-Over Requests
Drive 1	section with low traffic density and changing speed limit traffic jam	On parking area In motorway junction	1 × before highway intersection 1 × before exit 1 × bad lane markings
Drive 2	section with low traffic density and changing speed limit traffic jam	On parking area In construction site	1 × construction site 1 × exit 2 × bad lane markings
Drive 3	section with low traffic density and changing speed limit traffic jam	On parking area	1 × exit 1 × before moving roadworks (+traffic)
Drive 4	section with low traffic density and changing speed limit	On parking area In motorway junction In construction site	1 × construction site 1 × highway junction 1 × before exit 2 × bad lane markings
Drive 5	section with low traffic density and changing speed limit	On parking area In heavy rain	1 × exit 1 × moving roadworks 1 × heavy rain
Drive 6	section with low traffic density and changing speed limit	On parking area In heavy rain	1 × exit 1 × moving roadworks 1 × heavy rain

Table 2. Description of the analysed takeover scenarios.

Takeover Scenario	Drivers' Task	Occurrence in Drives	Speed Before TOR	During Situation
Highway intersection	Take driving task back Follow direction signs through intersection Two-lane changes are required for right way	1,3	120 km/h	Situationally adapted through intersection
Exit	Take driving task back Take exit lane Stop vehicle in car park	1,2,3,4	120 km/h	Situationally adapted on exit lane
Bad lane markings	Take driving task back Stay on lane through the section with bad lane marking (no traffic, no curvature)	1,2,4	120 km/h	80 km/h
Construction site	Take driving task back Slow down Drive through construction site	2,4	120 km/h	80 km/h
Roadworks + traffic	Take driving task back Check driving environment Change unto unblocked lane while taking rear traffic into account	3	120 km/h	120 km/h

The two other drives were longer (90 min) and more monotonous, one of them taking place at 6 am in the morning. Those two drives were included to study specific hypotheses on driver state which will be presented elsewhere (in preparation). The order of the drives was varied between participants to avoid sequence effects. The two monotonous drives always took place in the third and the fifth session.

2.3. Data Logging

Most methods used were defined in the common methodological approach within L3Pilot and that will also be used for the on-road tests of L3Pilot [24]. This specifically relates to the questionnaire developed within L3Pilot which assesses aspects like acceptance, perceived safety, trust, workload, etc. The questionnaire was designed for on-road tests where drivers have the opportunity to test an

L3/L4-system once. It consisted of a pre-drive questionnaire in which demographic information, as well as pre-experiences with in-vehicle systems, were collected. The post-drive questionnaire assessed the evaluation of the tested system through a mixture of standardized items (e.g., [25]) and items specifically tailored to the project questions of L3Pilot. The specifically developed questionnaire items mostly consisted of a statement with which the participants could agree or disagree on a 5-point scale (for example see Figure 2a). In the present study, the pre-drive questionnaire was administered once at the beginning of the first session. The full post-drive questionnaire was filled in after the 1st and the 6th session, a shortened version was used after session two to five. Directly after every TOR, drivers rated the criticality of the previous driving situation on a ten-point scale, ranging from harmless to uncontrollable with intermediate steps of unpleasant and dangerous (based on [26], see Figure 2b). The rating related to the TOR itself and to the following driving scenario (e.g., drive through a construction site).

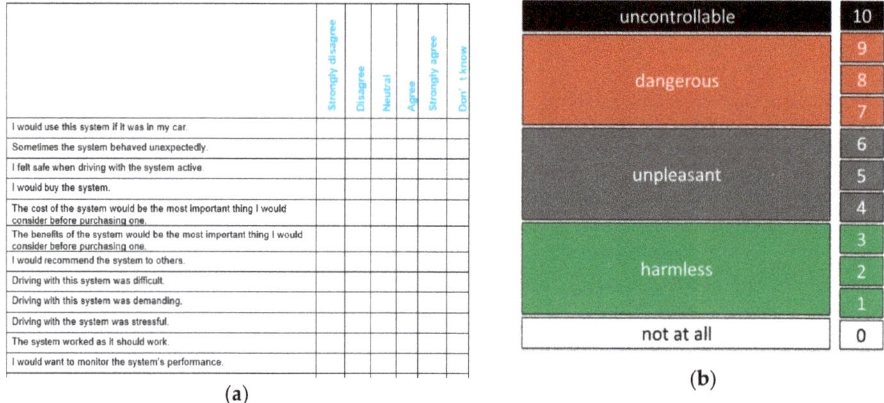

Figure 2. Example of questionnaire items used to assess concepts like acceptance and trust (**a**) and scale used to asses experienced criticality for takeover scenarios (**b**).

Furthermore, during all sessions a variety of objective parameters was logged:

- Signals from the driving simulator software that cover the areas of vehicle dynamics (v, ax, ay), state of the L3-system (TORs, system status), vehicle handling (brake pedal position, steering angle, hands-on detection) and vehicle environment (distance to other vehicles, lane position).
- Continuous video recording of the driver and the driving scenery.
- Continuous coding by the experimenter whether participants were engaged in NDRAs, whether the NDRA involved the hands (manual distraction, e.g., through browsing on a smartphone, holding food) and whether drivers closed their eyes for a longer time.
- Gaze and head direction, as well as eyelid-opening level, were logged with the 3-camera gaze tracker Smart Eye Pro® (SmartEye; Gothenburg, Sweden).

2.4. Procedure

In the introductory session, drivers were informed about the schedule for their test drives. Before each session, they knew the length of the oncoming trip and they were informed that they were free to prepare for the drive as they liked to. This meant for instance that they could bring something to read, something to eat or prepare other potential side tasks to fill the time of the automated drive. Besides being free to attend to side tasks as they liked while in automated mode, drivers were also free to use the system as they liked. This meant that they were allowed to override the system or deactivate it in situations where they preferred to drive manually. Table 3 gives an overview of the 6 experimental sessions.

Table 3. Overview of the content of the six sessions of the experiment. Results from the sessions in bold are included in this paper. The two other sessions are excluded to avoid a confusion of the effects of repeated usage and driver state.

Session	Content
Session 1—introductory session 90 min	Information on experiment and planned schedule Informed Consent Handing out of Pilot site questionnaire part 1 Introductory drive (10 min) Drive 1 (30 min) Post drive questionnaire (full version)
Session 2 45 min	Short pre-drive questionnaire Drive 2, Drive 3 or Drive 4 (30 min) Post drive questionnaire (short version)
Session 3 2 h 30 min	Short pre-drive questionnaire Drive 5 or Drive 6 (2 h) Post drive questionnaire (short version)
Session 4 45 min	Short pre-drive questionnaire Drive 2, Drive 3 or Drive 4 (30 min) Post drive questionnaire (short version)
Session 5 2 h 30 min	Short pre-drive questionnaire Drive 5 or Drive 6 (2 h) Post drive questionnaire (short version)
Session 6 90 min	Short pre-drive questionnaire Drive 2, Drive 3 or Drive 4 (30 min) Post drive questionnaire (full version)

2.5. Sample

The study was conducted with N = 31 drivers (mean age = 37, sd = 11.75); 58% of the sample were male. Nearly 70% of the sample have had their driving license for at least 10 years. In the pre-questionnaire, participants also stated on average that for them driving on a highway is neither difficult nor stressful, but that they also do not enjoy driving on motorways; 42% of the sample stated that they drive on a highway at least 1–2 times per week; 10% that they are stuck in traffic jams on highway with the same frequency of 1–2 times per week. All participants had completed an extensive training for the driving simulator before participating in the study in order to avoid learning effects and simulator sickness.

2.6. Data Analysis

To investigate drivers' performance in takeover scenarios, two approaches were chosen:

1. Reaction times are calculated that are defined as the duration between the start of the TOR and the first time point the analysed driver reaction was observed (eyes on road, hands on the steering wheel and deactivation of the AD).
2. Expert rating of takeover performance based on the video. For that the take-over controllability (TOC) rating was performed ([27], for more details see https://toc-rating.de/en/) which defines a standardized procedure for the evaluation of take-over situations. Relevant dimensions for evaluation are provided with defined criteria for the different rating categories. In the end, a final rating is given on a 10-point scale which corresponds to the categories shown in Table 4. Besides the overall rating, detailed information on the observed error types is provided by the method.

Reaction times and TOC-rating were only analysed for situations where drivers took control back after a takeover request was issued by the L3ADS.

Table 4. Scale for take-over controllability (TOC)-rating which was used to evaluate takeover performance.

1	2	3	4	5	6	7	8	9	10
Perfect	Good with imperfections		Reduced performance with driving errors			Endangerment, critical situations			Uncontrollable

For all questionnaire items, general agreement or disagreement was evaluated with single t-tests against zero (meaning neutral on the scale). Results are reported for the evaluation of the system after the first and after the sixth session.

To investigate the behavioural adaptations with repeated usage, the changes over experimental sessions were analysed for the following parameters:

- Agreement with various statements regarding the evaluation of the L3ADS, derived from post-drive questionnaires (see [24]).
- Subjective change of alertness over the drives measured with the Karolinska Sleepiness Scale (KSS, [28]) directly before and after the drives.
- The proportion of time the L3ADS was activated [%] in relation to the time it was available, derived from signals logged in SILAB®.
- The proportion of time with active L3ADS spent on NDRAs [%] and proportion of time with active L3ADS spent on NDRAs with active involvement of the hands [%], derived from table application.
- The proportion of driving time the gaze was on the road ([29], measured as percentage road centre, (PRC, %), [30]), derived from SmartEye signals.
- Subjective criticality of an experienced take-over situation (see [30]).
- Reaction times after a TOR occurred until eyes were on the road, hands on the steering wheel and until control was taken back (ADS deactivated) (sec) (see [30]).
- TOC rating measuring overall take over performance [27].

For statistical testing of the effect of repeated usage, repeated measures ANOVAs were calculated with time (session) as a within-subject factor. To avoid having the effects of repeated usage mixed with effects of drive state (which was experimentally influenced in the two monotonous drives) only the four shorter drives are included in the analysis of behavioural adaptation. These drives always took place in the first, second, fourth and sixth experimental sessions. In the result section, graphs show means and 95% confidence interval.

3. Results

3.1. Evaluation of the L3 Motorway ADS

For more general statements about the L3ADS, there is either a general agreement or disagreement (see Figure 3 and Table 5). Drivers state that they would use the system, recommend it and trust it. Furthermore, driving with L3ADS was rated as being comfortable and fun, drivers did not evaluate it as demanding, stressful or difficult and drivers felt safe while driving with the system. Therefore, drivers evaluate L3ADS positively.

A significant change of drivers' evaluation with repeated usage occurs for the statements "driving was stressful" ($F(3, 84) = 2.99$, $p = 0.03536$), "I felt safe driving with the system" ($F(3, 84) = 5.54$, $p = 0.00161$), "I trust the system" ($F(3, 81) = 3.87$, $p = 0.01221$), "I would use this system" ($F(3, 84) = 3.16$, $p = 0.02882$) and "Using the system was fun" ($F(3, 84) = 3.06$, $p = 0.03260$). Post-hoc tests show that with repeated usage, there is an increase in trust and driving safety going together with a decrease of subjective stress which is most pronounced during the fourth drive. Afterwards, there is again a decrease of expressed trust. Experienced fun is most pronounced during the second drive.

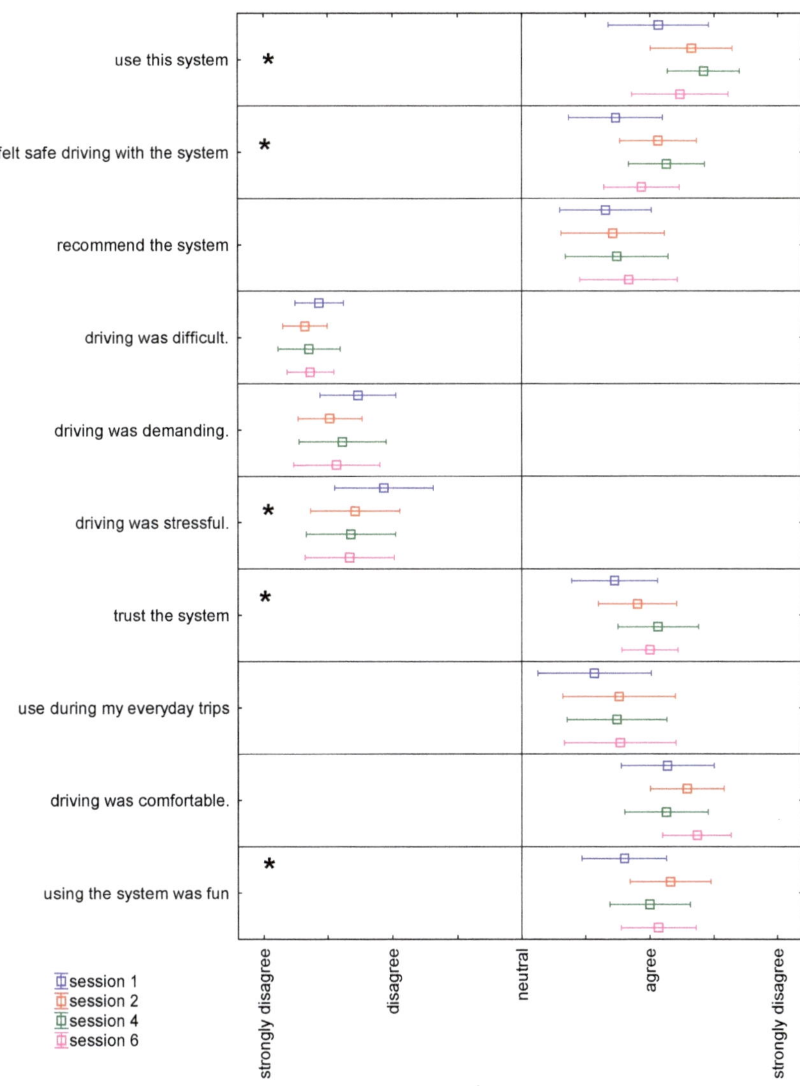

Figure 3. Drivers' agreement with general statements about the L3-motorway automated driving system (ADS) (L3ADS). * marks statements with a significant effect of the session.

Table 5. Results of t-tests evaluation agreement or disagreement with the questionnaire items for evaluation of the ADS. The table gives mean (m), standard deviation (sd), number of included participants (N), t-value, degrees of freedom (df) and p-value. Significant t-tests are marked in bold.

Item	Session	m	sd	N	t	df	p
driving was comfortable	1	2.138	0.953	29	6.43	28	**0.0000**
	6	2.367	0.718	30	10.42	29	**0.0000**
driving was demanding	1	−0.267	0.785	30	−8.84	29	**0.0000**
	6	−0.433	0.898	30	−8.75	29	**0.0000**
driving was difficult	1	−0.567	0.504	30	−17.03	29	**0.0000**
	6	−0.633	0.490	30	−18.25	29	**0.0000**
driving was stressful	1	−0.067	1.015	30	−5.76	29	**0.0000**
	6	−0.333	0.922	30	−7.92	29	**0.0000**
felt safe driving with the system	1	1.733	0.980	30	4.10	29	**0.0003**
	6	1.933	0.785	30	6.51	29	**0.0000**
recommend the system	1	1.655	0.936	29	3.77	28	**0.0008**
	6	1.833	1.020	30	4.48	29	**0.0001**
trust the system	1	1.724	0.882	29	4.42	28	**0.0001**
	6	2.000	0.587	30	9.33	29	**0.0000**
use during my everyday trips	1	1.567	1.194	30	2.60	29	**0.0146**
	6	1.767	1.165	30	3.60	29	**0.0012**
use this system	1	2.067	1.048	30	5.57	29	**0.0000**
	6	2.233	1.006	30	6.71	29	**0.0000**
using the system was fun	1	1.800	0.887	30	4.94	29	**0.0000**
	6	2.067	0.785	30	7.44	29	**0.0000**
driving would make me tired	1	1.821	1.124	28	3.87	27	**0.0006**
	6	1.767	1.223	30	3.43	29	**0.0018**
use the time to do other activities	1	2.069	0.998	29	5.77	28	**0.0000**
	6	2.500	0.682	30	12.04	29	**0.0000**
want to monitor the system's performance	1	1.600	0.932	30	3.53	29	**0.0014**
	6	1.400	1.192	30	1.84	29	0.0763
monitored environment more	1	0.800	1.424	30	−0.77	29	0.4479
	6	0.067	1.311	30	−3.90	29	**0.0005**
more aware of hazards	1	0.000	0.910	30	−6.02	29	**0.0000**
	6	−0.233	0.971	30	−6.95	29	**0.0000**
obvious why takeover requests occurred	1	2.733	0.521	30	18.23	29	**0.0000**
	6	2.625	0.576	24	13.83	23	**0.0000**
during takeovers I felt safe	1	2.167	0.986	30	6.48	29	**0.0000**
	6	2.208	0.833	24	7.11	23	**0.0000**
takeovers warned appropriately	1	2.467	0.629	30	12.78	29	**0.0000**
	6	2.542	0.509	24	14.84	23	**0.0000**
takeovers were with sufficient time	1	2.333	0.758	30	9.63	29	**0.0000**
	6	2.167	0.868	24	6.58	23	**0.0000**

3.2. Usage of L3 Motorway ADS

The overall positive evaluation of the system is reflected in system usage: 90% of the time the L3ADS is available it is actually activated (see Figure 4a). There is no change of system activation with repeated usage (F(3, 90) = 1.03, p = 0.38470). Instead, the increase of trust is reflected in a significant increase of engagement in NDRAs (F(3, 90) = 5.87, p = 0.00104) from 68% during the first session to about 80% in the following sessions. The significant increase in manual NDRAs (F(3, 90) = 7.95, p = 0.00009) is even more pronounced; from 32% of driving time in session one, over 40% in session two up to 59% in session four and 63% in session six.

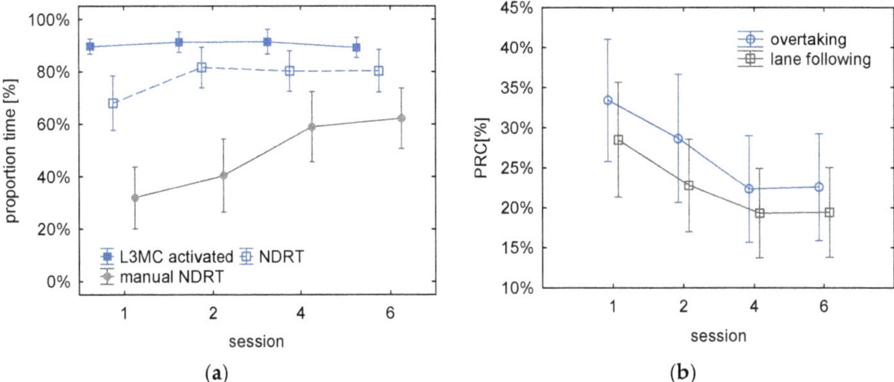

Figure 4. The proportion of time L3ADS was activated, drivers attended to non-driving related activities (NDRAs), drivers attended to NDRAs involving manual distraction (**a**) and drivers spent looking on the road (percentage road centre, PRC) during all time with L3ADS active and during time L3ADS was overtaking other vehicles (**b**).

With the increase of manual NDRAs while driving with the L3ADS activated, the proportion of glances directed to the road decreases (F(3, 90) = 5.79, p = 0.00115, see Figure 4a). There is a decrease between sessions one and two and a further decrease during session four. Then, PRC stays at a constant level. PRC decreases from 30% of the time with the system active in session one to 20% in sessions four and six. The decrease is similar for situations where the L3ADS overtakes other vehicles including lane changes and for situations where the L3ADS follows its own lane. However, during overtaking manoeuvres, drivers' gaze is direct on average during 5% more driving time to the road compared to lane following (F(1, 30) = 12.073, p = 0.00158, see Figure 4b). Therefore, with repeated usage of the L3ADS, the willingness of the drivers increases to engage in other activities and to draw attention away from the driving environment, but situational differences remain unchanged.

3.3. Driver State with L3 Motorway ADS

The measurable behavioural changes are reflected in the subjective evaluation as well (see Figure 5): over the sessions, drivers agreed significantly more strongly with the statement "I use the time to do other activities" (F(3, 78) = 6.38, p = 0.00063) and significantly less with the statement "I monitored the environment more than in manual driving" (F(3, 84) = 8.40, p = 0.00006). For both statements, the change is most pronounced after the first session.

Drivers agree significantly with the statement "driving with the system would make me tired" (see Table 5). This subjective impression is supported by the comparison of ratings of fatigue assessed with the KSS directly before and after the drives. There is a significant increase of fatigue (F(1, 26) = 17.71, p = 0.00027) of about 0.6 scale points on average for the four drives.

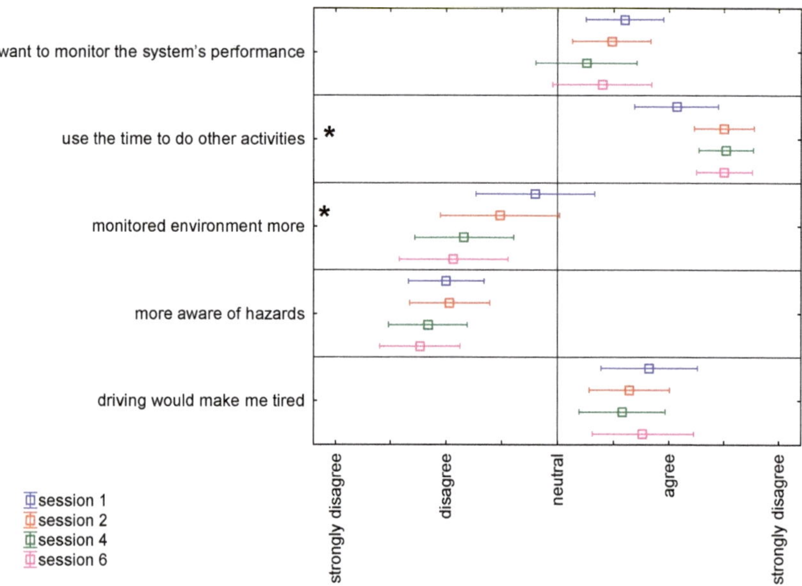

Figure 5. Drivers' agreement with statements about the effect of L3ADS on drivers' state. * marks statements with a significant effect of the session.

3.4. Take-Over Situations

Drivers agree significantly with the statements "during take-overs I felt safe", "it was obvious to me why take-over requests occurred", "take-overs were warned appropriately" and "take-overs were with sufficient time" (see Table 5). For none of the statements on takeover situations, there is a significant change in the evaluation with repeated usage.

Within the four drives, frequency and reasons of TORs varied. Overall, the majority of take-over situations are experienced as being harmless or unpleasant (see Figure 6a). N = 7 out of 433 situations are rated as dangerous, but in four of these situations, drivers took control back even before a TOR was issued by the system. Therefore, the rating mostly relates to the following driving situation, which was a highway intersection with traffic in the two most critical situations.

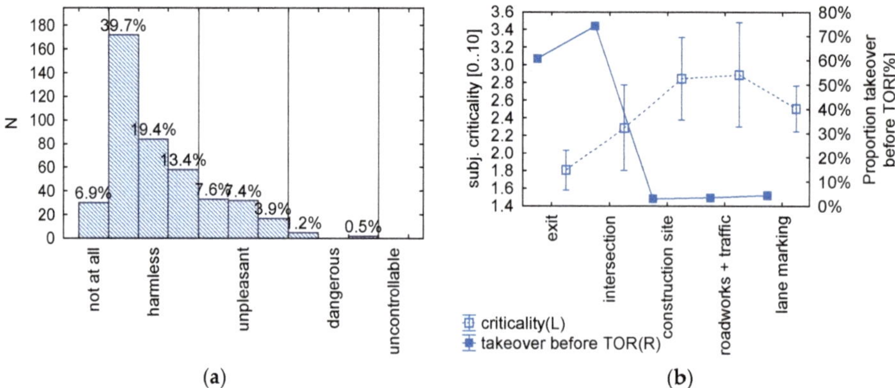

Figure 6. Experienced criticality of take-over situations (**a**) and criticality and proportion of take-over before take-over request (TOR) split by situation type (**b**).

As can be seen in Figure 6b, there are situations in which control is taken back quite frequently before a TOR actually occurred (exit and highway intersection) because these system limits are announced by the navigation system before a TOR. These situations are rated as less critical than situations without a pre-announcement like TORs before a construction site, before roadworks or because of missing lane markings (F(4, 108) = 8.12, p = 0.00001).

To analyse behavioural adaptation to TORs, take-over situations are averaged per driver and driving session separately for situations where drivers take control back before or after a TOR. For subjective criticality, there is a significant interaction between the type of take-over situation and the number of sessions (F(3, 100) = 3.20, p = 0.02671, see Figure 7). During the first session, experienced criticality is similar to situations where drivers take control back before and after a TOR. After the first session, situations are rated as less critical when the driver takes control back before the system issues a TOR. There is no change in the evaluation of situations where control is taken back after a TOR.

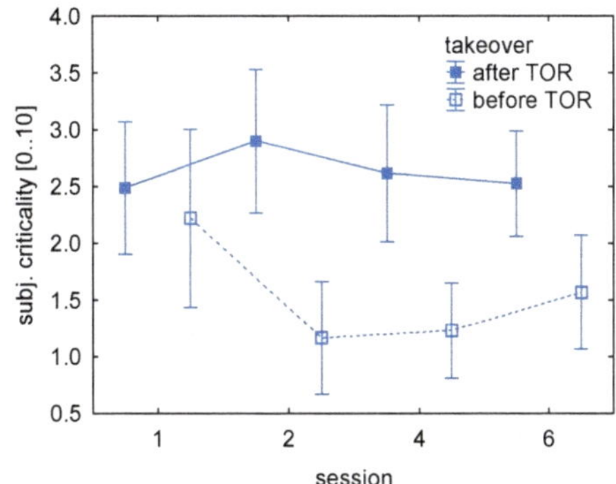

Figure 7. Experienced criticality in a situation where drivers took control back before and after a TOR.

For situations where drivers react after the TOR, TOC-rating and reaction times are analysed (see Figure 8). The time it takes until drivers look onto the road (eyes on-road) is shorter than one second for all sessions and it does not change with repeated usage (F(3, 72) = 0.26, p = 0.85355, Figure 8b, lowest parameter). It takes between two and three seconds until drivers put their hands on the wheel (Figure 8b, middle parameter) and between three and four seconds until the L3ADS is deactivated and the driver starts driving manually (Figure 8b, upmost parameter). For the time until drivers put their hands on the wheel there is a tendency (F(3, 87) = 2.51, p = 0.06424) and for the time until control is actually taken back there is a significant (F(3, 87) = 4.51, p = 0.00547) change over time. For both parameters, the effect is based on an increase of reaction times during the second session. This pattern resembles the results for the TOC-rating. Descriptively there is an increase in average TOC-rating in session 2, which means a worsened takeover performance. Nevertheless, this change is not significant (F(3, 87) = 1.3382, p = 0.26723). In all sessions, between 31% and 42% of all takeover reactions are rated either as perfect or good (on the scale 1–3) with the highest proportion during the first session and the lowest during the second. Between 56% and 69% of takeover reactions are evaluated as being with errors (on the scale 4–6), now vice version session one having the lowest and session two the highest proportion. Overall there is only one takeover scenario rated as being critical that occurred during session one.

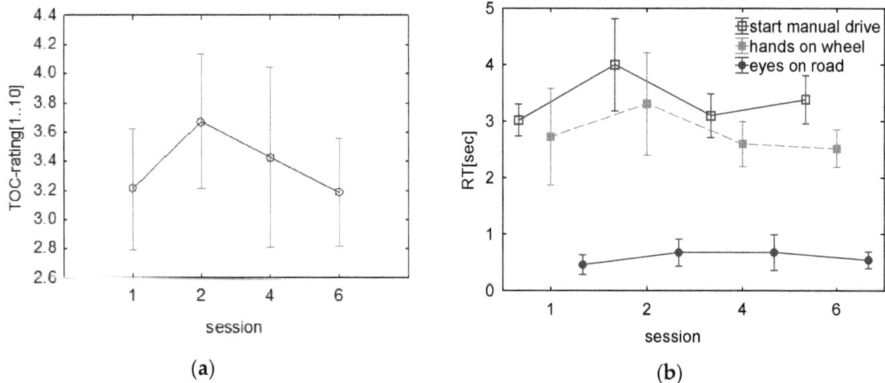

(a) (b)

Figure 8. Experienced criticality in a situation where drivers took control back before and after a TOR (**a**) and reaction times until eyes were on the road, hands were on the steering wheel and control was taken back (**b**).

The pattern of errors occurring in the takeover scenarios remains similar to repeated usage (see Figure 9). Most errors/imprecision rated relate to imprecise lateral control like jerky steering, too low lateral safety distance and crossing of lane markings. Furthermore, drivers frequently forget to use the indicator or use it too late. Errors in longitudinal control (like braking too strong or too late) and errors indicate problems on the decision level (e.g., missing, hesitant or wrong lane change) are rare.

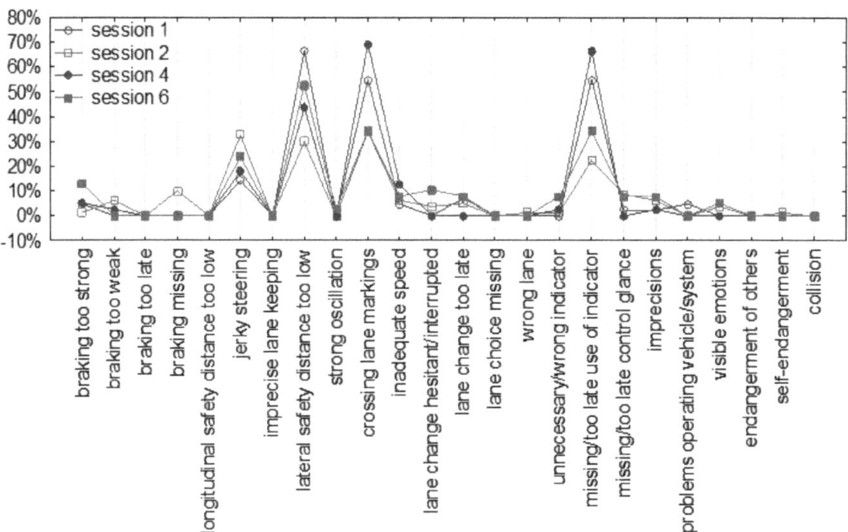

Figure 9. The proportion of takeover scenarios with the different types of errors/imprecision rated in the TOC-rating.

4. Discussion

In summary, several of the investigated measures change with repeated usage of the L3ADS:

- With repeated usage, drivers trust the function more and feel safer and less stressed.

- With repeated usage, drivers spent more time with the function active on NDRAs, especially on tasks that involve both hands.
- With repeated usage, drivers feel less the need to monitor the system and they direct less attention to the road.

In the course of the drives, there is no change in the proportionate time that the system is activated. This can be explained by the fact that during the first drive the usage is already very high, with the system being activated more than 80% of the time it is available. This level remains rather stable in the course of the six drives. Therefore, the growing trust in the system is reflected not in an increase of usage of the system but rather in an increase in the willingness to engage in NDRAs and let the system be unsupervised. For most measures, the main increase can be observed between the first and the fourth drive, the second drive ranging somewhere in between. For the proportion of time the gaze is directed to the road, there is a continuous decrease from session one to session four. During sessions four and six, the level remains stable. Therefore, subjective as well as objective measures indicate an increase of trust over the first four drives. Afterwards, no further behavioural change can be observed. The results are in-line with the model of Martens and Jenssen [12] that describes that after the first encounter where the driver first explores the system, a phase of learning starts. In this phase, the driver experiences the system behaviour in different situations or scenarios. Even though the timely dimensions stated in the model (1–6 h for the first encounter and 3–4 weeks for the learning phase) do not apply to the results of our study, the phases seem applicable.

Performance changes over time as assumed by the model of behavioural adaptation [12] were expected in terms of better reactions to TORs. In summary, drivers were able to handle TORs safely and easily within the available timeframe of 15 s. There is a small effect of repeated usage on reaction time to a TOR based on increased reaction times in the second drive which is reflected at least on a descriptive level in the TOC-rating. This pattern does not support the assumption of a learning effect in terms of a constant improvement of take-over performance. However, it has to be considered that the applied take-over situations were easy to handle. This is also reflected by an overall very low subjective criticality. Especially, situations that were designed such that drivers received a cue that a take-over situation would occur soon, e.g., the information from the navigation system that was given before the TOR was issued were rated as not critical. During the first session, drivers learn to use the pre-announcement to react without time pressure before reaching the system limit and to take control back even before a TOR is issued by the L3ADS.

For those take-over situations where drivers react after a TOR was issued, there is no change of experienced criticality over time. Probably, due to the time pressure after a TOR announces the on-coming end of ODD 15 s before it is actually reached and probably also due to the variability and the changing complexity of the oncoming driving situations, there is no change of experienced criticality over time. It might either be that the number of actual TORs experienced in the experimental drives was too low for such an adaptation to take place or that there is no room for adaptation because appropriate reaction and timing are largely pre-defined by the situation itself. For reactions after a TOR, reaction times for later parts of the reaction (hands-on the steering wheel and control taken back) are delayed during the second session. Whether this indicates a relevant but short change with regard to the concept of behavioural adaptation is questionable.

5. Conclusions

Investigating behavioural adaptation to ADSs poses high requirements for the study design. Simply comparing driving parameters when using the system with driving without the system as applied in studies on ADAS (see e.g., [7]) is not applicable for automated driving systems from SAE level 3 onwards. This would mean that manual driving behaviour is compared to a driving behaviour defined by the automation technology. An alternative approach is to investigate the drivers' behaviour in a timely perspective when interacting with the system. As described by Martens and Jenssen [12], drivers' behaviour when using an ADAS changes over time. Especially the phase of building trust in the system seems highly critical for explaining changes in the drivers' behaviour. Self-reported trust in the L3ADS in our study increased in the course of the driving sessions. Between the first and the fourth drive, an increase in trust in the system was evident. Along with increasing trust, a decrease in monitoring behaviour (decrease in PRC) and increasing engagement in NDRAs was observed. Even though the causal relation of this development is unclear, it can be assumed that drivers change their monitoring behaviour as well as their engagement in NDRAs due to their growing trust in the L3ADS. Furthermore, the observed changes are in line with the predictions of the theory of risk homeostasis [5]. The increase in subjective trust went along with an increase in perceived safety. The increased trust explains why drivers led their attention away from the driving environment and engaged in other activities. Therefore, it can be argued that the overall subjective risk was kept constant by the drivers.

It seems likely that the progress in behavioural adaptation varies for different aspects of using and handling an L3-system. Since driving with the activated system, seeing the system work and experiencing its advantages included the largest proportion of the total 8 h of driving time, the six sessions seem to be sufficient to investigate changes in drivers' attitudes and also in their decisions regarding handling the activated system and using the driving time. Compared to that, actual TORs are rare and short situations. Furthermore, they often lead to situations that require a situationally adapted reaction from the driver with little room for behavioural variations. It is likely that TORs were not frequent enough to study behavioural adaptation, especially because they were experienced as being harmless and manageable.

Regarding the different dimensions of behavioural adaptation discussed in the literature, a clear differentiation between cognitive changes and performance changes turned out to be difficult to capture for driving with L3 automation. This is mainly because the driving task is performed by automation most of the time, therefore the performance of the driver cannot be measured. What can be measured is the decision to activate the system and how the time with the system active is used. These are measures that to our understanding mirror the cognitive decisions of the driver. Also, for driving with ADAS, these two dimensions are probably the ones that interact most, because with mostly manual driving a decision (e.g., to attend to an NDRA) often directly impacts the measured driving performance (e.g., lane-keeping performance). With L3 ADSs, drivers' performance is only measurable in take-over situations where control is handed back to the driver. For situations with a pre-announcement of a system limit (e.g., due to the navigation system), experienced criticality decreases in parallel to other measures during the second session. To gain further insight into potential behavioural adaptation in takeover scenarios more research is needed. It needs to be investigated whether no behavioural adaptation to TORs occurs, e.g., due to the nature of takeover situations (time pressure, varying situational demands) or whether the number of TOR in our study was too low to observe behavioural adaptation.

The approach of operationalizing behavioural adaptation by comparing the driver's attitudes and behaviours over different points in time seems applicable to ADSs. Therefore, the driver's behaviour when using the system for the first time can be compared to the behaviour when using the system at a later point in time. The only question is: When do changes in behaviour occur? What is a reasonable period of usage to observe a change? The timely dimension of the five phases of behavioural adaptation to ADAS by Martens and Jenssen [12] is 1–2 years and was clearly not covered in the presented study. However, the results suggest that for the use of highly automated driving systems this process might be faster. Between the first and the fourth drive, an increase in subjectively reported trust, perceived safety and the willingness to use the system was evident (*attitudinal changes*). The engagement in NDRAs also increased in parallel with reported trust (*cognitive changes*). It seems that especially the learning phase (3–4 weeks) passes much faster since besides the system handling (activation/deactivation) there were mainly the system limits that had to be learned. Drivers experienced various system limits during the experimental drives. This might have been sufficient for "learning" the system. It is obvious that if drivers use the system for an extended period of time and experience the system in more diverse situations, drivers might adjust their behaviour at a later point in time (see *adjustment phase*, [12]).

However, it can be argued that behavioural adaptation to ADSs seems to occur faster than for ADAS. A longer-term user study on ADSs preferably in a real driving environment could yield more insights into further behavioural changes due to ADSs. Nevertheless, such a study requires that L3/L4-ADSs are on the market or at least available in a market-ready version. Such a study would also help to replicate the findings from the presented study.

Author Contributions: conceptualization, B.M. and J.W.; methodology, B.M. and J.W.; software, M.H.; analysis, B.M.; study conduction, J.W., A.L. and M.S.; writing—original draft preparation, B.M.; writing—review and editing, J.W. All authors have read and agreed to the published version of the manuscript.

Funding: The study is part of the research project L3Pilot (https://www.l3pilot.eu/), which receives funding from the European Union's Horizon 2020 research and innovation programme under grant agreement No 723051.

Acknowledgments: The research leading to these results has received funding from the European Commission Horizon 2020 program under the project L3Pilot, grant agreement number 723051. Responsibility for the information and views set out in this publication lies entirely with the authors. The authors would like to thank all partners within L3Pilot for their cooperation and valuable contribution.

Conflicts of Interest: The authors declare no conflict of interest.

References

1. SAE. *Taxonomy and Definitions for Terms Related to on-Road Motor Vehicle Automated Driving Systems*; Society of Automotive Engineers: Warrendale, PA, USA, 2018; Volume 3016, pp. 1–16.
2. Kyriakidis, M.; de Winter, J.C.; Stanton, N.; Bellet, T.; van Arem, B.; Brookhuis, K.; Martens, M.H.; Bengler, K.; Andersson, J.; Merat, N.; et al. A human factors perspective on automated driving. *Theor. Issues Ergon. Sci.* **2019**, *20*, 223–249. [CrossRef]
3. OECD. *Behavioural Adaptations to Changes in the Road Transport System 1990*; Organisation for Economic Co-Operation and Development: Paris, France, 1990.
4. Peltzman, S. The effects of automobile safety regulation. *J. Political Econ.* **1975**, *83*, 677–725. [CrossRef]
5. Wilde, G.J. The theory of risk homeostasis: Implications for safety and health. *Risk Anal.* **1982**, *2*, 209–225. [CrossRef]
6. Hollnagel, E.; Nåbo, A.; Lau, I.V. A systemic model for driver-in-control. In Proceedings of the Second International Driving Symposium on Human Factors in Driver Assessment, Training and Vehicle Design, Park City, UT, USA, 21–24 July 2003.
7. Rudin-Brown, C.M.; Parker, H.A. Behavioural adaptation to adaptive cruise control (ACC): Implications for preventive strategies. *Transp. Res. F* **2004**, *7*, 59–76. [CrossRef]
8. Dikmen, M.; Burns, C. Trust in autonomous vehicles: The case of Tesla Autopilot and Summon. In Proceedings of the IEEE International Conference on Systems, Man, and Cybernetics (SMC), Banff, AB, Canada, 5–8 October 2017.
9. Gold, C.; Körber, M.; Hohenberger, C.; Lechner, D.; Bengler, K. Trust in automation–Before and after the experience of take-over scenarios in a highly automated vehicle. *Procedia Manuf.* **2015**, *3*, 3025–3032. [CrossRef]
10. Naujoks, F.; Totzke, I. Behavioral adaptation caused by predictive warning systems—The case of congestion tail warnings. *Transp. Res. F Traffic Psychol. Behav.* **2014**, *26*, 49–61. [CrossRef]
11. Jamson, H.; Merat, N.; Carsten, O.M.; Lai, F.C. Behavioural changes in drivers experiencing highly-automated vehicle control in varying traffic conditions. *Transp. Res. C* **2013**, *30*, 116–125. [CrossRef]
12. Martens, M.H.; Jenssen, G.D. Behavioral adaptation and acceptance. In *Handbook of Intelligent Vehicles*; Springer: New York, NY, USA, 2012; pp. 117–138.
13. Parasuraman, R.; Riley, V. Humans and automation: Use, misuse, disuse, abuse. *Hum. Factors* **1997**, *39*, 230–253. [CrossRef]
14. Naujoks, F.; Wiedemann, K.; Schömig, N. The importance of interruption management for usefulness and acceptance of automated driving. In Proceedings of the 9th International Conference on Automotive User Interfaces and Interactive Vehicular Applications, Oldenburg, Germany, 24–27 September 2017.
15. Kyriakidis, M.; Happee, R.; de Winter, J.C. Public opinion on automated driving: Results of an international questionnaire among 5000 respondents. *Transp. Res. F* **2015**, *32*, 127–140. [CrossRef]

16. Vogelpohl, T.; Kühn, M.; Hummel, T.; Vollrath, M. Asleep at the automated wheel—Sleepiness and fatigue during highly automated driving. *Accid. Anal. Prev.* **2018**, *126*, 70–84. [CrossRef]

17. Saad, F.; Hjälmdahl, M.; Cañas, J.; Alonso, M.; Garayo, P.; Macchi, L.; Nathan, F.; Ojeda, L.; Papakostopoulos, V.; Panou, M.; et al. *Literature Review of Behavioural Effects*; AIDE deliverable D1. 2.1 2004; Information Society Technologies: Brussels, Belgium, 2004.

18. Patten, C.J. Behavioural adaptation to in-vehicle intelligent transport systems. In *Behavioural Adaptation and Road Safety: Theory, Evidence and Action*; CRC Press: Boca Raton, FL, USA, 2013; pp. 161–176.

19. Ajzen, I. Perceived behavioral control, self-efficacy, locus of control, and the theory of planned behavior 1. *J. Appl. Soc. Psychol.* **2002**, *32*, 665–683. [CrossRef]

20. Muir, B.M. Trust between humans and machines, and the design of decision aids. *Int. J. ManMach. Stud.* **1987**, *27*, 527–539. [CrossRef]

21. Large, D.R.; Burnett, G.; Morris, A.; Muthumani, A.; Matthias, R. A Longitudinal Simulator Study to Explore Drivers' Behaviour during Highly-Automated Driving. In Proceedings of the International Conference on Applied Human Factors and Ergonomics, Los Angeles, CA, USA, 17–21 July 2017.

22. Straßenverkehrsgesetz 2018, §1b Rechte und Pflichten des Fahrzeugführers bei Nutzung hoch- und vollautomatisierter Fahrfunktionen, (Germany). Available online: https://www.gesetze-im-internet.de/stvg/StVG.pdf (accessed on 3 October 2019).

23. Griffin, T.; Sauvaget, J.-L.; Geronimi, S.; Bolovinou, A.; Brouwer, T. Deliverable D4.1—Description and Taxonomy of Automated Driving Functions. Deliverable D3.3 of L3Pilot Project. 2019. Available online: https://l3pilot.eu/fileadmin/user_upload/Downloads/Deliverables/L3Pilot-SP4-D4.1-Description_and_taxonomy_of_AD_functions-v2.0_for_website.pdf (accessed on 17 February 2020).

24. Metz, B.; Rösener, C.; Louw, T.; Aitoniemi, E.; Bjorvatn, A.; Wörle, J.; Weber, H.; Torrao, G.A.; Silla, A.; Malin, F.; et al. Deliverable D3.3—Evaluation Methods; Deliverable D3.3 of L3Pilot Project. 2019. Available online: https://l3pilot.eu/fileadmin/user_upload/L3Pilot-SP3-D3.3_Evaluation_Methods-v1.0_DRAFT_for_website.pdf (accessed on 3 October 2019).

25. Van der Laan, J.D.; Heino, A.; De Waard, D. A simple procedure for the assessment of acceptance of advanced transport telematics. *Transp. Res. C* **1997**, *5*, 1–10. [CrossRef]

26. Neukum, A.; Lübbeke, T.; Krüger, H.-P.; Mayser, C.; Steinle, J. ACC-Stop&Go: Fahrerverhalten an funktionalen Systemgrenzen. In *5 Workshop Fahrerassistenzsysteme—FAS 2008*; Maurer, M., Stiller, C., Eds.; Fmrt: Karlsruhe, Germany, 2008; pp. 141–150.

27. Naujoks, F.; Wiedemann, K.; Schömig, N.; Jarosch, O.; Gold, C. Expert-based controllability assessment of control transitions from automated to manual driving. *MethodsX* **2018**, *5*, 579–592. [CrossRef] [PubMed]

28. Akerstedt, T.; Anund, A.; Axelsson, J.; Kecklund, G. Subjective sleepiness is a sensitive indicator of insufficient sleep and impaired waking function. *J. Sleep Res.* **2014**, *23*, 240–252. [CrossRef] [PubMed]

29. Wiedemann, K.; Naujoks, F.; Wörle, J.; Kenntner-Mabiala, R.; Kaussner, Y.; Neukum, A. Effect of different alcohol levels on take-over performance in conditionally automated driving. *Accid. Anal. Prev.* **2018**, *115*, 89–97. [CrossRef] [PubMed]

30. Victor, T. Keeping Eye and Mind on the Road. Ph.D. Thesis, Universitet Uppsala, Uppsala, Sweden, 2005.

Article

Measuring Drivers' Physiological Response to Different Vehicle Controllers in Highly Automated Driving (HAD): Opportunities for Establishing Real-Time Values of Driver Discomfort

Vishnu Radhakrishnan [1,*], Natasha Merat [1], Tyron Louw [1], Michael G. Lenné [2], Richard Romano [1], Evangelos Paschalidis [1], Foroogh Hajiseyedjavadi [1], Chongfeng Wei [1] and Erwin R. Boer [3]

[1] Institute of Transport Studies, University of Leeds, Leeds LS29JT, UK; N.Merat@its.leeds.ac.uk (N.M.); T.L.Louw@leeds.ac.uk (T.L.); R.Romano@leeds.ac.uk (R.R.); E.Paschalidis@leeds.ac.uk (E.P.); F.Hajiseyedjavadi@leeds.ac.uk (F.H.); chongfeng.wei@northumbria.ac.uk (C.W.);
[2] Seeing Machines Ltd., Canberra 2609, Australia; mike.lenne@seeingmachines.com
[3] Entropy Control, Inc., San Francisco, CA 94107, USA; erwinboer@entropycontrol.com
* Correspondence: mn16vr@leeds.ac.uk; Tel.: +44-74-69318182

Received: 2 July 2020; Accepted: 5 August 2020; Published: 8 August 2020

Abstract: This study investigated how driver discomfort was influenced by different types of automated vehicle (AV) controllers, compared to manual driving, and whether this response changed in different road environments, using heart-rate variability (HRV) and electrodermal activity (EDA). A total of 24 drivers were subjected to manual driving and four AV controllers: two modelled to depict "human-like" driving behaviour, one conventional lane-keeping assist controller, and a replay of their own manual drive. Each drive lasted for ~15 min and consisted of rural and urban environments, which differed in terms of average speed, road geometry and road-based furniture. Drivers showed higher skin conductance response (SCR) and lower HRV during manual driving, compared to the automated drives. There were no significant differences in discomfort between the AV controllers. SCRs and subjective discomfort ratings showed significantly higher discomfort in the faster rural environments, when compared to the urban environments. Our results suggest that SCR values are more sensitive than HRV-based measures to continuously evolving situations that induce discomfort. Further research may be warranted in investigating the value of this metric in assessing real-time driver discomfort levels, which may help improve acceptance of AV controllers.

Keywords: driver state; discomfort; psychophysiology; heart-rate variability (HRV); skin conductance response (SCR); highly automated driving (HAD)

---

1. Introduction

In the recent past, there has been an increasing interest in implementing vehicles with a range of advanced driver assistant systems (ADAS), fuelled by manufacturers' desire to introduce higher levels of vehicle automation capability [1]. The primary motivation for these implementations is their hypothesised provision of increased road safety, and enhanced mobility, accessibility, efficiency and comfort [2]. According to Carsten and Mertens [3], manufacturers have been using comfort as one of the main selling points for ADAS. Additionally, the comfort of the driver is considered to be a determining factor for the broader acceptance of the automated system [4]. Therefore, it can be argued that, if an automated system can measure driver comfort in real-time, it can adapt its driving style/behaviour to match the drivers' expectations accordingly, and thereby potentially increase acceptance. This could have the additional benefit of reducing unnecessary driver initiated takeovers, which can otherwise jeopardise the safety of the vehicle and its occupants [5]. This study, conducted as part of the HumanDrive project, considered the effect of a number of road and vehicle-based factors on driver comfort, investigating whether physiological metrics can be used to provide an objective measure of comfort, to help inform the design process when investigating the acceptance of future automated vehicles.

Currently, there is no unanimously agreed on definition of comfort. In a general context, Slater [6] (p. 158) described comfort as "a pleasant state of physiological, psychological and physical harmony between human being and the environment". In the context of driving, and especially highly automated driving (HAD), Beggiato et al. [7] (p. 446), defined comfort as "a subjective, pleasant state of relaxation resulting from confidence in safe vehicle operation which is achieved by the absence of uneasiness and distress". Beggiato et al. [7] further suggested this is still a rather broad definition of comfort, and is associated with other concepts, such as stress, mental workload, fear, motion sickness or anger, with stress and mental workload having the closest link to discomfort (i.e., lack of comfort). Siebert et al. [4] argued that it is easier to measure discomfort rather than comfort, since signs of discomfort tend to be more well-defined and pronounced, compared to the un-aroused relaxed state of comfort. Summala [8] proposed four factors that need to be maintained above a certain threshold to keep drivers within their "comfort zone" during manual driving. These are safety margins (to road edges, obstacles or other vehicles), vehicle-road system (accelerations, road geometry), rule-following (obeying traffic laws, maintaining speed limits) and good progress of the trip (meeting one's expectations for the pace or progress of the travel). However, assuming 100% performance of the automated system, Siebert et al. [4] noted that the rule-following factor for comfort is redundant in HAD, as the automated vehicle (AV) will almost certainly follow the rules, and that good progress of the trip is dependent on traffic conditions, rather than automation state in itself, assuming the route selected by the AV is similar to that in manual driving, where the navigation system decides/recommends the optimal route to be followed. Therefore, in this paper, we focus specifically on how factors that affect the safety margins, and vehicle-road system, affect driver discomfort, for manual and automated driving.

Summala [8], suggested that sufficient safety margins from potential hazards are required for a driver to feel safe and comfortable. Factors influencing these safety margins, and likely to increase driver discomfort, include situations which increase drivers' stress levels, such as navigating in crowded cities, interactions with other road users, or when passing another car/obstacle [9,10].

Comfort is affected by jerk and acceleration forces of the vehicle, with higher accelerations and jerks (in terms of both magnitude and frequency) associated with an increase in discomfort [11–13], and an increase in motion sickness [14]. Drivers tend to keep their lateral and longitudinal acceleration under 2 m/s^2 for a comfortable driving experience [15–17]. However, it should be noted that drivers' comfort threshold for lateral acceleration varies with respect to their velocity, with an increase in velocity resulting in lower threshold values for lateral acceleration [17,18]. Within the public transport domain, especially in railway systems, standard acceleration values are limited to under 1.47 m/s^2, and jerk values are kept under 0.6 m/s^3, to ensure passenger comfort [13,16,19]. However, the acceleration and jerk thresholds used in public transport systems consider both seated and standing passengers. Therefore, it may be permissible to have slightly higher thresholds in HAD, where passengers are typically seated. For instance, Eriksson and Svensson [20] suggested an acceleration and jerk threshold of under 2 m/s^2 and 0.9 m/s^3 respectively, to ensure a comfortable ride in HAD.

Because AVs are still in the prototype and testing phase, most individuals have not had a real-world experience of HAD. Therefore, our expectations of what constitutes a 'comfortable' experience during HAD can only be based on our current understanding of users' comfort in either manual driving, or in other surface transport modes. However, there are considerable differences between these modes, in terms of Summala's [8] proposed four factors, described above, making them difficult to compare to HAD. Thus, to assist with the development of more acceptable AVs, and to ensure user uptake of these systems in the future, it is of value to understand what particular features of an AV's manoeuvres are likely to enhance or diminish user discomfort. For example, humans try to minimise the jerk during manual driving, whereas most current ADAS features tend to have a relatively higher jerk, due to their preference to stay closer to the lane centre and unwillingness to cut corners, unlike human drivers. Thus, it is important to know if users would prefer, and feel more comfortable with, a more "human-like" AV controller, which favours manoeuvres that result in lower acceleration and jerk, over a more conventional AV controller, with very strict margins for optimal and accurate lane-keeping and vehicle velocities.

Studies on comfort in manual driving have used subjective measures, such as comfort questionnaires [21] and comfort scales [22]. Since comfort is highly subjective, it can be challenging to measure it accurately and reliably on a moment-to-moment basis. In a real-world HAD scenario, the driver may become annoyed if they are asked to rate their comfort levels time and again during the drive, especially when they have the option to engage in more appealing non-driving related activities. Thus, in HAD, there is a need for a non-intrusive, objective, discomfort detection system, which can ultimately be used to adapt the automated system's driving style, to ensure the driver is relaxed and at ease [7]. Physiological techniques are one example of such objective methods, which have been used in the past to assess driver state both in HAD [7] and manual driving [23,24]. Recent technological advancements have led to the development of non-intrusive physiological devices that measure heart rate variability (HRV) and electrodermal activity (EDA), such as wearable smart-band sensors like Empatica E4 [25] or Microsoft band 2 [7], and non-contact methods, such as those listed in [26]. Previously, studies have shown strong correlations between stress and workload, and users' HRV, and EDA. A general finding is that heart rate (HR) increases, and HRV (including the time-domain based metric of root mean square of successive differences in R-R intervals (RMSSD)) decreases, during periods of high stress or workload [10,27–29].

An EDA signal consists of the slow-changing tonic component called skin conductance level (SCL) and the rapidly changing phasic component, known as skin conductance response (SCR) [30]. SCRs are generally used to understand short-term fluctuations in the EDA signal, due to a short-term stimulus (for example, being startled or passing an obstacle), whereas SCL is used to understand the overall change in a person's skin conductance when the stimulus is spread over a longer period (for example, fatigue induced by driving for a long time). SCRs have a much shorter decay time than SCLs, and, hence, can more accurately capture differences in manipulations, without the need for recovery/resting periods in between [30,31]. In the context of driving, both SCL and SCRs have been shown to increase with an increase in stress and workload for a driver [10,23,32], and, thus, are associated with increases in discomfort [7]. Based on these findings, we analysed RMSSD, HR and SCR responses per minute (nSCR/min) in this study, as the objective physiological metrics of drivers' comfort.

Current Study

This study was undertaken as part of a 10-member consortium of the HumanDrive project, part-funded by the UK's Centre for Connected and Autonomous Vehicles (CCAV), via Innovate UK. The main aim of the project was to develop an advanced vehicle controller, which allowed the vehicle to perform a 'natural', human-like, driving style, using artificial intelligence (AI), and deep learning techniques. As outlined above, developing a human-like controller could potentially help with the broader acceptance of AVs, driven by a more natural driving style, which is familiar to the driver. Using manual driving data collected from 44 drivers in an earlier HumanDrive study, an aggregated model for human-like controllers, focusing on both vehicle safety and comfort, was developed for the present study (see also [33], for more details of the controllers). An environment-specific risk model was developed to guide the design of the experiments. The simulated drives were constructed to include risk elements present in the drive, based on road width and curvature, as well as on the presence of road-based furniture and obstacles, such as hedges of different heights, grass/asphalt verges, pedestrian refuges and parked-cars or roadworks (see [34] for more details). The development of this risk model was based on satisficing risk corridors, proposed by Boer [35], where a set of vehicle states are within acceptable bounds. The vehicle state includes velocity and lateral offset. The trajectory of the vehicle is always within this risk corridor and adopts a comfortable smoothness for the ride. The model holds that drivers' perceived risk level is based on minimum time to lane crossing, wherein the lateral position for the vehicle stays within the road boundaries [35]. Based on this model, two human-like AV controllers (SLOW and FAST, with the FAST controller having higher velocities than the SLOW controller) were developed, and compared to a conventional controller (LKAS), and drivers' replay of their own drive (see Section 2.3, for more details). To understand how the different physical characteristics of a drive can affect drivers' discomfort, our study exposed participants to a range of accelerations, induced by the four different AV controllers and manual driving. Participants experienced these controllers in two different road environments (rural and urban), which included a variety of road geometries, such as roads of different curvatures/width/speed limit, containing a range of road furniture/obstacles (parked cars, roadworks and pedestrian refuges). Previous studies on driver discomfort during HAD, such as Beggiato et al. [7], have focused on discrete situations causing discomfort, such as negotiating an intersection, exit ramp or an obstacle. In our study, we considered the effects of longer, repeated exposure to different road environment, human-like AV controllers and interactions with road furniture and obstacles, on drivers' discomfort. Drivers' HR and EDA data were compared to drivers' self-reported level of perceived discomfort for each road environment, which was measured in real-time, using a button pressing technique (see Section 3.2 for more details). We addressed the following research questions:

i.   How is driver discomfort, as measured by changes in physiological state (i.e., HRV and EDA), affected by the various controllers, and manual driving?

ii.   Do drivers' discomfort levels change, based on the behaviour of the different controllers, in the different road environments (rural and urban)?

iii.   Does the change in drivers' physiological state reflect their self-reported level of perceived discomfort during HAD?

2. Materials and Methods

2.1. Participants

In total, 24 participants (10 Female), each with a valid UK driving licence, took part in this driving simulator-based study. Their mean age was 43 ± 17 years, with a mean driving experience of 23 ± 18 years. All participants gave consent to take part in the study, in accordance with the rules and regulations of the University of Leeds ethics committee (LTTRAN-086) and were compensated with £50 for taking part in the study. Participants were pre-screened for physiological data collection and those with pre-existing heart conditions were not included in the study (as per [30,36]). In addition, participants were requested to avoid consuming food and beverages that had cardiac stimulants such as caffeine or alcohol for 24 h before they took part in the study.

2.2. Aparatus

The experiment was conducted in the full motion-based University of Leeds Driving Simulator (UoLDS), which consists of a Jaguar S-type cab housed in a 4 m diameter spherical projection dome with a 300-degree field-of-view projection system. The simulator also incorporates an 8 degree-of-freedom electrical motion system. This consists of a 500 mm stroke-length hexapod motion platform, carrying the 2.5 T payload of the dome and vehicle cab combination, and allowing movement in all six orthogonal degrees-of-freedom of the Cartesian inertial frame. Additionally, the platform is mounted on a railed gantry that allows a further 5 m of effective travel in surge and sway. Drivers' physiological data were collected using a Biopac MP35 data acquisition system at 1000 Hz, which consisted of ECG electrodes and an EDA sensor.

2.3. Study Design

The study used a within-participant design and included a short familiarisation drive for ~10 min. Each participant experienced five drives: a MANUAL drive, two with human-like AV controllers (SLOW and FAST), a replay of their manual drive (REPLAY) and one conventional lane-keeping assist-based AV controller (LKAS) which did not adapt its behaviour to road furniture, such as kerbs or hedges. Each drive consisted of two different road environments (rural and urban). The design of the drives and the road environments are discussed below.

2.3.1. Road Design

Each drive was 15.8 km long, and incorporated several situations that demanded greater attention and a shift in lateral position and speed, which could be deemed uncomfortable by the driver based on how it was negotiated, presented across two different road environments (rural and urban, see Figure 1). The speed limits, geometries, and obstacle locations, for each road are listed in Table 1 and Figure 2. The road design was similar across all drives except for LKAS, which did not include any obstacles, which were partly within the lane, such as roadworks or parked cars.

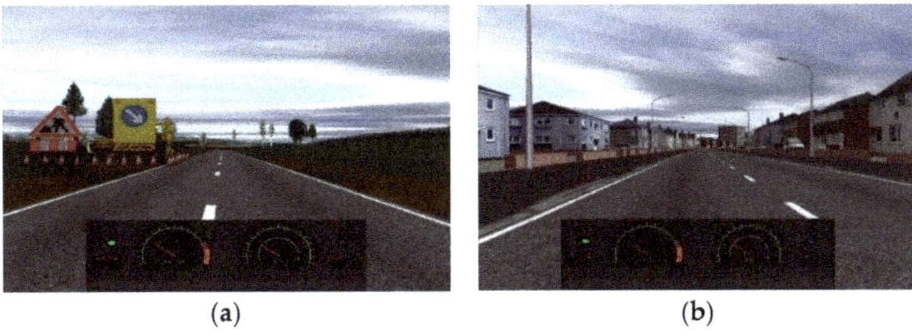

(a)   (b)

Figure 1. (**a**) Rural environment with roadworks; (**b**) urban environment.

Table 1. Road geometry and furniture across different segments (in the order they were experienced).

Segment	Obstacles	Environment	Speed Limit (mph)	Road Width (m)	Radius and Number of Curves 100 m	170 m	252 m	750 m
Segment 1	-	rural	60	7.3	-	2	3	-
Segment 2	4	rural	60	5.8	1	4	-	-
Segment 3	4	urban	40	7.3	-	-	-	5
Segment 4	-	rural	60	5.8	1	4	-	-
Segment 5	6	urban	40	7.3	-	-	-	5

Roads in the rural environments were narrower than those in the urban environments, except in the first segment, which was wider than the other two rural segments (see Table 1). We did this to assess whether a decrease in road-width increased discomfort within the same road environment. Overall, rural environments were designed to have narrower roads, tighter curves, and higher speed limits (and therefore, higher resultant acceleration), along with the presence of obstacles (parked-cars and roadworks, see Figure 1). These factors were designed to increase the attentional demand of the driver at varying degrees, which could possibly induce discomfort depending on how they were negotiated by the controllers, or drivers' individual manual driving style. There were more obstacles (parked-cars, roadworks, or pedestrian refuge, see Figure 2) in the urban environments (10), when compared to the rural environments (4), to investigate whether participants' discomfort increased with the number of obstacles.

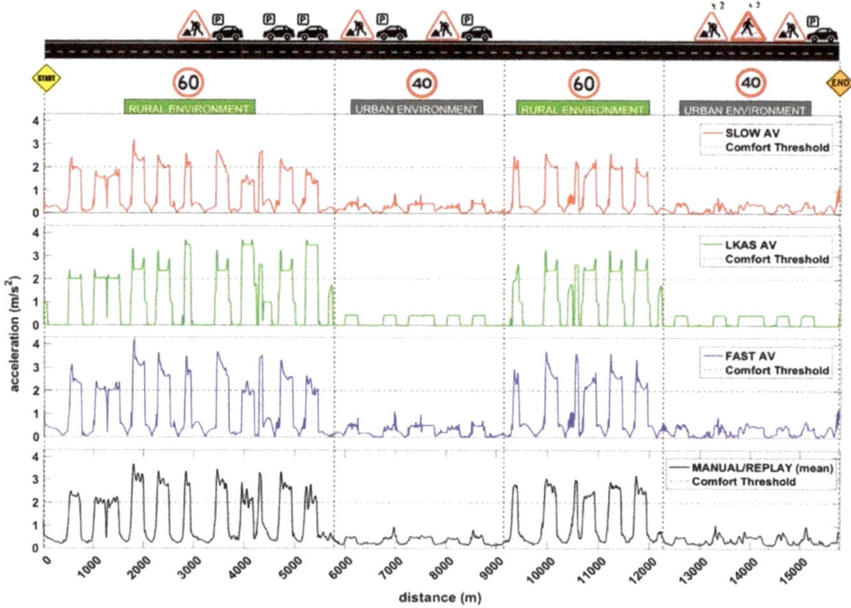

Figure 2. Resultant acceleration of the different controllers and manual driving, along with the location of obstacles across all drives, except LKAS.

2.3.2. Experimental Design

The five drives were counterbalanced, with the exception of the MANUAL drive, which was always the first drive for every participant, so that data could be collected for their REPLAY drive, although participants were not explicitly informed about this. As discussed in the Introduction, the SLOW and FAST controllers were modelled, based on data collected during manual driving across similar road segments in a previous HumanDrive study (see [34]). They were designed to mimic human-like driving, based on a risk model, which defined a range of acceptable vehicle states, such as velocity and lateral offset, depending on drivers' perceived risk levels in response to different road furnitures and features present in the drive, such as parked-cars or sharp curves. The FAST controller had higher velocities, compared to the SLOW controller, with a maximum difference of 4 m/s, and a minimum difference of 0.15 m/s. The driving data used to create the models (see [33]) showed that when driving at higher velocities, drivers' time to lane crossing (TLC) decreased, and, in order to maintain their preferred safety boundary, they moved further away from the road edge. Taking this knowledge into account, we increased the lateral offset of the FAST controller from the left edge of the road, at a rate of 5 cm for every 1 m/s increase in relative speed, compared to the SLOW controller. The LKAS controller was a simple lane-keeping assist controller, which had a constant velocity for most parts of the drive (at the speed limit for that section), except for when the vehicle had to negotiate a curve, or when it moved from an urban to rural environment (or vice-versa). The LKAS controller mostly kept to the lane centre (even when on curves). The objective of the design of the different drives with these controllers was to understand how discomfort was affected by factors such as manual and automated driving, the behaviour of the human-like AV controllers, a conventional lane-keeping controller and the controller based on one's own driving style. The different drives and their properties are shown in Figure 2, Tables 2 and 3, which show that the LKAS controller had the highest resultant acceleration (combined lateral and longitudinal accelerations) in rural environments, whereas the SLOW controller had the lowest resultant acceleration in rural environments. The 95th percentile of resultant acceleration and lateral jerk values across all the drives in rural environments was higher than the suggested comfort threshold value for acceleration and jerk (2 m/s^2 and 0.9 m/s^3, respectively, according to [20]), whereas it was well below this threshold across all drives in the urban environments. The resultant acceleration values were mainly governed by the lateral accelerations, as the longitudinal accelerations were minimal, and within the suggested comfort threshold for longitudinal acceleration, across both environments, for all controllers.

Table 2. The 95th percentile of resultant acceleration (in m/s^2) for different drives across different road environments.

	MANUAL	*SLOW*	*LKAS*	*FAST*	*REPLAY*
Rural	3.42	2.34	3.48	3.20	3.42
Urban	0.74	0.47	0.45	0.57	0.74

Table 3. The 95th percentile of absolute values of lateral jerk (m/s^3) for different drives across different road environments.

	MANUAL	*SLOW*	*LKAS*	*FAST*	*REPLAY*
Rural	2.27	1.38	1.71	2.13	2.27
Urban	0.66	0.83	0.19	0.83	0.66

2.4. Subjective Discomfort Rating (Button Presses)

For each of the automated drives, the participants heard 41 auditory beep triggers. These beeps were played immediately after the participants were exposed to any obstacles, changes in road furniture, changes in road curvature or changes in road environment. In response to these triggers, they were required to press one of two buttons on an Xbox handset, to state: *"Yes, I found the behaviour to be safe/natural/comfortable"* (right button) or *"No, I did not find the behaviour to be safe/natural/comfortable"* (left button). This response explicitly pertained to the behaviour of the car within a couple of seconds around the moment of the beep's occurrence. Additionally, participants were encouraged to give this binary input whenever they felt necessary, across each drive.

2.5. Procedure

Upon arrival, the participants were briefed with the description of the study, after which they were invited to sign a consent form, with an opportunity to ask questions. Three ECG electrodes were then attached to the participant's chest, and 2 EDA electrode bands were attached on the index and middle finger of their non-dominant hand. They then performed a manual familiarisation drive, where they could become accustomed to the simulator environment and vehicle controls. Participants were instructed to adhere to the posted speed limit and to obey the normal rules of the road. After each drive, the participants were given a 10-min break, during which they were asked to complete a set of subjective questionnaires relating to that drive and the controllers. The results of the subjective questionnaires are not within the scope of this paper and will not be reported here.

2.6. Data Analysis Tools

The ECG data was processed on Kubios HRV premium software [37]. EDA signals were pre-processed, and artefacts were removed using custom algorithms based on recommendations in [30] and [38], on MATLAB R2016a. The data were analysed using Ledalab v3.9 [39], a MATLAB-based software package.

2.7. Statistical Analysis

Statistical analysis was conducted on IBM SPSS Statistics 26. Shapiro Wilk's test, which showed that not all estimates across the independent variables were normally distributed, but, in general, the majority of the estimates (>75%) were normally distributed for each of the dependent variables used.

We judged the repeated measures ANOVA to be sufficiently robust to these issues, with only a small effect on Type I error rate [40]. For statistical significance, an α-value of 0.05 was used, and partial eta-squared was computed as an effect size statistic. Degrees of freedom were Greenhouse-Geisser corrected when Mauchly's test showed a violation of sphericity. Pair-wise comparisons with Bonferroni corrections were used to determine the differences in different drives and road segments. Pearson's correlation coefficient was used for any correlation analyses. Data from participants 24 and 14 were classified as outliers, and the data recorded from participants 10 and 15 were of poor quality, and, hence, these were discarded for RMSSD and HR analysis. Participant 12 did not respond to the instructions given for button presses, and participant 13 had an abnormally high rate of button presses. Therefore, these participants were not considered in the subjective button press analysis.

3. Results

Initially, the data were analysed for five separate segments (three in rural and two in urban environments) for each of the five drives, but results for physiological metrics, and the button presses, were not statistically different between the different segments, within the same environment. Therefore, the physiological and button press data across the three rural and 2 urban segments were aggregated for analysis, with the two independent variables being drive (MANUAL, SLOW, LKAS, FAST, REPLAY) and environment (rural and urban). The dependent variables were RMSSD, mean HR and nSCR/min.

3.1. Physiological Metrics

To understand how the behaviour of the AV controllers and manual driving affected drivers' physiological response, and discomfort, across the different road environments, we conducted a 5 (Drive: SLOW, LKAS, FAST, MANUAL and REPLAY) × 2 (Environment: rural, urban) repeated-measures ANOVA on all three physiological metrics (RMSSD, mean HR, nSCR/min). As discussed in the Introduction, previous research has shown that RMSSD values tend to decrease with an increase in discomfort, whereas mean HR and nSCR/min values tend to increase with an increase in discomfort [7,32].

There was a main effect of drive on RMSSD values, $F(2.4, 45.2) = 5.27$, $p = 0.006$, $\eta_p^2 = 0.22$, (Figure 3), with post-hoc tests showing significantly lower RMSSD values in the MANUAL drive, compared to the *LKAS* ($p = 0.007$) and FAST ($p = 0.008$) drives. No other significant differences were found between the drives. There was no effect of environment on RMSSD, or any interactions between drive and environment.

There was a main effect of drive on drivers' mean HR, $F(4, 76) = 6.81$, $p < 0.001$, $\eta_p^2 = 0.23$, (Figure 3), with post-hoc tests showing that drivers had significantly higher mean HR values in the MANUAL drive, compared with the FAST *drive* ($p = 0.001$). There were no significant differences between the other drives. There was no main effect of environment and no interactions between drive and environment.

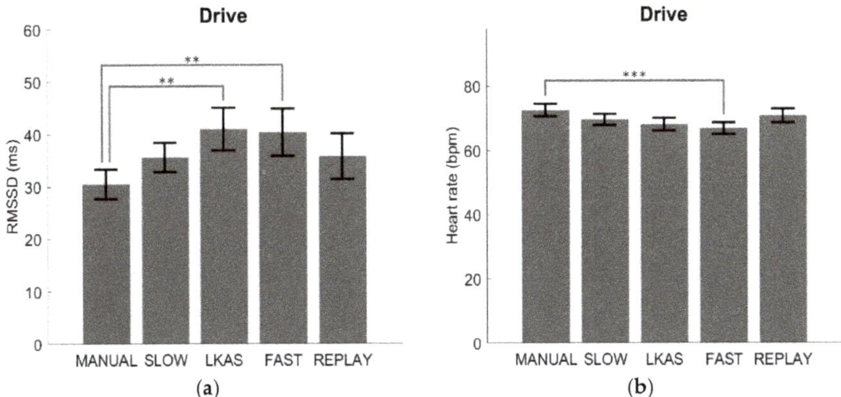

Figure 3. (**a**) Root mean square of successive differences (RMSSD) and (**b**) heart rate (HR) plots for drive. ** $p \leq 0.01$, *** $p \leq 0.001$. Error bars denote s.e.

There was a main effect of drive on nSCR/min, $F(4, 92) = 4.70$, $p = 0.002$, $\eta_p^2 = 0.17$, (Figure 4a), with post-hoc tests showing that there were significantly higher nSCRs/min in the MANUAL drive, compared to the SLOW ($p = 0.006$) and REPLAY drives ($p = 0.005$). There were no other significant differences. There was also a main effect of environment on drivers' nSCR/min, $F(1, 23) = 40.54$, $p < 0.001$, $\eta_p^2 = 0.64$, (Figure 4b), with higher values seen in the rural environments, than the urban environments ($p < 0.001$). An interaction between drive and environment, $F(4, 92) = 3.37$, $p = 0.013$, $\eta_p^2 = 0.13$, (Figure 4c) was also observed. Pairwise comparisons with Bonferroni corrections ($\alpha = 0.002$) revealed that, in the MANUAL drive, drivers had a significantly higher nSCR/min while driving in rural environments, compared to the urban environments ($p < 0.001$). Additionally, within the rural environments, drivers showed significantly higher nSCR/min values in the MANUAL drive, when compared to the SLOW ($p < 0.001$), FAST ($p < 0.001$) and REPLAY ($p = 0.001$) drives. Amongst the AV controllers, LKAS showed the largest reduction in nSCR/min values between rural and urban environments (20.3% reduction in mean nSCR/min from rural to urban).

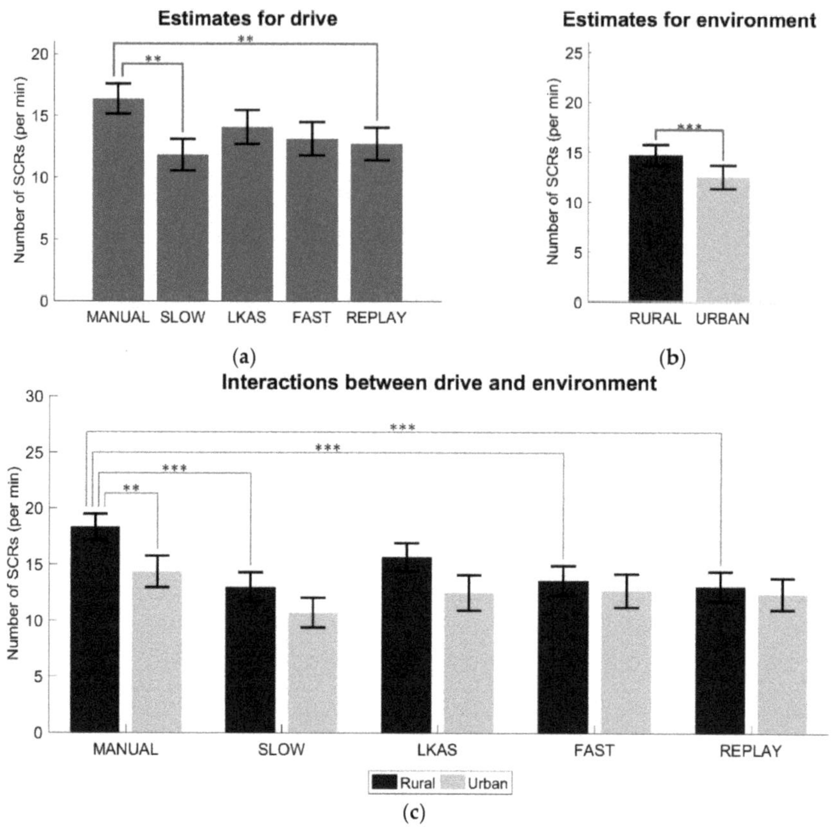

Figure 4. Number of skin conductance responses (SCRs) per minute (nSCR/min) for: (**a**) each drive; (**b**) across different environments; (**c**) and interaction effects. ** $p \leq 0.01$, *** $p \leq 0.001$. Error bars denote s.e.

3.2. Subjective Discomfort Ratings (Button Presses)

In the previous section, we reported a comparison of drivers' physiological state during each drive. However, physiological signals are sensitive to a wide range of stimuli, and are prone to individual differences. Therefore, care must be taken when interpreting a psychological construct, such as discomfort, using physiological measures only [7]. Hence, we used data from the button presses (see Section 2.4, in the Methods section) to establish whether the changes in physiological state correlated with the participants' overall subjective discomfort rating. Correlation analysis showed that button presses and nSCR/min were significantly positively correlated ($r(20) = 0.46$, $p = 0.04$).

To normalise the button press data across all participants, the percentage of NO presses was calculated in relation to the total number of presses, for each road environment, in each drive. A 4×2 repeated measures ANOVA was performed on the percentage of NO presses to assess discomfort, comparing the values across the four drives (SLOW, LKAS, FAST, and REPLAY) at two different road environments (rural and urban).

ANOVA results showed no main effect of drive on participants' button presses, but there was a main effect of environment, where drivers reported a significantly higher percentage of discomfort ratings in the rural, compared to the urban environment, $F(1, 21) = 9.83$, $p = 0.005$, $\eta_p^2 = 0.32$ (Figure 5a). This pattern is similar to that observed for drivers' nSCR/min values, above.

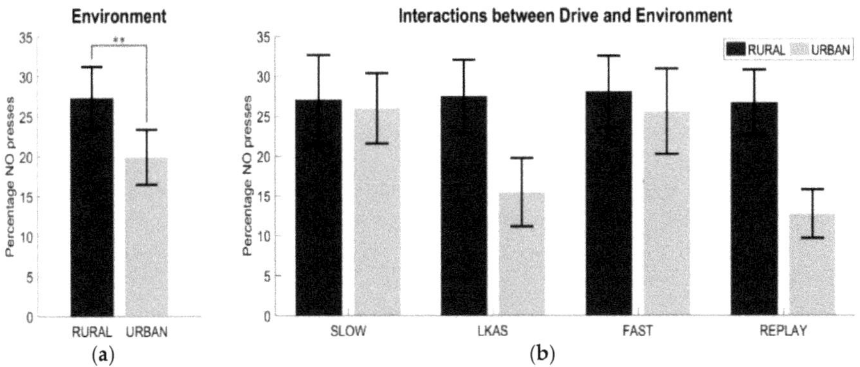

Figure 5. Percentage of NO presses: (**a**) across the two environments; (**b**) the interaction between these two factors is shown in the right graph. ** $p \leq 0.01$. Error bars denote s.e.

There was also an interaction effect, $F(3, 63) = 3.16$, $p = 0.031$, $\eta_p^2 = 0.13$ (Figure 5b). Pair-wise comparisons with Bonferroni corrections ($\alpha = 0.003125$) did not show any significant differences between any of the drives, in each environment. Discomfort ratings were similar across all the drives in the rural environment. However, there was a 43.8% and 52.3% reduction in mean discomfort ratings for LKAS and REPLAY drives, respectively, in the urban environment, compared to their respective values in the rural environment.

4. Discussion and Conclusions

This study investigated driver discomfort, from a physiological perspective, and sought to establish whether drivers' physiological state changes in line with the behaviour of different automated vehicle controllers. Drivers' response in manual driving was compared to four automated drives, with each navigating through a range of road geometries and speeds, associated with urban and rural road environments.

Physiological signals can be highly subjective, and therefore individuals may respond slightly differently to a particular stimulus. Additional care must be given whilst interpreting a physiological change to a psychological construct, as a range of constructs could initiate similar psychological responses [7]. In this study, participants were pre-screened for any physiological anomalies that could occur from usage of cardiac stimulants, exercise, or any medication that they were taking. Furthermore, for EDA analysis, we used nSCR/min instead of amplitude sum of each SCRs, and the former is less susceptible to individual differences such as thickness of skin, as each event related SCR is generally initiated as a response to a particular stimuli. This, and, given the fact that our study incorporated a within-subject design, additional standardisation techniques were not applied for processing RMSSD, mean HR and nSCR/min metrics.

Results showed lower RMSSD values, and higher mean HR and nSCR/min values, in the MANUAL drive, compared to at least one of the AV controllers. However, since drivers were not required to evaluate their own driving, by button presses in the MANUAL drive, it is not possible to conclude whether this difference in physiological metric between the MANUAL and automated drives reflects driver discomfort only, or rather, whether it is due to an increased physical and mental demand associated with the manual driving task, or both.

There were no significant main effects in either the physiological metrics, or button press data, between the four automated drives. This may be because overall, the drives had similar resultant acceleration profiles across the whole drive (see Figure 2.). We analysed physiological metrics and subjective button press data for each segment/environment, which were at least 2 min long. Hence, some of the instantaneous variations in controller behaviour may have produced opposing effects, which cancelled each other out when averaged across a larger time window. These findings are in agreement with [7], where the authors did not find any significant differences in physiological responses between their three automated drives (defensive, aggressive and replay of manual drive). Those authors attributed the lack of difference in physiological responses to high confidence interval bands in their analysis, where missing or opposite effects would have increased the confidence bands dramatically.

In contrast, there were some observable differences, both in terms of physiological metrics nSCR/min), and subjective button presses, for the two road environments, with the rural roads being significantly more uncomfortable than the urban environments. This increase in discomfort is likely attributed to the significantly higher resultant acceleration and jerk experienced in the rural environments, for all drives, which often crossed the 2 m/s^2 and 0.9 m/s^3 threshold for acceleration and jerk, respectively, for a comfortable driving experience, as suggested by [20]. In other words, the higher speed limits, narrower roads and tighter curves associated with the rural environments, seem to be the main cause of increased driver discomfort in this environment. Although more obstacles were present in the urban sections (10 vs. 4), it seems that the way these were negotiated by the vehicle in the rural sections (i.e., passed at a much higher velocity and on narrower roads), was a significant source of driver discomfort during rural environment. These findings are in line with those of [41], where the authors found higher levels of simulator sickness in high-velocity rural environments, when compared to city environments. These results also suggest that those developing automated vehicle controllers should focus on improving comfort, and thereby minimising jerk, when the vehicle is negotiating higher speed, higher acceleration, road geometries.

While the mean discomfort ratings and nSCR/min seemed to be quite similar across all AV controllers in the rural environments, these were particularly low for the urban section of the LKAS (as seen in both discomfort ratings and nSCR/min) and REPLAY (as seen in the discomfort ratings) drives. This is likely due to the absence of any obstacles in the LKAS drive, resulting in very little variations in velocity and lateral offset (and thus, resultant acceleration). With respect to the REPLAY drive, it is likely that participants visibly recognised their own driving style and preferred this familiar behaviour during the lower speed urban environment, where their comfort threshold for acceleration forces was not breached. This was also reflected in their subjective ratings. This recognition was indeed noted by some participants, after their REPLAY drive, although not formally recorded. There seems to be incongruence in participants' physiological indicator of discomfort and perceived level of discomfort during the REPLAY drive in urban environments, indicating a bias in rating one's own driving behaviour. These findings suggest that when the resultant acceleration and jerk experienced by the driver remains well below the comfort threshold, other factors that affect discomfort, such as familiarity of the drive or presence of obstacles, become more prominent and noticeable. In contrast, when the resultant acceleration and jerk values moves above the comfort threshold, it seemingly overshadows other determinants of driver discomfort. This warrants further research into understanding drivers' comfort threshold in terms of jerk and acceleration forces, and its impact on other factors that induce discomfort to the driver.

This study was conducted on a dynamic driving simulator (see Section 2.2 for more details), and the acceleration and jerk forces experienced by the participants would be similar to that in a real-world scenario. Since acceleration and jerk were two main factors affecting discomfort, we believe a drivers' feeling of discomfort due to these forces is quite similar in a simulator and real-world environment. Johnson et al. [42] conducted a study on effect of physiological responses in fixed-based simulator vs. real-world driving and concluded that while level of immersion is at an acceptable level to elicit presence and the trends observed in physiological data during simulated driving relative to real-world driving were quite similar, the absolute physiological responses for virtual and real-world environments were significantly different. There is also the possibility of different behavioural responses by drivers in simulator, when compared to a real-world driving situation [43]. This study incorporated conventional techniques and sensors to measure drivers' physiological data, which were intrusive in nature. However, recent technological advancements have led to non-intrusive [7,25] and even non-contact physiological sensor technologies [26], which need to be validated with on-road studies.

To conclude, there is a need to measure discomfort objectively, and in real-time, so that future AVs can adapt their driving behaviour and provide a more comfortable and pleasant driving experience for human occupants. The novelty of this study is in understanding and measuring the long-term effects of discomfort, across various road environments and a range of AV controllers, using physiological measures. This study suggests that, compared to HR variability measures, EDA-based SCR values are more sensitive to continuous changes in discomfort inducing stimuli, such as those experienced when a vehicle navigates through different geometric and speed-based scenarios. We observed a moderately positive correlation between participants' nSCR/min and their subjective rating of discomfort. Further research may, therefore, be warranted to investigate the value of this metric for assessing real-time driver discomfort levels, which may be useful when developing more acceptable controllers for future automated vehicles.

Author Contributions: Conceptualisation, N.M., V.R., T.L., E.R.B., R.R.; data curation, V.R.; formal analysis, V.R.; funding acquisition, N.M., R.R., E.R.B.; investigation, V.R., E.P., F.H.; methodology, V.R., N.M., R.R., E.R.B., C.W., F.H., E.P.; project administration, N.M., E.R.B., R.R.; software, V.R.; supervision, N.M., T.L., M.G.L.; validation, V.R., N.M., T.L.; visualisation, V.R.; writing—original draft preparation, V.R.; writing—review and editing, V.R., N.M., T.L., M.G.L. All authors have read and agreed to the published version of the manuscript.

Funding: The work described in this paper was undertaken as part of the HumanDrive project, which is co-funded by the Centre for Connected and Automated Vehicles (CCAV) and Innovate UK, the UK's innovation agency. The lead author's Ph.D. is funded by EPSRC CASE studentship in partnership with Seeing Machines Ltd.

Acknowledgments: This paper is published with kind permission from the HumanDrive consortium: Nissan, Hitachi, Horiba MIRA, Atkins Ltd., Aimsun Ltd., SBD Automotive, University of Leeds, Highways England, Cranfield University, and the Connected Places Catapult. The data collection for this paper was feasible due to the help and technical support provided by the University of Leeds Driving Simulator (UoLDS) team.

Conflicts of Interest: The funders had no role in the design of the study; in the collection, analyses, or interpretation of data; in the writing of the manuscript, or in the decision to publish the results.

References

1. *Taxonomy and Definitions for Terms Related to Driving Automation Systems for On-Road Motor Vehicles*; SAE International: Warrendale, PA, USA, 2018; p. J3016.
2. ERTRAC. *Automated Driving Roadmap*; European Road Transport Research Advisory Council: Brussels, Belgium, 2017.
3. Carsten, O.; Martens, M.H. How can humans understand their automated cars? HMI principles, problems and solutions. *Cogn. Technol. Work* **2019**, *21*, 3–20. [CrossRef]
4. Siebert, F.W.; Oehl, M.; Höger, R.; Pfister, H.R. Discomfort in Automated Driving—The Disco-Scale. In *HCI International 2013—Posters' Extended Abstracts*; Stephanidis, C., Ed.; Communications in Computer and Information Science; Springer: Berlin/Heidelberg, Germany, 2013; Volume 374, ISBN 978-3-642-39476-8.
5. Beggiato, M.; Hartwich, F.; Krems, J. Using Smartbands, Pupillometry and Body Motion to Detect Discomfort in Automated Driving. *Front. Hum. Neurosci.* **2018**, *12*, 338. [CrossRef] [PubMed]
6. Slater, K. The assessment of comfort. *J. Text. Inst.* **1986**, *77*, 157–171. [CrossRef]
7. Beggiato, M.; Hartwich, F.; Krems, J. Physiological correlates of discomfort in automated driving. *Transp. Res. Part F Traffic Psychol. Behav.* **2019**, *66*, 445–458. [CrossRef]
8. Summala, H. Modelling driver behaviour in automotive environments. In *Modelling Driver Behaviour in Automotive Environments: Critical Issues in Driver Interactions with Intelligent Transport Systems*; Cacciabue, P.C., Ed.; Springer: London, UK, 2007; pp. 189–207. ISBN 978-1-84628-618-6.
9. Cahour, B. Discomfort, affects and coping strategies in driving activity. In Proceedings of the ECCE 2008 (European Conference on Cognitive Ergonomics), Madeira, Portugal, 16–19 September 2008; pp. 45–53.
10. Healey, J.A.; Picard, R.W. Detecting stress during real-world driving tasks using physiological sensors. *IEEE Trans. Intell. Transp. Syst.* **2005**, *6*, 156–166. [CrossRef]
11. Wertheim, A.H.; Hogema, J.H. *Thresholds, Comfort and Maximum Acceptability of Horizontal Accelerations Associated with Car Driving*; TNO report TM-97-C003; TNO: Soesterberg, The Netherlands, 1997.

12. Beard, G.F.; Griffin, M.J. Discomfort caused by low-frequency lateral oscillation, roll oscillation and roll-compensated lateral oscillation. *Ergonomics* **2013**, *56*, 103–114. [CrossRef]

13. Martin, D.; Litwhiler, D. An investigation of acceleration and jerk profiles of public transportation vehicles. In Proceedings of the ASEE Annual Conference and Exposition, Pittsburgh, PA, USA, 22 June 2008.

14. Vogel, H.; Kohlhaas, R.; von Baumgarten, R.J. Dependence of motion sickness in automobiles on the direction of linear acceleration. *Eur. J. Appl. Physiol. Occup. Physiol.* **1982**, *48*, 399–405. [CrossRef]

15. Moon, S.; Yi, K. Human driving data-based design of a vehicle adaptive cruise control algorithm. *Veh. Syst. Dynam.* **2008**, *46*, 661–690. [CrossRef]

16. Bae, I.; Moon, J.; Seo, J. Toward a comfortable driving experience for a self-driving shuttle bus. *Electronics* **2019**, *8*, 943. [CrossRef]

17. Bosetti, P.; Da Lio, M.; Saroldi, A. On the human control of vehicles: An experimental study of acceleration. *Eur. Transp. Res. Rev.* **2014**, *6*, 157–170. [CrossRef]

18. Levison, W.H.; Campbell, J.L.; Kludt, K.; Bittner, A.; Harwood, D.W.; Hutton, J.; Gilmore, D.; Howe, J.G.; Chrstos, J.P.; Allen, R.W.; et al. *Development of a Driver Vehicle Module for the Interactive Highway Safety Design Model*; FHWA report FHWA-HRT-08-019; Federal Highway Administration: McLean, VA, USA, 2007.

19. Powell, J.P.; Palacín, R. Passenger Stability Within Moving Railway Vehicles: Limits on Maximum Longitudinal Acceleration. *Urban Rail Transit* **2015**, *1*, 95–103. [CrossRef]

20. Eriksson, J.; Svensson, L. *Tuning for Ride Quality in Autonomous Vehicle Application to Linear Quadratic Path Planning Algorithm*; Uppsala University: Uppsala, Sweden, 2015.

21. Thakurta, K.; Koester, D.; Bush, N.; Bachle, S. *Evaluating Short and Long Term Seating Comfort*; SAE Technical Paper 950144; SAE International: Warrendale, PA, USA, 1995.

22. Myers, A.M.; Paradis, J.A.; Blanchard, R.A. Conceptualizing and Measuring Confidence in Older Drivers: Development of the Day and Night Driving Comfort Scales. *Arch. Phys. Med. Rehabil.* **2008**, *89*, 630–640. [CrossRef] [PubMed]

23. Mehler, B.; Reimer, B.; Coughlin, J.; Dusek, J. Impact of Incremental Increases in Cognitive Workload on Physiological Arousal and Performance in Young Adult Drivers. *Transp. Res. Rec. J. Transp. Res. Board* **2009**, *2138*, 6–12. [CrossRef]

24. Lal, S.K.L.; Craig, A. Driver fatigue: Electroencephalography and psychological assessment. *Psychophysiology* **2002**, *39*, 313–321. [CrossRef] [PubMed]

25. McCarthy, C.; Pradhan, N.; Redpath, C.; Adler, A. Validation of the Empatica E4 wristband. In Proceedings of the 2016 IEEE EMBS International Student Conference (ISC), Ottawa, ON, Canada, 29–31 May 2016; Institute of Electrical and Electronics Engineers Inc.: Piscataway, NJ, USA, 2016; pp. 1–4. [CrossRef]

26. Kranjec, J.; Beguš, S.; Geršak, G.; Drnovšek, J. Non-contact heart rate and heart rate variability measurements: A review. *Biomed. Signal Process. Control* **2014**, *13*, 102–112. [CrossRef]

27. Mehler, B.; Reimer, B.; Wang, Y. A comparison of heart rate and heart rate variability indices in distinguishing single-task driving and driving under secondary cognitive workload. In Proceedings of the Proceedings of the Sixth International Driving Symposium on Human Factors in Driver Assessment, Training and Vehicle Design, Olympic Valley, CA, USA, 27–30 July 2011; Public Policy Center, University of Iowa: Iowa City, IA, USA, 2011; pp. 590–597.

28. Orsila, R.; Virtanen, M.; Luukkaala, T.; Tarvainen, M.; Karjalainen, P.; Viik, J.; Savinainen, M. Perceived mental stress and reactions in heart rate variability—A pilot study among employees of an electronics company. *Int. J. Occup. Saf. Ergon.* **2008**, *14*, 275–283. [CrossRef]

29. Cinaz, B.L.; Marca, R.; Arnrich, B.; Tröster, G. Monitoring of mental workload levels during an everyday life office-work scenario. *Personal and Ubiquitous Computing* **2013**, *17*, 229–239. [CrossRef]

30. Braithwaite, J.J.; Watson, D.G.; Jones, R.; Rowe, M. *A Guide for Analysing Electrodermal Activity & Skin Conductance Responses (SCRs) for Psychophysiological Experiments*; Behavioural Brain Sciences Centre, University of Birmingham: Birmingham, UK, 2015.

31. Dawson, M.E.; Schell, A.M.; Filion, D.L. The Electrodermal System. In *Handbook of Psychophysiology*, 4th ed.; Cacioppo, J.T., Tassinary, L.G., Berntson, G.G., Eds.; Cambridge University Press: Cambridge, UK, 2016; pp. 159–181. ISBN 9781107415782.

32. Foy, H.J.; Chapman, P. Mental workload is reflected in driver behaviour, physiology, eye movements and prefrontal cortex activation. *Appl. Ergon.* **2018**, *73*, 90–99. [CrossRef]

33. Hajiseyedjavadi, F.; Merat, N.; Romano, R.; Paschalidis, E.; Boer, E. Effect of Environmental and Individual Differences on Subjective Evaluation of Human-Like and Conventional Automated Vehicle Controllers. 2020, Unpublished work.

34. Louw, T.; Hajiseyedjavadi, F.H.; Jamson, H.; Romano, R.; Boer, E.; Merat, N. The Relationship between Sensation Seeking and Speed Choice in Road Environments with Different Levels of Risk. In Proceedings of the Tenth International Driving Symposium on Human Factors in Driver Assessment, Training and Vehicle Design, Santa Fe, NM, USA, 24–27 June 2019; pp. 29–35.

35. Boer, E.R. Satisficing Curve Negotiation: Explaining Drivers' Situated Lateral Position Variability. *IFAC-PapersOnLine* **2016**, *49*, 183–188. [CrossRef]

36. Laborde, S.; Mosley, E.; Thayer, J.F. Heart rate variability and cardiac vagal tone in psychophysiological research—Recommendations for experiment planning, data analysis, and data reporting. *Front. Psychol.* **2017**, *8*, 213. [CrossRef]

37. Tarvainen, M.P.; Niskanen, J.P.; Lipponen, J.A.; Ranta-aho, P.O.; Karjalainen, P.A. Kubios HRV—Heart rate variability analysis software. *Comput. Methods Prog. Biomed.* **2014**, *113*, 210–220. [CrossRef] [PubMed]

38. Kikhia, B.; Stavropoulos, T.G.; Andreadis, S.; Karvonen, N.; Kompatsiaris, I.; Sävenstedt, S.; Pijl, M.; Melander, C. Utilizing a wristband sensor to measure the stress level for people with dementia. *Sensors* **2016**, *16*, 1989. [CrossRef] [PubMed]

39. Benedek, M.; Kaernbach, C. A continuous measure of phasic electrodermal activity. *J. Neurosci. Methods* **2010**, *190*, 80–91. [CrossRef] [PubMed]

40. Blanca, M.J.; Alarcón, R.; Arnau, J.; Bono, R.; Bendayan, R. Datos no normales: ¿es el ANOVA una opción válida? *Psicothema* **2017**, *29*, 552–557. [CrossRef]

41. Mourant, R.R.; Thattacherry, T.R. Simulator sickness in a virtual environments driving simulator. In Proceedings of the XIVth Triennial Congress of the International Ergonomics Association and 44th Annual Meeting of the Human Factors and Ergonomics Association, Ergonomics for the New Millennium, San Diego, CA, USA, 29 July 2000; SAGE Publications: Los Angeles, CA, USA, 2000; pp. 534–537.

42. Johnson, M.J.; Chahal, T.; Stinchcombe, A.; Mullen, N.; Weaver, B.; Bédard, M. Physiological responses to simulated and on-road driving. *Int. J. Psychophysiol.* **2011**, *81*, 203–208. [CrossRef]

43. Ekanayake, H.B.; Backlund, P.; Ziemke, T.; Ramberg, R.; Hewagamage, K.P.; Lebram, M. Comparing Expert Driving Behavior in Real World and Simulator Contexts. *Int. J. Comput. Games Technol.* **2013**, *2013*. [CrossRef]

MDPI

St. Alban-Anlage 66

4052 Basel

Switzerland

Tel. +41 61 683 77 34

Fax +41 61 302 89 18

www.mdpi.com

Information Editorial Office

E-mail: information@mdpi.com

www.mdpi.com/journal/information